AF294510

Gunther Schmidt   Thomas Ströhlein

# Relations and Graphs

## Discrete Mathematics for Computer Scientists

With 203 Figures

Springer-Verlag
Berlin Heidelberg New York
London Paris Tokyo
Hong Kong Barcelona
Budapest

*Authors*

Gunther Schmidt
Fakultät für Informatik, Universität der Bundeswehr München
Werner-Heisenberg-Weg 39, W-8014 Neubiberg, FRG

Thomas Ströhlein
Fakultät für Informatik, Technische Universität München
Postfach 20 24 20, W-8000 München 2, FRG

*Editors*

Wilfried Brauer
Fakultät für Informatik, Technische Universität München
Arcisstrasse 21, W-8000 München 2, FRG

Grzegorz Rozenberg
Institute of Applied Mathematics and Computer Science
University of Leiden, Niels-Bohr-Weg 1, P.O. Box 9512
2300 RA Leiden, The Netherlands

Arto Salomaa
The Academy of Finland
Department of Mathematics, University of Turku
SF-20 500 Turku, Finland

Improved, extended and translated (in cooperation with Tilmann Würfel) from the German version *Relationen und Graphen* (ISBN 3-540-50304-8) which appeared in 1989 in the Springer series *Mathematik für Informatiker*, edited by F. L. Bauer

CR Subject Classification (1991): G.2, B.6, F.3.1-2, D.2.4, I.2.2

Mathematics Subject Classification (1991): 00-01, 00A06, 03B70, 03G15, 05-01, 05C50, 68Q55, 68Q60, 68R10

ISBN-13: 978-3-642-77970-1     e-ISBN-13: 978-3-642-77968-8
DOI: 10.1007/978-3-642-77968-8



The authors compiled the text on a Macintosh with the T$_E$X program Textures; figures were arranged by Canvas and MacPaint; print-outs were provided by a Hewlett-Packard LaserJet IIISi.

45/3140 - 5 4 3 2 1 0 - Printed on acid-free paper

# Preface

Applications of mathematics in other sciences often take the form of a network of relations between certain objects. But in this situation one seldom does "calculations", in contrast to problems involving mathematical analysis, although a sufficiently developed theory of relations together with a practicable calculus has been in existence for a long while.

The mathematical treatment of relations is said to have its origin in Aristotle, but its modern story starts with the contributions of George Boole, Augustus de Morgan, and Charles S. Peirce. Their work was continued in a systematic way notably by Ernst Schröder who around 1890 published his three volumes of *Algebra der Logik*. In the preface he wrote[1]: "The merit is particularly due to Mr. Peirce for having built the bridge between the older, merely verbal treatment of that discipline and the new, mathematical one." This new treatment was promoted by Leopold Löwenheim who, in view of the recently arisen paradoxes in set theory, suggested "schröderizing" the whole of mathematics.

Half a century later Alfred Tarski, Jacques Riguet, and J. C. C. McKinsey, among others, laid the foundations for the modern calculus of relations, but there has been no easily accessible textbook on the subject as yet. It is our aim to present the theory of relations and to give many applications of their calculus. The subjects to be treated include graphs, combinatorics (e.g., the matching problem), games such as chess and nim, the basics of the relational data base concept, and an extensive discussion of verification and correctness for programs. But we shall also give applications to more abstract areas such as the principle of transfinite induction.

Reading this book requires few prerequisites. The student who has absorbed the introductory mathematics courses in his or her field should be able to read the bulk of every chapter. The later sections of some chapters may be harder, though. The number of applications and examples may even make the book a suitable text for nonexperts.

In conjunction with the manuscript the formula manipulation system RALF has been developed over the past years. It comprises by now all the formulas derived in the text, and a number of proofs have been checked with it. The system RELVIEW, in addition, allows fast and flexible calculations with Boolean matrices; product spaces together with their projection relations as well as function spaces can easily be manipulated.

We have not attempted to give a complete bibliography. The original literature on relation algebra often differs with regard to notation and assumptions so as to render quick comparison difficult. Moreover, much of this literature pursues

---

[1] „Namentlich gebührt Herrn Peirce das Verdienst, die Brücke von den älteren, bloß verbalen Behandlungen jener Disziplin zu der neuen rechnerisch zu Werke gehenden geschlagen zu haben."

goals more purely mathematical in nature than the applications we have in mind. Nevertheless, we have taken care to mention all the relevant literature which precedes our approach or which we took as a model, including some items of historical interest.

The writing of this book extended over a number of years during which we enjoyed the support of students, friends, and colleagues to whom we wish to express our sincere gratitude.

After the German version appeared in 1989 we benefitted from Tilmann Würfel's steady help in the tedious process of translating this book into English. Natalia Schmidt—in addition to gracefully enduring her spouse during that time—was the first to read the emerging text.

Our thanks also go to the following colleagues for their stimulating support and comments: Ludwig Bayer, Rudolf Berghammer, Thomas Gritzner, Jürgen Janas, Wolfram Kahl, and Peter Kempf.

Help and constructive criticism from several researchers from abroad who have read various parts of the book is gratefully acknowledged: Jules Desharnais, Roger Maddux, Ali Mili, Jacques Riguet, and Iain A. Stewart.

We are indebted to Friedrich L. Bauer for accompanying the emergence of this book over the years with stimulation and scientific impetus. We are grateful to the editors of the EATCS monographs for including the book in this series. We thank Springer-Verlag for their agreeable cooperation, in particular Hans Wössner, and not least J. Andrew Ross for his careful copy editing.

Munich, August 1992

Gunther Schmidt<br>Thomas Ströhlein

# Table of Contents

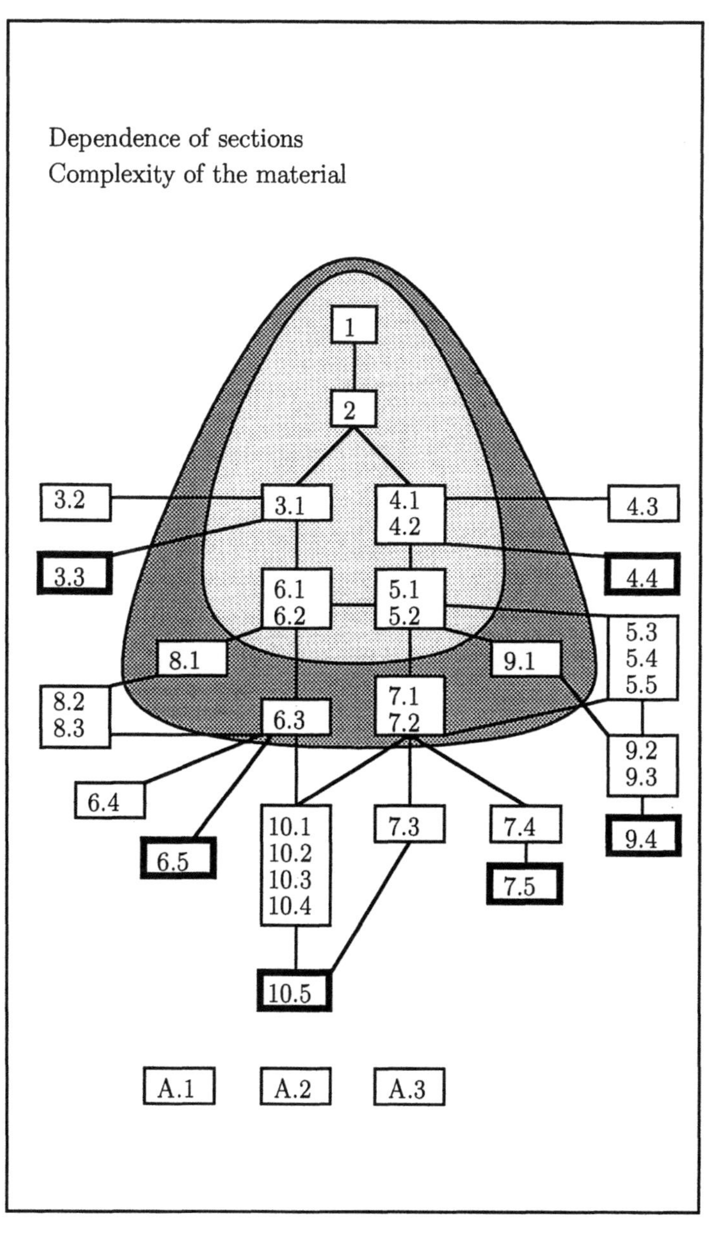

Dependence of sections
Complexity of the material
1
2
3.2
3.1
4.1
4.2
4.3
3.3
4.4
6.1
6.2
5.1
5.2
5.3
5.4
5.5
8.1
9.1
8.2
8.3
7.1
7.2
6.3
9.2
9.3
6.4
9.4
6.5
10.1
10.2
10.3
10.4
7.3
7.4
7.5
10.5
A.1
A.2
A.3

# 1. Sets

In this chapter we introduce relations on a set. For that purpose we first recall some well-known facts from set theory and explain our notation. A set $M$ is a collection of well-defined objects called its *elements*. We write $x \in X$ if $x$ is an element of the set $X$. The symbol $\emptyset$ denotes the *empty set*. The set of all elements which have the property $E$ is written as $M = \{\, x \mid E(x) \,\}$. The power set of a set $X$ is denoted by $\mathbf{2}^X$; so $M$ is an element of $\mathbf{2}^X$ if and only if $M$ is a subset of $X$.

*Union* and *intersection* of two sets $M, N$ are denoted by $M \cup N$ and $M \cap N$, respectively. If $M$ is contained in $N$ we use the symbol $\subset$ for *inclusion*, $M \subset N$. The *complement* of a subset $M$ with respect to a set $X$ is denoted by $X \backslash M$ or, if $X$ is tacitly given, by $\overline{M}$. The operations of union and intersection are

$$
\begin{array}{ll}
\textbf{associative} & (M \cup N) \cup P = M \cup (N \cup P), \\
 & (M \cap N) \cap P = M \cap (N \cap P), \\
\textbf{commutative} & M \cup N = N \cup M, \quad M \cap N = N \cap M, \\
\textbf{distributive} & M \cup (N \cap P) = (M \cup N) \cap (M \cup P), \\
 & M \cap (N \cup P) = (M \cap N) \cup (M \cap P), \\
\textbf{absorptive} & M \cap (M \cup N) = M, \quad M \cup (M \cap N) = M.
\end{array}
$$

(These are the laws of a *distributive lattice*.) Union and intersection are duals of one another. By the absorption law, these operations are also

$$
\begin{array}{ll}
\textbf{idempotent} & M \cup M = M, \quad M \cap M = M.
\end{array}
$$

The inclusion $M \subset N$ can also be interpreted as another way of expressing $M \cap N = M$ or $M \cup N = N$. The power set $\mathbf{2}^X$ is ordered by inclusion, i.e., $\subset$ satisfies the following conditions. It is

$$
\begin{array}{lll}
\textbf{reflexive} & M \subset M, & \\
\textbf{antisymmetric} & M \subset N,\ N \subset M & \implies \quad M = N, \\
\textbf{transitive} & M \subset N,\ N \subset P & \implies \quad M \subset P.
\end{array}
$$

Often, there are pairs of elements $M, N$ which are *incomparable*, in the sense that neither $M \subset N$ nor $N \subset M$.

So far, we have emphasized the algebraic operations of set theory. For finite sets such as $X = \{\, a, b \,\}$ one can set up composition tables as in Fig. 1.1.

We may also start out with the ordering by inclusion $\subset$ of elements $M, N$ of the power set $\mathbf{2}^X$ of $X$. (The last table in Fig. 1.1 indicates when the relation $\subset$ holds for the example above.) Using the inclusion, one can express union, and

| $.\cup.$ | $\emptyset$ | $\{a\}$ | $\{b\}$ | $\{a,b\}$ |
|---|---|---|---|---|
| $\emptyset$ | $\emptyset$ | $\{a\}$ | $\{b\}$ | $\{a,b\}$ |
| $\{a\}$ | $\{a\}$ | $\{a\}$ | $\{a,b\}$ | $\{a,b\}$ |
| $\{b\}$ | $\{b\}$ | $\{a,b\}$ | $\{b\}$ | $\{a,b\}$ |
| $\{a,b\}$ | $\{a,b\}$ | $\{a,b\}$ | $\{a,b\}$ | $\{a,b\}$ |

| $.\cap.$ | $\emptyset$ | $\{a\}$ | $\{b\}$ | $\{a,b\}$ |
|---|---|---|---|---|
| $\emptyset$ | $\emptyset$ | $\emptyset$ | $\emptyset$ | $\emptyset$ |
| $\{a\}$ | $\emptyset$ | $\{a\}$ | $\emptyset$ | $\{a\}$ |
| $\{b\}$ | $\emptyset$ | $\emptyset$ | $\{b\}$ | $\{b\}$ |
| $\{a,b\}$ | $\emptyset$ | $\{a\}$ | $\{b\}$ | $\{a,b\}$ |

| $\bar{\phantom{x}}$ | $\emptyset$ | $\{a\}$ | $\{b\}$ | $\{a,b\}$ |
|---|---|---|---|---|
| | $\{a,b\}$ | $\{b\}$ | $\{a\}$ | $\emptyset$ |

| $.\subset.$ | $\emptyset$ | $\{a\}$ | $\{b\}$ | $\{a,b\}$ |
|---|---|---|---|---|
| $\emptyset$ | 1 | 1 | 1 | 1 |
| $\{a\}$ | 0 | 1 | 0 | 1 |
| $\{b\}$ | 0 | 0 | 1 | 1 |
| $\{a,b\}$ | 0 | 0 | 0 | 1 |

**Fig. 1.1** Set-theoretic operations

intersection as follows:

$$M \cup N := \text{least subset of } X \text{ containing } M \text{ and } N;$$

$$M \cap N := \text{greatest subset of } X \text{ contained in } M \text{ and in } N.$$

So we are forming the *least upper bound* or *supremum* and the *greatest lower bound* or *infimum*, respectively, of two sets with respect to the ordering given by inclusion. Accordingly, we define for an arbitrary (possibly infinite) set of subsets $\mathcal{A} \subset \mathbf{2}^X$:

$$\sup \mathcal{A} := \text{least subset of } X \text{ containing all subsets in } \mathcal{A};$$

$$\inf \mathcal{A} := \text{greatest subset of } X \text{ contained in all subsets in } \mathcal{A}.$$

If $\mathcal{A}$ consists of the subsets $A_1 := \{a,b,c\}$, $A_2 := \{b,f\}$ and $A_3 := \{a,b,f\}$ then we have $\sup \mathcal{A} = A_1 \cup A_2 \cup A_3 = \{a,b,c,f\}$ and $\inf \mathcal{A} = A_1 \cap A_2 \cap A_3 = \{b\}$. It is possible to determine arbitrary suprema and infima in $\mathbf{2}^X$ in this descriptive fashion, even if there are infinitely many sets to be compared.

The complement is characterized by the following conditions which hold for every subset $M$ of $X$:

$$\overline{M} \cup M = X, \qquad \overline{M} \cap M = \emptyset.$$

Moreover, we have

$$\emptyset \cup M = M, \qquad \emptyset \cap M = \emptyset,$$
$$X \cup M = X, \qquad X \cap M = M.$$

Forming the complement $\overline{M}$ of a set $M$ is an

**involution**     $M = \overline{\overline{M}}.$

In connection with union and intersection there are the laws of

**de Morgan**     $\overline{M \cup N} = \overline{M} \cap \overline{N}, \quad \overline{M \cap N} = \overline{M} \cup \overline{N},$

$$\overline{\sup\{\,M \mid M \in \mathcal{A}\,\}} = \inf\{\,N \mid \overline{N} \in \mathcal{A}\,\},$$

$$\overline{\inf\{\,M \mid M \in \mathcal{A}\,\}} = \sup\{\,N \mid \overline{N} \in \mathcal{A}\,\}.$$

The inclusion $M \subset N$ is equivalent to each of the following statements:

$$\overline{M} \cup N = X, \quad M \cap \overline{N} = \emptyset, \quad M \cap N = M, \quad M \cup N = N.$$

(Union, intersection, and complement therefore satisfy the laws of a *complete Boolean lattice*). In the ordered set $\mathbf{2}^X$ the empty set $\emptyset$ is the least and the set $X$ is the greatest element. (In a general lattice such elements are also called *zero* and *universal element*, respectively).

An important Boolean lattice is the lattice of truth-values $\mathbb{B} = \{\,0,1\,\}$ whose operations are $\vee$ (disjunction, "or", Latin "vel"), $\wedge$ (conjunction, "and", Latin "et"), and $^{-}$ (negation, "not", Latin "non"). Figure 1.2 indicates the ordering of the subsets of $X = \{1,2,3\}$. There, a set is understood to be contained in those above it which are connected to it by a path of lines.

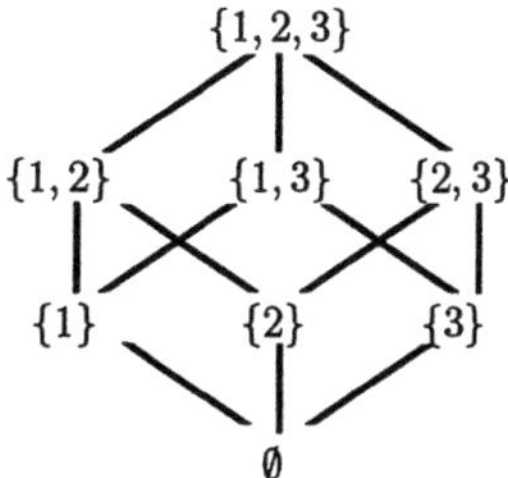

**Fig. 1.2** A power set ordered by inclusion

In the ordering by inclusion, the singleton sets $\{x\}$, $x \in X$, are closest to the empty set. For every nonempty subset $M$, $M \subset \{x\}$ implies $M = \{x\}$. So $\{x\}$ has no proper nonempty subsets. Such sets are called *atoms*. Distinct atoms have empty intersection. Moreover, we mention the following property:

$$a \text{ atom}, \ a \subset M \cup N \implies a \subset M \text{ or } a \subset N.$$

We add to our collection of operations on sets the difference of two sets $A, B$, defined as $A \backslash B := A \cap \overline{B}$. We then have $A \cup (A \backslash B) = A$. The symmetric version,

$$A + B := (\overline{A} \cap B) \cup (A \cap \overline{B}),$$

is called the *symmetric difference* of $A$ and $B$ (or symmetric sum) and is of greater interest. In the Boolean lattice of truth-values it is called the "exclusive or". One easily verifies

$$A + B = (A \cup B) \cap \overline{A \cap B},$$

as well as the laws of

$$
\begin{array}{ll}
\text{\textit{associativity}} & (A + B) + C = A + (B + C), \\
\text{\textit{commutativity}} & A + B = B + A, \\
\cap\text{\textit{-distributivity}} & A \cap (B + C) = (A \cap B) + (A \cap C).
\end{array}
$$

By associativity one may write $A+B+C$. In Fig. 1.3 a few symmetric differences are visualized.

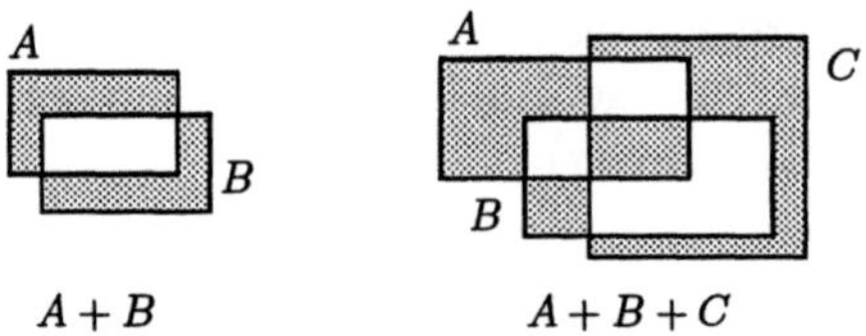

$A+B$          $A+B+C$

**Fig. 1.3** Symmetric differences

We indicate some further important rules without proof:

$$A + A = \emptyset, \quad A + \emptyset = A, \quad A + \overline{A} = X, \quad A + X = \overline{A},$$
$$A + (A + B) = B, \quad A + B = \emptyset \iff A = B,$$
$$A + B = \overline{\overline{A} + B} = \overline{A} + \overline{B}.$$

The *Cartesian product* $M \times N$ of two sets $M, N$ is the set of all (ordered) pairs having one component in $M$ and the other in $N$. Considered as an operation on subsets of a given set, this product is no longer an internal operation.

# 2. Homogeneous Relations

Relations between elements of the same set are called homogeneous; they are
the easier ones to investigate. We defer the study of heterogeneous relations,
i.e., those between elements of different sets, to Chap. 4. The homogeneous case
already presents many of the essential features and is notationally simpler. In
Sect. 2.1 we discuss the Boolean operations. They deal only with subsets of some
given set since union, intersection, and complement of relations are not dependent
on their nature as sets of pairs. This latter feature, however, gives rise to two
new operations: transposition and composition of relations which are treated in
Sects. 2.2 and 2.3. There we make the transition from the algebra of sets to the
algebra of relations. Section 2.4 contains a number of new concepts concerning
subsets, points, and orderings which are taken up again in the appendix.

## 2.1 Boolean Operations on Relations

We now consider relations. In ordinary language this word is used to indicate an
aspect that connects two or more things. Here is the mathematical definition in
terms of set theory:

**2.1.1 Definition.** Let $V$ be a set[1]. A **(homogeneous) relation** $R$ on $V$ is
a subset of the Cartesian product $V \times V$. Elements $x, y \in V$ are said to be in
relation $R$ if $(x, y) \in R$. $\qquad\qquad\square$

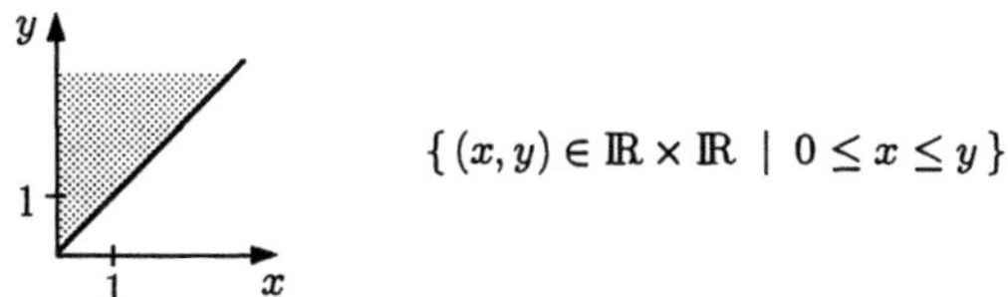

$$\{\, (x, y) \in \mathbb{R} \times \mathbb{R} \mid 0 \leq x \leq y \,\}$$

**Fig. 2.1.1** A relation as a subset

The relation "is a brother of" is defined on a set of persons; the relation "less
than" can be defined on the set $\mathbb{N}$ of natural numbers. A relation either does, or
does not, hold between two elements; 3 is less than 5, but it is not true that 4 is
less than 2. By this alternative, a relation partitions the set of pairs of elements
of the given set $V$ into those for which the relation holds, and those for which
it does not. The relation "is succeeded by" on $\mathbb{N}$ corresponds to the subset

$$S := \{(0, 1), (1, 2), (2, 3), \ldots\} \subset \mathbb{N} \times \mathbb{N},$$

---

[1] We will often speak of a relation $R$; however, we will almost never name or mention the
underlying set $V$.

for 0 is succeeded by 1, 1 by 2, etc. Figure 2.1.1 shows the relation $0 \leq x \leq y$ on the real numbers as a subset of the plane.

Relations on finite sets can also be represented by tables, matrices, or correspondences, as shown in Fig. 2.1.2 for the relation "congruent modulo 3" on the set $\{1, 2, \ldots, 7\} \subset \mathbb{N}$.

|   | 1 | 2 | 3 | 4 | 5 | 6 | 7 |
|---|---|---|---|---|---|---|---|
| 1 | × |   | × |   |   | × |   |
| 2 |   | × |   | × |   |   | × |
| 3 |   |   | × |   | × |   |   |
| 4 | × |   | × |   |   | × |   |
| 5 |   | × |   | × |   |   | × |
| 6 |   |   | × |   | × |   |   |
| 7 | × |   | × |   |   | × |   |

|   | 1 | 2 | 3 | 4 | 5 | 6 | 7 |
|---|---|---|---|---|---|---|---|
| 1 | 1 | 0 | 0 | 1 | 0 | 0 | 1 |
| 2 | 0 | 1 | 0 | 0 | 1 | 0 | 0 |
| 3 | 0 | 0 | 1 | 0 | 0 | 1 | 0 |
| 4 | 1 | 0 | 0 | 1 | 0 | 0 | 1 |
| 5 | 0 | 1 | 0 | 0 | 1 | 0 | 0 |
| 6 | 0 | 0 | 1 | 0 | 0 | 1 | 0 |
| 7 | 1 | 0 | 0 | 1 | 0 | 0 | 1 |

|   |   |
|---|---|
| 1 | $\{1, 4, 7\}$ |
| 2 | $\{2, 5\}$ |
| 3 | $\{3, 6\}$ |
| 4 | $\{1, 4, 7\}$ |
| 5 | $\{2, 5\}$ |
| 6 | $\{3, 6\}$ |
| 7 | $\{1, 4, 7\}$ |

**Fig. 2.1.2** A relation given by tables and by a Boolean matrix

Relations and graphs are closely related. Any homogeneous relation can be interpreted as the *transition relation* of a graph and vice versa. One usually represents a relation $B$ on a set $V$ graphically as follows: the elements of $V$ are plotted as points and two points $x, y \in V$ are joined by an oriented line segment (*arrow*) having $x$ as its *tail* and $y$ as its *head* precisely when $(x, y) \in B$. Sometimes, both these points are misleadingly called "endpoints" of the line segment. A relation depicted in such a way is called a graph, more precisely a directed 1-graph[2].

**2.1.2 Definition.** A **1-graph** (graph in short) $G = (V, B)$ consists of a set $V$ of **points** (also vertices, nodes) and a relation $B \subset V \times V$, the **associated relation**. The **order** of the graph $G$ is the cardinality $|V|$ of the set of points; $G$ is termed **finite** if its order is.                    □

A **loop** is an arrow whose head and tail coincide; in Fig. 2.1.3 a loop $(b, b)$ is attached to point $b$.

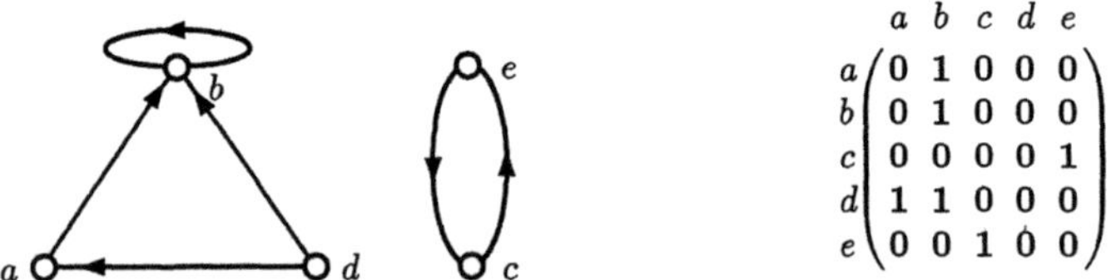

Fig. 2.1.3 part, Boolean matrix:

|   | a | b | c | d | e |
|---|---|---|---|---|---|
| a | 0 | 1 | 0 | 0 | 0 |
| b | 0 | 1 | 0 | 0 | 0 |
| c | 0 | 0 | 0 | 0 | 1 |
| d | 1 | 1 | 0 | 0 | 0 |
| e | 0 | 0 | 1 | 0 | 0 |

**Fig. 2.1.3** A 1-graph and its associated relation as a Boolean matrix

---

[2] The notion 1-graph originates from the fact that elements for which the relation holds are joined by exactly *one* arrow. (If $s$ parallel arrows are allowed, then the term $s$-graph is used.) In this chapter, arrows are not studied as independent objects but are used for visualizing relationships.

In a programming language such as Pascal a data structure for storing a finite relation can be given by an array; the same applies to 1-graphs. In general, this can be a wasteful method of storage. Moreover, it seldom occurs that a relation must be stored in the computer with its set of points staying the same during the entire period of investigation. In more instances the set may increase or decrease in the course of time (node dynamical situation), requiring sophisticated data structures.

Relations are sets, so we can consider their intersection, union, complement, and inclusion. With respect to these four operations the set of all relations on the set $V$, namely the powerset $2^{V \times V}$, is a complete Boolean lattice. In order to make clear when we are not working with arbitrary sets, we use the following modified notation for the related operations on relations $R, S$:

$$
\begin{aligned}
\text{union} \qquad & R \sqcup S = \{\, (x,y) \mid (x,y) \in R \vee (x,y) \in S \,\}, \\
\text{supremum} \qquad & \sup\{\, R \mid R \in \mathcal{S} \,\} = \{\, (x,y) \mid \exists R \in \mathcal{S} : (x,y) \in R \,\}, \\
\text{intersection} \qquad & R \sqcap S = \{\, (x,y) \mid (x,y) \in R \wedge (x,y) \in S \,\}, \\
\text{infimum} \qquad & \inf\{\, R \mid R \in \mathcal{S} \,\} = \{\, (x,y) \mid \forall R \in \mathcal{S} : (x,y) \in R \,\}, \\
\text{complement} \qquad & \overline{R} = \{\, (x,y) \mid (x,y) \notin R \,\}, \\
\text{inclusion} \qquad & R \subset S \iff \forall x,y : [(x,y) \in R \rightarrow (x,y) \in S].
\end{aligned}
$$

We are using the symbols $(\vee, \wedge)$, $(\cup, \cap)$, and $(\sqcup, \sqcap)$ for the two Boolean operations on truth values, subsets, and relations, respectively. The **empty relation** $\emptyset \subset V \times V$ is denoted[3] by $O$ (zero relation), and the entire Cartesian product $V \times V$ is called the **universal relation** $L$. Thus we have, for example,

$$
L = \overline{R} \sqcup S \iff R \subset S.
$$

A graph is said to be **empty** or **complete**, if its associated relation is zero or universal, respectively. If the associated relation $B$ of a graph $G$ is contained in that of another graph $G'$ (on the same set of points), then the graph $G$ is called a **partial graph** of $G'$. A **subgraph** $(V_1, B_1)$ of a given graph has a possibly smaller set $V_1 \subset V$ of points which, however, satisfy all the relationships of the given graph. A point is called **isolated** if it is not related to any other point.

Fig. 2.1.4 Zero and universal relations

Concerning the graph on the empty set, there is a paper by F. Harary and R. C. Read entitled "Is the null-graph a pointless concept?".

---

[3] Notations $O$ and $L$ should be used only if it is clear on which set $V$ they are defined; else one should rather use $O_V$ and $L_V$ for clarity.

Particular relations[4] on $V$ are the identity and nonidentity (diversity) of elements

$$I = \{ (x,y) \mid x = y \} \subset V \times V,$$
$$\overline{I} = \{ (x,y) \mid x \neq y \} \subset V \times V.$$

They are represented by the unit matrix and its Boolean complement, respectively; see Fig. 2.1.5. We shall see later that the identity is the neutral element for the multiplication or composition of relations.

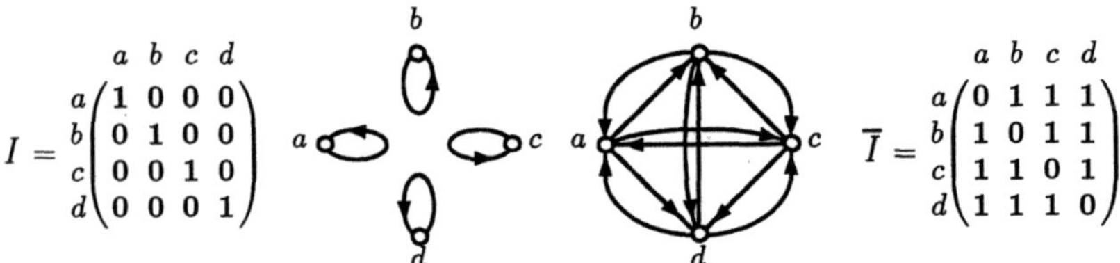


$$I = \begin{array}{c} \\ a \\ b \\ c \\ d \end{array} \begin{array}{cccc} a & b & c & d \\ \left( 1 \right. & 0 & 0 & 0 \\ 0 & 1 & 0 & 0 \\ 0 & 0 & 1 & 0 \\ 0 & 0 & 0 & \left. 1 \right) \end{array} \qquad \overline{I} = \begin{array}{c} \\ a \\ b \\ c \\ d \end{array} \begin{array}{cccc} a & b & c & d \\ \left( 0 \right. & 1 & 1 & 1 \\ 1 & 0 & 1 & 1 \\ 1 & 1 & 0 & 1 \\ 1 & 1 & 1 & \left. 0 \right) \end{array}$$

**Fig. 2.1.5** Identity and nonidentity

The identity relation is helpful for defining the following simple properties of relations. When dealing with graphs, one often uses the term "loop-free" instead of "irreflexive".

**2.1.3 Definition.** Let $R$ be a homogeneous relation. Then we define:

$R$ **reflexive**    $:\!\!\Longleftrightarrow$    $I \subset R$    $\Longleftrightarrow$    $R \sqcup I = R$    $\Longleftrightarrow$    $\forall x : (x,x) \in R;$

$R$ **irreflexive**    $:\!\!\Longleftrightarrow$    $R \subset \overline{I}$    $\Longleftrightarrow$    $R \sqcap \overline{I} = R$    $\Longleftrightarrow$    $\forall x : (x,x) \in \overline{R}.$
(loop-free)    $\square$

The above equivalences are obvious. The minimum enlargement of a relation necessary to make it reflexive is our first instance of a closure operator. We shall see more such examples in Sects. 2.2, 3.1, and 4.4.

**2.1.4 Definition.** The **reflexive closure** of a homogeneous relation $R$:

$$h_{\mathbf{refl}}(R) := \inf \{ H \mid H \supset R \text{ and } H \text{ reflexive} \} = R \sqcup I. \qquad \square$$

The above infimum has a *descriptive* character as it is the intersection of *all* relations in question, whereas the supremum as the union of two concrete relations is *constructive*. Both notions yield the same result: $H \supset R$ and $H \supset I$ imply $H \supset R \sqcup I$; so $R \sqcup I$ is a lower bound for the set of all reflexive relations $H \supset R$ and is therefore contained in $h_{\mathbf{refl}}(R)$. Conversely, $S := R \sqcup I$ is a reflexive relation that contains $R$; thus it occurs among the relations $H$ and therefore contains their infimum. We shall take up lattice-theoretic reasoning like this again, in a systematic way, in the appendix.

---

[4] Again, if it is not tacitly understood that $V$ is the underlying set, then $I_V$ and $\overline{I}_V$ could be used.

Incidentally, an algebra of homogeneous relations also carries a ring structure where the addition is $+$, the multiplication is $\sqcap$, and $L$ is the multiplicative unit element. Moreover, this multiplication is *idempotent*. If, conversely, such a **Boolean ring** is given with addition $+$ and multiplication $\sqcap$, then one can define the operations

$$\text{union} \qquad A \sqcup B := (A + B) + (A \sqcap B) \quad \text{and}$$
$$\text{complement} \qquad \overline{A} := A + L$$

which make the ring a Boolean lattice. (Compare with Chap. 1.)

## 2.2 Transposition of a Relation

A relation can be transposed—this is a characteristic trait of relation algebra as opposed to mere subset algebra. We define the **transpose** or **converse** of a relation $R$ as follows

$$R^{\mathsf{T}} := \{\, (x, y) \mid (y, x) \in R \,\}.$$

In Fig. 2.2.1 we explain this notion using two ways of representing a relation. So the relation $R^{\mathsf{T}}$ holds between $x$ and $y$ precisely when $R$ holds between $y$ and $x$.

$$
\begin{array}{c|c}
a & \{b\} \\
b & \{d, e\} \\
c & \{e\} \\
d & \{d\} \\
e & \emptyset
\end{array}
\qquad
\begin{array}{c|c}
a & \emptyset \\
b & \{a\} \\
c & \emptyset \\
d & \{b, d\} \\
e & \{b, c\}
\end{array}
\qquad
\begin{array}{c}
\begin{array}{ccccc} a & b & c & d & e \end{array} \\
\begin{array}{c|ccccc}
a & 0 & 1 & 0 & 0 & 0 \\
b & 0 & 0 & 0 & 1 & 1 \\
c & 0 & 0 & 0 & 0 & 1 \\
d & 0 & 0 & 0 & 1 & 0 \\
e & 0 & 0 & 0 & 0 & 0
\end{array}
\end{array}
\qquad
\begin{array}{c}
\begin{array}{ccccc} a & b & c & d & e \end{array} \\
\begin{array}{c|ccccc}
a & 0 & 0 & 0 & 0 & 0 \\
b & 1 & 0 & 0 & 0 & 0 \\
c & 0 & 0 & 0 & 0 & 0 \\
d & 0 & 1 & 0 & 1 & 0 \\
e & 0 & 1 & 1 & 0 & 0
\end{array}
\end{array}
$$

Tables for $R$ and $R^{\mathsf{T}}$       Boolean matrices for $R$ and $R^{\mathsf{T}}$

**Fig. 2.2.1** The transpose of a relation

When representing this relation as a 1-graph in Fig. 2.2.2, transposing amounts to reversing the arrows.

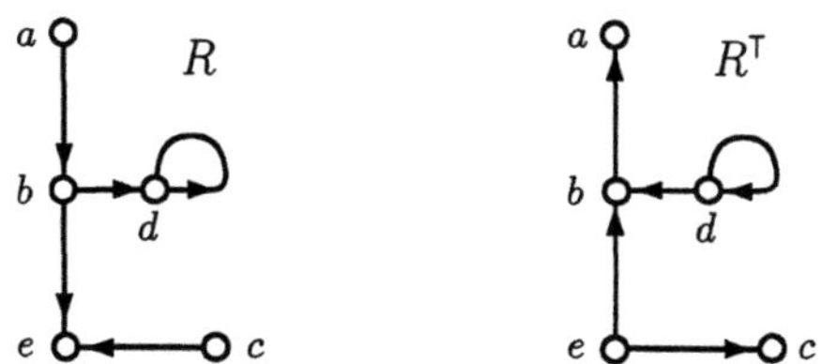


**Fig. 2.2.2** Transposing a 1-graph

Transposition is an involution and commutes with complementation:

$$(R^{\mathsf{T}})^{\mathsf{T}} = R, \qquad \overline{R}^{\mathsf{T}} = \overline{R^{\mathsf{T}}}.$$

We prove the second formula: Starting with the elements of $\overline{R}^{\mathsf{T}}$ we have for all $x, y$

$$(x, y) \in \overline{R}^{\mathsf{T}} \iff (y, x) \in \overline{R}$$
$$\iff (y, x) \notin R \iff (x, y) \notin R^{\mathsf{T}} \iff (x, y) \in \overline{R^{\mathsf{T}}}.$$

Transposition is compatible with the ordering by inclusion and commutes with union, intersection, and the formation of suprema and infima.

$$R \subset S \iff R^{\mathsf{T}} \subset S^{\mathsf{T}},$$
$$(R \sqcup S)^{\mathsf{T}} = R^{\mathsf{T}} \sqcup S^{\mathsf{T}}, \quad (R \sqcap S)^{\mathsf{T}} = R^{\mathsf{T}} \sqcap S^{\mathsf{T}},$$
$$[\,\sup\{\, R \mid R \in \mathcal{A} \,\}]^{\mathsf{T}} = \sup\{\, S \mid S^{\mathsf{T}} \in \mathcal{A} \,\},$$
$$[\,\inf\{\, R \mid R \in \mathcal{A} \,\}]^{\mathsf{T}} = \inf\{\, S \mid S^{\mathsf{T}} \in \mathcal{A} \,\}.$$

Identity, zero and universal relations stay the same when transposed:

$$I^{\mathsf{T}} = I, \quad O^{\mathsf{T}} = O, \quad L^{\mathsf{T}} = L.$$

In the sequel we shall make use of these simple rules for transposition[5] without expressly mentioning them.

Next we recall the notions of symmetry for homogeneous relations which we shall then reexpress using the transpose:

$$R \text{ symmetric} \quad :\iff \forall x, y : [(x,y) \in R \to (y,x) \in R];$$
$$R \text{ asymmetric} \quad :\iff \forall x, y : [(x,y) \in R \to (y,x) \notin R];$$
$$R \text{ antisymmetric} :\iff \forall x, y : [x \neq y \to \{(x,y) \notin R \vee (y,x) \notin R\}].$$

Antisymmetry can be expressed, in an equivalent way, as follows; it is therefore also called the **identitive** law:

$$\forall x, y : [\{(x,y) \in R \wedge (y,x) \in R\} \to x = y].$$

These notions can easily be expressed in terms of relations; the theorem below could also serve as a definition.

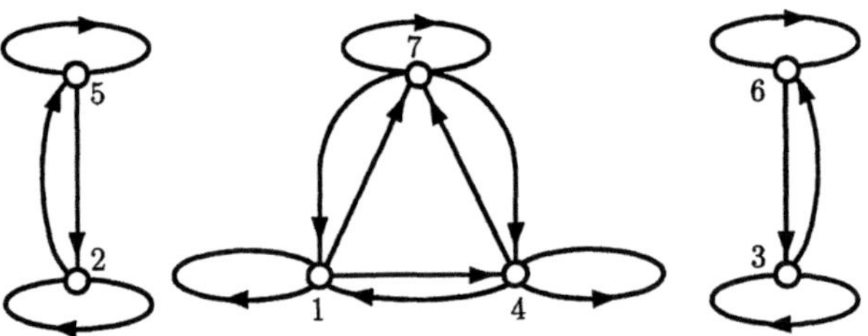

**Fig. 2.2.3** Reflexive and symmetric relation as a graph

**2.2.1 Proposition.** Let $R$ be a homogeneous relation. Then we have

$$R \text{ symmetric} \quad \iff \quad R^{\mathsf{T}} \subset R \quad \iff \quad R^{\mathsf{T}} = R;$$
$$R \text{ asymmetric} \quad \iff \quad R \sqcap R^{\mathsf{T}} \subset O \quad \iff \quad R^{\mathsf{T}} \subset \overline{R};$$
$$R \text{ antisymmetric} \iff \quad R \sqcap R^{\mathsf{T}} \subset I \quad \iff \quad R^{\mathsf{T}} \subset \overline{R} \sqcup I. \qquad \square$$

The 1-graph in Fig. 2.2.3 represents the symmetric relation from Fig. 2.1.2. But this simple example already lacks clarity, because points are joined by couples

---

[5] In Appendix A.2 we show how to derive these rules algebraically from a few axioms, without resorting to their set-theoretic formulation.

of arrows pointing in opposite directions. For drawing symmetric relations one therefore prefers "undirected" graphs, as in Fig. 2.2.4, where the (undirected) *edges* represent pairs of arrows pointing in opposite directions.

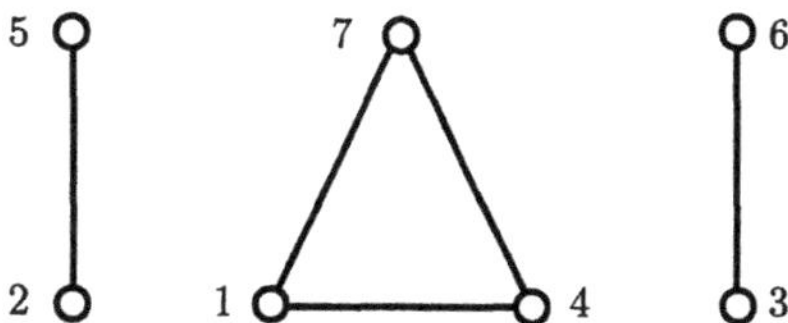

**Fig. 2.2.4** Irreflexive and symmetric relation as an undirected graph

In a loop-free graph only distinct points are joined by an edge. In general, it may be more convenient to express reflexivity in a graph by marking the nodes, rather than drawing loops. A symmetric and irreflexive relation is called an **adjacency**.

**2.2.2 Definition.** A **simple graph** (graph in short) $G = (V, \Gamma)$ consists of a set $V$ of points and an irreflexive, symmetric relation $\Gamma$ on $V$, called the **adjacency** relation of the graph. So $\Gamma = \Gamma^{\mathsf{T}} \subset \overline{I}$ holds. Two points $x, y$ are called **adjacent**, if they are related by $\Gamma$, i.e., if $(x, y) \in \Gamma$. Given a 1-graph $(V, B)$, we call $\left(V, \overline{I} \sqcap (B \sqcup B^{\mathsf{T}})\right)$ its **associated simple graph**.     □

The irreflexive relation $\Gamma$ is sometimes written as a set of two-element subsets of points. If $|V| = n$ and $\Gamma = \overline{I}$ is the set of *all* two-element subsets of $V$, then $(V, \Gamma)$ is called the **$n$-clique**. In Fig. 2.2.5 a simple example is given.

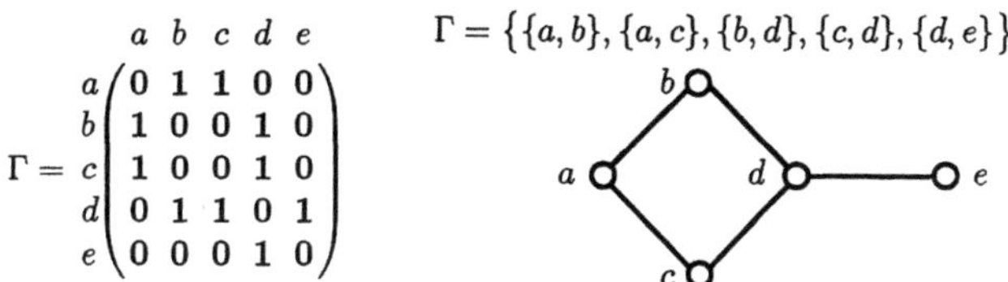

$$\Gamma = c \begin{pmatrix} & a & b & c & d & e \\ a & 0 & 1 & 1 & 0 & 0 \\ b & 1 & 0 & 0 & 1 & 0 \\ c & 1 & 0 & 0 & 1 & 0 \\ d & 0 & 1 & 1 & 0 & 1 \\ e & 0 & 0 & 0 & 1 & 0 \end{pmatrix}$$

$$\Gamma = \{\{a, b\}, \{a, c\}, \{b, d\}, \{c, d\}, \{d, e\}\}$$

**Fig. 2.2.5** Representations of a simple graph

Figure 2.2.6 represents the adjacency relation for the 11 states of the Federal Republic of Germany as of 1989. Two (distinct) states are joined by an edge precisely when they have a border line in common.

Every relation is contained in a symmetric one, namely, the universal relation $L$. As in the case of the reflexive closure (Def. 2.1.4), there is the symmetric closure, and, again, our definition has a descriptive and an, obviously equivalent, constructive version:

**2.2.3 Definition.** The **symmetric closure** of a homogeneous relation $R$ is

$$h_{\text{symm}}(R) := \inf\{H \mid H \supset R, H \text{ symmetric}\} = R \sqcup R^{\mathsf{T}}.$$     □

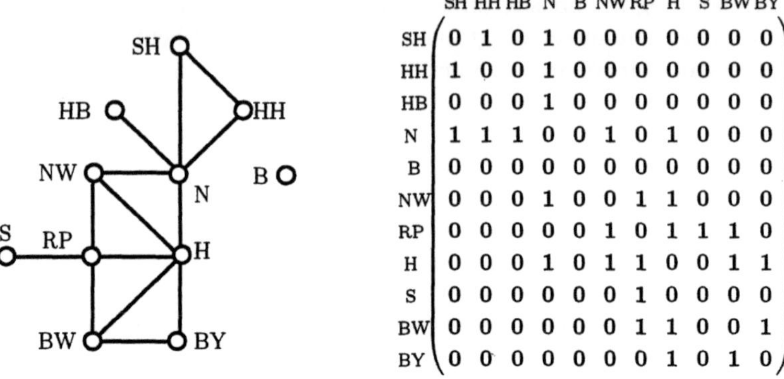

| | SH | HH | HB | N | B | NW | RP | H | S | BW | BY |
|---|---|---|---|---|---|---|---|---|---|---|---|
| SH | 0 | 1 | 0 | 1 | 0 | 0 | 0 | 0 | 0 | 0 | 0 |
| HH | 1 | 0 | 0 | 1 | 0 | 0 | 0 | 0 | 0 | 0 | 0 |
| HB | 0 | 0 | 0 | 1 | 0 | 0 | 0 | 0 | 0 | 0 | 0 |
| N | 1 | 1 | 1 | 0 | 0 | 1 | 0 | 1 | 0 | 0 | 0 |
| B | 0 | 0 | 0 | 0 | 0 | 0 | 0 | 0 | 0 | 0 | 0 |
| NW | 0 | 0 | 0 | 1 | 0 | 0 | 1 | 1 | 0 | 0 | 0 |
| RP | 0 | 0 | 0 | 0 | 0 | 1 | 0 | 1 | 1 | 1 | 0 |
| H | 0 | 0 | 0 | 1 | 0 | 1 | 1 | 0 | 0 | 1 | 1 |
| S | 0 | 0 | 0 | 0 | 0 | 0 | 1 | 0 | 0 | 0 | 0 |
| BW | 0 | 0 | 0 | 0 | 0 | 0 | 1 | 1 | 0 | 0 | 1 |
| BY | 0 | 0 | 0 | 0 | 0 | 0 | 0 | 1 | 0 | 1 | 0 |

**Fig. 2.2.6** Adjacency of the states of the Federal Republic of Germany as of 1989

We now define connexity (not to be confused with connectedness) and later express it by means of the transpose:

$$R \text{ connex} \quad :\!\Longleftrightarrow \quad \forall x,y : (x,y) \in R \lor (y,x) \in R;$$

$$R \text{ semiconnex} \quad :\!\Longleftrightarrow \quad \forall x,y : [x \neq y \to \{ (x,y) \in R \lor (y,x) \in R \}].$$

With respect to a connex relation $R$, every pair of points is either in $R$ or in $R^{\mathsf{T}}$. A good example is provided by a complete championship series of games where a result of win or loss has been entered in a table for every pair of teams. Therefore an antisymmetric and connex relation is sometimes called a **tournament**. For $R$ semiconnex, this is required for pairs of distinct points only. Hence a reflexive, semiconnex relation is connex. We formulate, without proof, the relation-algebraic characterization of these properties:

**2.2.4 Proposition.** Let $R$ be a homogeneous relation. Then

$$R \text{ connex} \quad \Longleftrightarrow \quad L \subset R \sqcup R^{\mathsf{T}} \Longleftrightarrow \overline{R} \subset R^{\mathsf{T}} \quad \Longleftrightarrow \overline{R} \text{ asymmetric};$$

$$R \text{ semiconnex} \Longleftrightarrow \overline{I} \subset R \sqcup R^{\mathsf{T}} \Longleftrightarrow \overline{R} \subset R^{\mathsf{T}} \sqcup I \Longleftrightarrow \overline{R} \text{ antisymmetric.} \quad \Box$$

If we call a pair of points $(x,y)$ **collateral (independent, incomparable)** when it satisfies $x \neq y$, $(x,y) \notin R$, $(y,x) \notin R$, then we can say that a relation without collateral pairs of points is semiconnex.

## 2.3 The Product of Two Relations

The composition of relations is a generalization of the composition of mappings. In contrast to mappings, however, relations associate *sets* of values with elements. These sets may be empty, or contain one, or more, elements. So, by composing the relations $R$ and $S$, we associate with the element $x$ the union of all sets of values associated by $S$ with elements $y$ such that $(x, y) \in R$. When representing relations by graphs, composition amounts to linking consecutive arrows, see Fig. 2.3.1.

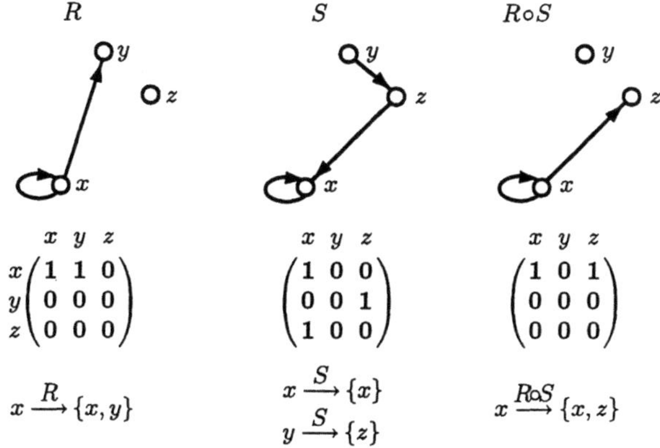

**Fig. 2.3.1** Composition of relations

**2.3.1 Definition.** Let $R, S \subset V \times V$ be relations. Their **product** $R, S \subset V \times V$ is given by

$$R \circ S := \{\, (x, z) \in V \times V \mid \exists y \in V : (x, y) \in R \wedge (y, z) \in S \,\}. \qquad \square$$

Accordingly, one also speaks of **multiplication** or **composition** of relations. The product $R \circ S$ is also written as $RS$ or $R \,;\, S$. The semicolon was introduced by E. Schröder (1895), and, indeed, the operational symbol for the product has a meaning similar to the semicolon used as a separating particle in programming languages. After the preliminaries we will simply write $RS$ for the product and $R^2, R^3 \ldots$ for the powers of $R$. In the example in Fig. 2.3.1 we have $(x, z) \in R \circ S$ because

$$\exists i \in V : (x, i) \in R \wedge (i, z) \in S,$$

(namely, $i = y$) which is tantamount to

$$[(x, x) \in R \wedge (x, z) \in S] \vee [(x, y) \in R \wedge (y, z) \in S] \vee [(x, z) \in R \wedge (z, z) \in S].$$

When multiplying relations it is convenient to utilize their matrix representation because the result can then be obtained by rewriting the formula for matrix multiplication in a Boolean way: replace addition "+" and multiplication "·" for a field by disjunction "∨" and conjunction "∧" for the 2-element Boolean lattice $\mathbb{B} = \{\, 0, 1 \,\}$ of truth values. Form the conjunction element by element, of the $x$-row of $R$ with the $z$-column of $S$, and then form the disjunction of all these results with respect to $y$:

$$(R_{xx} \wedge S_{xz}) \vee (R_{xy} \wedge S_{yz}) \vee (R_{xz} \wedge S_{zz}) = \exists i : R_{xi} \wedge S_{iz} = (R \circ S)_{xz}.$$

Compare the result obtained to the formula for the ordinary matrix product,

$$(R_{xx} \cdot S_{xz}) + (R_{xy} \cdot S_{yz}) + (R_{xz} \cdot S_{zz}) = \sum_{i \in V} R_{xi} \cdot S_{iz} = (R \cdot S)_{xz}$$

where, in the first formula, we put $R_{xy} = 1 \in \mathbb{B}$ if and only if $(x, y) \in R$.

As for multiplication and addition of numbers, we use the convention that $\circ$ binds more strongly than $\sqcup$ and $\sqcap$. So, for example, $Q \circ R \sqcap S = (Q \circ R) \sqcap S$.

The set $\mathcal{R} = \mathcal{R}(V)$ of all relations on a set $V$ carries two structures: With respect to the set-theoretic operations, $(\mathcal{R}, \sqcup, \sqcap, \bar{\phantom{x}})$ is a complete, atomic, Boolean lattice with zero element $O$, universal element $L$, and ordering by inclusion $\subset$ (see Chap. 1). Secondly, $(\mathcal{R}, \circ)$ is a semigroup with unit element $I$; we formulate this as a proposition:

**2.3.2 Proposition** (*Properties of multiplication*).

i)    The multiplication of relations is associative:
$$(Q \circ R) \circ S = Q \circ (R \circ S).$$

ii)    The identity $I$ acts as unit element for multiplication, i.e.:
$$I \circ R = R \circ I = R. \qquad \square$$

This fact, as well as Prop. 2.3.3 below, is easily established by looking at the quantifier form of the expressions in question. Figure 2.3.2 serves as an illustration for our next proposition.

$$R = \begin{pmatrix} 0 & 0 & 0 & 0 \\ 0 & 1 & 0 & 0 \\ 0 & 0 & 0 & 1 \\ 0 & 1 & 0 & 1 \end{pmatrix} \quad L \circ R = \begin{pmatrix} 0 & 1 & 0 & 1 \\ 0 & 1 & 0 & 1 \\ 0 & 1 & 0 & 1 \\ 0 & 1 & 0 & 1 \end{pmatrix} \quad R \circ L = \begin{pmatrix} 0 & 0 & 0 & 0 \\ 1 & 1 & 1 & 1 \\ 1 & 1 & 1 & 1 \\ 1 & 1 & 1 & 1 \end{pmatrix}$$

**Fig. 2.3.2** Tarski rule

**2.3.3 Proposition** (*Tarski rule*[6]).  $L \circ R \circ L = L$    for all $R \neq O$.    $\square$

Composition and transposition are connected by the following formula which is familiar from matrix calculus and easy to prove:

$$(R \circ S)^{\mathsf{T}} = S^{\mathsf{T}} \circ R^{\mathsf{T}}.$$

---

[6] We gave this name in honour of Alfred Tarski (1911–1983) who revitalized relational algebra in 1941. In his axiomatic considerations he observed that the rule has to do with the "simplicity" of the algebra. He also gave an equivalent formulation,
$$R \circ L = L \quad \text{or} \quad L \circ \overline{R} = L \quad \text{for all } R,$$
of the rule which is investigated in Exercise 2.3.5.

The following equivalences for containments of relations are already present[7] in E. Schröder's work (1895) and have been rediscovered several times since then—this may be due to the fact that M. H. Stone proved that every Boolean lattice can be represented as a lattice of subsets only as late as 1936. Before that time things developed separately. Concerning subset lattices, L. Löwenheim, for instance, invented an "area calculus" based on geometrical, rather than logical, arguments.

The interplay between composition, transposition, and complement, with respect to containment, is given by the following:

**2.3.4 Proposition** (*Schröder equivalences*). Given relations $Q, R, S$, one has

$$Q \circ R \subset S \quad\Longleftrightarrow\quad Q^{\mathsf{T}} \circ \overline{S} \subset \overline{R} \quad\Longleftrightarrow\quad \overline{S} \circ R^{\mathsf{T}} \subset \overline{Q}.$$

**Proof**: The assertion is obviously equivalent to

$$\overline{A}^{\mathsf{T}} \circ \overline{B}^{\mathsf{T}} \subset C \quad\Longleftrightarrow\quad \overline{B}^{\mathsf{T}} \circ \overline{C}^{\mathsf{T}} \subset A \quad\Longleftrightarrow\quad \overline{C}^{\mathsf{T}} \circ \overline{A}^{\mathsf{T}} \subset B,$$

where the relations $A, B, C$ are cyclically permuted. The statement $\overline{A}^{\mathsf{T}} \circ \overline{B}^{\mathsf{T}} \subset C$ written explicitly runs as follows

$$\forall x, y : [\{ \exists z : (x, z) \in \overline{A}^{\mathsf{T}} \wedge (z, y) \in \overline{B}^{\mathsf{T}} \} \to (x, y) \in C].$$

After replacing the implication $a \to b$ by $\overline{a} \vee b$, de Morgan's rule yields

$$\forall x, y : \forall z : (x, z) \notin \overline{A}^{\mathsf{T}} \vee (z, y) \notin \overline{B}^{\mathsf{T}} \vee (x, y) \in C$$

which is equivalent to

$$\forall x, y, z : (z, x) \in A \vee (y, z) \in B \vee (x, y) \in C.$$

The other two cases will lead to the same result, up to a permutation of the disjunction terms. $\qquad\square$

The transition from one containment to another can be memorized as follows: Transpose the first (or second) relation and complement, and permute the other two.

There is no analog of Schröder's equivalences for the case where $Q \subseteq R \circ S$. Exercise 2.3.4 shows how much more special this case is. Nor can $\overline{R \circ S}$ be calculated by algebraic laws.

As an example we now apply the Schröder equivalences to some types of kinship. Denote by $B, F, M$ the relations "is a brother of", "is the father of", and "is the mother of", respectively; in addition let $G$ denote "is a godfather of" (see Figs. 2.3.3 and 2.3.4). Then $P := F \sqcup M$ means "is a parent of" and $B \circ P$ means "is an uncle of". Let us assume that uncles in our family traditionally become godfathers, so that we have $B \circ P \subset G$, while allowing other types of godfathership, too.

By Schröder's equivalence applied to our family the formula $B^{\mathsf{T}} \circ \overline{G} \subset \overline{P}$ holds true and reads as follows: If there is a family member $x$ who is a brother of $z$, but not a godfather of $y$, then $z$ cannot be a parent of $y$—of course not, because

---

[7]  As we learned from Roger Maddux, while preparing the English translation, this result was already known to C. S. Peirce and even earlier to A. de Morgan in 1864 as "Theorem K".

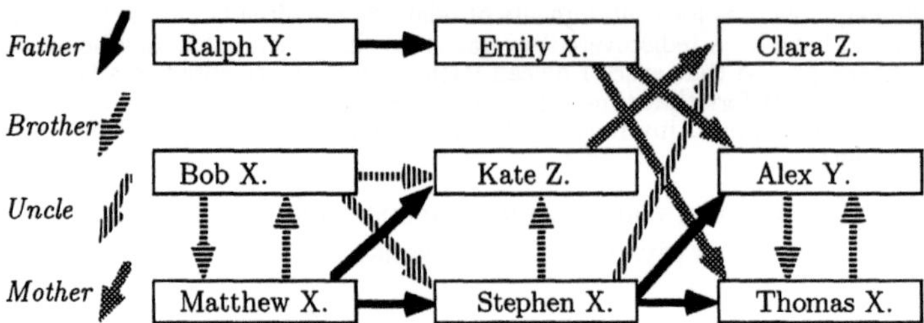


**Fig. 2.3.3** An instance of Schröder equivalence

otherwise, contrary to family tradition, $x$ would be an uncle of $y$ without being $y$'s godfather. We feel the formal manipulation is shorter and safer than figuring everything out using quantifiers. The kinships from Fig. 2.3.3 are represented by matrices in Fig. 2.3.4.

$$
F = \begin{array}{c c}
 & \begin{array}{c c c c c c c c c} R & B & M & E & K & S & C & A & T \end{array} \\
\begin{array}{c} R \\ B \\ M \\ E \\ K \\ S \\ C \\ A \\ T \end{array} &
\left(\begin{array}{c c c c c c c c c}
0 & 0 & 0 & 1 & 0 & 0 & 0 & 0 & 0 \\
0 & 0 & 0 & 0 & 0 & 0 & 0 & 0 & 0 \\
0 & 0 & 0 & 0 & 1 & 1 & 0 & 0 & 0 \\
0 & 0 & 0 & 0 & 0 & 0 & 0 & 0 & 0 \\
0 & 0 & 0 & 0 & 0 & 0 & 0 & 0 & 0 \\
0 & 0 & 0 & 0 & 0 & 0 & 0 & 1 & 1 \\
0 & 0 & 0 & 0 & 0 & 0 & 0 & 0 & 0 \\
0 & 0 & 0 & 0 & 0 & 0 & 0 & 0 & 0 \\
0 & 0 & 0 & 0 & 0 & 0 & 0 & 0 & 0
\end{array}\right)
\end{array}
\qquad
M = \begin{array}{c}
\begin{array}{c c c c c c c c c} R & B & M & E & K & S & C & A & T \end{array} \\
\left(\begin{array}{c c c c c c c c c}
0 & 0 & 0 & 0 & 0 & 0 & 0 & 0 & 0 \\
0 & 0 & 0 & 0 & 0 & 0 & 0 & 0 & 0 \\
0 & 0 & 0 & 0 & 0 & 0 & 0 & 0 & 0 \\
0 & 0 & 0 & 0 & 0 & 0 & 0 & 1 & 1 \\
0 & 0 & 0 & 0 & 0 & 0 & 1 & 0 & 0 \\
0 & 0 & 0 & 0 & 0 & 0 & 0 & 0 & 0 \\
0 & 0 & 0 & 0 & 0 & 0 & 0 & 0 & 0 \\
0 & 0 & 0 & 0 & 0 & 0 & 0 & 0 & 0 \\
0 & 0 & 0 & 0 & 0 & 0 & 0 & 0 & 0
\end{array}\right)
\end{array}
\qquad
B \circ P = \begin{array}{c}
\begin{array}{c c c c c c c c c} R & B & M & E & K & S & C & A & T \end{array} \\
\left(\begin{array}{c c c c c c c c c}
0 & 0 & 0 & 0 & 0 & 0 & 0 & 0 & 0 \\
0 & 0 & 0 & 0 & 1 & 1 & 0 & 0 & 0 \\
0 & 0 & 0 & 0 & 0 & 0 & 0 & 0 & 0 \\
0 & 0 & 0 & 0 & 0 & 0 & 0 & 0 & 0 \\
0 & 0 & 0 & 0 & 0 & 0 & 0 & 0 & 0 \\
0 & 0 & 0 & 0 & 0 & 0 & 1 & 0 & 0 \\
0 & 0 & 0 & 0 & 0 & 0 & 0 & 0 & 0 \\
0 & 0 & 0 & 0 & 0 & 0 & 0 & 0 & 0 \\
0 & 0 & 0 & 0 & 0 & 0 & 0 & 0 & 0
\end{array}\right)
\end{array}
$$

$$
B = \begin{array}{c c}
 & \begin{array}{c c c c c c c c c} R & B & M & E & K & S & C & A & T \end{array} \\
\begin{array}{c} R \\ B \\ M \\ E \\ K \\ S \\ C \\ A \\ T \end{array} &
\left(\begin{array}{c c c c c c c c c}
0 & 0 & 0 & 0 & 0 & 0 & 0 & 0 & 0 \\
0 & 0 & 1 & 0 & 0 & 0 & 0 & 0 & 0 \\
0 & 1 & 0 & 0 & 0 & 0 & 0 & 0 & 0 \\
0 & 0 & 0 & 0 & 0 & 0 & 0 & 0 & 0 \\
0 & 0 & 0 & 0 & 0 & 0 & 0 & 0 & 0 \\
0 & 0 & 0 & 0 & 1 & 0 & 0 & 0 & 0 \\
0 & 0 & 0 & 0 & 0 & 0 & 0 & 0 & 0 \\
0 & 0 & 0 & 0 & 0 & 0 & 0 & 0 & 1 \\
0 & 0 & 0 & 0 & 0 & 0 & 0 & 1 & 0
\end{array}\right)
\end{array}
\qquad
G = \left(\begin{array}{c c c c c c c c c}
0 & 0 & 0 & 0 & 0 & 1 & 0 & 0 & 0 \\
0 & 0 & 0 & 0 & 1 & 1 & 0 & 0 & 0 \\
0 & 0 & 0 & 0 & 0 & 0 & 0 & 0 & 0 \\
0 & 0 & 0 & 0 & 0 & 0 & 0 & 0 & 0 \\
0 & 0 & 0 & 0 & 0 & 0 & 0 & 0 & 0 \\
0 & 0 & 0 & 0 & 0 & 0 & 1 & 0 & 0 \\
0 & 0 & 0 & 0 & 0 & 0 & 0 & 0 & 0 \\
0 & 0 & 0 & 0 & 0 & 0 & 0 & 0 & 0 \\
0 & 0 & 0 & 0 & 0 & 0 & 0 & 0 & 0
\end{array}\right)
\qquad
B^{\mathsf{T}} \circ \overline{G} = \left(\begin{array}{c c c c c c c c c}
0 & 0 & 0 & 0 & 0 & 0 & 0 & 0 & 0 \\
1 & 1 & 1 & 1 & 1 & 1 & 1 & 1 & 1 \\
1 & 1 & 1 & 1 & 0 & 0 & 1 & 1 & 1 \\
0 & 0 & 0 & 0 & 0 & 0 & 0 & 0 & 0 \\
1 & 1 & 1 & 1 & 1 & 1 & 0 & 1 & 1 \\
0 & 0 & 0 & 0 & 0 & 0 & 1 & 0 & 0 \\
0 & 0 & 0 & 0 & 0 & 0 & 0 & 0 & 0 \\
1 & 1 & 1 & 1 & 1 & 1 & 1 & 1 & 1 \\
1 & 1 & 1 & 1 & 1 & 1 & 1 & 1 & 1
\end{array}\right)
$$

**Fig. 2.3.4** Boolean matrices for Fig. 2.3.3

There is a variant[8] of the Schröder equivalences, introduced by J. Riguet as "formule de Dedekind", which is very helpful in the calculus of relations; P. Lorenzen calls it the "Bund-Axiom"; see GERICKE 83.

**2.3.5 Corollary** (*Dedekind formula, J. Riguet 1948*). Given $Q, R$ and $S$, one has

$$Q \circ R \sqcap S \ \subset\ (Q \sqcap S \circ R^{\mathsf{T}}) \circ (R \sqcap Q^{\mathsf{T}} \circ S).$$

**Proof:** A "component free" proof can be given using the Schröder equivalences (see Exercise 2.3.3). Here we argue as follows:

$$(x, y) \in Q \circ R \sqcap S \quad\Longleftrightarrow\quad [\exists z : (x, z) \in Q \wedge (z, y) \in R] \wedge (x, y) \in S$$

---

[8] In abstract relation algebra this variant can be used as an axiom and is then an equivalent substitute for Schröder's equivalence; see Exercise 2.3.3.

$$\begin{aligned}
\Longleftrightarrow \quad & \exists z : [(x,z) \in Q \wedge (x,y) \in S \wedge (y,z) \in R^{\mathsf{T}}] \wedge \\
& \qquad [(z,y) \in R \wedge (z,x) \in Q^{\mathsf{T}} \wedge (x,y) \in S] \\
\Longrightarrow \quad & \exists z : [(x,z) \in Q \wedge [\exists w : (x,w) \in S \wedge (w,z) \in R^{\mathsf{T}}]] \wedge \\
& \qquad (z,y) \in R \wedge [\exists u : (z,u) \in Q^{\mathsf{T}} \wedge (u,y) \in S] \\
\Longleftrightarrow \quad & \exists z : [(x,z) \in Q \wedge (x,z) \in S \circ R^{\mathsf{T}}] \wedge [(z,y) \in R \wedge (z,y) \in Q^{\mathsf{T}} \circ S] \\
\Longleftrightarrow \quad & \exists z : [(x,z) \in Q \sqcap S \circ R^{\mathsf{T}}] \wedge [(z,y) \in R \sqcap Q^{\mathsf{T}} \circ S] \\
\Longleftrightarrow \quad & (x,y) \in (Q \sqcap S \circ R^{\mathsf{T}}) \circ (R \sqcap Q^{\mathsf{T}} \circ S). \qquad \qquad \square
\end{aligned}$$

The Dedekind formula does not involve the complement and can be advantageous for giving proofs in the form of a chain of containments. If we apply

$$B \circ P \sqcap G \subset (B \sqcap G \circ P^{\mathsf{T}}) \circ (P \sqcap B^{\mathsf{T}} \circ G)$$

to the relations of our example above, we get the following interpretation: If $x$ is an uncle and a godfather of $y$, then there is a family member $z$ so that $x$ is a brother of $z$ and has a godchild $w$ with $z$ as a parent; and moreover, $z$ is a parent of $y$, and there is a brother $u$ of $z$ who is a godfather of $y$. Apparently, we did not assume every uncle to be a godfather.

We now prove the essential laws for multiplication.

**2.3.6 Proposition** (*Laws for multiplication*).

i)
$$R \circ O = O \circ R = O;$$

ii)   Monotonicity of multiplication
$$P \subset Q, \quad R \subset S \quad \Longrightarrow \quad P \circ R \subset Q \circ S;$$

iii)   $\sqcap$-subdistributivity of multiplication
$$Q \circ (R \sqcap S) \subset Q \circ R \sqcap Q \circ S,$$
$$Q \circ \inf \{\, R \mid R \in \mathcal{A} \,\} \subset \inf \{\, S \mid S = Q \circ R \text{ and } R \in \mathcal{A} \,\};$$

iv)   $\sqcup$-distributivity of multiplication
$$Q \circ (R \sqcup S) = Q \circ R \sqcup Q \circ S,$$
$$Q \circ \sup \{\, R \mid R \in \mathcal{A} \,\} = \sup \{\, S \mid S = Q \circ R \text{ and } R \in \mathcal{A} \,\}.$$

For multiplication by $Q$ from the right, (iii, iv) hold accordingly.

**Proof**[9]: i) Since $L$ contains every other relation, we have $R^{\mathsf{T}} \circ L \subset L$; the Schröder equivalence yields $R \circ O \subset O \Longleftrightarrow R^{\mathsf{T}} \circ L \subset L$.

ii) Since $P \circ S \subset P \circ S$, Schröder implies $P^{\mathsf{T}} \circ \overline{P \circ S} \subset \overline{S} \subset \overline{R}$ as well as monotonicity with respect to the second factor, $P \circ R \subset P \circ S$. Similarly, one deduces $P \circ S \subset Q \circ S$ and thus gets $P \circ R \subset Q \circ S$.

iii) $Q \circ (R \sqcap S) \subset Q \circ R \sqcap Q \circ S$, since $Q \circ (R \sqcap S) \subset Q \circ R$ and $Q \circ (R \sqcap S) \subset Q \circ S$. Infinite intersections are handled likewise: We have $\inf \{\, R \mid R \in \mathcal{A} \,\} \subset W$ for

---

[9] Note that the transposition rules from Sect. 2.2 will not be used for the proof; see the axiomatic considerations of Appendix A.2.

every $W \in \mathcal{A}$ and so, by (ii), $Q \circ \inf \{ R \mid R \in \mathcal{A} \} \subset Q \circ W =: S$. Therefore, the left-hand side is contained in the intersection of all these right-hand sides.

iv) By monotonicity, $Q \circ R \subset Q \circ (R \sqcup S)$, $Q \circ S \subset Q \circ (R \sqcup S)$; so $Q \circ R \sqcup Q \circ S \subset Q \circ (R \sqcup S)$. On the other hand, by Schröder, we have $Q^{\mathsf{T}} \circ \overline{Q \circ R} \subset \overline{R}$ and $Q^{\mathsf{T}} \circ \overline{Q \circ S} \subset \overline{S}$ which, by (iii), yields $Q^{\mathsf{T}} \circ \overline{Q \circ R \sqcup Q \circ S} = Q^{\mathsf{T}} \circ (\overline{Q \circ R} \sqcap \overline{Q \circ S}) \subset Q^{\mathsf{T}} \circ \overline{Q \circ R} \sqcap Q^{\mathsf{T}} \circ \overline{Q \circ S} \subset \overline{R} \sqcap \overline{S} = \overline{R \sqcup S}$. Again by Schröder we obtain $Q \circ (R \sqcup S) \subset Q \circ R \sqcup Q \circ S$. Suprema can be treated in a similar fashion.     $\square$


In order to see that $\sqcap$-distributivity does not hold in general, we look at Boolean matrices with two or more rows:

$$O = L \circ O = L \circ (I \sqcap \overline{I}) \subset L \circ I \sqcap L \circ \overline{I} = L \sqcap L = L.$$

As a straightforward consequence of Propositions 2.3.3 and 2.3.6, the universal relation $L$ is annihilated by the zero relation $O$ only. Thus

$$R \circ L = O \quad \Longleftrightarrow \quad R = O.$$

From now on we write $RS$ instead of $R \circ S$ for the product of two relations.


## Residues and Symmetric Quotient

The material of this section serves only for external references and may therefore be omitted at the first reading. Guided by the similarity between matrices over a field and relations, one might go a step further and ask: Under which circumstances can "linear" equations for relations be solved? This question was dealt with by R. D. Luce in 1952; he rediscovered what we call Schröder equivalence and proved the following:


**2.3.7 Proposition** (*Solving linear equations*). Let $R$ and $S$ be relations. Then

$$RX = S \text{ is solvable in } X \quad \Longleftrightarrow \quad S \subset \overline{R\overline{R^{\mathsf{T}}\overline{S}}} \quad \Longleftrightarrow \quad S = \overline{R\overline{R^{\mathsf{T}}\overline{S}}}.$$

If solutions exist, then every solution is contained in the special solution $\overline{R^{\mathsf{T}}\overline{S}}$.

**Proof:** If $X$ is a relation with $RX = S$, then $RX \subset S$ is by Schröder equivalent to $R^{\mathsf{T}}\overline{S} \subset \overline{X}$ and to $X \subset \overline{R^{\mathsf{T}}\overline{S}}$. This yields $S = RX \subset \overline{R\overline{R^{\mathsf{T}}\overline{S}}}$. The reverse containment $\overline{R\overline{R^{\mathsf{T}}\overline{S}}} \subset S$ is equivalent to $R^{\mathsf{T}}\overline{S} \subset R^{\mathsf{T}}\overline{S}$ and, therefore, is always true.     $\square$


We now discuss some less familiar operations on relations which are, however, closely related to composition. Occasionally, the operational symbol "$\dagger$" is used—it can already be found in Schröder, 1890, and is reminiscent of "$;$", that is, "$.$" and "$,$", which in turn was used by Schröder for the relation product. The operational symbols "$\cdot\,.$" and "$.\,\cdot$" are used by G. Birkhoff.

**2.3.8 Definition.** Let $R$ and $S$ be relations.

i) $S \dagger R := \overline{\overline{S}\,\overline{R}}$;

ii) $S \,.\!\cdot\, R := \overline{R^{\mathsf{T}}\overline{S}} = \overline{R}^{\mathsf{T}} \dagger S = \sup\{X \mid RX \subset S\}$ **right residue** of $S$ by $R$;

iii) $S \,\cdot\!.\, R := \overline{\overline{S}\,R^{\mathsf{T}}} = S \dagger \overline{R}^{\mathsf{T}} = \sup\{Y \mid YR \subset S\}$ **left residue** of $S$ by $R$. $\square$

The right residue constitutes the *largest* solution to $RX \subset S$ and is therefore also the largest solution to $RX = S$ if this equation is solvable (cf. Prop. 2.3.7). The left residue has also been named "weakest prespecification" in HOARE, HE 86, denoted by $R/S$—it has arisen in connection with methods of program verification.

It is plausible that the double negation in (i) has the effect of exchanging distributivity for subdistributivity, and conversely. We now gather some further results of this sort without proving them.

**2.3.9 Proposition.** i) The operation $\dagger$ is $\sqcap$-distributive:

$$(Q \sqcap R) \dagger S = (Q \dagger S) \sqcap (R \dagger S),$$
$$Q \dagger (R \sqcap S) = (Q \dagger R) \sqcap (Q \dagger S).$$

ii) The operation $\dagger$ is isotonic: $\quad P \subset Q, \; R \subset S \;\Longrightarrow\; P \dagger R \subset Q \dagger S$;

iii) $$\overline{I} \dagger R = R \dagger \overline{I} = R;$$

iv) $$\overline{I} \,\cdot\!.\, \overline{R} = R^{\mathsf{T}}.$$

v) The following rules of distributivity are only shown in one version

$$(P \sqcap Q) \,.\!\cdot\, R = (P \,.\!\cdot\, R) \sqcap (Q \,.\!\cdot\, R),$$
$$P \,.\!\cdot\, (Q \sqcup R) = (P \,.\!\cdot\, Q) \sqcap (P \,.\!\cdot\, R);$$

vi) $$PQ \subset R \;\Longleftrightarrow\; Q \subset R \,.\!\cdot\, P \;\Longleftrightarrow\; P \subset R \,\cdot\!.\, Q;$$

vii) $$(PQ) \,.\!\cdot\, P \supset Q, \qquad (PQ) \,\cdot\!.\, Q \supset P. \qquad\qquad \square$$

Equation (iv) can serve to rewrite relation algebra by replacing the transpose by the left residue. Schröder's equivalences are then to be replaced by (vi). However, we will not pursue this any further. The following symmetrized version of residues will be used later.

**2.3.10 Definition.** If $R$ and $S$ are given relations, we call

$$\mathbf{syq}(R,S) := \overline{R^{\mathsf{T}}\overline{S}} \sqcap \overline{\overline{R}^{\mathsf{T}}S} = \sup\{\, X \mid RX \subset S \text{ and } XS^{\mathsf{T}} \subset R^{\mathsf{T}} \,\}$$

the **symmetric quotient**[10]. $\square$

---

[10] The word "quotient" is partly justified by the fact that $R\,\mathbf{syq}(R,S) \subset R\overline{R^{\mathsf{T}}\overline{S}} \subset S$. We shall see in Sect. 4.4 under what condition there is equality. Note that the first version of $\mathbf{syq}(R,S)$,

Let us show that the two expressions given in the definition are equal. Let $W := \overline{R^\mathsf{T}\overline{S}} \sqcap \overline{\overline{R}^\mathsf{T} S}$. Clearly, $RX \subset S$ and $XS^\mathsf{T} \subset R^\mathsf{T}$ is equivalent to $X \subset \overline{R^\mathsf{T}\overline{S}}$ and $X \subset \overline{\overline{R}^\mathsf{T} S}$ so that any such $X$ satisfies $X \subset W$. On the other hand, we have already seen (proof of Prop. 2.3.7) that $W$ satisfies $RW \subset S$, the first defining property. Also, $WS^\mathsf{T} \subset \overline{\overline{R}^\mathsf{T} S}S^\mathsf{T} \subset R^\mathsf{T}$. So $W$ is indeed the largest relation satisfying the two conditions and is thus the supremum of all these relations.

### Abstract Relation Algebras

We have deduced Propositions 2.3.6 and 2.3.7 from Propositions 2.3.2, 2.3.3, and 2.3.4 by arguments involving no element relationship $(x, y) \in R$. In the exercises we require similar proofs for Dedekind's formula 2.3.5. Proofs not involving quantifiers are usually shorter and more transparent, hence also easier to check and to memorize. Moreover, this type of proof is capable of being performed by a machine.

We define an (abstract) relation algebra to be an algebraic structure that satisfies the laws of an atomic, complete, Boolean lattice, the properties of multiplication (2.3.2), the Tarski rule (2.3.3), and the Schröder equivalences (2.3.4). By contrast, when we consider a given set $V$ and relations on it, we refer to concrete relations. We call a proof **relation-algebraic**, if it involves only the laws of a relation algebra, and their consequences, such as (2.3.6), but makes no use of the matrix-like structure of $R \subset V \times V$. Results proved in such a way hold true in every relation algebra, hence also in those constructed by the procedure of (7.5.14).

### Exercises

**2.3.1**   Prove that $(I \sqcap RR^\mathsf{T})R = R = R(I \sqcap R^\mathsf{T}R)$ for every relation $R$.

**2.3.2**   Prove by relation-algebraic techniques

$$Q\overline{\overline{RS}} \subset \overline{\overline{QRS}} \quad \text{and} \quad \overline{Q\overline{R}S} \subset \overline{\overline{QRS}}.$$

**2.3.3**   Deduce the Dedekind formula $QR \sqcap S \subset (Q \sqcap SR^\mathsf{T})(R \sqcap Q^\mathsf{T}S)$ relation-algebraically. Hint: In the product $QR$ the factor $Q$ should be split with respect to $SR^\mathsf{T}$ and its complement; ($R$ with respect to $Q^\mathsf{T}S$).

**2.3.4**   In 1952, R. D. Luce proved the existence of $X$ with the properties

$$X \subset S, \quad Q \subset LX \quad \text{and} \quad QX^\mathsf{T} \subset R$$

to be sufficient for $Q \subset RS$. Prove this using the Dedekind formula.

**2.3.5**   Show that in a relation algebra the Tarski rule

$$R \neq O \implies LRL = L \quad \text{for all } R$$

holds precisely when the following rule is satisfied

$$SL = L \quad \text{or} \quad L\overline{S} = L \quad \text{for all } S.$$

---

although symmetric with respect to $R$ and $S$, is not, in general, a symmetric relation. For $R = S$ this construction was introduced by J. Riguet under the name of "noyau".

**2.3.6**   Show that the relations $R, R^\mathsf{T} R$, and $R R^\mathsf{T}$ can only vanish simultaneously:

$$R \neq O \quad \Longleftrightarrow \quad R^\mathsf{T} R \neq O.$$

**2.3.7**   Show that every relation contained in the identity is idempotent:

$$R \subset I \quad \Longrightarrow \quad R^2 = R.$$

**2.3.8**   Show that for any $R$ we have $\overline{RL} = \overline{RLL}$.

**2.3.9**   Prove that $P^\mathsf{T} Y Q^\mathsf{T} \subset X \quad \Longleftrightarrow \quad P \overline{X} Q \subset \overline{Y}$.

**2.3.10**   Prove that $PQ \subset \overline{I} \quad \Longleftrightarrow \quad QP \subset \overline{I}$.

**2.3.11**   Prove that $PQ \sqcap R = P(P^\mathsf{T} R \sqcap Q) \sqcap R$.

## 2.4 Subsets and Points

Relations can be used for characterizing subsets of a set. To this end associate with a subset $U \subset V$ the relation

$$u := \{\, (x,y) \mid x \in U \,\} \subset V \times V.$$

It imposes a condition only on the first component of pairs. The relation $u$ corresponding to the subset $U = \{\, b,c \,\} \subset V = \{\, a,b,c,d \,\}$ is represented by a matrix in Fig. 2.4.1.

$$u = \begin{array}{c} \\ a \\ b \\ c \\ d \end{array}\begin{array}{c} a\ \ b\ \ c\ \ d \\ \begin{pmatrix} 0 & 0 & 0 & 0 \\ 1 & 1 & 1 & 1 \\ 1 & 1 & 1 & 1 \\ 0 & 0 & 0 & 0 \end{pmatrix} \end{array} \begin{pmatrix} 0 \\ 1 \\ 1 \\ 0 \end{pmatrix}$$

**Fig. 2.4.1** A subset as a relation

This matrix is "row-constant", and it is easy to see that such matrices are characterized by $u = uL$. So there is a one-to-one correspondence between subsets $U \subset V$ and relations $R$ on $V$ subject to $R = RL$. One can abbreviate the row-constant matrix by the vector on the right. Relations then correspond to matrices, and subsets to vectors. As in matrix calculus, we write capitals for relations and lowercase letters for vectors.

**2.4.1 Definition.**   A relation $x$ satisfying $x = xL$ is called a **subset given by a predicate** or a **(Boolean) vector**.                                    $\square$

For any relation $R$ the construct $x := RL$ is a vector which, by analogy to mappings, describes the **domain** of $R$:

$$(i,j) \in RL \quad \Longleftrightarrow \quad \exists k : (i,k) \in R.$$

If $x, y$ are vectors, then so are $\overline{x}$, $x \sqcup y$, and $x \sqcap y$, but not necessarily $x^\mathsf{T}$. In particular, $\overline{RL} = \overline{RLL}$ (see Exercise 2.3.8). Therefore, vectors form a sublattice which is actually an ideal in the lattice of all relations, since $Rx$ is a vector for every relation $R$ and vector $x$. We now prove two frequently applied formulas describing the effect of the relation product on vectors.

Intersecting a Boolean matrix with a vector such as $RL$ amounts to **singling out** certain rows and replacing the entries of all the other rows by zeros. Similarly, $LR$ will single out certain columns.

**2.4.2 Proposition.** The following formulas hold for arbitrary relations $Q, R, S$.

i)
$$[Q \sqcap RL]S = QS \sqcap RL;$$

ii)
$$[Q \sqcap (RL)^{\mathsf{T}}]S = Q[S \sqcap RL].$$

**Proof:** i) $(Q \sqcap RL)S \subset QS \sqcap RLS \subset QS \sqcap RL$
$$\subset (Q \sqcap RLS^{\mathsf{T}})(S \sqcap Q^{\mathsf{T}}RL) \subset (Q \sqcap RL)S.$$
ii) $[Q \sqcap (RL)^{\mathsf{T}}]S \sqcap L \subset [Q \sqcap (RL)^{\mathsf{T}} \sqcap LS^{\mathsf{T}}][S \sqcap (Q^{\mathsf{T}} \sqcap RL)L]$
$$\subset Q(S \sqcap RL) = Q(S \sqcap RL) \sqcap L$$
$$\subset [Q \sqcap L(S \sqcap RL)^{\mathsf{T}}][S \sqcap RL \sqcap Q^{\mathsf{T}}L] \subset [Q \sqcap (RL)^{\mathsf{T}}]S. \qquad \square$$

Of course, one also has $(Q \sqcap \overline{RL})S = QS \sqcap \overline{RL}$, as well as $(Q \sqcap \overline{RL}^{\mathsf{T}})S = Q(S \sqcap \overline{RL})$.

Equation (i) means that singling out rows in a product can also be performed by first doing this for the first factor and then multiplying by the second. Transposing in (i) and renaming the relations yields

$$Q[S \sqcap (RL)^{\mathsf{T}}] = QS \sqcap (RL)^{\mathsf{T}},$$

i.e., the singling out of columns of a product is equivalent to doing this for the second factor only. In (ii), the singling out of columns in the first factor of a product is equivalent to singling out the corresponding rows of the second factor. (ii) is invariant with respect to transposition. Figure 2.4.2 gives an example.

$[Q \sqcap RL] \circ S = Q \circ S \sqcap RL:$

$$\left[\begin{pmatrix} 0 & 0 & 0 & 1 \\ 1 & 0 & 1 & 0 \\ 0 & 0 & 0 & 0 \\ 1 & 0 & 1 & 0 \end{pmatrix} \sqcap \begin{pmatrix} 1 & 1 & 1 & 1 \\ 1 & 1 & 1 & 1 \\ 1 & 1 & 1 & 1 \\ 0 & 0 & 0 & 0 \end{pmatrix}\right] \circ \begin{pmatrix} 0 & 1 & 1 & 1 \\ 1 & 0 & 1 & 1 \\ 0 & 0 & 0 & 1 \\ 1 & 0 & 1 & 1 \end{pmatrix} = \begin{pmatrix} 0 & 0 & 0 & 1 \\ 1 & 0 & 1 & 0 \\ 0 & 0 & 0 & 0 \\ 1 & 0 & 1 & 0 \end{pmatrix} \circ \begin{pmatrix} 0 & 1 & 1 & 1 \\ 1 & 0 & 1 & 1 \\ 0 & 0 & 0 & 1 \\ 1 & 0 & 1 & 1 \end{pmatrix} \sqcap \begin{pmatrix} 1 & 1 & 1 & 1 \\ 1 & 1 & 1 & 1 \\ 1 & 1 & 1 & 1 \\ 0 & 0 & 0 & 0 \end{pmatrix}$$

$[Q \sqcap (PL)^{\mathsf{T}}] \circ S = Q \circ [S \sqcap PL]:$

$$\left[\begin{pmatrix} 0 & 0 & 0 & 1 \\ 1 & 0 & 1 & 0 \\ 0 & 0 & 0 & 0 \\ 1 & 0 & 1 & 0 \end{pmatrix} \sqcap \begin{pmatrix} 1 & 1 & 0 & 0 \\ 1 & 1 & 0 & 0 \\ 1 & 1 & 0 & 0 \\ 1 & 1 & 0 & 0 \end{pmatrix}\right] \circ \begin{pmatrix} 0 & 1 & 1 & 1 \\ 1 & 0 & 1 & 1 \\ 0 & 0 & 0 & 1 \\ 1 & 0 & 1 & 1 \end{pmatrix} = \begin{pmatrix} 0 & 0 & 0 & 1 \\ 1 & 0 & 1 & 0 \\ 0 & 0 & 0 & 0 \\ 1 & 0 & 1 & 0 \end{pmatrix} \circ \left[\begin{pmatrix} 0 & 1 & 1 & 1 \\ 1 & 0 & 1 & 1 \\ 0 & 0 & 0 & 1 \\ 1 & 0 & 1 & 1 \end{pmatrix} \sqcap \begin{pmatrix} 1 & 1 & 1 & 1 \\ 1 & 1 & 1 & 1 \\ 0 & 0 & 0 & 0 \\ 0 & 0 & 0 & 0 \end{pmatrix}\right]$$

**Fig. 2.4.2** Singling out rows before and after multiplication

Single elements of $V$, interpreted as one-element subsets, can be characterized via relations as follows:

**2.4.3 Definition.** We say that a relation

$$x \text{ is a } \textbf{point} \quad :\Longleftrightarrow \quad x = xL, \ xx^{\mathsf{T}} \subset I, \ x \neq O. \qquad \square$$

Indeed, a vector $x$ such that $xx^\mathsf{T} \subset I$ can have at most one element since $xx^\mathsf{T} \subset I$ implies

$$\forall i,j : [\{ \exists k : (i,k) \in x \wedge (k,j) \in x^\mathsf{T}\} \to (i,j) \in I]$$

what can also be written as

$$\forall i,j : [\{ \exists k : (i,k) \in x \wedge (j,k) \in x \} \to i = j].$$

$$\begin{pmatrix} 0 & 0 & 0 & 0 & 0 \\ 1 & 1 & 1 & 1 & 1 \\ 1 & 1 & 1 & 1 & 1 \\ 0 & 0 & 0 & 0 & 0 \\ 1 & 1 & 1 & 1 & 1 \end{pmatrix} \qquad \begin{pmatrix} 0 & 0 & 0 & 0 & 0 \\ 0 & 0 & 0 & 0 & 0 \\ 1 & 1 & 1 & 1 & 1 \\ 0 & 0 & 0 & 0 & 0 \\ 0 & 0 & 0 & 0 & 0 \end{pmatrix} \qquad \begin{pmatrix} 0 \\ 0 \\ 1 \\ 0 \\ 0 \end{pmatrix}$$

**Fig. 2.4.3** Subset, point, and vector

Since $x = xL$ has constant rows, there is at most one row with nonzero entries. Nonemptiness is part of the assumption. The following propositions deal with the "algebraic" manipulation of elements of $V$ interpreted as relations.

**2.4.4 Proposition.** i) $\quad R \subset Sx \quad \Longleftrightarrow \quad Rx^\mathsf{T} \subset S, \quad$ if $x$ is a point.

ii) $\quad y \subset Sx \quad \Longleftrightarrow \quad x \subset S^\mathsf{T}y, \quad$ if $x,y$ are points.

**Proof:** i) We use $R \subset Sx$ together with $xx^\mathsf{T} \subset I$ and monotonicity to obtain $Rx^\mathsf{T} \subset Sxx^\mathsf{T} \subset S$. In the other direction, Tarski's rule and the equivalence $Rx^\mathsf{T} \subset S \Longleftrightarrow \overline{S}x \subset \overline{R}$ are used to prove

$$L = LxL = Lx = \overline{S}x \sqcup Sx \subset \overline{R} \sqcup Sx,$$

yielding $R \subset Sx$. ii) From (i) we have $y \subset Sx \Longleftrightarrow yx^\mathsf{T} \subset S$, which, transposed, gives $xy^\mathsf{T} \subset S^\mathsf{T}$. The proof is completed by employing (i) again. $\qquad \square$

We now show that points are atoms (see Fig. 2.4.4). In addition, we give algebraic criteria for two points to be equal or different.

$$x = \begin{array}{c} \\ a \\ b \\ c \end{array}\!\!\begin{array}{c} a\ b\ c \\ \begin{pmatrix} 0 & 0 & 0 \\ 0 & 0 & 0 \\ 1 & 1 & 1 \end{pmatrix} \end{array} \qquad y^\mathsf{T} = \begin{array}{c} \\ a \\ b \\ c \end{array}\!\!\begin{array}{c} a\ b\ c \\ \begin{pmatrix} 0 & 1 & 0 \\ 0 & 1 & 0 \\ 0 & 1 & 0 \end{pmatrix} \end{array} \qquad xy^\mathsf{T} = \begin{array}{c} \\ a \\ b \\ c \end{array}\!\!\begin{array}{c} a\ b\ c \\ \begin{pmatrix} 0 & 0 & 0 \\ 0 & 0 & 0 \\ 0 & 1 & 0 \end{pmatrix} \end{array}$$

**Fig. 2.4.4** Atomicity of points

**2.4.5 Proposition.** i) Every point is an atom among the vectors, i.e., if $x$ is a point and $y$ is a vector with $O \neq y \subset x$, then necessarily $x = y$.

ii) If $x,y$ are points, then:

$$x \neq y \quad \Longleftrightarrow \quad x \subset \overline{y} \quad \Longleftrightarrow \quad xy^\mathsf{T} \subset \overline{I} \quad \Longleftrightarrow \quad x^\mathsf{T}y = O;$$
$$x = y \quad \Longleftrightarrow \quad x \subset y \quad \Longleftrightarrow \quad xy^\mathsf{T} \subset I \quad \Longleftrightarrow \quad x^\mathsf{T}y = L.$$

**Proof:**   i) Since $Ly = LyL = L$ by Tarski's rule, we may use $xx^\mathsf{T} \subset I$ to obtain

$$x = Ly \sqcap x \subset (L \sqcap xy^\mathsf{T})(y \sqcap L^\mathsf{T}x) \subset xy^\mathsf{T}y \subset xx^\mathsf{T}y \subset y.$$

ii) We begin by considering the first equivalences on the left-hand sides of the two lines. Since $x \neq y$, we cannot have $x \subsetneq y$ because otherwise, by (i), $x = y$. So $x \sqcap \overline{y} \neq O$. Again by (i), $x \sqcap \overline{y} = x$. The first equivalence in line two follows from (i). In order to establish the remaining equivalences, write $x \subset \overline{y}$ or $x \subset y$ as $Ix \subset \overline{y}$ and $\overline{I}y = \overline{y} \subset \overline{x}$, respectively, for the third version, and as $xL \subset \overline{y}$ and $Lx^\mathsf{T} \subset y^\mathsf{T}$, for the fourth version. To the first three of the four inclusions at hand apply Schröder's equivalences, to the fourth, Prop. 2.4.4. It remains to establish $\overline{I}y = \overline{y}$. Here, "$\subset$" is equivalent to $yy^\mathsf{T} \subset I$, and "$\supset$" follows by Prop. 2.4.4 from $Iy \subset y$ which is equivalent to $\overline{y}y^\mathsf{T} \subset \overline{I}$. $\qquad\square$

**2.4.6 Proposition** (*Point axiom*).  For every relation $R \neq O$ there exist two points $x, y$ such that $xy^\mathsf{T} \subset R$.

**Proof**[11]:  If $R \neq O$ is a relation on $V$ then there are elements $x, y \in V$ such that $(x, y) \in R$. Interpreting these elements relation-algebraically as points, we have $xy^\mathsf{T} \subset R$. $\qquad\square$

It is not hard to see that relations of the form $xy^\mathsf{T}$ ($x, y$ points) are always atoms in a relation algebra. The proof is left as Exercise 2.4.2.

**2.4.7 Proposition.**  Every nonzero vector contains a point.

**Proof:**  Let $O \neq v = vL$ be some vector. Then by Prop. 2.4.6 there are points $x$, $y$ with $xy^\mathsf{T} \subset v$. This yields $x = xL = xy^\mathsf{T}y \subset vy \subset vL = v$. $\qquad\square$

The following proposition, too, is dependent on the point axiom. In order to get an impression of what may happen if the point axiom is not valid, one can think of parallel processes which cannot be observed simultaneously. In a situation like that there are no algebraic means of identifying intermediate stages of the procedure as precise points.

**2.4.8 Proposition** (*Intermediate point theorem*).  For points $x$, $y$ and relations $R$, $S$ we have

$$x \subset RSy \quad \Longleftrightarrow \quad \text{There exists a point } z \text{ with } x \subset Rz \text{ and } z \subset Sy.$$

---

[11] The proof of this proposition constitutes a break of style. The preceding propositions were proved in a *relation-algebraic* way, i.e., by algebraic reasoning using the catalogue of laws set up in Sect. 2.1. Except when discussing examples, we never had to resort to the original definition of a relation as a set of pairs. Proposition 2.4.6 *cannot* be proved in this fashion. There are algebraic structures satisfying all the statements on relations proved so far, but where Prop. 2.4.6 does not hold. In Sect. 7.5 a procedure for constructing such examples will be given. So one can work with a set-up of relation algebras which does, or does not, contain Prop. 2.4.6 as an independent **axiom**.

**Proof:** Direction "$\Longleftarrow$" is trivial. In order to prove "$\Longrightarrow$" we start by showing that $x \subset RSy = R(Sy \sqcap \overline{\overline{R}^{\mathsf{T}}x}) \sqcup R(Sy \sqcap \overline{R}^{\mathsf{T}}x)$ can be strengthened to $x \subset R(Sy \sqcap \overline{R}^{\mathsf{T}}x)$: Because of $xx^{\mathsf{T}} \subset I$ we have $xx^{\mathsf{T}}\overline{R} \subset \overline{R}$ and $R\overline{R}^{\mathsf{T}}x \subset \overline{x}$, yielding $R(Sy \sqcap \overline{R}^{\mathsf{T}}x) \subset \overline{x}$. Since $Sy \sqcap \overline{\overline{R}^{\mathsf{T}}x}$ is a nonempty vector (otherwise $x = O$) we can apply the preceding theorem and choose some point $z \subset Sy \sqcap \overline{\overline{R}^{\mathsf{T}}x}$. Now $z \subset Sy$, fulfilling one of the requirements, and also $z \subset \overline{\overline{R}^{\mathsf{T}}x}$. The latter leads to $x^{\mathsf{T}}\overline{R} \subset \overline{z}^{\mathsf{T}}$ and $xz^{\mathsf{T}} \subset R$, which by Prop. 2.4.4.i is equivalent to the other requirement $x \subset Rz$. $\qquad\square$

**Remark.** In Def. 2.1.2 we introduced relations as subsets of $V \times V$ with $V$ an explicitly given set which, however, will lose its role when we deal with abstract relation algebra in the sense of Appendix A.2. With this intention in mind we shall always work with the relation-algebraic properties of points. Even in those few instances where it is necessary to assume the validity of the point axiom we will stick to relation-algebraic reasoning, although some simple results may then look overformalized.

## Predecessors, Successors and Neighbors

When dealing with subsets, especially of finite sets, their numbers are of interest. Relations on *finite* sets are typically represented by graphs, so we return now to the language of 1-graphs or simple graphs.

**2.4.9 Definition.** If the vector $x$ stands for a finite subset, then we denote by

$$|x| \quad \text{the number of elements of the subset } x.$$

$\qquad\square$

Given a point $x$, we consider the set of points which are tails of arrows ending in $x$, as well as the set of points that are heads of arrows originating in $x$. So we give the following

**2.4.10 Definition.** Let $B$ denote the associated relation of a graph and let $x$ be a vector. Then we denote by

$$Bx \quad \text{the set of \textbf{predecessors} of } x,$$
$$B^{\mathsf{T}}x \quad \text{the set of \textbf{successors} of } x.$$

When $x$ is a point, we call

$$d_p(x) := |Bx| \quad \text{the \textbf{predecessor degree} of } x,$$
$$d_s(x) := |B^{\mathsf{T}}x| \quad \text{the \textbf{successor degree} of } x.$$

$\qquad\square$

Often the set $Bx \sqcap \overline{x}$ of the **proper predecessors** and the set $B^{\mathsf{T}}x \sqcap \overline{x}$ of the **proper successors** of the point $x$ are needed.

In Fig. 2.1.3, for instance, we have $Bd = \emptyset$ and $B^{\mathsf{T}}d = \{a, b\}$, so $d$ has predecessor degree $d_p(d) = 0$ and successor degree $d_s(d) = 2$.

In general, we have:

$$xy^\mathsf{T} \subset R^\mathsf{T} R \quad \Longleftrightarrow \quad x \text{ and } y \text{ have a common predecessor,}$$
$$xy^\mathsf{T} \subset R R^\mathsf{T} \quad \Longleftrightarrow \quad x \text{ and } y \text{ have a common successor.}$$

As an example, $R R^\mathsf{T} \subset I$ means that no two distinct elements have a successor in common. In many cases, weakened versions of certain assumptions suffice already. A typical such situation is to require that a property hold only for pairs of points having a predecessor in common, rather than for all pairs of points. "Completeness" ($L \subset R$) can be weakened to "local completeness" ($R^\mathsf{T} R \subset R$); being "connex" ($L \subset R \sqcup R^\mathsf{T}$), to being "locally connex" ($R^\mathsf{T} R \subset R \sqcup R^\mathsf{T}$). Later, in connection with orderings, we shall encounter the properties of being "directed" ($L \subset R R^\mathsf{T}$) and "locally directed" or "rhombically confluent" ($R^\mathsf{T} R \subset R R^\mathsf{T}$). Notions of this sort are interrelated in many ways, especially in the presence of transitivity or with regard to the identity $I$.

**2.4.11 Definition.** Let $x$ be a point in a graph with associated relation $B$. Then we say that

$$x \text{ \textbf{carries a loop}} \quad :\Longleftrightarrow \quad x \subset Bx \quad \Longleftrightarrow \quad x \subset B^\mathsf{T} x \quad \Longleftrightarrow \quad xx^\mathsf{T} \subset B$$
$$\Longleftrightarrow \quad x \text{ is its own predecessor and successor.} \qquad \square$$

Using the terminology for graphs, the associated relation $B$ is called *loop-free* [12] if it is irreflexive, i.e., if $B \subset \overline{I}$ (see Def. 2.1.3).

By Prop. 2.4.4, the various characterizations of a loop-carrying point are equivalent. Since a loop-carrying point is its own predecessor and successor it must to be counted for its predecessor and successor degrees. In Fig. 2.1.3 the loop-carrying point $b$ has the only proper predecessors $a$ and $d$; it is, however, its own (improper) predecessor and also the only (improper) successor of itself [13]. A loop-carrying point which is unrelated, by relation $B$, to any other point is called *isolated* with respect to the associated adjacency $\Gamma := \overline{I} \sqcap (B \sqcup B^\mathsf{T})$.

**2.4.12 Definition.** For $\Gamma$ an adjacency relation and $x$ a vector we call

$$\Gamma x \text{ the set of \textbf{neighbors} of } x.$$

For one-element sets $x$ we define

$$d_0(x) := |\Gamma x| \text{ the \textbf{simple degree} of the point } x. \qquad \square$$

---

[12] Note that we have not defined "loops" as mathematical objects proper. In spite of this, we are using the notions of "carrying a loop" for a point and of "loop-free" for the associated relation.

[13] A graph which is loop-free in the sense of Def. 2.1.3 cannot contain a loop-carrying point $x$ because $xx^\mathsf{T} \subset B \subset \overline{I}$ would contradict $xx^\mathsf{T} \subset I$. By the point axiom, a *non*loop-free graph contains points $x, y$ such that $xy^\mathsf{T} \subset B \sqcap I \neq O$. By Prop. 2.4.5, then $x = y$ which, together with $xx^\mathsf{T} \subset B \sqcap I \subset B$, yields a loop-carrying point. One should be cautious, however, because there are relation algebras for which the point axiom (Prop. 2.4.6) does not hold; they may contain relations $B \not\subset \overline{I}$, (corresponding to *non*loop-free graphs) *without* loop-carrying points. If the point axiom is valid, however, then loop-freeness and the nonexistence of loop-carrying points are equivalent.

Point $H$ in Fig. 2.2.6 has five neighboring points, namely, N, NW, RP, BW, and BY. So H has simple degree $d_0 = 5$.

Since an occurrence of $xy^\top \subset \Gamma$ contributes to the simple degree $d_0(x)$ as well as to $d_0(y)$, the sum of all simple degrees must be even.

**2.4.13 Proposition.** In a finite, simple graph the number of points of odd degree is even.                                                          $\square$

In Sect. 5.4 we shall classify points according to the number of their neighbors. Points having no neighbor are termed isolated. The set of all isolated points is $\overline{\Gamma L}$. A graph contains *only* isolated points exactly when $\Gamma = O$. Graphs containing *no* isolated point are characterized by $\Gamma L = L$.

## Exercises

**2.4.1** Prove $\quad R \subset I \implies \overline{RL} = (I \sqcap \overline{R})L$.

**2.4.2** Show that $xy^\top$ is an atomic relation if $x, y$ are points.

**2.4.3** Show that points $x, y$ are uniquely determined by the product term $xy^\top$:

$$xy^\top = uv^\top \quad \Longleftrightarrow \quad xL = uL \quad \text{and} \quad yL = vL.$$

**2.4.4** Define the so-called demonic composition $R\,;;\,S := RS \sqcap \overline{R\overline{SL}}$ and prove that it is an associative operation.

# 2.5 References

See also the references in the appendix.


GERICKE H: *Theorie der Verbände.* Bibliographisches Institut, Mannheim, 1963.

HARARY F, READ RC: *Is the null-graph a pointless concept?* In: Bari RA, Harary F (eds.): *Graphs and Combinatorics.* Proc. of the Capital Conf. on Graph Theory and Combinatorics at the G. Washington Univ., June 18–22, 1973, Lecture Notes in Computer Science **406**, Springer, Berlin, 1974

HOARE CAR, HE JIFENG: *The weakest prespecification, Part 1.* Fund. Inform. (4) **9** (1986) 51–84.

LÖWENHEIM L: *Über Möglichkeiten im Relativkalkül.* Math. Ann. **76** (1915) 447–470.

LORENZEN P: *Über die Korrespondenzen einer Struktur.* Math. Z. **60** (1954) 61–65.

LUCE RD: *A note on boolean matrix theory.* Proc. Amer. Math. **3** (1952) 382–388.

DE MORGAN A: *On the Syllogism: IV, and on the Logic of Relations.* Transactions of the Cambridge Philosophical Society **10** (1864) 331–358; Reprint: DE MORGAN A: *On the Syllogism, and Other Logical Writings,* edited, with an Introduction, by Peter Heath. Routledge & Kegan Ltd., London, 1966.

RIGUET J: *Relations binaires, fermetures, correspondances de Galois.* Bull. Soc. Math. France **76** (1948) 114–155.

SCHRÖDER E: *Algebra der Logik.* Vol. 3, Teubner, Leipzig, 1895.

STONE MH: *The theory of representations for boolean algebras.* Trans. Amer. Math. Soc. **40** (1936) 37–111.

TARSKI A: *On the calculus of relations.* J. Symbolic Logic **6** (1941) 73–89.

# 3. Transitivity

The notion of transitivity, introduced in Sect. 3.1, is fundamental for orderings and equivalence relations. In Sect. 3.2 the transitive closure is defined, which is a closure operation like the ones we have encountered already in two earlier instances. We then discuss a well-known algorithm for constructing the transitive closure. Related algorithms, appropriately adapted, are widely used today as resolution procedures in logic programming.

In this book, we started out with orderings and lattices as given mathematical concepts to treat relations algebraically, but did not investigate them in greater depth themselves. They served merely as tools. We now focus on orderings as relations and on lattices. We therefore deviate from the usual infix notation. Section 3.3 deals with maxima, bounds, and extrema from the relation-algebraic point of view. That section may be omitted at a first reading.

The discreteness of orderings will be analyzed in detail in Sect. 6.5 only, because it requires the concepts of path and reachability yet to be developed.

## 3.1 Orderings and Equivalence Relations

The formal apparatus for handling relations established in the preceding chapter will now be applied to specific types of relations. The following set of properties of relations is basic to the fundamental concepts of *equivalence*, *ordering*, and *strict-ordering*.

**3.1.1 Definition.** We call a homogeneous relation

$$R \ \textbf{transitive} \quad :\Longleftrightarrow \quad R^2 \subset R$$
$$\Longleftrightarrow \quad \forall x, y, z : [(x,y) \in R \wedge (y,z) \in R \to (x,z) \in R]. \quad \square$$

As in the proof of Proposition 2.3.4 the relational and the predicate-logical way of expressing things are easily seen to be equivalent:

$$RR \subset R$$
$$\Longleftrightarrow \quad \forall x, z : [(\exists y : (x,y) \in R \wedge (y,z) \in R) \to (x,z) \in R]$$
$$\Longleftrightarrow \quad \forall x, z : [\forall y : (x,y) \notin R \vee (y,z) \notin R] \vee (x,z) \in R$$
$$\Longleftrightarrow \quad \forall x, y, z : (x,y) \notin R \vee (y,z) \notin R \vee (x,z) \in R$$
$$\Longleftrightarrow \quad \forall x, y, z : [((x,y) \in R \wedge (y,z) \in R) \to (x,z) \in R].$$

Combining transitivity with the various properties of symmetry from Sect. 2.2 yields the notions of quasiordering and equivalence. To see that their relation-

algebraic and predicate logic forms are equivalent one can proceed as above, so
we omit the proofs.

**3.1.2 Definition.** We call a homogeneous relation

$$R \; \textbf{preorder} \quad :\Longleftrightarrow \quad R \text{ transitive and reflexive}$$
$$\Longleftrightarrow \quad R^2 \subset R, \; I \subset R \quad \Longleftrightarrow \quad R^2 = R, \; I \subset R.$$

$$R \; \textbf{equivalence} \quad :\Longleftrightarrow \quad R \text{ transitive, symmetric and reflexive}$$
$$\Longleftrightarrow \quad R^2 \subset R \subset R^{\mathsf{T}}, \; I \subset R.$$

A preorder is often called a **quasiordering**.                                         $\square$

An equivalence relation $R$ on the set $V$ yields a partitioning of $V$ into equivalence classes. The **equivalence class** of $x$ with respect to $R$ is the set
$[x]_R := \{ y \in V \mid (x,y) \in R \}$ consisting of all elements of $V$ related to $x$ by $R$
or $R$-equivalent to $x$; it can also be described by the vector $Rx = R^{\mathsf{T}}x$. As is
well-known one has $x \in [x]_R$, and $[x]_R = [y]_R$, if $y \in [x]_R$ and $[x]_R \cap [y]_R = \emptyset$, if
$y \notin [x]_R$. Thus, in terms of vectors, $Rx = Ry$ or $Rx \sqcap Ry = O$ for every couple
of points $x, y$.

The set of equivalence classes is called the **quotient set** $V/R$. Figure 3.1.1
indicates the partition defined by the equivalence relation $R$ from Fig. 2.2.3 as
shaded areas. Conversely, every partition of a set gives rise to an equivalence
relation. Its Boolean matrix has block diagonal form if suitably arranged.

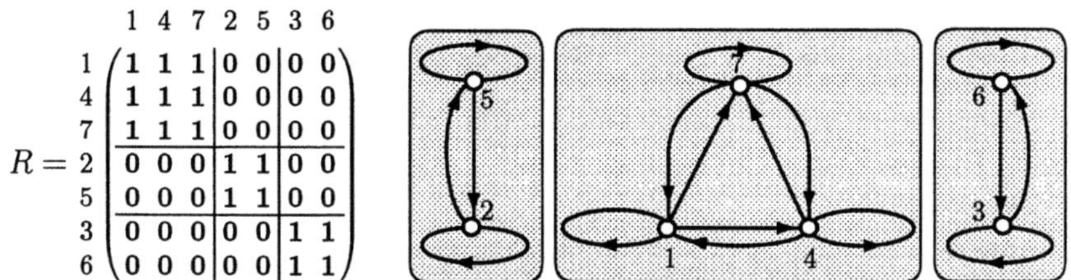

Fig. 3.1.1 Equivalence relation and equivalence classes

There are various ways of denoting preorders, stemming from different fields of
application: one often says, $x$ "is a predecessor of" $y$, $x$ "comes before" $y$,
$x \le y$. If $x$ precedes $y$ and $y$ precedes $z$, then $y$ is usually said "to lie between"
$x$ and $z$.

A symmetric preorder is an equivalence; given two elements $x, y$, the equivalence relation checks whether $x$ precedes $y$ and $y$ precedes $x$, or whether $x$ and
$y$ are incomparable (collateral).

In a preorder two distinct elements can be related to one another in both
directions.

Lists of results for sports championships can sometimes define a preorder. For example, Fig. 3.1.2 gives the outcome of the women's high jump competition in the 1987 track and field world championship. Positions four to six and seven to eight are held by several participants, respectively. The order in which the names of the holders of a given position are printed in the newspaper is arbitrary since it is not always justified by the numerical results.

| 1. Kostadinova | BUL | 2.09 m |
| 2. Bykova | URS | 2.04 m |
| 3. Beyer | GDR | 1.99 m |
| 4. Costa | CUB | 1.96 m |
| Kosizina | URS | 1.96 m |
| Redetzky[1] | FRG | 1.96 m |
| 7. Issajeva | BUL | 1.93 m |
| Awdejenko | URS | 1.93 m |

**Fig. 3.1.2** Preorder in sports

Place number four is occupied by a *set* of athletes whose athletic performance in that event afforded them equal rank. For the sake of uniformity one should therefore interpret the entries in places one, two, and three as one-element *sets*. As a result, we have put an ordering, not on a group of individual athletes, but on a set of classes consisting of athletes ranked equal[2].

So we see that identifying elements which are mutually comparable in a preorder always results in an order relation.

In Fig. 3.1.3 an example of a reflexive relation is given that does not contain a largest equivalence relation.

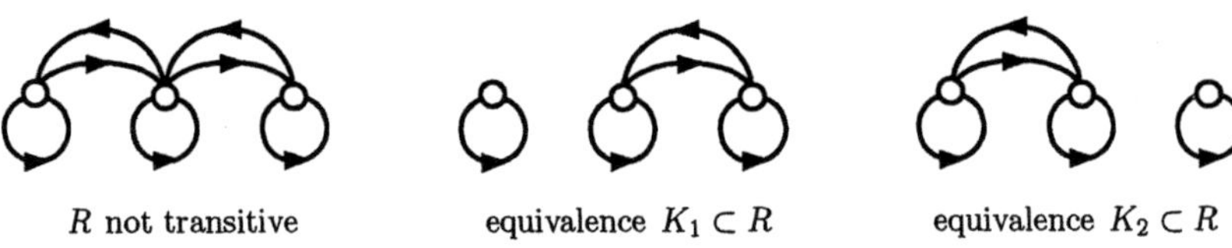

$R$ not transitive    equivalence $K_1 \subset R$    equivalence $K_2 \subset R$

**Fig. 3.1.3** A relation not containing a *greatest* equivalence

**3.1.3 Proposition.** The relation $R \sqcap R^\mathsf{T}$ is the greatest equivalence contained in a preorder $R$.

**Proof:** $R \sqcap R^\mathsf{T}$ is symmetric and reflexive, and it is transitive because of

$$(R \sqcap R^\mathsf{T})^2 \subset R^2 \sqcap (R^\mathsf{T})^2 = R \sqcap R^\mathsf{T}.$$

On the other hand, for every equivalence $K$ contained in $R$ we have $K = K^\mathsf{T} \subset R^\mathsf{T}$, so that $K \subset R \sqcap R^\mathsf{T}$. □

We now define orderings and strict-orderings. They differ from preorders with regard to pairs of elements allowed to be related in both directions. There is no restriction for preorders. In an ordering, two such elements must be equal; in a strict-ordering they must not even occur.

---

[1] Ranking no. 1 in 1991 with 2.05 m under her new name Henkel, GER.

[2] In most athletic disciplines such situations do not occur. Ambiguity in ranking is eliminated by secondary criteria such as chance, number of failed attempts, or lighter body weight, so that the results can be properly ordered.

To facilitate graphic representation of orderings one often omits the arrowheads, with the convention in mind that edges are always pointing upward. This, however, does not apply to horizontal connections in preorders, cf. Fig. 3.1.4.

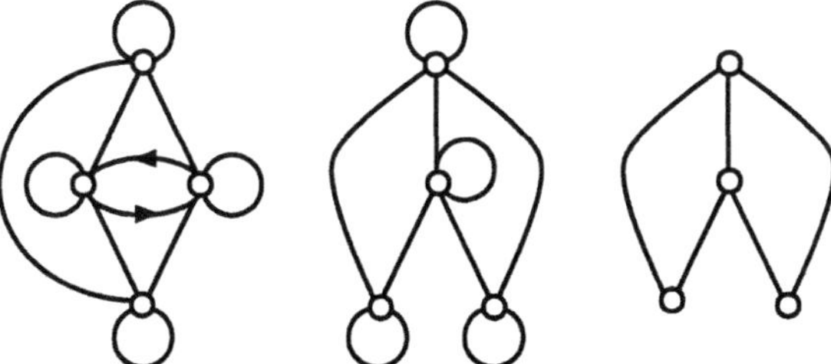

**Fig. 3.1.4** Preorder, ordering and strict-ordering using an abbreviated representation

**3.1.4 Definition.** We call a homogeneous relation

$$R \text{ ordering} \quad :\Longleftrightarrow \quad R \text{ transitive, antisymmetric, reflexive}$$
$$\Longleftrightarrow \quad R^2 \subset R, \quad R \sqcap R^\mathsf{T} \subset I, \quad I \subset R;$$

$$R \text{ strict-ordering} \quad :\Longleftrightarrow \quad R \text{ transitive and asymmetric}$$
$$\Longleftrightarrow \quad R^2 \subset R, \quad R \sqcap R^\mathsf{T} \subset O$$
$$\Longleftrightarrow \quad R^2 \subset R, \quad R^\mathsf{T} \subset \overline{R}$$
$$\Longleftrightarrow \quad R \text{ transitive and irreflexive}$$
$$\Longleftrightarrow \quad R^2 \subset R, \quad R \subset \overline{I}.$$

An ordering is often more precisely termed a **reflexive ordering**, and a strict-ordering an **irreflexive ordering**, regardless of the fact that a strict-ordering is *not* an ordering. □

Examples of order relations are divisibility on the set of natural numbers, inclusion of subsets in the powerset of a given set. A strict-ordering $R$ can be characterized in many ways. The two characterizations involving asymmetry are equivalent by a Boolean argument. From this we deduce irreflexivity splitting $R$ into $(R \sqcap R^2)$ and $(R \sqcap \overline{R^2})$. Clearly, $R \sqcap R^2 \subset R^2 \subset \overline{I}$ because, by Schröder's equivalence, $RR \subset \overline{I}$ is equivalent to $R^\mathsf{T} I \subset \overline{R}$. The second piece $R \sqcap \overline{R^2}$ is contained in $\overline{I}$ because $I \sqcap R \subset R^2$ always holds (see Exercise 2.3.7). Conversely, $RR \subset R$ and $R \subset \overline{I}$ imply $RR \subset \overline{I}$ and hence $R^\mathsf{T} I \subset \overline{R}$. Thus asymmetry $R^\mathsf{T} \subset \overline{R}$ has been established.

To a *reflexive* order relation, such as "$\leq$" on $\mathbb{N}$, one can associate the strict-ordering in the sense of "$<$". It is straightforward to see that every reflexive ordering $R$ induces the strict-ordering $\overline{I} \sqcap R$, and that every strict-ordering $R$ has the reflexive counterpart $I \sqcup R$. The irreflexive part $\overline{I} \sqcap R$ of any ordering $R$ is a strict-ordering, and the reflexive extension $I \sqcup R$ is an ordering for any strict-ordering $R$.

So orderings and strict-orderings come in pairs, and they require two different signs to denote them. It has become customary to denote order relations by infix symbols such as $\leq, \subseteq, \preceq$, etc., and their associated strict-orderings by the related symbols $<, \subset, \prec$, respectively. If letters are employed to denote order relations then we write $C$ for $\subset$ and $E$ for $\subseteq$ (because $E$ is reminiscent of $\subseteq$, and $C$ of $\subset$).

Thus the following relationships hold throughout this chapter

$$E = I \sqcup C \quad \text{and} \quad C = E \sqcap \overline{I}.$$

**3.1.5 Definition.** We call a relation

| | | |
|---|---|---|
| $E$ **linear ordering** | $:\Longleftrightarrow$ | $E$ ordering with $L = E \sqcup E^{\mathsf{T}}$, |
| $C$ **linear strict-ordering** | $:\Longleftrightarrow$ | $C$ strict-ordering with $\overline{I} = C \sqcup C^{\mathsf{T}}$. |

In accordance with Prop. 2.2.4, linear orderings are also termed **connex orderings**, and linear strict-orderings, **semiconnex strict-orderings**.      □

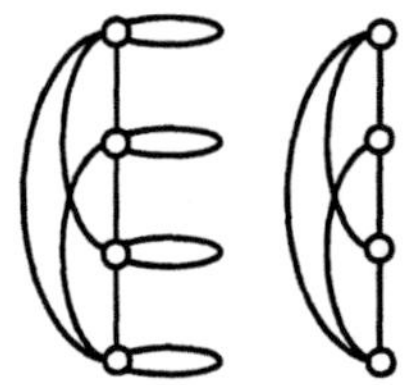

**Fig. 3.1.5** Connex ordering and
semiconnex strict-ordering

To a connex ordering corresponds a full upper triangular matrix, after suitably arranging rows and columns. A connex ordering cannot contain distinct *parallel strands*; in the reflexive case any two elements are comparable, in the irreflexive case, any two distinct elements are comparable. Antisymmetry now reads $C^{\mathsf{T}} \subset \overline{E}$. Any connex ordering satisfies $C^{\mathsf{T}} = \overline{E}$.

In a strictly ordered set such as $\mathbb{N}$ one may ask whether there are immediate successors or predecessors. These, however, do not exist in general, not for example in the reals or the rational numbers. For immediate successors and predecessors to exist, a certain discreteness condition for the ordering suffices: Given two points $a, b$, an ordered sequence of intermediate points between them cannot be refined ad infinitum. We will discuss this topic with greater precision in Sect. 6.5. Such a discrete ordering can already be *generated* by its immediate-successor relationship, which goes by the name of **Hasse diagram**. Figure 3.1.6 shows a discrete strict-ordering and its associated Hasse diagram. This phenomenon can also occur in more general transitive relations, generated by some relation they contain. See Fig. 3.2.1 for an example.

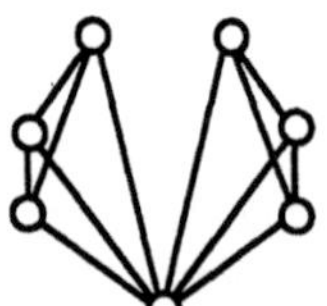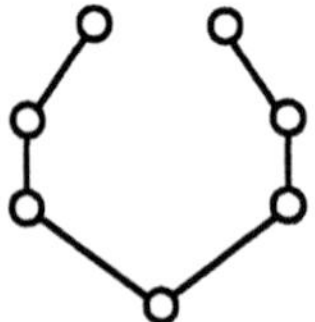

**Fig. 3.1.6** Strict-ordering and associated Hasse diagram

Even for small examples such as Fig. 3.1.6, it is economical to represent the ordering by the Hasse diagram of its associated strict-ordering. The matrix corresponding to the Hasse diagram of a linear strict-ordering consists of an upper off-diagonal of ones, and of zeros elsewhere.

## Compatibility with Equivalence Relations

We now prove some rules which are useful for calculations involving equivalence relations; the first one deals with the effect of multiplication by an equivalence relation with regard to complementation and intersection.

**3.1.6 Proposition.** Let $\Theta$ be an equivalence and let $A, B$ be arbitrary relations.

i)
$$\Theta\overline{\Theta A\Theta}\Theta = \overline{\Theta A\Theta}.$$

ii)
$$(A\Theta \sqcap B)\Theta = A\Theta \sqcap B\Theta = (A \sqcap B\Theta)\Theta.$$

**Proof:** i) Direction "$\supset$" is obvious. In the other direction, we have $\Theta A\Theta\Theta^{\mathsf{T}} \subset \Theta A\Theta$ from which we conclude, using Schröder's equivalences several times,

$$\overline{\Theta A\Theta}\Theta \subset \overline{\Theta A\Theta}, \quad \Theta^{\mathsf{T}}\Theta A\Theta = \Theta A\Theta \subset \overline{\overline{\Theta A\Theta}\Theta}, \quad \Theta\overline{\Theta A\Theta}\Theta \subset \overline{\Theta A\Theta}.$$

ii) $(A\Theta\sqcap B)\Theta \subset A\Theta^2\sqcap B\Theta = A\Theta\sqcap B\Theta$

$\subset (A\sqcap B\Theta\Theta^{\mathsf{T}})(\Theta\sqcap A^{\mathsf{T}}B\Theta) \subset (A\sqcap B\Theta)\Theta \subset A\Theta\sqcap B\Theta^2$

$= B\Theta\sqcap A\Theta \subset (B\sqcap A\Theta\Theta^{\mathsf{T}})(\Theta\sqcap B^{\mathsf{T}}A\Theta) \subset (B\sqcap A\Theta)\Theta.$ □

More about products involving equivalence relations is contained in the following proposition, which should be compared to Prop. 4.2.2.iv.

**3.1.7 Proposition.** Let $\Theta, \Omega$ be equivalences. If $G^{\mathsf{T}}\Theta G \subset \Omega$, $F\Omega \subset G\Omega$ and $FL \supset GL$, then $F\Omega = G\Omega$.

**Proof:** $G\Omega = GL \sqcap G\Omega \subset FL \sqcap G\Omega \subset (F \sqcap G\Omega L^{\mathsf{T}})(L \sqcap F^{\mathsf{T}}G\Omega)$

$\subset FF^{\mathsf{T}}G\Omega \subset F(F\Omega)^{\mathsf{T}}G\Omega \subset F(G\Omega)^{\mathsf{T}}G\Omega \subset F\Omega G^{\mathsf{T}}G\Omega$

$\subset F\Omega G^{\mathsf{T}}\Theta G\Omega \subset F\Omega\Omega\Omega \subset F\Omega.$ □

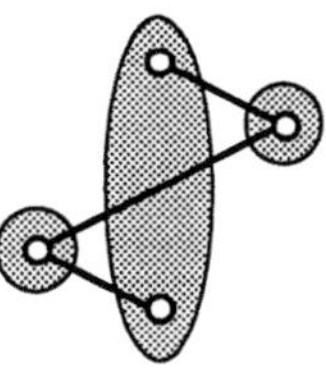

**Fig. 3.1.7** Ordering and equivalence which are incompatible

If a set carries an order relation as well as an equivalence relation, the question arises of whether the ordering induces an ordering on the equivalence classes. Of course, this will not be possible in general—see Fig. 3.1.7 for a counterexample. The following proposition gives a compatibility condition guaranteeing that the quotient set inherits the ordering. In Fig. 3.1.8 we give an example where the preorder $R := S^*$ is generated by $S$.

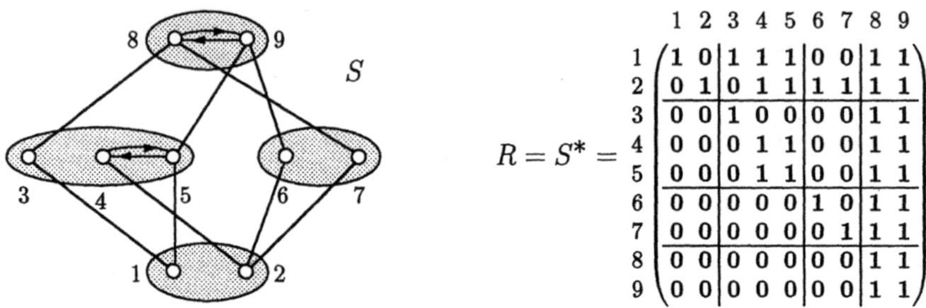

Fig. 3.1.8 Preorder and equivalence which are compatible

$$R = S^* = \begin{pmatrix} & 1 & 2 & 3 & 4 & 5 & 6 & 7 & 8 & 9 \\ 1 & 1 & 0 & 1 & 1 & 1 & 0 & 0 & 1 & 1 \\ 2 & 0 & 1 & 0 & 1 & 1 & 1 & 1 & 1 & 1 \\ 3 & 0 & 0 & 1 & 0 & 0 & 0 & 0 & 1 & 1 \\ 4 & 0 & 0 & 0 & 1 & 1 & 0 & 0 & 1 & 1 \\ 5 & 0 & 0 & 0 & 1 & 1 & 0 & 0 & 1 & 1 \\ 6 & 0 & 0 & 0 & 0 & 0 & 1 & 0 & 1 & 1 \\ 7 & 0 & 0 & 0 & 0 & 0 & 0 & 1 & 1 & 1 \\ 8 & 0 & 0 & 0 & 0 & 0 & 0 & 0 & 1 & 1 \\ 9 & 0 & 0 & 0 & 0 & 0 & 0 & 0 & 1 & 1 \end{pmatrix}$$

**3.1.8 Proposition.** Let $R$ be a preorder and let $\Theta$ be an equivalence relation, subject to the conditions

$$R\Theta \subset \Theta R, \quad R\Theta \sqcap \Theta R^\mathsf{T} \subset \Theta.$$

Then $\Theta R \Theta$ is a preorder, and $\Theta R \Theta \sqcap \Theta R^\mathsf{T} \Theta = \Theta$.

**Proof:** The relation $\Theta R \Theta$ is obviously reflexive and also transitive as a consequence of $\Theta R \Theta \Theta R \Theta \subset \Theta R \Theta R \Theta \subset \Theta \Theta R R \Theta \subset \Theta R \Theta$. Furthermore,

$$\Theta \subset \Theta R \Theta \sqcap \Theta R^\mathsf{T} \Theta$$
$$= \Theta(R\Theta \sqcap \Theta R^\mathsf{T}\Theta) = \Theta(R\Theta \sqcap \Theta R^\mathsf{T})\Theta \quad \text{(cf. Prop. 3.1.6.ii)}$$
$$\subset \Theta\Theta\Theta \subset \Theta. \qquad \qquad \Box$$

In other words, relation $\Theta$ is the largest equivalence relation contained in $\Theta R \Theta$ (cf. Prop. 3.1.3). The ordering of the quotient set $V/\Theta$ which can be read off from $\Theta R \Theta$, is given by

$$[x]_\Theta \leq [y]_\Theta \quad :\Longleftrightarrow \quad \exists u, v \in V : x\Theta u \wedge uRv \wedge v\Theta y.$$

In particular, $\Theta := R \sqcap R^\mathsf{T}$ obviously satisfies the hypotheses in Prop. 3.1.8. Therefore, the preorder $R$ induces an ordering on the equivalence classes of $R \sqcap R^\mathsf{T}$.


## Exercise

**3.1.1** Prove that $F\Theta = \Theta$ for an equivalence $\Theta$ with $L = FL$ and $F \subset \Theta$.


# 3.2 Closures and Closure Algorithms

Recovering an order relation from its Hasse diagram can be viewed as a *closure operation* to the effect of completing a relation with respect to certain properties. Sometimes (as for the reflexive and the symmetric closures of Chap. 2) a closure can in addition be given in a constructive way.

**3.2.1 Definition.** Given a homogeneous relation $R$, we call

i) $\quad R^+ := \sup_{i \geq 1} R^i = \inf \{ H \mid R \subset H, \ H \ \text{transitive} \}$
$$\text{the } \textbf{transitive closure} \text{ of } R,$$

ii) $\quad R^* := \sup_{i \geq 0} R^i = \inf \{ H \mid R \subset H, \ H \ \text{reflexive and transitive} \}$
$$\text{the } \textbf{reflexive-transitive closure} \text{ of } R. \qquad \square$$

Let us recall that these suprema and infima do indeed exist because we are working in a complete Boolean lattice. The infima, denoted by $J$ for the time being, are transitive: They constitute a lower bound for every $H$ satisfying the conditions considered, and therefore we have $JJ \subset HH \subset H$ for every $H$. So $JJ$ is another lower bound for every $H$ and hence is contained in the greatest lower bound $J$.

In order to see that the transitive closure can be written as an infimum as well as a supremum, we first show that the supremum is contained in the infimum $J$. Since every $H$ is transitive, we have $R^i \subset H^i \subset H$. So all the $R^i$ are lower bounds, hence $R^i \subset J$ for every $i$, and $J$ contains the supremum of the set of relations $R, R^2, \ldots$. On the other hand, since multiplication distributes over suprema, the supremum is transitive, is therefore one of the candidates $H$, and so contains their infimum $J$. We note a modified characterization which is often more convenient and whose proof we give as Exercise 3.2.3:

$$R^+ = \inf \{ H \mid R \subset H, \ RH \subset H \}.$$

The following can also be useful:

$$R \ \text{transitive} \iff R^+ \subset R \iff R^+ = R.$$

Moreover, the two transitive closures of a relation $R$ satisfy

$$R^+ = RR^*, \quad R^* = I \sqcup R^+.$$

We now introduce a similar closure operation for the purpose of getting equivalence relations.

**3.2.2 Definition.** For a homogeneous relation $R$ we define the

$$\textbf{equivalence closure} \ \ h_{\text{equiv}}(R) := \inf \{ H \mid R \subset H, \ H \ \text{equivalence} \}$$

as the lower bound of all enclosing equivalence relations. $\qquad \square$

Clearly, $h_{\text{equiv}}(R)$ is an equivalence relation and contains $R$. The conditions imposed on the relations $H$ are preserved when taking the infimum. Therefore $h_{\text{equiv}}(R)$ is the uniquely determined, smallest equivalence relation containing $R$. It can be expressed in terms of the reflexive-transitive closure:

**3.2.3 Proposition.** $\qquad\qquad h_{\text{equiv}}(R) = (R \sqcup R^{\mathsf{T}})^*.$

**Proof:** We have $h_{\text{equiv}}(R) \subset (R \sqcup R^{\mathsf{T}})^*$, since $(R \sqcup R^{\mathsf{T}})^*$ is an equivalence containing $R$. Now let $H$ be an arbitrary equivalence with $R \subset H$. Then $R^\mathsf{T} \subset H^\mathsf{T} = H$ and therefore $R \sqcup R^\mathsf{T} \subset H$, so that $H$, as a reflexive and transitive relation, contains the reflexive-transitive closure $(R \sqcup R^\mathsf{T})^*$. This holds for every such $H$ and consequently does so for their infimum $h_{\mathrm{equiv}}(R)$. $\qquad\square$

The closure operations considered so far have the following in common: They are mappings $h\colon \mathcal{B} \longrightarrow \mathcal{B}$ of some complete Boolean lattice $\mathcal{B}$ into itself and are easily seen to have the following properties: *extensive* $R \subset h(R)$, *isotonic* $R \subset S \implies h(R) \subset h(S)$ and *idempotent* $h(h(R)) = h(R)$. In general lattice theory a **closure operation** is defined by these three properties.

## Transitively Irreducible Kernels

We are now interested in the question, "Which relations $Q \subset R$ have the same transitive closure as $R$?" Therefore we consider the set

$$\mathcal{G}_R := \{ Q \mid Q \subset R,\ Q^+ = R^+ \},$$

the elements of which we call **generators.** For the sake of economy we are interested in generators that are as small as possible. As can be seen from Fig. 3.2.1 a smallest such relation need not exist; there are, however, such relations which are minimal with respect to inclusion.

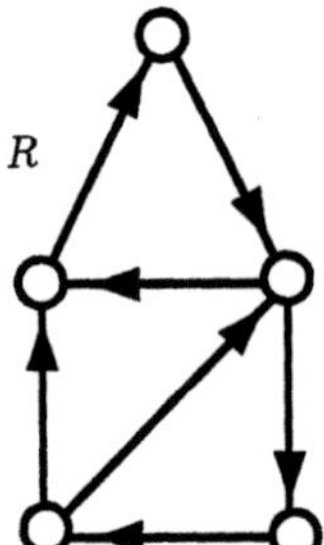

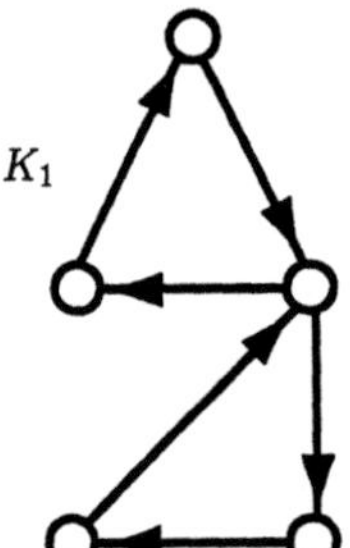

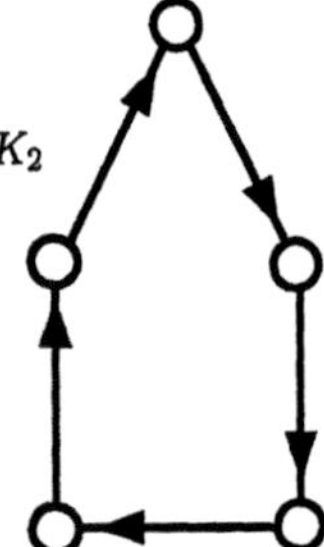


**Fig. 3.2.1** Transitively irreducible kernels

**3.2.4 Definition.** Given a relation $R$, we call a relation

$$K \text{ transitively irreducible kernel of } R \quad :\Longleftrightarrow \quad \begin{cases} K \subset R, \quad K^+ = R^+, \\ Q \subset K,\ Q^+ = R^+ \text{ implies } Q = K. \end{cases} \quad\square$$

Transitively irreducible kernels being minimal elements in $\mathcal{G}_R$ are by no means uniquely determined. We now ask when there is a *least* element, which will then, of course, be unique. The proposition below gives a hint as to its existence. There is an easy way to describe a lower bound for all transitively irreducible kernels: Given a transitive relation $R$, suppress all pairs $a, c$ for which there is a point $b$ such that $(a, b) \in R$ and $(b, c) \in R$. The resulting relation, however, is often much too small and may even be the zero relation. Additional properties are needed.

**3.2.5 Proposition.** Let $R$ be transitive. The relation $H_R := R \sqcap \overline{R^2}$ is contained in every generator of $R$.

**Proof:** Consider a generator $Q$ and define $W := H_R \sqcap \overline{Q}$. Then $W \subset R$ by definition of $H_R$. On the other hand $W \subset \overline{R^2}$ which, together with $R = R^+ = Q^+$ and $Q \subset R$, implies that

$$W \subset \overline{R^2} = \overline{RR} \subset \overline{QQ^+}.$$

Therefore $W \subset \overline{Q} \sqcap \overline{QQ^+} = \overline{Q \sqcup QQ^+} = \overline{Q^+} = \overline{R}$ so that $W \subset R \sqcap \overline{R} = O$, yielding $H_R \subset Q$. $\qquad\square$

This discussion will be resumed in Proposition 6.5.6 where we will show that, for a discrete relation $R$, $H_R$ is actually the smallest generator.

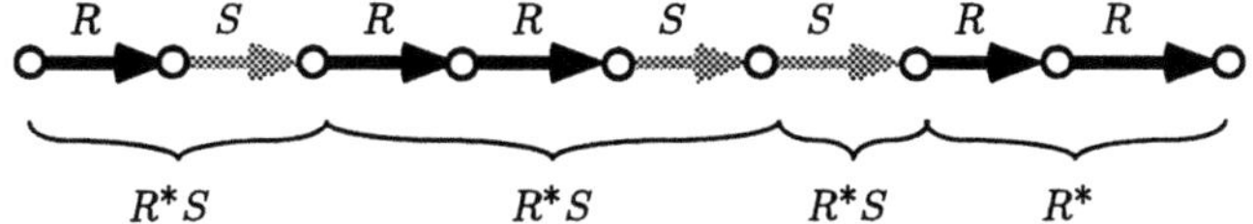


**Fig. 3.2.2** Transitive closure $(R \sqcup S)^* = (R^*S)^*R^*$ of a union

We collect several formulas concerning transitive closures:

**3.2.6 Proposition.** For the transitive closure of a relation the following holds:

i) $\quad (R^*)^\mathsf{T} = (R^\mathsf{T})^*, \qquad\qquad (R^+)^\mathsf{T} = (R^\mathsf{T})^+;$

ii) $\quad (R \sqcup S)^* = (R^*S)^*R^*, \quad (R \sqcup S)^+ = R^+ \sqcup (R^*S)^+R^*;$

iii) $\quad R^*S^* \subset (R \sqcup S)^*.$

**Proof:** i) is a consequence of $(R^\mathsf{T})^i = (R^i)^\mathsf{T}$. ii) Sorting the occurrences of $S$ we have $(R \sqcup S)^* = \sup_{i \geq 0}(R \sqcup S)^i = \sup_{i \geq 0}(R^*S)^iR^*$. Therefore

$$
\begin{aligned}
(R \sqcup S)^+ &= (R \sqcup S)(R \sqcup S)^* = (R \sqcup S)(R^*S)^*R^* \\
&= R(I \sqcup (R^*S)^+)R^* \sqcup S(R^*S)^*R^* \\
&= R^+ \sqcup RR^*S(R^*S)^*R^* \sqcup S(R^*S)^*R^* = R^+ \sqcup (R^+S \sqcup S)(R^*S)^*R^* \\
&= R^+ \sqcup (R^*S)(R^*S)^*R^* = R^+ \sqcup (R^*S)^+R^*.
\end{aligned}
$$

iii) $\quad (R \sqcup S)^* = (S^*R)^*S^* = IS^* \sqcup (S^*R)^+S^* \supset IS^* \sqcup R^+S^* = R^*S^*. \qquad\square$

## Closure Algorithms

We will now apply the formulas above in order to analyze algorithms for finding the transitive closure. We represent relations by their graphs and matrices. If a relation $R$ on a finite set is represented by a Boolean matrix, then calculating its transitive closure given by $R^+ = \sup_{1 \leq i \leq n} R^i$ seems to require considerable effort. Forming the product of two matrices by standard procedures requires a computational effort of the order $O(n^3)$. So for the first $n$ powers of $R$ one needs $O(n^4)$ operations (these first $n$ powers of $R$ suffice by Exercise 3.2.6). Working with the formula $R^+ = \inf\{X \mid R \sqcup RX \subset X\}$ would be even more costly because here $2^{n^2}$ relations are to be compared. A first improvement in this direction is provided by a *layer algorithm* which is based on the formula in Exercise 3.2.6.iv; it reduces the effort to $O(n^3 \log n)$ and is suitable for certain applications.

The following considerations are fundamental for effectively calculating the transitive closure $R^+$. They deal with preserving transitivity *locally*, i.e., at some element $x$, and are frequently attributed to S. Warshall but are in fact due to B. Roy.

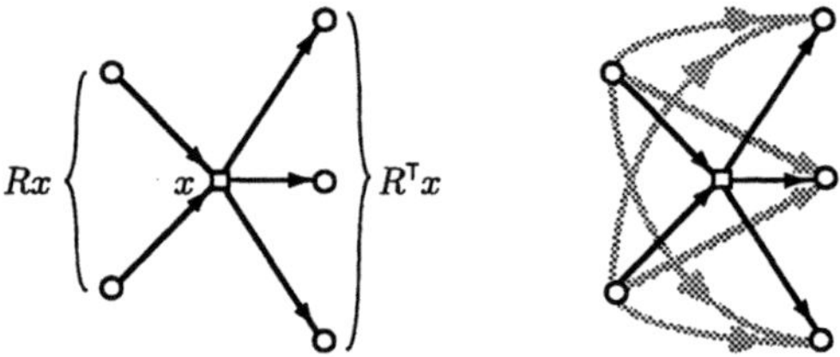

Fig. 3.2.3 Basic operation of the Roy-Warshall algorithm

In each step an additional element $x$ is chosen, and its predecessors $Rx$ are joined to its successors $R^\mathsf{T}x$. So partial transitivity is established in the sense that every pair of consecutive arrows joining each other at $x$ gives rise to a single new arrow. After this step involving $x$, $R$ is enlarged by precisely $Rx(R^\mathsf{T}x)^\mathsf{T} = Rxx^\mathsf{T}R \subset R^2$.

The next proposition shows that our construction of local transitivity suffices for calculating $R^+$ (see Fig. 3.2.4).

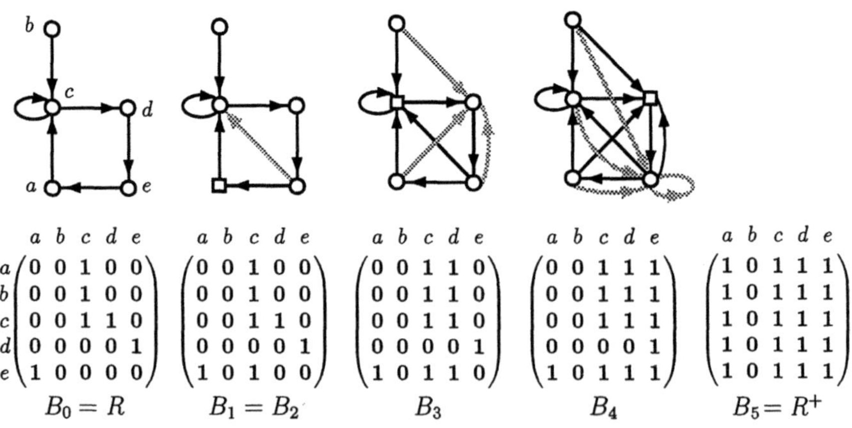

$$
\begin{array}{c|ccccc}
 & a & b & c & d & e \\
\hline
a & 0 & 0 & 1 & 0 & 0 \\
b & 0 & 0 & 1 & 0 & 0 \\
c & 0 & 0 & 1 & 1 & 0 \\
d & 0 & 0 & 0 & 0 & 1 \\
e & 1 & 0 & 0 & 0 & 0
\end{array}
\qquad
\begin{array}{ccccc}
a & b & c & d & e \\
0 & 0 & 1 & 0 & 0 \\
0 & 0 & 1 & 0 & 0 \\
0 & 0 & 1 & 1 & 0 \\
0 & 0 & 0 & 0 & 1 \\
1 & 0 & 1 & 0 & 0
\end{array}
\qquad
\begin{array}{ccccc}
a & b & c & d & e \\
0 & 0 & 1 & 1 & 0 \\
0 & 0 & 1 & 1 & 0 \\
0 & 0 & 1 & 1 & 0 \\
0 & 0 & 0 & 0 & 1 \\
1 & 0 & 1 & 1 & 0
\end{array}
\qquad
\begin{array}{ccccc}
a & b & c & d & e \\
0 & 0 & 1 & 1 & 1 \\
0 & 0 & 1 & 1 & 1 \\
0 & 0 & 1 & 1 & 1 \\
0 & 0 & 0 & 0 & 1 \\
1 & 0 & 1 & 1 & 1
\end{array}
\qquad
\begin{array}{ccccc}
a & b & c & d & e \\
1 & 0 & 1 & 1 & 1 \\
1 & 0 & 1 & 1 & 1 \\
1 & 0 & 1 & 1 & 1 \\
1 & 0 & 1 & 1 & 1 \\
1 & 0 & 1 & 1 & 1
\end{array}
$$

$$B_0 = R \qquad\qquad B_1 = B_2 \qquad\qquad B_3 \qquad\qquad B_4 \qquad\qquad B_5 = R^+$$

Fig. 3.2.4 Intermediate states of the Roy-Warshall algorithm

**3.2.7 Proposition.** Let $x_1, \ldots, x_n$ be the points of the finite graph associated with $R$. Then iterating

$$B_0 := R, \quad B_i := B_{i-1} \sqcup B_{i-1} x_i x_i^\mathsf{T} B_{i-1}, \ i = 1, \ldots, n,$$

leads to the transitive closure $R^+ = B_n$.

**Proof:** For the sake of simplicity we assume $R$ is reflexive. Hence $R^+ = R^*$. We have the sequence of matrices $B_0 \subset B_1 \subset \ldots$ and we must show that $B_n = \sup_{1 \leq i \leq n} B_i = R^+$. Since the points satisfy $x_i x_i^\mathsf{T} \subset I$, we get first of all that $B_i \subset B_{i-1} \sqcup B_{i-1}^2 \subset \ldots \subset R^+$ and hence $B_n \subset R^+$.

In order to establish the converse inclusion we put

$$E_i := (R \sup_{1 \leq m \leq i} x_m x_m^\mathsf{T})^*(I \sqcup R).$$

So $E_i$ represents, roughly speaking, that part of the transitivity conditions stemming from the first $i$ points used as intermediate points in the sense of Fig. 3.2.3. Since $\sup_{1 \leq i \leq n} x_i x_i^\mathsf{T} = I$ it follows that $\sup_i E_i = (R\,I)^*(I \sqcup R) = R^*$. So it remains to verify that $\sup_i E_i \subset B_n$. We show inductively that $E_i \subset B_i$ for all $1 \leq i \leq n$. Clearly $E_0 = O \subset B_0$. The induction argument is as follows:

$$\begin{aligned}
E_i &= (R \sup_{1 \leq m \leq i-1} x_m x_m^\mathsf{T} \sqcup R x_i x_i^\mathsf{T})^*(I \sqcup R) \\
&= [(R \sup_{1 \leq m \leq i-1} x_m x_m^\mathsf{T})^* R x_i x_i^\mathsf{T}]^*(R \sup_{1 \leq m \leq i-1} x_m x_m^\mathsf{T})^*(I \sqcup R) \\
&\subset (E_{i-1} x_i x_i^\mathsf{T})^* E_{i-1} = (I \sqcup E_{i-1} x_i x_i^\mathsf{T}) E_{i-1}, \\
&\qquad\qquad \text{because} \quad (A x x^\mathsf{T})^* = I \sqcup A x x^\mathsf{T} \quad \text{for every subset } x \\
&\subset B_{i-1} \sqcup B_{i-1} x_i x_i^\mathsf{T} B_{i-1} = B_i, \qquad \text{by induction hypothesis.} \qquad \square
\end{aligned}$$

If this closure operation is performed on a computer where the memory reserved for the matrix $R$ can be "overwritten", then the above iterative process is to be implemented as

$$\textbf{for } i := 1 \textbf{ step } 1 \textbf{ until } n \textbf{ do } \ R := R \sqcup R x_i x_i^\mathsf{T} R.$$

Here, $R x_i$ and $R^\mathsf{T} x_i$ stand for the $i$th column or row, respectively. We now formulate our result in pseudo-code:

**3.2.8 Algorithm** (*B. Roy 1959; S. Warshall 1962*). Let $R$ be a Boolean $n \times n$-matrix. The following program converts $R$ into its transitive closure $R^+$:

```
for i := 1 step 1 until n do
    for j := 1 step 1 until n do
        if R(j, i) then
            for k := 1 step 1 until n do
                R(j, k) := R(j, k) ∨ R(i, k)
```

This program requires $O(n^2)$ memory bits and carries out $O(n^3)$ Boolean $\vee$-operations. $\qquad \square$

This is not, however, the best method theoretically, although it is the most common one. For symmetric $R$, one can apply the variant of the Roy-Warshall algorithm where the inner loop is already terminated after reaching the first true or nonzero $R(j, i)$. The computational effort then reduces to $O(n^2)$. Asymptotically faster algorithms for the nonsymmetric case can be based on the following result, which is an instance of the frequently used principle of successively halving the problem (for computing purposes).

**3.2.9 Proposition.** The transitive closure of a matrix which is subdivided into four parts with quadratic diagonal matrices is given by

i) $\begin{pmatrix} A & O \\ C & O \end{pmatrix}^+ = \begin{pmatrix} A^+ & O \\ CA^* & O \end{pmatrix}$,    $\begin{pmatrix} O & B \\ O & D \end{pmatrix}^+ = \begin{pmatrix} O & BD^* \\ O & D^+ \end{pmatrix}$;

ii) $\begin{pmatrix} A & B \\ C & D \end{pmatrix}^+ = \begin{pmatrix} A^+ \sqcup A^*BE^*CA^* & A^*BE^* \\ E^*CA^* & E^+ \end{pmatrix}$, with $E := CA^*B \sqcup D$.

**Proof:** i) $\begin{pmatrix} A & O \\ C & O \end{pmatrix}^2 = \begin{pmatrix} A^2 & O \\ CA & O \end{pmatrix}$; $\begin{pmatrix} A & O \\ C & O \end{pmatrix}^3 = \begin{pmatrix} A^3 & O \\ CA^2 & O \end{pmatrix}$    etc.

ii) In order to be able to use (i) and Prop. 3.2.6 one starts from

$$\begin{pmatrix} A & B \\ C & D \end{pmatrix}^+ = \left[ \begin{pmatrix} A & O \\ C & O \end{pmatrix} \sqcup \begin{pmatrix} O & B \\ O & D \end{pmatrix} \right]^+ \qquad \square$$

Thus the task of calculating the transitive closure of a $2n \times 2n$-matrix can be reduced to doing it for the two $n \times n$-matrices $A^*, E^*$, at the expense, though, of a certain number of additional multiplications. Using an idea of V. Strassen, this additional effort can be kept asymptotically small.

It should be mentioned that algorithms of a quite similar structure have long been in use in numerical analysis. A closer look at an intermediate state of the Roy-Warshall algorithm, i.e., relation $B_i$ from Prop. 3.2.7, shows that it is equal to the matrix

$$\begin{pmatrix} A^+ & A^*B \\ CA^* & D \sqcup CA^*B \end{pmatrix}$$

after the first columns of $\begin{pmatrix} A & B \\ C & D \end{pmatrix}$ have been worked out. Compare this to the intermediate state

$$\begin{pmatrix} A^{-1} & -A^{-1}B \\ +CA^{-1} & D - CA^{-1}B \end{pmatrix} \begin{pmatrix} U \\ Y \end{pmatrix} = \begin{pmatrix} X \\ V \end{pmatrix}$$

of a Gauss-Jordan algorithm with diagonal pivot choice, applied to the system of linear equations

$$\begin{pmatrix} A & B \\ C & D \end{pmatrix} \begin{pmatrix} X \\ Y \end{pmatrix} = \begin{pmatrix} U \\ V \end{pmatrix}.$$

**Exercises**

**3.2.1** Show that every relation $R$ contained in $I$ is symmetric, and that every asymmetric relation is irreflexive.

**3.2.2** Prove that a relation $R$ is an equivalence if and only if it is reflexive and satisfies $RR^{\mathsf{T}} \subset R$.

**3.2.3** Show that for every homogeneous relation $R$

i) $\inf\{H \mid R \sqcup RH \subset H\} = \inf\{H \mid R \sqcup RH = H\} = R^{+}$;
ii) $\inf\{H \mid S \sqcup RH \subset H\} = R^{*}S$.

**3.2.4** Prove the following statement: For every relation $Q$ which is contained in an equivalence relation $\Xi$ and for any other relation $R$ the equation $Q(R \sqcap \Xi) = QR \sqcap \Xi$ holds. (Compare this result to the modular law of lattice theory.)

**3.2.5** Prove the following statements:

i) For every relation $R$, the construct $\overline{R^{\mathsf{T}}\overline{R}} = R \mathbin{.\cdot} R$ is reflexive and transitive; see Definition 2.3.8.

ii) For every reflexive and transitive relation $A$ there exists a relation $R$ such that $A = \overline{R^{\mathsf{T}}\overline{R}}$.

**3.2.6** Use the pigeon-hole principle to prove that for an arbitrary Boolean $n \times n$-matrix $R$:

i) $R^{n} \subset (I \sqcup R)^{n-1}$.
ii) $R^{*} = \sup_{0 \leq i < n} R^{i}$.
iii) $R^{+} = \sup_{0 < i \leq n} R^{i}$.
iv) $(I \sqcup R)(I \sqcup R^{2})(I \sqcup R^{4})(I \sqcup R^{8}) \ldots (I \sqcup R^{2^{\lfloor \log n \rfloor}}) = R^{*}$.

# 3.3 Extrema, Bounds, and Suprema

When dealing with an ordered set one typically wants to find greatest or maximal points for a given subset. We consider two different types of orderings in this chapter: firstly, the ordering on a set of points, given by some relation $E$, and secondly the ordering of relations themselves, by containment $R \subset S$. We will refer to these two situations as the object level and the metalevel, respectively. We will now analyze bounds, least or greatest bounds, and extrema in a relation-algebraic way, i.e., at the metalevel. The results of this section are not required for the later chapters.

We first look at "extremal" elements of a subset; these are either maximal or minimal ones, i.e., they have no proper successor or predecessor in the subset, respectively.

**3.3.1 Definition.** We consider the irreflexive version $C$ of a given ordering together with a subset $t$ and a point $x$. Then we call

i)  $x$ **maximal element of** $t$   $:\Longleftrightarrow$   $x \subset t \subset \overline{C^\mathsf{T}x}$

   $\Longleftrightarrow$   $x \subset t$ and $xt^\mathsf{T} \subset \overline{C}$

   $\Longleftrightarrow$   Point $x$ belongs to the set $t$ and is not less than any other point of $t$.

$\quad\mathbf{max}(t) := t \sqcap \overline{Ct}$ set of the maxima of $t$.

ii)  $x$ **minimal element of** $t$   $:\Longleftrightarrow$   $x \subset t \subset \overline{Cx}$

   $\Longleftrightarrow$   $x \subset t$ and $tx^\mathsf{T} \subset \overline{C}$

   $\Longleftrightarrow$   Point $x$ belongs to the set $t$ and is not greater than any point of $t$.

$\quad\mathbf{min}(t) := t \sqcap \overline{C^\mathsf{T}t}$ set of the minima of $t$.

We will write $\mathbf{max}_C(t)$ instead of $\mathbf{max}(t)$ if necessary.   $\square$


The equivalence of the first two versions of the definition follows from Proposition 2.4.4.i. Then $x \subset \mathbf{max}(t) = t \sqcap \overline{Ct}$ is equivalent to $x \subset t$ and $x \subset \overline{Ct}$. The latter means $Ct \subset \overline{x}$ and $xt^\mathsf{T} \subset \overline{C}$, so that $\mathbf{max}(t)$ indeed contains all the maximal points.

The property of being maximal for a point $x$ of a subset $t$ can be viewed as an *inner* property of $t$ endowed with the given ordering, whereas bounds for $t$ can lie outside of $t$. Of course, extremal elements may or may not exist, and need not be uniquely determined. We shall give a condition for the existence of maximal elements in connection with the discussion of progressive boundedness in Definition 6.3.2.

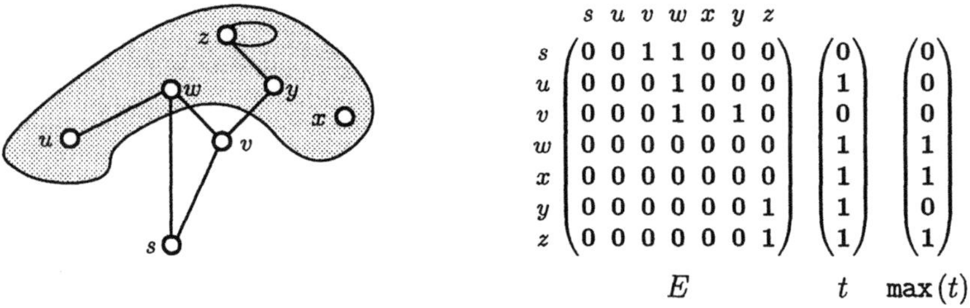

Fig. **3.3.1** Maximal elements of a relation

As for subsets $t$, one can form maximal elements column-wise for a relation $X$ by $\mathbf{max}(X) := X \sqcap \overline{CX}$. Many of the results below remain valid, and interesting identities such as $\mathbf{max}(E) = I$ arise.

The definition of extremal elements does not require any special conditions on the relation $C$, in particular, $C$ need not be a strict-ordering. In Fig. 3.3.1, where the arrows are understood to point upwards, the relation $E$ is not transitive, hence not an ordering. In such a case, too, the set $\mathbf{max}(t) = t \sqcap \overline{(E \sqcap \overline{I})t}$ consists of all the points of $t$ having no proper successor in $t$.

## Bounds

Apart from extremal elements, the bounds of a subset are of interest.

**3.3.2 Definition.** Let an ordering $E$ be given together with the associated strict-ordering $C = E \sqcap \overline{I}$, and let $t$ be a subset and $x$ a point. We call

i)  $x$ **upper bound** of $t$ $\quad :\Longleftrightarrow\quad t \subset Ex$

$\qquad\qquad\qquad\qquad\quad \Longleftrightarrow\quad t \sqcap \overline{x} \subset Cx \quad\Longleftrightarrow\quad tx^{\mathsf{T}} \subset E$

$\qquad\qquad\quad \Longleftrightarrow\quad$ Every point of $t$ different from $x$ is smaller than $x$.

$\qquad \mathrm{ubd}(t) := \overline{\overline{E}^{\mathsf{T}}t}$ set of all upper bounds of $t$.

ii)  $x$ **lower bound** of $t$ $\quad :\Longleftrightarrow\quad t \subset E^{\mathsf{T}}x$

$\qquad\qquad\qquad\qquad\quad \Longleftrightarrow\quad t \sqcap \overline{x} \subset C^{\mathsf{T}}x \quad\Longleftrightarrow\quad tx^{\mathsf{T}} \subset E^{\mathsf{T}}$

$\qquad\qquad\quad \Longleftrightarrow\quad$ Every point of $t$ different from $x$ is greater than $x$.

$\qquad \mathrm{lbd}(t) := \overline{\overline{E}t}$ set of all lower bounds of $t$.

If the ordering is to be specified, we will write $\mathrm{ubd}_E$ and $\mathrm{lbd}_E$. $\qquad\qquad\square$

As a consequence of Proposition 2.4.4 the formal variants of the definition are equivalent. Furthermore, $x \subset \mathrm{ubd}(t) = \overline{\overline{E}^{\mathsf{T}}t}$ is equivalent to $\overline{E}^{\mathsf{T}}t \subset \overline{x}$ and to $xt^{\mathsf{T}} \subset E^{\mathsf{T}}$, so that the set of upper bounds is indeed the union of the single upper bounds.

Obviously, $\mathrm{ubd}(O) = L$, i.e., every point is an upper bound of the empty set. On the other hand, $\mathrm{ubd}(L) = \overline{\overline{E}^{\mathsf{T}}L}$. Since $E$ is antisymmetric, there should be at most one upper bound of all points. Indeed, $y := \mathrm{ubd}(L)$ is injective, so it satisfies one of the conditions characterizing a point:

$$yy^{\mathsf{T}} \subset \overline{\overline{E}^{\mathsf{T}}LL} \sqcap \overline{L\overline{E}^{\mathsf{T}}L}^{\mathsf{T}} \subset E^{\mathsf{T}} \sqcap E \subset I.$$

Using transitivity and the identities $E\overline{E}^{\mathsf{T}} = \overline{E}^{\mathsf{T}} = \overline{E}^{\mathsf{T}}E$ for an ordering, one obtains immediately

$$\mathrm{ubd}(E) = E^{\mathsf{T}}, \quad \mathrm{lbd}(E^{\mathsf{T}}) = E,$$
$$E^{\mathsf{T}}\,\mathrm{ubd}(t) = \mathrm{ubd}(t) = \mathrm{ubd}(Et), \quad E\,\mathrm{lbd}(t) = \mathrm{lbd}(t) = \mathrm{lbd}(E^{\mathsf{T}}t).$$

Again, we might drop the requirement on $E$ to be an ordering and demand instead that an upper bound (lower bound) be an immediate successor (predecessor) of *all* points of a given set $t$. For a relation $R$ that is not necessarily reflexive, we could introduce

$$\mathrm{ubd}(t) = \overline{\overline{R \sqcup I}^{\mathsf{T}}t} \quad \text{and} \quad \mathrm{lbd}(t) = \overline{\overline{R \sqcup I}t},$$

and thus obtain quite similar results. Figure 3.3.2 contains examples of upper bounds of a relation that is not an ordering.

In the following proposition we prove further important properties of $\mathrm{ubd}$ and $\mathrm{lbd}$ which allow us to treat these functionals in a more general lattice-theoretic context, as will be done in the appendix.

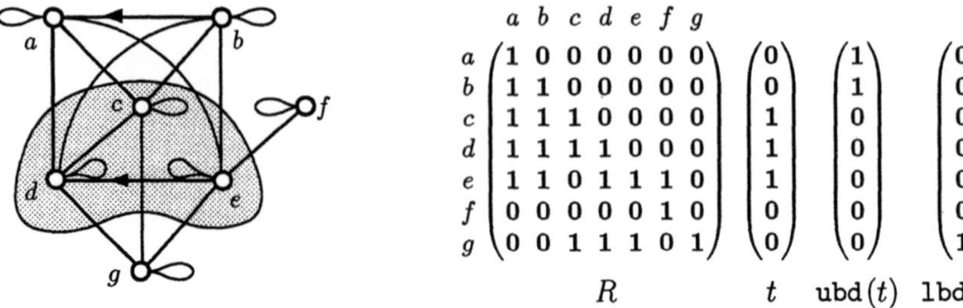

$$\begin{array}{c|ccccccc} & a & b & c & d & e & f & g \\\hline a & 1 & 0 & 0 & 0 & 0 & 0 & 0 \\ b & 1 & 1 & 0 & 0 & 0 & 0 & 0 \\ c & 1 & 1 & 1 & 0 & 0 & 0 & 0 \\ d & 1 & 1 & 1 & 1 & 0 & 0 & 0 \\ e & 1 & 1 & 0 & 1 & 1 & 1 & 0 \\ f & 0 & 0 & 0 & 0 & 0 & 1 & 0 \\ g & 0 & 0 & 1 & 1 & 1 & 0 & 1 \end{array} \qquad \begin{pmatrix}0\\0\\1\\1\\1\\0\\0\end{pmatrix} \quad \begin{pmatrix}1\\1\\0\\0\\0\\0\\0\end{pmatrix} \quad \begin{pmatrix}0\\0\\0\\0\\0\\0\\1\end{pmatrix}$$

$$R \qquad\qquad\qquad t \quad \mathrm{ubd}(t) \quad \mathrm{lbd}(t)$$

**Fig. 3.3.2** Upper and lower bounds

**3.3.3 Proposition.** Let the bound functionals ubd and lbd be given.

i)  For any two subsets $t, t'$ we have the equivalence

$$t \subset \mathrm{lbd}(t') \quad\Longleftrightarrow\quad t' \subset \mathrm{ubd}(t).$$

ii)  ubd and lbd are antitonic and, if applied one after the other, extensive:

$$t \subset \mathrm{ubd}(\mathrm{lbd}(t)), \quad t \subset \mathrm{lbd}(\mathrm{ubd}(t)).$$

**Proof:** i) $t \subset \overline{\overline{E}t'} \iff \overline{E}t' \subset \bar{t} \iff \overline{E}^{\mathsf{T}} t \subset \bar{t'} \iff t' \subset \overline{\overline{E}^{\mathsf{T}}t}.$

ii) It is trivial that the functionals are antitonic. The rest is obtained by Schröder's equivalence from $\mathrm{ubd}(t) \subset \overline{\overline{E}^{\mathsf{T}}t}$ and $\mathrm{lbd}(t) \subset \overline{\overline{E}t}$.  □

Exercise 3.3.1 shows that

$$\mathrm{ubd}(t) = \mathrm{ubd}(\mathrm{lbd}(\mathrm{ubd}(t))), \quad \mathrm{lbd}(t) = \mathrm{lbd}(\mathrm{ubd}(\mathrm{lbd}(t))).$$

Here, as in the proof of Proposition 3.3.3, it is not necessary that $E$ be an ordering. So our reasoning above applies to relations in general.

Pairs of functions with properties (i) and (ii) (which are actually equivalent) are called **Galois correspondences** or Galois connections; they occur in many other areas, too. In a Galois correspondence any one member is uniquely determined by the other. See Appendix A.3 for similar questions.

We now give two more relationships for the functionals ubd and lbd.

**3.3.4 Proposition.** Given a subset $t \neq O$ we have

$$\mathrm{ubd}(t) \subset E^{\mathsf{T}}t, \qquad \mathrm{lbd}(t) \subset Et.$$

**Proof:** From $t \neq O$ it follows that $L = Lt = \overline{E}^{\mathsf{T}}t \sqcup E^{\mathsf{T}}t$; hence $\mathrm{ubd}(t) \subset E^{\mathsf{T}}t$.  □

As we observed earlier, $\mathrm{ubd}(L)$ is equal to $O$ or to a single point depending on whether there does not, or does, exist a unique upper bound. This can be sharpened to an analogous result for $t \sqcap \mathrm{ubd}(t)$. So *within* the point set $t$ there can be at most one upper bound for $t$:

$$[t \sqcap \mathrm{ubd}(t)][t \sqcap \mathrm{ubd}(t)]^{\mathsf{T}} \subset t[\mathrm{ubd}(t)]^{\mathsf{T}} \sqcap \mathrm{ubd}(t)t^{\mathsf{T}} \subset E \sqcap E^{\mathsf{T}} \subset I.$$

If that point exists then it is uniquely determined and is called the greatest element of $t$. In Exercise 3.3.4 results of this kind can be obtained for $E$ not necessarily antisymmetric. This is achieved by taking $\max(t) \sqcap \mathtt{ubd}(t)$ instead of $t \sqcap \mathtt{ubd}(t)$.

**3.3.5 Definition.** We consider a point $x$ together with a subset $t$ in context with an ordering $E$ and call

i) $x$ a **greatest point** of $t$ $\quad :\Longleftrightarrow\quad x \subset t \subset Ex$

$\qquad\qquad\qquad\qquad\qquad\quad \Longleftrightarrow\quad x$ is an upper bound and a point of $t$.

$\quad \mathtt{gre}(t) := t \sqcap \mathtt{ubd}(t)$  set of greatest points of $t$.

ii) $x$ a **least point** of $t$ $\qquad :\Longleftrightarrow\quad x \subset t \subset E^{\mathsf{T}}x$

$\qquad\qquad\qquad\qquad\qquad\quad \Longleftrightarrow\quad x$ is a lower bound and a point of $t$.

$\quad \mathtt{lea}(t) := t \sqcap \mathtt{lbd}(t)$  set of least points of $t$. $\hfill\square$

If a set $t$ contains a greatest point, then this is clearly the only maximal element, for $t \subset E\,\mathtt{gre}(t)$, provided $\mathtt{gre}(t) \neq O$. For the following to hold it is essential that $E$ be reflexive and antisymmetric.

**3.3.6 Proposition.** Given an ordering, the sets of greatest or least points of a given subset $t$ satisfy:

i) $\mathtt{gre}(t) = \max(t) \sqcap \mathtt{ubd}(t), \qquad \mathtt{gre}(t) \neq O \quad \Longrightarrow \quad \mathtt{gre}(t) = \max(t);$

ii) $\mathtt{lea}(t) = \min(t) \sqcap \mathtt{lbd}(t), \qquad \mathtt{lea}(t) \neq O \quad \Longrightarrow \quad \mathtt{lea}(t) = \min(t).$

**Proof** for (i): It suffices to show $\mathtt{gre}(t) \subset \max(t)$. By antisymmetry we have $C = E \sqcap \overline{I} \subset \overline{E}^{\mathsf{T}}$, so we may proceed by writing

$$\overline{\mathtt{gre}(t)} \sqcup \max(t) = \overline{t} \sqcup \overline{\mathtt{ubd}(t)} \sqcup \max(t) = \overline{t} \sqcup \overline{E}^{\mathsf{T}}t \sqcup (t \sqcap \overline{Ct})$$
$$\supset \overline{t} \sqcup Ct \sqcup (t \sqcap \overline{Ct}) \supset \overline{t} \sqcup t = L.$$

This result is used in order to show:

$$t \subset E\,\mathtt{gre}(t) = (C \sqcup I)\,\mathtt{gre}(t) \subset Ct \sqcup \mathtt{gre}(t);$$

thus $\max(t) = t \sqcap \overline{Ct} \subset \mathtt{gre}(t)$. $\hfill\square$

An ordering $E_W$, where every nonempty subset $t$ contains a least element, is called a **well-ordering**; it is characterized by $t \neq O \Longrightarrow \mathtt{lea}(t) \neq O$. Suppose a set is given with an ordering $E$ and associated strict-ordering $C$. Suppose also that the set carries some well-ordering $E_W$. The relation $F := C \sqcap \overline{CE_W^{\mathsf{T}}}$ then associates with each point that point among the greater ones with respect to $E$ which is least with respect to $E_W$.

## Least Upper and Greatest Lower Bounds

We are now concerned with order-related concepts involving the complement of $t$. We define the supremum as the least of the upper bounds, and the infimum accordingly.

**3.3.7 Definition.** Given an ordering and a subset $t$, we call

i) $\mathtt{lub}(t) := \mathtt{lea}(\mathtt{ubd}(t))$

$\quad\quad\quad = \mathtt{ubd}(t) \sqcap \mathtt{lbd}(\mathtt{ubd}(t))$  the set of **least upper bounds**.

$\quad$ If $x := \mathtt{lub}(t) \neq O$, $x$ is called the least upper bound or **supremum** of $t$.

ii) $\mathtt{glb}(t) := \mathtt{gre}(\mathtt{lbd}(t))$

$\quad\quad\quad = \mathtt{lbd}(t) \sqcap \mathtt{ubd}(\mathtt{lbd}(t))$  the set of **greatest lower bounds**.

$\quad$ If $x := \mathtt{glb}(t) \neq O$, $x$ is called the greatest lower bound or **infimum** of $t$. $\square$

As a consequence of $\mathtt{lub}(O) = \mathtt{lea}(\mathtt{ubd}(O)) = \mathtt{lea}(L)$, the set of upper bounds of the empty set is the set of smallest elements which contains at most one element. For a point $x$ we have

$$\mathtt{lub}(x) = \overline{\overline{E^{\mathsf{T}}}x} \sqcap \overline{\overline{E}\,\overline{E^{\mathsf{T}}}x} = E^{\mathsf{T}}x \sqcap \overline{\overline{E}\,E^{\mathsf{T}}x} = E^{\mathsf{T}}x \sqcap \overline{\overline{E}x}$$

$$= E^{\mathsf{T}}x \sqcap Ex = (E^{\mathsf{T}} \sqcap E)x = Ix = x.$$

Furthermore $\mathtt{lub}(E) = I$ and $\mathtt{lub}(t) = \mathtt{lub}(Et)$ since $\mathtt{ubd}(t) = \mathtt{ubd}(Et)$.

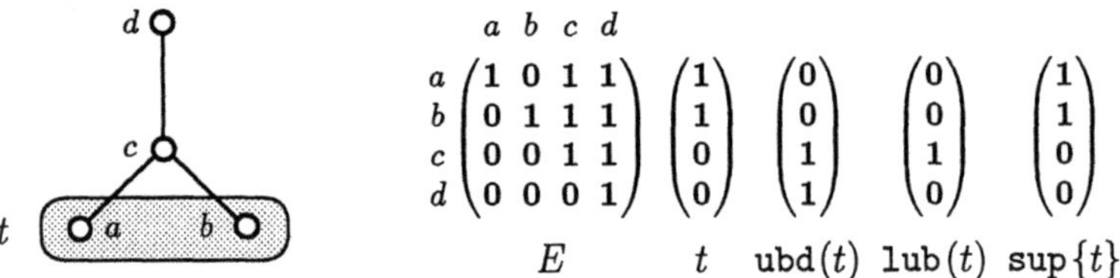


$$
\begin{array}{c}
\begin{array}{cccc} a & b & c & d \end{array} \\
\begin{array}{c} a \\ b \\ c \\ d \end{array}
\begin{pmatrix} 1 & 0 & 1 & 1 \\ 0 & 1 & 1 & 1 \\ 0 & 0 & 1 & 1 \\ 0 & 0 & 0 & 1 \end{pmatrix}
\begin{pmatrix} 1 \\ 1 \\ 0 \\ 0 \end{pmatrix}
\begin{pmatrix} 0 \\ 0 \\ 1 \\ 1 \end{pmatrix}
\begin{pmatrix} 0 \\ 0 \\ 1 \\ 0 \end{pmatrix}
\begin{pmatrix} 1 \\ 1 \\ 0 \\ 0 \end{pmatrix} \\
\quad E \quad\quad\quad\quad t \quad \mathtt{ubd}(t) \;\; \mathtt{lub}(t) \;\; \mathtt{sup}\{t\}
\end{array}
$$

**Fig. 3.3.3** Comparing $\mathtt{lub}$ and $\mathtt{sup}$

We take care to distinguish between $\mathtt{lub}$ and $\mathtt{sup}$. The $\mathtt{sup}$ is reserved for the supremum of a set of relations in a relation algebra ordered by inclusion $\subset$; it belongs to our metalevel. For example, the transitive closure is defined as

$$\mathtt{sup}\{\, X \mid R \subset X,\ XX \subset X \,\};$$

it is a *relation*. Whereas $\mathtt{lub}(t)$ denotes the supremum of a subset $t$ with respect to a given order relation $E$; it belongs to the object level:

$$\mathtt{lub}(t) = \overline{\overline{E^{\mathsf{T}}}t} \sqcap \overline{\overline{E}\,\overline{E^{\mathsf{T}}}t}.$$

The result is the zero relation or one point. Compare Fig. 3.3.3 where we apply $\mathtt{lub}$ to the subset $t$, and $\mathtt{sup}$ to the set $\{t\}$ consisting solely of the relation $t$.

By definition, $\mathtt{lub}(t)$ is the empty set or the one-element set consisting of the least upper bound. The operators $\mathtt{gre}$ and $\mathtt{lub}$ are related by the following proposition.

**3.3.8 Proposition.** Let a subset $t$ be given.

i)   $\mathtt{gre}(t) = t \sqcap \mathtt{lub}(t),$ $\qquad\qquad\qquad$ $\mathtt{lea}(t) = t \sqcap \mathtt{glb}(t);$

ii)  $\mathtt{gre}(t) \neq O \implies \mathtt{gre}(t) = \mathtt{lub}(t),$ $\quad$ $\mathtt{lea}(t) \neq O \implies \mathtt{lea}(t) = \mathtt{glb}(t).$

**Proof:** i) $\mathtt{gre}(t) = t \sqcap \mathtt{ubd}(t)$ and $t \sqcap \mathtt{lub}(t) = t \sqcap \mathtt{ubd}(t) \sqcap \mathtt{lbd}(\mathtt{ubd}(t))$ coincide since $t \subset \mathtt{lbd}(\mathtt{ubd}(t))$, by Prop. 3.3.3.
ii) Every $\mathtt{lub}(t) \neq O$ is a point; furthermore $O \neq \mathtt{gre}(t) \subset \mathtt{lub}(t)$. For a point $x$ it has been proved in Prop. 2.4.5.i that $O \neq y \subset x$ implies $y = x$. $\qquad\qquad$ $\square$


By (i), a least upper bound for the set $t$, which belongs to $t$, is its greatest element; so there is no greatest element if the least upper bound does not lie in $t$, or does not exist at all.

Since $\mathtt{lub}(t) = \mathtt{lub}(Et)$, the least upper bound does not change when $t$ is enlarged by the set of predecessors of elements of $t$ with respect to the ordering $E$. It follows that (also when $\mathtt{lub}(t) = O$)

$$\mathtt{ubd}(t) = \mathtt{ubd}(\mathtt{lbd}(\mathtt{ubd}(t))) = \overline{\overline{E}^{\mathsf{T}} \, \mathtt{lbd}(\mathtt{ubd}(t))} \subset \overline{\overline{E}^{\mathsf{T}} \, \mathtt{lub}(t)} = \mathtt{ubd}(\mathtt{lub}(t)).$$

As a simple consequence of the definition we have:


**3.3.9 Proposition.** Let a subset $t$ be given. Then the following holds:

i)    $t \subset \mathtt{lbd}(\mathtt{lub}(t)),$ $\qquad\qquad$ $t \subset \mathtt{ubd}(\mathtt{glb}(t));$

ii)   $\mathtt{lub}(t) \neq O \implies t \subset E\,\mathtt{lub}(t),$ $\quad$ $\mathtt{glb}(t) \neq O \implies t \subset E^{\mathsf{T}}\mathtt{glb}(t);$

iii)  $t \neq O \implies \mathtt{lub}(t) \subset E^{\mathsf{T}}t,$ $\qquad$ $t \neq O \implies \mathtt{glb}(t) \subset Et;$

iv)   $\mathtt{lub}(t) = \mathtt{glb}(\mathtt{ubd}(t)),$ $\qquad$ $\mathtt{glb}(t) = \mathtt{lub}(\mathtt{lbd}(t)).$

**Proof:** i) follows from $\mathtt{lub}(t) \subset \mathtt{ubd}(t)$ and Prop. 3.3.3.i. ii) is a special case of (i) which can be proved using Prop. 2.4.4.i, since $\mathtt{lub}(t) \neq O$ is a point. iii) can be shown using $\mathtt{lub}(t) \subset \mathtt{ubd}(t)$ and Prop. 3.3.4. iv) means in full

$$\mathtt{ubd}(t) \sqcap \mathtt{lbd}(\mathtt{ubd}(t)) = \mathtt{lbd}(\mathtt{ubd}(t)) \sqcap \mathtt{ubd}(\mathtt{lbd}(\mathtt{ubd}(t))),$$

which can be proved using the remark following Prop. 3.3.3. $\qquad\qquad$ $\square$


We now prove the main properties of bounds.


**3.3.10 Proposition.** Let subsets $t$ and $t'$ be given.

i)  $t \subset \mathtt{lbd}(t') \implies \mathtt{lub}(t) \subset \mathtt{lbd}(t');$
    $t \subset \mathtt{ubd}(t') \implies \mathtt{glb}(t) \subset \mathtt{ubd}(t').$

ii) For a point $x$ the following holds
    $t \subset Ex \implies \mathtt{lub}(t) \subset Ex, \quad t \subset E^{\mathsf{T}}x \implies \mathtt{glb}(t) \subset E^{\mathsf{T}}x.$

**Proof:** i) $\mathtt{lub}(t) \subset \mathtt{lbd}(\mathtt{ubd}(t)) \subset \mathtt{lbd}(\mathtt{ubd}(\mathtt{lbd}(t'))) = \mathtt{lbd}(t')$ since these functionals are antitonic. ii) If $x$ is a point then $\mathtt{lbd}(x) = \overline{\overline{Ex}} = Ex$.    □

The following example shows that for (ii) to hold true the point property of $x$ is essential. Let $t = x = O$. Then $t \subset Ex$, but in general $\mathtt{lub}(O) = \mathtt{ubd}(O) \sqcap \mathtt{lbd}(\mathtt{ubd}(O)) = L \sqcap \mathtt{lbd}(L) = \mathtt{lbd}(L) = \overline{\overline{EL}} \neq O$. It does not suffice to suppose just $x \neq O$, as can be seen by letting $t := x := (0,1)$ be an open interval of the ordered set of real numbers.

An ordering is particularly nice to work with when $\mathtt{lub}$ and $\mathtt{glb}$ always exist. At the metalevel this is in fact the case for the ordering by inclusion $\subset$ of relations. We now formulate this condition for the object level:

**3.3.11 Definition.** We call an ordering

i) $E$ a **complete lattice**    $:\Longleftrightarrow$    $O \neq \mathtt{lub}(t)$  for all  $t$
        $\Longleftrightarrow$    $O \neq \mathtt{glb}(t)$  for all  $t$
        $\Longleftrightarrow$    $\mathtt{glb}(t) \neq O \neq \mathtt{lub}(t)$  for all  $t$
        $\Longleftrightarrow$    Every subset $t$ has a least upper and a greatest lower bound.

ii) $E$ a **complete $\mathtt{lub}$-semilattice**
    $:\Longleftrightarrow$    $O \neq \mathtt{lub}(t)$  for all  $t \neq O$.
    $\Longleftrightarrow$    Every nonempty subset $t$ has a least upper bound.

iii) $E$ a **complete $\mathtt{glb}$-semilattice**
    $:\Longleftrightarrow$    $O \neq \mathtt{glb}(t)$  for all  $t \neq O$.
    $\Longleftrightarrow$    Every nonempty subset $t$ has a greatest lower bound.    □

The various statements in (i) are equivalent by Prop. 3.3.9.iv. Clearly, any complete lattice is also a complete $\mathtt{lub}$-semilattice, as well as a complete $\mathtt{glb}$-semilattice. Our next proposition is known as the Least Upper Bound Theorem. It asserts that in a complete $\mathtt{glb}$-semilattice *least upper bounds* exist, provided there are *upper bounds*.

**3.3.12 Proposition** (*Least Upper Bound Theorem*). In a complete $\mathtt{glb}$-semilattice $E$, for every subset $t$ the following is true:

i)    $\qquad\qquad\qquad \mathtt{ubd}(t) \neq O \quad \Longrightarrow \quad \mathtt{lub}(t) \neq O;$

ii)    $\qquad\qquad\qquad t \subset E\,\mathtt{ubd}(t) \quad \Longrightarrow \quad t \subset E\,\mathtt{lub}(t).$

**Proof:** i) Due to Prop. 3.3.9.iv, $\mathtt{lub}(t) = \mathtt{glb}(\mathtt{ubd}(t))$ holds and we apply Prop. 3.3.11.iii to the subset $\mathtt{ubd}(t) \neq O$. (ii) is a consequence of (i) and Prop. 3.3.9.ii.    □

We omit the dual result for $\mathtt{lub}$-semilattices.

## Exercises

**3.3.1** Prove that for arbitrary relations $R, S$ the identity $\overline{\overline{RS}} = \overline{\overline{R}\ \overline{\overline{R^{\mathsf{T}}}\ \overline{\overline{RS}}}}$ holds. Use Def. 3.3.2 in order to give a formulation in terms of $\mathtt{lbd}$ and $\mathtt{ubd}$ .

**3.3.2** Let $R$ be an arbitrary relation which is not necessarily an ordering. By analogy with Def. 3.3.2, define $\mathtt{ubd}_R(X) := \overline{\overline{R^{\mathsf{T}}}X}$ and $\mathtt{lbd}_R(X) := \overline{\overline{R}X}$, and prove (cf. Prop. 3.3.3.ii)

$$\mathtt{ubd}_R(X) = \mathtt{ubd}_R(\mathtt{lbd}_R(\mathtt{ubd}_R(X))).$$

**3.3.3** For an ordering $E$ and an arbitrary relation $X$ prove $E^{\mathsf{T}}\overline{EX} = \overline{EX}$.

**3.3.4** Define for an arbitrary relation $S$ the formation of a "greatest element"

$$G(X) := \mathtt{ubd}(X) \sqcap \mathtt{max}(X) = \overline{\overline{S \sqcup I}^{\mathsf{T}}X} \sqcap [X \sqcap \overline{(S \sqcap \overline{I})X}],$$

and show that there is at most one such element:

$$G(X)[G(X)]^{\mathsf{T}} \subset I.$$

**3.3.5** For points $x, y$, the condition $\mathtt{ubd}(x) = \mathtt{ubd}(y)$ implies $x = y$.

**3.3.6** Prove that $\mathtt{ubd}(E^{\mathsf{T}}) = \mathtt{gre}(E^{\mathsf{T}})$.

**3.3.7** Prove that $\mathtt{lub}(X \sqcup Y) = \mathtt{lub}(\mathtt{lub}(X) \sqcup \mathtt{lub}(Y))$.

**3.3.8** Show that $L = L\,\mathtt{lub}(X) \implies \mathtt{ubd}(X) = \mathtt{ubd}(\mathtt{lub}(X))$.

**3.3.9** An ordering $E$ is called *directed* if $L = EE^{\mathsf{T}}$. Prove $\mathtt{max}(L) = \mathtt{gre}(L)$ for any directed $E$.

## 3.4 References

BODEWIG E: *Bericht über die verschiedenen Methoden zur Lösung eines Systems linearer Gleichungen mit reellen Koeffizienten. IV.* Nederl. Akad. Wetensch. Proc. Ser. A. **51** (1948) 53–64. Also: Indag. Math. **10** (1948) 24–35.

GIVE'ON Y: *Lattice matrices.* Inform. and Control **7** (1964) 477–484.

ROY B: *Transitivité et connexité.* C. R. Acad. Sci. Paris **249** (1959) 216.

STRASSEN V: *Gaussian elimination is not optimal.* Numer. Math. **13** (1969) 354–356.

WARSHALL S: *A theorem on boolean matrices.* J. Assoc. Comput. Mach. **9** (1962) 11–12.

# 4. Heterogeneous Relations

We now pass from homogeneous to heterogeneous relations. In terms of matrices this amounts to passing from square matrices to general rectangular ones. The general principles stay the same as before, but when multiplying, joining, or intersecting heterogeneous relations one has to make sure that these operations are defined.

In Sect. 4.1 we introduce heterogeneous relations and show how they can be represented by bipartitioned graphs. Section 4.2 contains the relation-algebraic treatment of univalent and totally defined relations, and of the property of being a function. There we also give the decomposition of a relation into its univalent and its multivalent part.

Beside binary relations which are the chief topic of this book there are $n$-ary relations. Section 4.3 establishes a connection with $n$-ary relations used for relational data bases. Section 4.4 deals with difunctionality. It provides preparation for the rather involved algebraic investigation of comparing functions by their course of values and can be skipped at the first reading.

## 4.1 Bipartite Graphs

So far we have only been concerned with homogeneous relations, i.e., with subsets of a Cartesian product $V \times V$ which we also represented by square Boolean matrices. We now generalize this concept to relations relating elements of two different sets. In applications in computer science these sets can be of different *sort* or *type*.

**4.1.1 Definition.** Given possibly different sets $X$ and $Y$, a subset $R$ of $X \times Y$ is called a (**heterogeneous**) **relation** between $X$ and $Y$. $\qquad\square$

For representing a heterogeneous relation one can use a rectangular Boolean matrix and also a bipartitioned graph. Representations by hypergraphs will be discussed in Sect. 5.3.

Most of the calculus of homogeneous relations developed in Chap. 2 extends to heterogeneous relations:

- The Boolean operations from Sect. 2.1 carry over completely: The set of relations between two sets $X, Y$ is the complete lattice of subsets of $X \times Y$. Of course, it would not make much sense for our applications to intersect a relation $R \subset X \times Y$ with an $S \subset Y \times X$ when $X \neq Y$.

- Transposing a relation $R \subset X \times Y$ leads to $R^\mathsf{T} \subset Y \times X$, but the rules for conversion from Sect. 2.2 still apply. We will not consider, e.g., $R \sqcup R^\mathsf{T}$ unless $X = Y$.
- The product of two relations is defined, provided the underlying sets or sorts match: if $R \subset X \times Y$ and $S \subset Y \times Z$ then $RS \subset X \times Z$. The results of Sect. 2.3 remain true when the matching condition above is satisfied. For example, the products $RR^\mathsf{T} \subset X \times X$ and $R^\mathsf{T}R \subset Y \times Y$ are always defined.

We retain our notation, $O$ for the empty relation, $L$ for the universal relation, and $I$ for the identity, in the heterogeneous case. Thus, for $R \subset X \times Y$, there is a left identity $I \subset X \times X$ and a right identity $I \subset Y \times Y$. If necessary, one can specify the underlying sets and write more precisely $I_X R = R$, $R I_Y = R$, and $O_{XY} \subset R \subset L_{XY}$.

$$\underset{I}{\begin{pmatrix}1&0&0&0&0\\0&1&0&0&0\\0&0&1&0&0\\0&0&0&1&0\\0&0&0&0&1\end{pmatrix}}\underset{R}{\begin{pmatrix}1&0&0\\1&1&1\\0&1&0\\0&1&1\\0&0&0\end{pmatrix}} = \underset{R}{\begin{pmatrix}1&0&0\\1&1&1\\0&1&0\\0&1&1\\0&0&0\end{pmatrix}} = \underset{R}{\begin{pmatrix}1&0&0\\1&1&1\\0&1&0\\0&1&1\\0&0&0\end{pmatrix}}\underset{I}{\begin{pmatrix}1&0&0\\0&1&0\\0&0&1\end{pmatrix}}$$

**Fig. 4.1.1** Left identity and right identity of a Boolean matrix

A heterogeneous relation $R \subset X \times Y$ is commonly represented as in Fig. 4.1.2: The points of $X$ and $Y$, respectively, are aligned opposite to each other, and are joined by arrows according to $R$. This object can be made a 1-graph by extending $R$ to $V = \{a, b, c, d, e, x, y, z\} = X \cup Y$ and taking its associated graph $B$. The corresponding matrix contains a lot of zeros and is therefore less economical to store than $R$. One may also want to consider, as in Sect. 8.3, arrows that point from $Y$ to $X$. We make this precise in the following definition.

$$R = \begin{array}{c|ccc} & x & y & z\\ \hline a & 1&0&0\\ b&1&1&1\\ c&0&1&0\\ d&0&1&1\\ e&0&0&0\end{array}$$

$$B = \begin{array}{c|ccccc|ccc} & a&b&c&d&e&x&y&z\\ \hline a&0&0&0&0&0&1&0&0\\ b&0&0&0&0&0&1&1&1\\ c&0&0&0&0&0&0&1&0\\ d&0&0&0&0&0&0&1&1\\ e&0&0&0&0&0&0&0&0\\ \hline x&0&0&0&0&0&0&0&0\\ y&0&0&0&0&0&0&0&0\\ z&0&0&0&0&0&0&0&0\end{array}$$

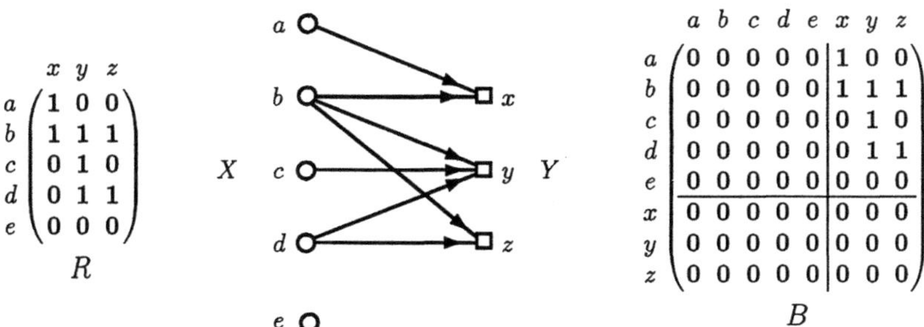

**Fig. 4.1.2** Possible representations of a heterogeneous relation

**4.1.2 Definition.** A quadruple $(X, Y, R, S)$ is called a **bipartitioned graph**, if

i) $X$ and $Y$ are nonempty disjoint point sets (**left** and **right** respectively),

ii) $R \subset X \times Y$ and $S \subset Y \times X$ are heterogeneous relations, the arrows of which lead from left to right and from right to left, respectively. □

If the bipartitioned graph is symmetric, $R = S^\mathsf{T}$, or if we are only interested in an "undirected" graph, we restrict the description to the constituents of the triple $(X, Y, Q)$ with $Q := R \sqcup S^\mathsf{T}$.

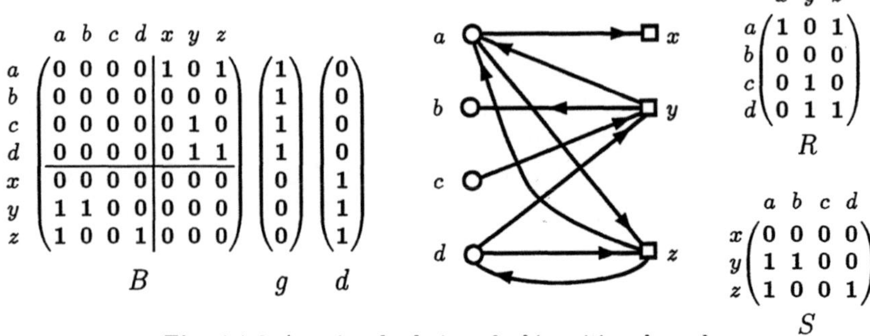

Fig. 4.1.3 Associated relation of a bipartitioned graph

Thus, a bipartitioned graph $(X, Y, R, S)$ leads to a 1-graph $(V, B)$ with (total) point set $V = X \cup Y$ and an associated matrix of special form

$$B = \begin{pmatrix} O & R \\ S & O \end{pmatrix}.$$

We decide to use "$g$" and "$d$" for "left" and "right" combining this mnemonically with (French:) gauche, droit. Then the separation of the (total) point set is given by two vectors $g$, $d$ fulfilling

$$\overline{g} = d, \quad g^\mathsf{T} g = L, \quad d^\mathsf{T} d = L.$$

Of course[1], $g, d$ also fulfil $g \sqcap d = O$, $g \sqcup d = L$, $g^\mathsf{T} d = O$ and $d^\mathsf{T} g = O$, and $B \subset gd^\mathsf{T} \sqcup dg^\mathsf{T}$.

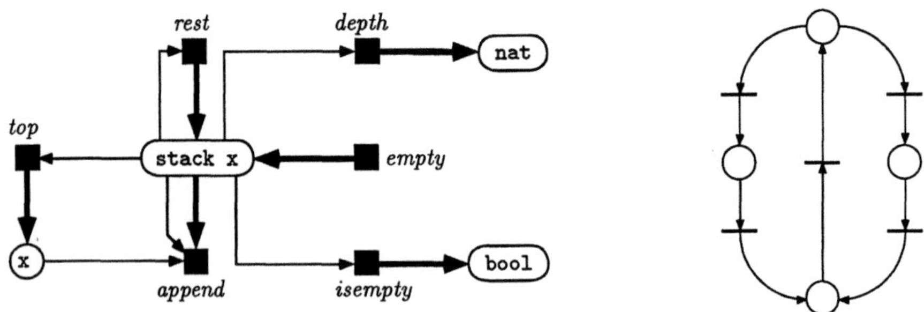


Fig. 4.1.4 Signature diagram and Petri net

Particularly important examples of bipartitioned graphs in computer science are **Petri nets** and **signature diagrams**. Signature diagrams enable us to represent

---

[1] So far we cannot express $B$ relation-algebraically in terms of $R, S, g, d$; nor $R$ and $S$ in terms of $B, g, d$. We shall achieve this, however, in Sect. 5.2, using the results of Sect. 4.2.

functionality and type of $n$-ary operations of an abstract data type. In a Petri net the points to the left or to the right are called **places** or **transitions** (or hurdles), respectively. But in practice the points will not be arranged in a strict left-right pattern, so as to be able to draw the mappings in a natural way. Figure 4.1.4 gives two such bipartitioned graphs.

Given an arbitary graph one may want to draw it as a bipartitioned graph in order to make its graphical representation clearer. When this is possible, by suitably arranging the points, then the associated matrix can be decomposed as in Fig. 4.1.3 and one obtains the characterization by $g$ and $d$ as above; in this situation the original graph is called **bipartite** according to a **bipartitioning** of the set of points. In Fig. 4.1.5 two possible bipartitionings for a bipartite graph are given.

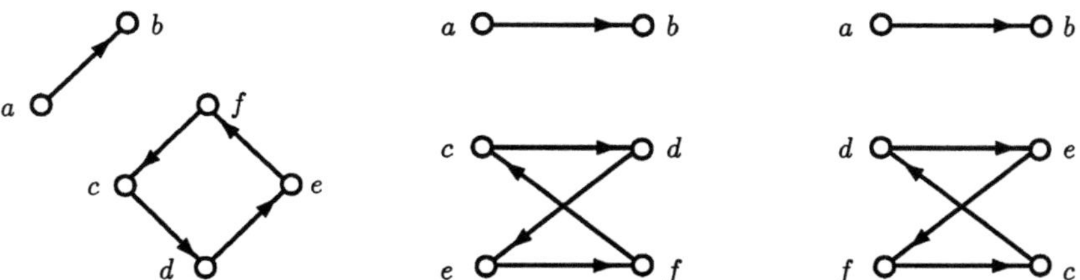

**Fig. 4.1.5** Bipartite graph with different bipartitionings

When determining the set of predecessors of a point set, a bipartitioned graph behaves as follows:

**4.1.3 Proposition.** The associated relation $B$ of a graph bipartitioned with $g, d$ together with a point set $x$ always fulfils

$$B(x \sqcap g) = Bx \sqcap d, \qquad B(x \sqcap d) = Bx \sqcap g.$$

**Proof:** For reasons of symmetry, we only prove the first equation. Since

$$B(x \sqcap g) \subset Bg \subset (gd^{\mathsf{T}} \sqcup dg^{\mathsf{T}})g = gd^{\mathsf{T}}g \sqcup dg^{\mathsf{T}}g = O \sqcup dL = d$$

we have "$\subset$". The opposite direction "$\supset$" follows directly from

$$Bx \sqcap d \subset (B \sqcap dx^{\mathsf{T}})(x \sqcap B^{\mathsf{T}}d) \subset B(x \sqcap [gd^{\mathsf{T}} \sqcup dg^{\mathsf{T}}]d) = B(x \sqcap g). \qquad \square$$

The concept of bipartition can be generalized to $n$-partitions which are called colorings of the set of points. The points of an "$n$-partite" graph can be colored with at most $n$ colors in such a way that no two related points have the same color ($n$-point-coloring).

## 4.2 Functions and Mappings

Of special interest are heterogeneous relations that are *univalent* or *total* in the following sense: A relation $R \subset X \times Y$ is called univalent if it associates to every $x \in X$ *at most* one $y \in Y$, i.e.,

$$\forall x \in X \; \forall y, z \in Y : [(x,y) \in R \wedge (x,z) \in R] \rightarrow y = z;$$

$R$ is called total if it associates to every $x \in X$ *at least* one $y \in Y$, i.e.,

$$\forall x \in X \; \exists y \in Y : (x,y) \in R.$$

An $R$ that is both univalent and total is a *mapping* or a *totally defined function*, a fundamental concept of mathematics.

**4.2.1 Definition.** If $R$ is a relation, we call

$$
\begin{aligned}
R \; \textbf{total} \quad &:\Longleftrightarrow \quad L = RL \quad \Longleftrightarrow \quad I \subset RR^{\mathsf{T}} \quad \Longleftrightarrow \quad \overline{R} \subset R\overline{I} \\
&\Longleftrightarrow \quad \text{For all } S, \text{ from } SR = O \text{ follows } S = O; \\[4pt]
R \; \textbf{univalent} \quad &:\Longleftrightarrow \quad R^{\mathsf{T}}R \subset I \quad \Longleftrightarrow \quad R\overline{I} \subset \overline{R}; \\
&\quad\text{also: functional or right-univalent} \\[4pt]
R \; \textbf{surjective} \quad &:\Longleftrightarrow \quad R^{\mathsf{T}} \text{ total} \\
&\Longleftrightarrow \quad L = LR \quad \Longleftrightarrow \quad I \subset R^{\mathsf{T}}R \quad \Longleftrightarrow \quad \overline{R} \subset \overline{I}R \\
&\Longleftrightarrow \quad \text{For all } S, \text{ from } RS = O \text{ follows } S = O; \\[4pt]
R \; \textbf{injective} \quad &:\Longleftrightarrow \quad R^{\mathsf{T}} \text{ univalent} \\
&\Longleftrightarrow \quad RR^{\mathsf{T}} \subset I \quad \Longleftrightarrow \quad \overline{I}R \subset \overline{R}; \\[4pt]
R \; \textbf{mapping} \quad &:\Longleftrightarrow \quad R \text{ total and univalent} \quad \Longleftrightarrow \quad R\overline{I} = \overline{R}. \qquad \Box
\end{aligned}
$$

Let us explain the various conditions for a relation $R \subset X \times Y$ to be total. $L = RL$ means that with *every* $x \in X$ there is associated at least one $y \in Y$; so the domain of $R$ is all of $X$. Starting with an element of $X$ and applying $R$ and then $R^{\mathsf{T}}$ should at least yield the original element: $RR^{\mathsf{T}} \supset I$. If an element $x$ is *not* related, by $R$, to some element $y$, then $x$ must be related to some *other* element $z$: $\overline{R} \subset R\overline{I}$. The last condition says that $R$ is annihilated from the left by the zero relation only[2]. 

The four conditions for totality are easily seen to be equivalent. The equivalence of the first two follows from Dedekind's formula using monotonicity:

$$RR^{\mathsf{T}} = (R \sqcap IL^{\mathsf{T}})(L \sqcap R^{\mathsf{T}}I) \supset RL \sqcap I = L \sqcap I = I; \quad L = IL \subset RR^{\mathsf{T}}L \subset RL.$$

Equivalence of the first and the third variant is proved by

$$\overline{R} \subset R\overline{I} \quad \Longleftrightarrow \quad L = R \sqcup R\overline{I} = RI \sqcup R\overline{I} = RL.$$

Finally, we prove that a relation $R$ fulfilling $L \subset RL$ may only be annihilated

---

[2] This is reminiscent of linear independence. The condition is formulated by quantifying over all relations, so it no longer belongs to first-order theory.

by a relation $S$ if $S = O$. Of course, $SR \subset O$ precisely when $LR^\mathsf{T} \subset \overline{S}$ and, furthermore, precisely when $RL \subset \overline{S}^\mathsf{T}$; so $S$ must vanish as a consequence of $L \subset RL \subset \overline{S}^\mathsf{T}$. We now consider the opposite direction. We always have $L^\mathsf{T}R^\mathsf{T} \subset (RL)^\mathsf{T}$, so that $\overline{RL}^\mathsf{T} R \subset O$. As $R$ is annihilated, we have $\overline{RL}^\mathsf{T} \subset O$. This implies $L \subset RL$.

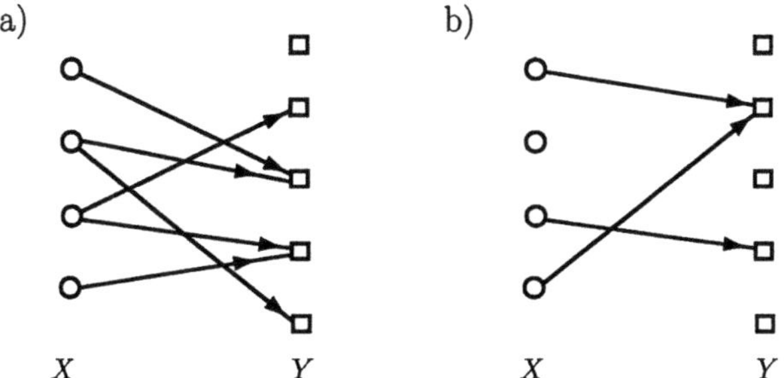



**Fig. 4.2.1** Total relation and univalent relation

We now consider the two univalency conditions. If we look at predecessors of $y \in Y$ along relation $R$ and then at the successors of these predecessors, we get at most the $y$ we started with. So $R^\mathsf{T}R \subset I$. The second form of the condition follows from Schröder's equivalences.

A univalent relation $R \subset X \times Y$ is also called a **(partially defined) function**, a **right-univalent** relation, or a **functional** relation (cf. RIGUET 48 or CHIN, TARSKI 51). For the term "injective" one also says **left-univalent**. Mappings or (totally defined) functions, on the other hand, associate with every element of $X$ *precisely* one element of $Y$. The usual notation is $f_R\colon X \longrightarrow Y$, or $X \xrightarrow{f_R} Y$ and $f_R(x) = y$ for $(x, y) \in R$. This notation is frequently used for partially defined functions, too: One writes $f_R\colon X \longrightarrow Y$ and puts

$$f_R(x) = \begin{cases} y, & \text{if } (x,y) \in R; \\ \text{undefined}, & \text{if } (x,z) \notin R \text{ for all } z. \end{cases}$$

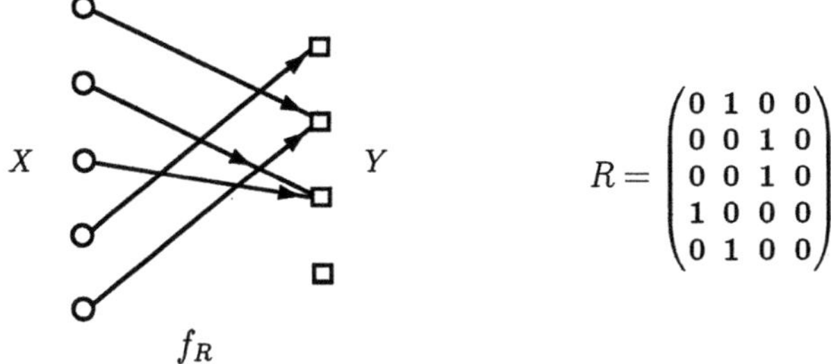



$$R = \begin{pmatrix} 0 & 1 & 0 & 0 \\ 0 & 0 & 1 & 0 \\ 0 & 0 & 1 & 0 \\ 1 & 0 & 0 & 0 \\ 0 & 1 & 0 & 0 \end{pmatrix}$$

**Fig. 4.2.2** A mapping $f_R$ as a total and univalent relation $R$

The order in which we write the product of two relations is opposite to the usual way of denoting the composition of two mappings: $(x, y) \in RS$ corresponds to

$f_S(f_R(x)) = y$. With this in mind, our relation-algebraic formulas translate into well-known statements on mappings. The two defining conditions for a mapping, $I \subset RR^\mathsf{T}$ and $R^\mathsf{T}R \subset I$ translate into $A \subset f_R^{-1}(f_R(A))$ and $f_R(f_R^{-1}(U)) \subset U$ when applying the mapping to subsets. A mapping can even be characterized by the single relation-algebraic equation $R\overline{I} = \overline{R}$.

The following laws hold for mappings and, more generally, for univalent relations.

### 4.2.2 Proposition

| | | | |
|---|---|---|---|
| i) | $Q, R$ univalent | $\implies$ | $QR$ univalent; |
| ii) | $Q$ univalent | $\iff$ | $Q(R \sqcap S) = QR \sqcap QS$ for all $R, S$; |
| iii) | $Q$ univalent | $\iff$ | $RQ \sqcap S = (R \sqcap SQ^\mathsf{T})Q$ for all $R, S$; |

iv) $\left. \begin{array}{l} R \subset Q,\ Q \text{ univalent} \\ RL \supset QL \end{array} \right\} \implies R = Q;$

v) $\quad Q$ univalent $\quad\iff\quad Q\overline{R} = QL \sqcap \overline{QR}$ for all $R$.

**Proof:** We use Dedekind's formula and $\sqcap$-subdistributivity:

i) $\quad (QR)^\mathsf{T}QR = R^\mathsf{T}Q^\mathsf{T}QR \subset R^\mathsf{T}R \subset I.$

ii) $\quad Q(R \sqcap S) \supset (Q \sqcap QSR^\mathsf{T})(R \sqcap Q^\mathsf{T}QS) \supset QR \sqcap QS \supset Q(R \sqcap S).$

For "$\impliedby$" take $R := \overline{I}, S := I$:
$$Q\overline{I} \sqcap Q = Q\overline{I} \sqcap QI = Q(\overline{I} \sqcap I) = QO = O \implies Q\overline{I} \subset \overline{Q}.$$

iii) $\quad (R \sqcap SQ^\mathsf{T})Q \subset RQ \sqcap SQ^\mathsf{T}Q$
$$\subset RQ \sqcap S \subset (R \sqcap SQ^\mathsf{T})(Q \sqcap R^\mathsf{T}S) \subset (R \sqcap SQ^\mathsf{T})Q.$$

For "$\impliedby$" take $R := L, S := I$.

iv) $\quad Q = QL \sqcap Q \subset RL \sqcap Q, \qquad$ since $Q \subset QL$
$$\subset (R \sqcap QL^\mathsf{T})(L \sqcap R^\mathsf{T}Q) \subset RR^\mathsf{T}Q \subset RQ^\mathsf{T}Q \subset R.$$

v) For "$\impliedby$" take $R := I$. "$\implies$" is proved by:
$$Q\overline{R} \subset \overline{QR} \quad\iff\quad Q^\mathsf{T}QR \subset R;$$
$$QL \sqcap \overline{QR} \subset Q\overline{R} \quad\iff\quad L = \overline{QL} \sqcup (QR \sqcup Q\overline{R}). \qquad \square$$

The product of univalent relations is again univalent by (i), so the composition of two partially defined functions is again a partially defined function. Whereas multiplication is, in general, only subdistributive with respect to intersection (Prop. 2.3.6 iii), we have shown in (ii) that the distributive law holds when multiplying from the left by a univalent relation.

This is reflected by $f_Q^{-1}(U) \cap f_Q^{-1}(V) = f_Q^{-1}(U \cap V)$, the well-known rule for mappings: "the intersection of inverse images is equal to the inverse image of the intersection", the validity of which does *not* require the entire mapping property.

On the other hand, when multiplying by a univalent relation $Q$ from the right, one cannot expect the distributive law to hold in general—it is easy to give counterexamples. However, in many instances one encounters situation (iii)

which can be viewed as a substitute for distributivity. In terms of mappings—
actually we have given the proof for partial functions—equation (iii) reads as
follows: $f_Q(U) \cap V = f_Q\left(U \cap f_Q^{-1}(V)\right)$ or "the intersection of $V$ and the image
of $U$ is equal to the image of the intersection of the inverse image of $V$ and of
$U$".

By (iv), a relation $R$ which is contained in, and has the same domain as, some
univalent relation $Q$ must be equal to $Q$ (cf. Prop. 3.1.7). Statement (v) yields
$f_Q^{-1}(\overline{U}) = f_Q^{-1}(L) \cap \overline{f_Q^{-1}(U)}$ or "the inverse image of the complement is equal to
the complement of the inverse image intersected with the domain".

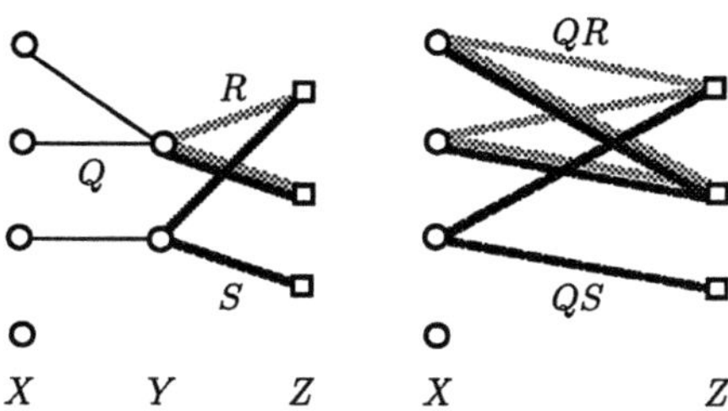


**Fig. 4.2.3** $Q(R \sqcap S) = QR \sqcap QS$  for a univalent relation $Q$

The following is a generalization of Prop. 2.4.4.i for transposed functions instead
of points. It reflects the rule "$U \subset f_Q^{-1}(V) \iff f_Q(U) \subset V$".

**4.2.3 Proposition.** If $F$ is a mapping, then arbitrary relations $R, S$ satisfy

$$R \subset SF^{\mathsf{T}} \iff RF \subset S.$$

**Proof:** Using $F^{\mathsf{T}}F \subset I$ and $FL = L$, the proof is completely analogous to that
of Prop. 2.4.4.i.                                                               $\square$

We add three more facts about mappings, total relations, and univalent relations
which show that the defining conditions $R\overline{I} = \overline{RI}$, $R\overline{I} \supset \overline{RI}$, $R\overline{I} \subset \overline{RI}$ remain
valid when replacing the identity $I$ by an arbitrary relation $S$.

**4.2.4 Proposition.** 

i)   $R$ **total** $\iff$ $R\overline{S} \supset \overline{RS}$ for all $S$;

ii)  $R$ **univalent** $\iff$ $R\overline{S} \subset \overline{RS}$ for all $S$;

iii) $R$ **mapping** $\iff$ $R\overline{S} = \overline{RS}$ for all $S$.

**Proof:** i) $L = RL = R(S \sqcup \overline{S}) = RS \sqcup R\overline{S} \iff \overline{RS} \subset R\overline{S}$.
ii) $R^{\mathsf{T}}R \subset I \implies R^{\mathsf{T}}RS \subset S \iff R\overline{S} \subset \overline{RS}$.
iii) is a consequence of (i, ii).                                               $\square$

Statement (iii) is the rule "$f_R^{-1}(\overline{U}) = \overline{f_R^{-1}(U)}$", i.e., "the inverse image of the
complement of a set is equal to the complement of the inverse image of this set".
So this is a special case of Prop. 4.2.2.v for (totally defined) mappings.

## Cardinality

If $R$ is a relation between finite sets, then let

$$\| R \| := \max_z |\{\, y \mid (y, z) \in R \,\}|$$
$$= \text{maximum number of elements in relation to an element.}$$

$\| R \|$ is then the maximum number of entries of $\mathbf{1}$ in a column. For a vector $x$ the number of rows of $\mathbf{1}$ is given by $|x|$. The following properties of these norm functions or absolute values are easy to prove:

**4.2.5 Proposition.** The following inequalities hold for a relation $R$ and a vector $x$, which are both finite:

i) $\qquad\qquad |Rx| \leq \| R \| \, |x|;$

ii) $\qquad\qquad \| R^{\mathsf{T}} \| \leq 1, \quad |R^{\mathsf{T}}x| \leq |x|, \quad$ if $R$ is univalent;

iii) $\qquad\qquad \| R^{\mathsf{T}} \| = 1, \quad |R^{\mathsf{T}}x| \leq |x|, \quad$ if $R$ is a mapping;

iv) $\qquad\qquad \| R^{\mathsf{T}} \| = 1, \quad |R^{\mathsf{T}}x| = |x|, \quad$ if $R$ is an injective mapping. $\qquad\square$

The notion of **cardinality** is also extended to infinite sets. A set $X$ is said to have cardinality less than or equal to that of a set $Y$ (written as $|X| \leq |Y|$), if there exists a relation $R$ between $X$ and $Y$ which is injective and total, i.e., $RR^{\mathsf{T}} = I$. Similarly, $X$ is said to have cardinality greater than or equal to that of $Y$ (written as $|X| \geq |Y|$), if there exists a univalent and surjective relation $R$ between $X$ and $Y$, i.e., $R^{\mathsf{T}}R = I$. This is, of course, only a preorder for sets.

Two sets are said to have the same cardinality if there exists a relation $R$ between them such that $R$ and $R^{\mathsf{T}}$ are mappings. Having the same cardinality constitutes an equivalence relation for sets.

A relation that is injective and surjective is called **bijective**. For mappings between finite sets of the same cardinality, the properties of being injective, surjective, or bijective are equivalent. Such mappings are also called **permutations**, or permutation matrices if viewed as relations. In order to justify our claim we prove:

$$RR^{\mathsf{T}} = I \iff \exists m : R^m = I, \ R^{\mathsf{T}} = R^{m-1}, \qquad \text{if } R \text{ is finite.}$$

For the nontrivial implication "$\Longrightarrow$" we use the fact that there are only finitely many distinct powers of $R$. So by the pigeon-hole principle, $R^t = R^s$ with $t > s$, say. Repeated multiplication by $R^{\mathsf{T}}$ and using $RR^{\mathsf{T}} = I$ yields $R^m = I$ with $m := t - s$. So $R^{m-1}R = I$ and therefore

$$R^{\mathsf{T}} = I R^{\mathsf{T}} = R^{m-1} RR^{\mathsf{T}} = R^{m-1}.$$

For the foundations of set theory it is of importance to know whether saying "$X$ and $Y$ have the same cardinality" is equivalent to "$X$ has cardinality less than or equal to that of $Y$, and vice versa".

Let us illustrate this by an example. Suppose an injective mapping $\iota$ from $\{a_1, a_2, \ldots\}$ to $\{b_1, b_2, \ldots\}$ is given, as well as an injective mapping $\kappa$ in the opposite direction (see Fig. 4.2.4). We want to construct, in an explicit way, a bijective mapping between these two sets. This can be accomplished by "alternating" between $\iota$ and $\kappa^{\mathsf{T}}$. Specifically, a subset $F$ is to be found such that $\lambda := (\iota \sqcap F) \sqcup (\kappa^{\mathsf{T}} \sqcap \overline{F})$ becomes a bijective mapping. In order that the $\lambda$ pieced together in this fashion has the desired property, it is necessary that the images $\kappa F$ of $F$ under $\kappa^{\mathsf{T}}$ all are images $\iota^{\mathsf{T}} F$ of $F$ under $\iota$. So the condition is $\kappa F \subset \iota^{\mathsf{T}} F$.

**4.2.6 Proposition** (*E. Schröder, F. Bernstein*). Let an injective mapping $\iota$ from $X$ to $Y$ and an injective mapping $\kappa$ from $Y$ to $X$ be given. Then there exists an injective and in addition *surjective* mapping $\lambda$ from $X$ to $Y$.

**Proof:** We consider the set $\mathcal{A}$ of relations between $X$ and $Y$,

$$\mathcal{A} := \{ A \mid \kappa A \subset \iota^{\mathsf{T}} A \} = \{ A \mid A \subset \overline{\kappa^{\mathsf{T}} \overline{\iota^{\mathsf{T}} A}} \},$$

described using two obviously equivalent conditions. It is clear that $O$ belongs to $\mathcal{A}$. Furthermore, $\mathcal{A}$ is closed with respect to set union. So $F := \sup \mathcal{A}$ belongs to $\mathcal{A}$ and, of course, is the greatest element of $\mathcal{A}$. One easily sees that

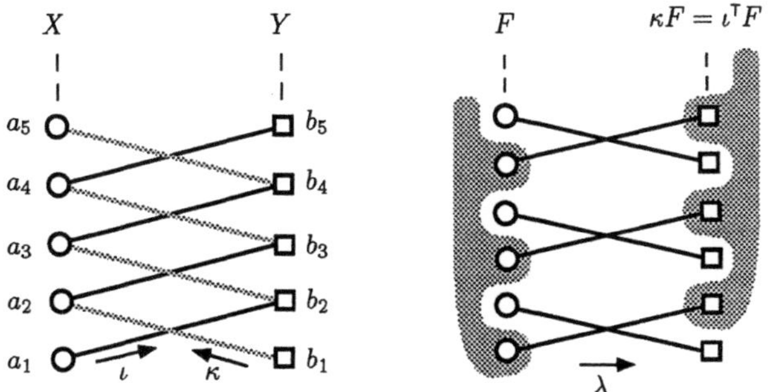


**Fig. 4.2.4** Comparing cardinalities

together with $A$ the relation $A' := \overline{\kappa^{\mathsf{T}} \overline{\iota^{\mathsf{T}} A}}$ containing $A$ belongs to $\mathcal{A}$; therefore $F = \overline{\kappa^{\mathsf{T}} \overline{\iota^{\mathsf{T}} F}}$. Furthermore $F = FL$, since $F \subset FL$ and since $\kappa F \subset \iota^{\mathsf{T}} F$ implies $\kappa FL \subset \iota^{\mathsf{T}} FL$. Finally we have $\iota \kappa F = \iota \kappa \kappa^{\mathsf{T}} \overline{\iota^{\mathsf{T}} F} = \iota \kappa \kappa^{\mathsf{T}} \overline{\iota^{\mathsf{T}} F} = \iota \iota^{\mathsf{T}} F = F$ so that $\kappa^{\mathsf{T}} \iota^{\mathsf{T}} F = \kappa^{\mathsf{T}} \iota^{\mathsf{T}} \iota \kappa F \subset F$. Using this vector $F$ we define

$$\lambda := (\iota \sqcap F) \sqcup (\kappa^{\mathsf{T}} \sqcap \overline{F})$$

composing the relations $\iota$ and $\kappa^{\mathsf{T}}$.

The proof that $\lambda$ is indeed a bijective mapping is left to the reader.     □

## Injective Mappings and Subsets

Having the necessary notions for mappings at our disposal, we return to subsets and predicates. To start with an example, consider the sets $X = \{1, 2, 3, 4, 5, 6\}$

and $Z = \{\underline{2}, \underline{4}, \underline{6}\}$ and associate with every element of $Z$ the corresponding element of $X$, thus defining a total injective mapping; see Fig. 4.2.5.

Fig. **4.2.5** Subset given by injection

We have defined the set $Z$ by letting its elements correspond to certain elements in $X$; it is thus as independent a set as $X$ is from which we copied it. In our earlier Definition 2.4.1 we viewed subsets as indicated in Fig. 4.2.6 below.

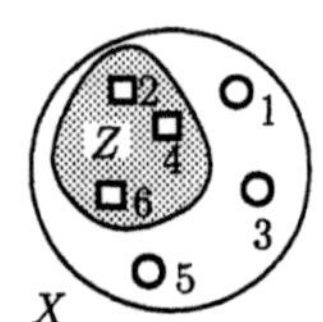

Fig. **4.2.6** Subset given by a predicate

Here subset $Z$ is regarded as a "dependent" set, owing its existence only to its elements being selected by some predicate. In the homogeneous relation algebra of all the relations on $X$ there are only relations that correspond to Boolean 6×6-matrices. So this relation algebra cannot contain relations on $Z$ because $3 \times 3$-matrices do not occur. Though our restriction to the given relation algebra may seem artificial, it is essential for our constructive approach to computer related set theory.

The following definition makes our concept of "independent" subset precise.

**4.2.7 Definition.** Given an injective mapping $\iota$, we call $\iota$ a **subset given by injection**. ☐

A transition between the two concepts of subsets, Definitions 2.4.1 and 4.2.7, is in general possible *in one direction only*. Clearly, there is an element-by-element correspondence between the subset given by injection $\iota$ and the vector $z := \iota^{\mathsf{T}} L$ representing the subset given by $\iota$ in the relation algebra. But conversely, given a vector $z$, it is not evident at all that there exists an appropriate injection $\iota$ *within the given* relation algebra.

## Univalent and Multivalent Part

A relation may associate with certain elements precisely one element and with others several or none. This distinction yields a decomposition of the relation into two parts as follows; the terminology is justified by the next proposition.

**4.2.8 Definition.** Given an arbitrary relation $R$, we call

i) $\qquad\qquad \text{unp}(R) := R \sqcap \overline{\overline{R}\overline{I}}$ the **univalent part** of $R$,

ii) $\qquad\qquad \text{mup}(R) := R \sqcap \overline{R}\overline{I}$ the **multivalent part** of $R$. $\qquad\qquad\square$

So every relation splits into its univalent and its multivalent part resulting in a disjunction:

$$R = \text{unp}(R) \sqcup \text{mup}(R), \quad \text{unp}(R) \sqcap \text{mup}(R) = O.$$

$R \sqcap \overline{\overline{R}\overline{I}}$ is characterized as follows: An element $x$ is related to $y$ by the univalent part of $R$ if $x$ is related to $y$ by $R$, but to no $z$ different from $y$. So in this case, $y$ is the unique element associated with $x$ by $R$. If on the other hand $R$ associates with $x$ two distinct elements $y$ and $z$, then row $x$ belongs to the multivalent part of $R$. See Fig. 4.2.7.

The complement $\overline{I}$ of the identity $I$ describes distinctness. We now prove some results involving the univalent and the multivalent parts of relations where we make use of $\overline{I}$.

$$
\begin{pmatrix} 0 & 0 & 1 & 0 \\ 1 & 0 & 0 & 1 \\ 1 & 0 & 0 & 0 \\ 0 & 0 & 0 & 0 \\ 1 & 1 & 1 & 1 \end{pmatrix}
\quad
\begin{pmatrix} 0 & 1 & 1 & 1 \\ 1 & 0 & 1 & 1 \\ 1 & 1 & 0 & 1 \\ 1 & 1 & 1 & 0 \end{pmatrix}
\quad
\begin{pmatrix} 1 & 1 & 0 & 1 \\ 1 & 1 & 1 & 1 \\ 0 & 1 & 1 & 1 \\ 0 & 0 & 0 & 0 \\ 1 & 1 & 1 & 1 \end{pmatrix}
\quad
\begin{pmatrix} 0 & 0 & 1 & 0 \\ 0 & 0 & 0 & 0 \\ 1 & 0 & 0 & 0 \\ 0 & 0 & 0 & 0 \\ 0 & 0 & 0 & 0 \end{pmatrix}
\quad
\begin{pmatrix} 0 & 0 & 0 & 0 \\ 1 & 0 & 0 & 1 \\ 0 & 0 & 0 & 0 \\ 0 & 0 & 0 & 0 \\ 1 & 1 & 1 & 1 \end{pmatrix}
$$

$$R \qquad\qquad \overline{I} \qquad\qquad \overline{R}\overline{I} \qquad\qquad \text{unp}(R) \qquad\qquad \text{mup}(R)$$

**Fig. 4.2.7** Univalent and multivalent part of a relation

**4.2.9 Proposition.** The following holds for the univalent part $\text{unp}(R)$:

i) $\text{unp}(R)$ is univalent: $\qquad [\text{unp}(R)]^{\mathsf{T}} \text{unp}(R) \subset I$.

ii) $\text{unp}$ is an idempotent functional: $\quad \text{unp}(\text{unp}(R)) = \text{unp}(R)$.

iii) The functional $\text{unp}$ is decreasing as long as the domain of the argument remains unaltered:
$$\left.\begin{array}{r} R \subset S \\ RL \supset SL \end{array}\right\} \quad \Longrightarrow \quad \text{unp}(R) \supset \text{unp}(S).$$

iv) $\text{mup}(R)$ and $\text{unp}(R)$ are "orthogonal" in the following sense:
$$[\text{mup}(R)]^{\mathsf{T}} \text{unp}(R) = O, \quad R^{\mathsf{T}} \text{unp}(R) = [\text{unp}(R)]^{\mathsf{T}} \text{unp}(R).$$

**Proof:** i) $[\text{unp}(R)]^{\mathsf{T}} \text{unp}(R) = (R \sqcap \overline{\overline{R}\overline{I}})^{\mathsf{T}}(R \sqcap \overline{\overline{R}\overline{I}}) \subset R^{\mathsf{T}} \overline{\overline{R}\overline{I}} \subset I$.

ii) "$\subset$" follows from the definition. On the other hand, $\overline{R}\overline{I} \subset \overline{(R \sqcap \overline{\overline{R}\overline{I}})}\overline{I}$, so the inverted inclusion holds, too.

iii) $R \subset S \implies \overline{R}\overline{I} \subset \overline{S}\overline{I} \implies \overline{\overline{S}\overline{I}} \subset \overline{\overline{R}\overline{I}}$ gives the first half of the result $\text{unp}(S) \subset \overline{\overline{R}\overline{I}}$. Furthermore, $\text{unp}(S) = S \sqcap \overline{\overline{S}\overline{I}} = S \sqcap \overline{\overline{S}\overline{I}} \sqcap SL \subset S \sqcap \overline{\overline{S}\overline{I}} \sqcap RL = (S \sqcap \overline{\overline{S}\overline{I}} \sqcap RI) \sqcup (S \sqcap \overline{\overline{S}\overline{I}} \sqcap R\overline{I}) \subset R \sqcup (\overline{\overline{R}\overline{I}} \sqcap \overline{R}\overline{I}) = R \sqcup O = R$.

iv) $[\operatorname{mup}(R)]^{\mathsf{T}}\operatorname{unp}(R) = (R \sqcap R\overline{I})^{\mathsf{T}}(R \sqcap \overline{R\overline{I}})$ may on the one hand be estimated by $\dots \subset R^{\mathsf{T}}\overline{R\overline{I}} \subset I$, and on the other by $\dots \subset (R\overline{I})^{\mathsf{T}}\overline{R\overline{I}} \subset \overline{I}$. Therefore, this product equals $O$. $\qquad\qquad\square$

Statement (iii) can also be seen as follows: In $\operatorname{unp}(R)$ those rows of $R$ that contain more than one $\mathbf{1}$ are replaced by rows of zeros. If a given relation $R$ is enlarged by successively replacing $\mathbf{0}$ by $\mathbf{1}$, then three situations can occur: Firstly, a $\mathbf{1}$ is entered into a row already containing two or more $\mathbf{1}$'s; this row is then made a zero-row in $\operatorname{unp}(R)$ before as well as afterwards. Secondly, a $\mathbf{1}$ is added to a row containing exactly one coefficient equal to $\mathbf{1}$; this row is then again made $\mathbf{0}$, and the result decreases. Thirdly, a $\mathbf{1}$ is put into a row of zeros which enlarges the domain of $R$. Statement (iii) constitutes a generalization of Prop. 4.2.2.iv.

There are corresponding assertions for the multivalent part:

**4.2.10 Proposition.** The following holds for the multivalent part $\operatorname{mup}(R)$ of a relation $R$:

i)   $\operatorname{mup}$ is an increasing functional:   $R \subset S \implies \operatorname{mup}(R) \subset \operatorname{mup}(S)$.

ii)  $\operatorname{mup}$ is an idempotent functional:   $\operatorname{mup}(\operatorname{mup}(R)) = \operatorname{mup}(R)$.

iii)   $$\operatorname{mup}(R)L = \operatorname{mup}(R)\overline{I}.$$

iv)   $$\operatorname{mup}(R)L = \operatorname{mup}(R)\overline{x}\quad\text{for every point } x.$$

v)   $$R^{\mathsf{T}}\operatorname{mup}(R) = [\operatorname{mup}(R)]^{\mathsf{T}}\operatorname{mup}(R).$$

**Proof:** i) together with the direction "$\subset$" of (ii) are trivial. $\operatorname{mup}(R) = R\overline{I} \sqcap R \subset (R \sqcap R\overline{I})(\overline{I} \sqcap R^{\mathsf{T}}R) \subset (R \sqcap R\overline{I})\overline{I} = \operatorname{mup}(R)\overline{I}$ gives direction "$\supset$" of (ii).

iii) $\operatorname{mup}(R)L = \operatorname{mup}(R)(I \sqcup \overline{I}) = \operatorname{mup}(R) \sqcup \operatorname{mup}(R)\overline{I} = \operatorname{mup}(R)\overline{I}$ using the result (ii) already obtained.

iv) "$\supset$" is trivial. Furthermore $\operatorname{mup}(R)L = \operatorname{mup}(R)(x \sqcup \overline{x}) = \operatorname{mup}(R)x \sqcup \operatorname{mup}(R)\overline{x}$, so it remains to prove

$$\begin{aligned}
\operatorname{mup}(R)x &= (R\overline{I} \sqcap R)x = R\overline{I}x \sqcap Rx \\
&= R\overline{x} \sqcap Rx && \text{following Props. 4.2.2.ii and 4.2.4.iii} \\
&\subset (R \sqcap Rx\overline{x}^{\mathsf{T}})(\overline{x} \sqcap R^{\mathsf{T}}Rx) \\
&\subset (R \sqcap R\overline{I})\overline{x} = \operatorname{mup}(R)\overline{x}, && \text{since always } I\overline{x} \subset \overline{x}.
\end{aligned}$$

v) is a direct consequence of Prop. 4.2.9.iv. $\qquad\qquad\square$

It is clear that the multivalent part of a relation is an increasing functional, for entering an additional $\mathbf{1}$ into a Boolean matrix has the effect of converting a row with exactly one entry of $\mathbf{1}$ into one with two $\mathbf{1}$'s which then belongs to the multivalent part, whereas the status of rows with at least two entries of $\mathbf{1}$ is not affected. Note that (iii) and (iv) are also true for $1 \times 1$-matrices although $\overline{I} = O$ in this case.

The following results on univalent and multivalent parts will be needed later in various contexts. These results deal in essence with an elementary way of counting in terms of the quantities of "zero", "one", and "more than one". Interpreting the outcome as nonexistence, unique or nonunique existence may answer questions frequently asked in mathematics. We also see here that such problems can often be formulated in linear relation-algebraic terms, so that explicit counting is not necessary.

Regardless of their individual shape the rows of a Boolean matrix can be marked according to whether they vanish, contain exactly one, or more than one, entry of $1$.

**4.2.11 Proposition.** For an arbitrary relation $R$,

$$L = \overline{RL} \sqcup \operatorname{unp}(R)L \sqcup \operatorname{mup}(R)L$$

is a decomposition into disjoint components.

**Proof:** Of course, $L = \overline{RL} \sqcup RL$ is a disjoint decomposition. Obviously, $RL = \operatorname{unp}(R)L \sqcup \operatorname{mup}(R)L$, and from orthogonality (Prop. 4.2.9.iv) we have $\operatorname{unp}(R)L \sqcap \operatorname{mup}(R)L = O$. $\qquad\square$

$$
\begin{pmatrix} 0 & 0 & 0 & 0 \\ 1 & 0 & 0 & 0 \\ 0 & 0 & 0 & 0 \\ 0 & 1 & 0 & 0 \\ 1 & 0 & 1 & 0 \\ 0 & 0 & 1 & 1 \end{pmatrix}
\quad
\begin{pmatrix} 0 & 0 & 0 & 0 \\ 1 & 0 & 0 & 0 \\ 0 & 0 & 0 & 0 \\ 0 & 1 & 0 & 0 \\ 0 & 0 & 0 & 0 \\ 0 & 0 & 0 & 0 \end{pmatrix}
\quad
\begin{pmatrix} 0 & 0 & 0 & 0 \\ 0 & 0 & 0 & 0 \\ 0 & 0 & 0 & 0 \\ 0 & 0 & 0 & 0 \\ 1 & 0 & 1 & 0 \\ 0 & 0 & 1 & 1 \end{pmatrix}
\quad
\begin{pmatrix} 1 \\ 0 \\ 1 \\ 0 \\ 0 \\ 0 \end{pmatrix}
\quad
\begin{pmatrix} 0 \\ 1 \\ 0 \\ 1 \\ 0 \\ 0 \end{pmatrix}
\quad
\begin{pmatrix} 0 \\ 0 \\ 0 \\ 0 \\ 1 \\ 1 \end{pmatrix}
$$

$$R \qquad\qquad \operatorname{unp}(R) \qquad\qquad \operatorname{mup}(R) \qquad\quad \overline{RL} \qquad \operatorname{unp}(R)L \qquad \operatorname{mup}(R)L$$

**Fig. 4.2.8** The decomposition $L = \overline{RL} \sqcup \operatorname{unp}(R)L \sqcup \operatorname{mup}(R)L$

Figure 4.2.8 shows this situation for a Boolean matrix. Of particular interest are the extreme cases of this decomposition when one of the three components vanishes. If two components vanish, then the third must coincide with the vector $L$. We discuss these extreme cases systematically and characterize them by conditions to be imposed on $R$. Characterizations (i), (iv), and (v) have already arisen in Definition 4.2.1.

**4.2.12 Proposition.** For a relation $R$ the following holds

$$
\begin{array}{clccl}
\text{i)} & \overline{RL} = O & \Longleftrightarrow & R \text{ total;} \\
\text{ii)} & \overline{RL} = L & \Longleftrightarrow & R = O; \\
\text{iii)} & \operatorname{unp}(R)L = O & \Longleftrightarrow & R \subset R\overline{I} \quad\Longleftrightarrow\quad RL \subset R\overline{I}; \\
\text{iv)} & \operatorname{unp}(R)L = L & \Longleftrightarrow & R \text{ mapping;} \\
\text{v)} & \operatorname{mup}(R)L = O & \Longleftrightarrow & R \text{ univalent;} \\
\text{vi)} & \operatorname{mup}(R)L = L & \Longleftrightarrow & R\overline{I} = L.
\end{array}
$$

The **proof** is omitted. $\qquad\square$

Situation (iii) could be referred to as a *nowhere univalent* relation, and situation (vi), as an *everywhere multivalent* relation.

### Exercises

**4.2.1** Prove $\operatorname{mup}(R) \sqcup \operatorname{mup}(S) \sqcup \operatorname{mup}(R+S) = \operatorname{mup}(R \sqcup S)$.

**4.2.2** Show that $\operatorname{unp}(R)\overline{I} = \overline{R} \sqcap \operatorname{unp}(R)L$ for arbitrary $R$.

**4.2.3** Prove that $F \fatsemi G$ is a mapping if $F$ and $G$ are.

## 4.3 $n$-ary Relations in Data Bases

So far we have dealt with binary (or dyadic) relations. In practice relations relating three or more objects also occur, such as the following ternary (homogeneous) relation,

$$R = \{(\text{Oscar M., Isabel M., Ulysses M.}), (\text{Oscar M., Eve M., Claire Z.}),$$
$$(\text{Ulysses M., Barbara M., Sandra Z.}), (\text{Walter Z., Claire Z., Thomas Z.}),$$
$$(\text{Thomas Z., Sandra Z., Edward Z.})\} \subset P^3,$$

which is given by a set of triples (father, mother, child) on the set

$$P := \{\text{Oscar M., Eve M., Isabel M., Walter Z., Claire Z., Ulysses M.,}$$
$$\text{Barbara M., Thomas Z., Sandra Z., Edward Z.}\}$$

of persons. It is much harder to visualize such a **ternary** relation. Tables, graphs, or matrices are two-dimensional in nature, so using them for $n$-ary relations can become quite complicated. On the other hand, representing information structured in this fashion is a task typical of computer science. It can be helpful to make use of *functional dependencies*; in the example above, for instance, one can associate with each child its (uniquely determined) father and mother.

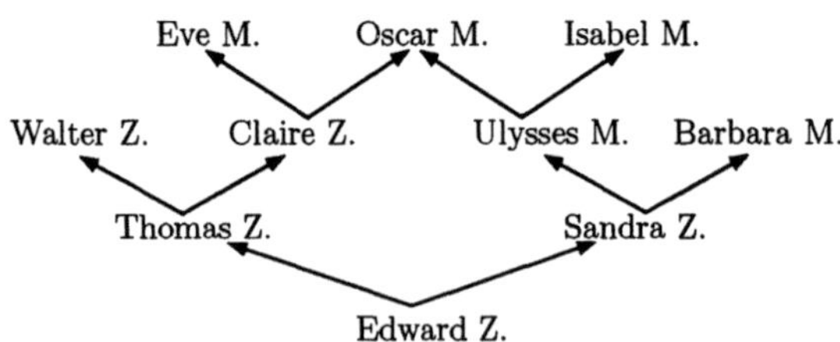


**Fig. 4.3.1** Representation of a ternary relation

So this example has been represented by means of binary relations.

Another representation is provided by the three projections $\pi_i \colon P^3 \longrightarrow P$, $i = 1, 2, 3$; $\pi_1$ associates with each triple in $R$, the father, $\pi_2$ the mother, and $\pi_3$ the child. Of course, $P^3$ may be a very large set, and $R \subset P^3$ only a small part thereof. So for reasons of economy, one only represents the restrictions $\pi_i|_R$ which, as binary relations, appear as follows:

$$\pi_1|_R \approx \text{father} = \{((\text{O M, I M, U M}), \text{O M}), ((\text{O M, E M, C Z}), \text{O M}),$$
$$((\text{U M, B M, S Z}), \text{U M}), ((\text{W Z, C Z, T Z}), \text{W Z}),$$
$$((\text{T Z, S Z, E Z}), \text{T Z})\} \subset R \times P$$

$$\pi_2|_R \approx \text{mother} = \{((\text{O M, I M, U M}), \text{I M}), ((\text{O M, E M, C Z}), \text{E M}),$$
$$((\text{U M, B M, S Z}), \text{B M}), ((\text{W Z, C Z, T Z}), \text{C Z}),$$
$$((\text{T Z, S Z, E Z}), \text{S Z})\} \subset R \times P$$

$$\pi_3|_R \approx \text{child} = \{((\text{O M, I M, U M}), \text{U M}), ((\text{O M, E M, C Z}), \text{C Z}),$$
$$((\text{U M, B M, S Z}), \text{S Z}), ((\text{W Z, C Z, T Z}), \text{T Z}),$$
$$((\text{T Z, S Z, E Z}), \text{E Z})\} \subset R \times P.$$

One writes this in a more compact way as in Fig. 4.3.2. There, the rows such as (Oscar M., Isabel M., Ulysses M.) constitute the domain of the three mappings. The image of the above row under the mapping *father* is then Oscar M.

| *father* | *mother* | *child* |
|---|---|---|
| Oscar M. | Eve M. | Claire Z. |
| Oscar M. | Isabel M. | Ulysses M. |
| Ulysses M. | Barbara M. | Sandra Z. |
| Walter Z. | Claire Z. | Thomas Z. |
| Thomas Z. | Sandra Z. | Edward Z. |

**Fig. 4.3.2** The ternary relation of Fig. 4.3.1 represented as a table

In Fig. 4.3.3 we consider the relation $V :=$ "was at some time married to" between the set $M$ of men and the set $W$ of women, and compare the two representations of $V$, as a matrix and as a table. On the left hand side, the columns of the matrix are labelled by *elements* of $W$, and the rows, by *elements* of $M$. On the right hand side, the columns of the table are headed by the words "husband" or "wife", which, for the matrix, indicate only the *collection* of all rows or all columns, respectively.

| *wife* / *husband* | Eve M. | Isabel M. | Barbara M. | Claire Z. | Sandra Z. |
|---|---|---|---|---|---|
| Oscar M. | × | × | | | |
| Ulysses M. | | | × | | |
| Walter Z. | | | | × | |
| Thomas Z. | | | | | × |
| Edward Z. | | | | | |

| *husband* | *wife* |
|---|---|
| Oscar M. | Eve M. |
| Oscar M. | Isabel M. |
| Ulysses M. | Barbara M. |
| Walter Z. | Claire Z. |
| Thomas Z. | Sandra Z. |

**Fig. 4.3.3** Matrix and table forms of a binary relation

We now sketch a unifying terminology introduced by E. F. Codd in 1970 in his investigations on a relational data base model. For further details we refer the reader to the extensive literature on relational data bases.

Clearly one has to distinguish between an **attribute** (such as "mother") and its representatives. Here, the attribute "mother" identifies those persons in the set $P$ that are listed in the relation $R$ in place two. In the example, Eve M., Isabel M., Barbara M., Claire Z. and Sandra Z. appear as mothers. The same attribute could yield completely different representatives if $P$ or $R$ were changed.

When creating an additional relation for a data base, one proceeds by first fixing a set $\mathcal{A} = \{ A_1, \ldots, A_n \}$ of attributes, i.e., *identifiers*, together with their respective domains. By doing this a **relation type** is defined. In a given concrete situation the labels of the domains are followed by sets $W(A_i)$, $1 \le i \le n$, called **domains**.

Often the *set* $\mathcal{A}$ of attributes, or **list of attributes, combination of attributes**, is denoted by merely writing down its elements; so it is common to write $\mathcal{A} = A_1 \ldots A_n$. A **relation** $R$ is then a subset of $W(A_1) \times \ldots \times W(A_n)$ which is also abbreviated by $W(\mathcal{A})$. Attributes can be interpreted as projections, associating with every $n$-tuple in $W(A_1) \times \ldots \times W(A_n)$ its respective component in $W(A_i)$. Our attribute "mother" associates with the triple (Oscar M., Eve M., Claire Z.) the value Eve M.

In books on data bases one can often find a remark to the effect that the order of the attributes does not matter. Mathematically this means the following: Every $n$-tuple is a mapping

$$t: \{ A_1, \ldots, A_n \} \longrightarrow \bigcup_{i=1}^{n} W(A_i)$$

of the set of the $n$ attributes into the union of their ranges, subject to the condition that $t(A_i) \in W(A_i)$.

According to this definition of relations, based on projections, the two tables in Fig. 4.3.4, where rows and columns are interchanged, represent the *same* relation.

| $X$ | $Y$ | $Z$ |
|---|---|---|
| $x$ | $y$ | $z$ |
| $u$ | $v$ | $w$ |

| $Y$ | $X$ | $Z$ |
|---|---|---|
| $v$ | $u$ | $w$ |
| $y$ | $x$ | $z$ |

**Fig. 4.3.4** A ternary relation with attributes $X$, $Y$ and $Z$

The set of relations, interpreted as mappings, is denoted by $\mathcal{R}(\mathcal{A})$.

A set of relation types is called a relational data base **scheme**. In general, the lists of attributes of the relation types of a data base scheme will overlap. A given set of relations of such a system is called **relational data base**.

## Set Operations for $n$-ary Relations

Since we introduced relations as certain sets, the usual set operations are defined for them:

| | |
|---|---|
| union | $R \cup S;$ |
| intersection | $R \cap S,$ |
| difference | $R - S.$ |

Here it is understood that $R$ and $S$ are "compatible" in the sense that they have the same number of arguments, and that the ranges of their attributes match.

| $X$ | $Y$ | $Z$ |
|---|---|---|
| $x$ | $y$ | $z$ |
| $u$ | $x$ | $w$ |

$R$

| $X$ | $Y$ | $Z$ |
|---|---|---|
| $x$ | $y$ | $z$ |
| $u$ | $v$ | $w$ |

$S$

| $X$ | $Y$ | $Z$ |
|---|---|---|
| $x$ | $y$ | $z$ |
| $u$ | $x$ | $w$ |
| $u$ | $v$ | $w$ |

$R \cup S$

| $X$ | $Y$ | $Z$ |
|---|---|---|
| $x$ | $y$ | $z$ |

$R \cap S$

| $X$ | $Y$ | $Z$ |
|---|---|---|
| $u$ | $x$ | $w$ |

$R - S$

**Fig. 4.3.5** Set operations for relations

Clearly, $R \cap S = R - (R - S)$ holds. Similarly the complement of any relation is defined in the usual way; for reasons of cardinality, however, it is not introduced as a proper data base construct. Consider, for instance, a relation coming with the attributes *employee's name* and *salary*. The above set operations suffice for handling all the elementary data base operations such as "insert", "delete", and "update", without having to resort to the complement.

## Special Data Base Operations

We now come to the operations on data base relations proper. We shall work with relations as sets, as well as with their characterization in terms of mappings. It is useful to bear in mind that the mathematical concept serves as a user interface, as it were, behind which the realization can be created in various ways by the data base system.

**Projection**     $p_{\mathcal{X}} \colon \mathcal{R}(\mathcal{A}) \longrightarrow \mathcal{R}(\mathcal{X}), \quad \mathcal{X} \subset \mathcal{A}$

In connection with *binary* relations, projections occur at most in a very rudimentary form: Given a relation $R \subset X \times Y$, one can consider the set $RL$ of elements of $X$ affected by $R$, and similarly the set $R^{\mathsf{T}}L$ of elements of $Y$ affected by $R$.

In the $n$-ary case, let a set $\mathcal{A}$ of attributes be given together with a subset $\mathcal{X} \subset \mathcal{A}$. If we write somewhat sloppily $\mathcal{A} = A_1 \ldots A_n$, then $\mathcal{X} = A_{i_1} \ldots A_{i_m}$ which, however, need not be a consecutive subsequence of the former; but we are assuming that the attributes $A_{i_1} \ldots A_{i_m}$, and hence the indices $i_1 \ldots i_m$, are all distinct. Consider now a relation $R \subset W(\mathcal{A})$ as a set of $n$-tuples over the ranges $W(A_1), \ldots, W(A_n)$. By the projection onto the components $i_1, \ldots, i_m$ we understand the set of those $m$-tuples $r' = (r'_{i_1}, \ldots, r'_{i_m})$ for which an $n$-tuple $r$ exists such that $r'$ is a subsequence of $r$ with $r_{i_\mu} = r'_{i_\mu}$ for $\mu = 1, \ldots, m$. If $R \in \mathcal{R}(\mathcal{A})$ is represented as a set of mappings indexed by the attributes in $\mathcal{A}$, then the projection of $R$ onto $\mathcal{X}$ is the set of mappings of $R$ whose indices are in $\mathcal{X}$. This restricted projection is denoted by $R|_{\mathcal{X}}$, $p_{i_1 \ldots i_m}(R)$, or $p_{A_{i_1} \ldots A_{i_m}}(R)$ or $p_{\mathcal{X}}(R)$; sometimes also by $p_{\mathcal{X}} \colon \mathcal{R}(\mathcal{A}) \longrightarrow \mathcal{R}(\mathcal{X})$.

**Cartesian Product**     $\times \colon \mathcal{R}(\mathcal{A}) \times \mathcal{R}(\mathcal{L}) \longrightarrow \mathcal{R}(\mathcal{A} \,\dot\cup\, \mathcal{L})$

Let $R$ and $S$ be relations of type $\mathcal{A}$ with $n$ attributes and of type $\mathcal{L}$ with $m$ attributes, respectively. If we consider $R \subset W(\mathcal{A})$ and $S \subset W(\mathcal{L})$, then $R \times S$

$\subset W(\mathcal{A}) \times W(\mathcal{L})$ is just the set of all $(n+m)$-tuples $(r, s)$, where $r$ is an $n$-tuple from $R$ and $s$ is an $m$-tuple from $S$. If we view $R \in \mathcal{R}(\mathcal{A})$ and $S \in \mathcal{R}(\mathcal{L})$ then $R \times S \in \mathcal{R}(\mathcal{A} \mathbin{\dot{\cup}} \mathcal{L})$ is the set of those mappings for which $(R \times S)|_{\mathcal{A}} = R$ and $(R \times S)|_{\mathcal{L}} = S$ (the symbol $\dot{\cup}$ denotes the disjoint union). When dealing with a relation type we always assume that we know how to distinguish between different attributes. If an attribute $X$ occurs in $\mathcal{A}$ as well as in $\mathcal{L}$, then it gives rise to two different attributes in $\mathcal{A} \mathbin{\dot{\cup}} \mathcal{L}$ which have to be denoted differently, e.g., as $X.\mathcal{A}$ and $X.\mathcal{L}$.

**Selection**    $s_B \colon \mathcal{R}(\mathcal{A}) \longrightarrow \mathcal{R}(\mathcal{A})$

Given a condition $B$ which may, or may not, hold for the $n$-tuples of a relation $R$, then

$$s_B(R) := \{\, r \in R \mid B(r) \,\}$$

is the relation formed by those $n$-tuples which satisfy $B$. In Fig. 4.3.6 a typical example is given.

$$R = \begin{array}{c|c|c} X & Y & Z \\ \hline 1 & 2 & 3 \\ 1 & 4 & 3 \end{array} \qquad\qquad S = \begin{array}{c|c} A & B \\ \hline 10 & 11 \\ -1 & 12 \end{array}$$

| X | Z |
|---|---|
| 1 | 3 |

| X | Y | Z | A | B |
|---|---|---|---|---|
| 1 | 2 | 3 | 10 | 11 |
| 1 | 2 | 3 | -1 | 12 |
| 1 | 4 | 3 | 10 | 11 |
| 1 | 4 | 3 | -1 | 12 |

| X | Y | Z |
|---|---|---|
| 1 | 2 | 3 |

| X | Y | Z | A | B |
|---|---|---|---|---|
| 1 | 2 | 3 | 10 | 11 |
| 1 | 4 | 3 | 10 | 11 |

$p_{XZ}(R)$         $R \times S$         $s_{Y=2}(R)$         $R[Y < A]S$

**Fig. 4.3.6** Projection, Cartesian product, selection and join

**Quotient**    $\div \colon \mathcal{R}(\mathcal{A}) \longrightarrow \mathcal{R}(\mathcal{A} - \mathcal{H})$

Let $R$ be a relation of type $\mathcal{A}$ and $S$ a relation of type $\mathcal{H} \subset \mathcal{A}$. The quotient $R \div S$ of $R$ divided by $S$ is the relation of type $\mathcal{L} := \mathcal{A} - \mathcal{H}$ defined by

$$R \div S := \{\, b \in W(\mathcal{L}) \mid \forall s \in S : (b, s) \in R \,\}$$
$$= p_{\mathcal{L}}(R) - p_{\mathcal{L}}([p_{\mathcal{L}}(R) \times S] - R).$$

The quotient can also be formed if $W(\mathcal{H}) \subset W(\mathcal{A})$ is given, instead of $\mathcal{H} \subset \mathcal{A}$.

**Join**    $[X \Theta Y] \colon \mathcal{R}(\mathcal{A}) \times \mathcal{R}(\mathcal{L}) \longrightarrow \mathcal{R}(\mathcal{A} \mathbin{\dot{\cup}} \mathcal{L})$

Let $R$ and $S$ be relations of types $\mathcal{A}$ and $\mathcal{L}$, respectively. Further, let $X$ be an attribute from $\mathcal{A}$ and $Y$ an attribute from $\mathcal{L}$ whose domains $W(X) = W(Y)$ coincide. Finally, let $\Theta$ be a comparison operator on this common domain $W(X)$. We say that a tuple from $R \times S$ satisfies condition $X \Theta Y$, if its components, with respect to the attributes $X, Y$, are in relation $\Theta$. The $\Theta$-**join** (or equijoin in case $\Theta$ is an equality) with respect to the attributes $X$ and $Y$ is then the result of the selection according to this condition:

$$R[X \Theta Y]S := s_{X \Theta Y}(R \times S).$$

$$R = \begin{array}{c|c|c|c|c} A & B & C & X & Y \\\hline x & y & z & 1 & 2 \\ x & y & z & 3 & 4 \\ y & z & z & 3 & 4 \\ u & v & y & 1 & 2 \\ u & v & y & 3 & 4 \\ x & y & y & 2 & 3 \end{array} \qquad S = \begin{array}{c|c} X & Y \\\hline 1 & 2 \\ 3 & 4 \end{array} \qquad R \div S = \begin{array}{c|c|c} A & B & C \\\hline x & y & z \\ u & v & y \end{array}$$

$$R = \begin{array}{c|c|c|c} X & A & B & Y \\\hline x & 1 & 2 & y \\ y & 1 & 3 & y \\ z & 2 & 3 & y \end{array} \qquad S = \begin{array}{c|c|c} A & B & Z \\\hline 1 & 2 & u \\ 1 & 2 & v \\ 1 & 3 & w \\ 3 & 2 & y \end{array} \qquad R \bowtie S = \begin{array}{c|c|c|c|c} X & A & B & Y & Z \\\hline x & 1 & 2 & y & u \\ x & 1 & 2 & y & v \\ y & 1 & 3 & y & w \end{array}$$

**Fig. 4.3.7** Quotient and natural join

**Natural join**     $\bowtie \colon \mathcal{R}(\mathcal{A}) \times \mathcal{R}(\mathcal{L}) \longrightarrow \mathcal{R}(\mathcal{A} \cup \mathcal{L})$

Let $\mathcal{H} := \mathcal{A} \cap \mathcal{L}$ be the set of the attributes $X$ which occur both for relations $R$ of type $\mathcal{A}$ and for relations $S$ of type $\mathcal{L}$, and which therefore have to be marked differently as $X.\mathcal{A}$ and $X.\mathcal{L}$ in the Cartesian product. The natural join of $R$ and $S$ is then defined as

$$R \bowtie S := p_{\mathcal{A} \cup \mathcal{L}} \left( s_{R.\mathcal{H} = S.\mathcal{H}}(R \times S) \right),$$

where $\mathcal{A} \cup \mathcal{L}$ stands for the attributes of $R \times S$ without the redundant $X.\mathcal{L}$, $X \in \mathcal{H}$. For the condition "$\forall X \in \mathcal{H} : R.X = S.X$" we write "$R.\mathcal{H} = S.\mathcal{H}$" for short. So $R \bowtie S$ consists of those tuples from $R \times S$ which have the same entries at *all* the attributes that $R$ and $S$ have in common, i.e., that lie in $\mathcal{H}$, with the redundant part being deleted.

We have tried to give an idea of how the above operations are interrelated; a similar treatment of this topic can be found in PÖSCHEL, KALUŽNIN 79. The **generalized product** is a derived operation: it is defined as a "convolution" of the $i$-th with the $j$-th component:

$$R \circ_{ij} S := p_{1,n+m} s_{i=j}(R \times S),$$

where $R$ and $S$ are relations with $n$ and $m$ arguments, respectively. When $R$ and $S$ are both binary, then their ordinary product is given by

$$R \circ S = p_{1,4} \left( s_{2=3}(R \times S) \right).$$

## Functional Dependency

Let $\mathcal{A}$ be the set of attributes of some relation type and let $\emptyset \neq \mathcal{X}, \mathcal{Y} \subset \mathcal{A}$ be subsets of attributes. When designing a data base scheme, relations may be such that

$$p_{\mathcal{X}}^{\mathsf{T}} p_{\mathcal{Y}} \colon \mathcal{R}(\mathcal{X}) \longrightarrow \mathcal{R}(\mathcal{Y})$$

is a partially defined function. Think of the attribute "mother" which is a univalent function since everybody has only one mother. However, not every person represented in the data base will have his mother listed in the data base, too. In such a situation one will add this a priori to the relation types in the initial set-up. One then says that $\mathcal{Y}$ is **functionally dependent** on $\mathcal{X}$, symbolically $\mathcal{X} \longrightarrow \mathcal{Y}$ (or more precisely $\mathcal{X} \xrightarrow{A} \mathcal{Y}$ because when the same attributes occur in another relation type they need no longer be functionally dependent). So functional dependency means the following: if two tuples of $R$ have the same $\mathcal{X}$-entries, then their $\mathcal{Y}$-entries agree, too. If $R$ is a binary relation with two attributes $\mathcal{X}$ and $\mathcal{Y}$, then functional dependency is tantamount to $R$ being univalent.

Moreover, $\mathcal{Y}$ is called **completely functionally dependent** on $\mathcal{X}$, denoted by $\mathcal{X} \dashrightarrow \mathcal{Y}$, if in addition, there is no functional dependency of the kind $\mathcal{X}' \longrightarrow \mathcal{Y}$ for any proper subcombination of attributes $\mathcal{X}' \subset \mathcal{X}$. Even if the relation type does not come equipped with some functional dependency in the first place, it may happen that for some given relation such a dependency is in existence, by chance, as it were. This situation, however, might change as the data base is being inserted into or updated. So one should not make such a dependency part of the formal treatment of the data base. Clearly, for any combination of attributes there is a minimum such combination (the so-called "key attributes") upon which it is functionally dependent. One may want to work with a small number of basic functional dependencies, and possibly derive others from these basic ones only. One then applies the following **Armstrong rules** for functional dependency:

1. Given a combination of attributes, all of its subcombinations are functionally determined.

2. From the functional dependency $\mathcal{X} \longrightarrow \mathcal{Y}$, deduce the functional dependency $\mathcal{X} \cup \mathcal{Z} \longrightarrow \mathcal{Y} \cup \mathcal{Z}$.

3. From the functional dependencies $\mathcal{X} \longrightarrow \mathcal{Y}$ and $\mathcal{Y} \longrightarrow \mathcal{Z}$, deduce the functional dependency $\mathcal{X} \longrightarrow \mathcal{Z}$.

These rules are complete in the sense that *all* functional dependencies can be derived from them. For any nondeducible functional dependency one can construct an example of a data base that contains the original functional dependencies but not the nondeducible. Functional dependencies are difficult to establish; usually they are postulated in accordance with the specific properties of the given date base scheme and are put in evidence in addition to the relation type. So functional dependencies are part of the overall concept, providing the advantage of effective implementation but possibly restricting the representation of future information.

### Decomposition of a Relation Type

Given a union $\mathcal{X} \cup \mathcal{Y} = \mathcal{A}$ which need not be disjoint, the relation types $\mathcal{R}(\mathcal{X})$, $\mathcal{R}(\mathcal{Y})$ constitute a **decomposition** of the type $\mathcal{R}(\mathcal{A})$. For every relation $R \in \mathcal{R}(\mathcal{A})$ one then has

$$R \subset p_{\mathcal{X}}(R) \bowtie p_{\mathcal{Y}}(R).$$

It may happen that the containment above is actually an equality for all relations satisfying a set $\mathcal{F}$ of functional dependencies. In this situation the decomposition determined by $\mathcal{R}(\mathcal{X})$ and $\mathcal{R}(\mathcal{Y})$ is said to be **lossless** with respect to $\mathcal{F}$; alternatively, the (natural) join of $\mathcal{R}(\mathcal{X})$ and $\mathcal{R}(\mathcal{Y})$ is said to be lossless with respect to $\mathcal{F}$. In a data base, one would prefer to use "smaller" relation types instead of the "big" type $\mathcal{R}(\mathcal{A})$. If these smaller types constitute a lossless decomposition of $\mathcal{R}(\mathcal{A})$, then the original information can be restored completely. Decomposing a relation type in a data base can be highly desirable, but it creates the problem of keeping the data consistent. There are fast algorithms for the testing of that decomposition property.

## 4.4 Difunctionality

In Definition 2.3.10 we introduced for homogeneous relations $A, B$ their symmetric quotient

$$\mathbf{syq}(A, B) = \overline{A^\mathsf{T} \overline{B}} \sqcap \overline{\overline{A}^\mathsf{T} B}.$$

This concept can be extended for suitable heterogeneous relations: Let $A$ be a relation between $X$ and $Y$, and $B$ a relation between $X$ and $Z$. The symmetric quotient $\mathbf{syq}(A, B)$ is then defined as above and is a relation between $Y$ and $Z$. The following properties of $\mathbf{syq}$ also hold for heterogeneous relations.

**4.4.1 Proposition.** i) $\quad \mathbf{syq}(\overline{A}, \overline{B}) = \mathbf{syq}(A, B)$;

ii) $\qquad\qquad\qquad \mathbf{syq}(B, A) = [\mathbf{syq}(A, B)]^\mathsf{T}$;

iii) $\qquad\qquad\quad A\,\mathbf{syq}(A, A) = A$;

iv) $\qquad\qquad\qquad\quad I \subset \mathbf{syq}(A, A)$;

v) $\qquad\qquad\mathbf{syq}(A, B) \subset \mathbf{syq}(CA, CB) \quad$ for every $C$;

vi) $\qquad\quad F\,\mathbf{syq}(A, B) = \mathbf{syq}(AF^\mathsf{T}, B) \quad$ for every mapping $F$.

**Proof:** (i) and (ii) are trivial. The first part "$\subset$" of (iii) has already been shown in the footnote to Def. 2.3.10. The second part "$\supset$", together with (iv), holds since $I \subset \overline{\overline{A}^\mathsf{T} A}$ for all relations $A$.

v) $CB \subset CB \iff C^\mathsf{T} \overline{CB} \subset \overline{B} \implies (CA)^\mathsf{T} \overline{CB} \subset \overline{A^\mathsf{T} B}$. The second case is proved analogously. vi) Following Prop. 4.2.4.iii, we have $F\overline{S} = \overline{FS}$ for a mapping $F$ and arbitrary $S$, so that

$$F\,\mathbf{syq}(A, B) = F\overline{A^\mathsf{T} \overline{B}} \sqcap F\overline{\overline{A}^\mathsf{T} B} = \overline{FA^\mathsf{T} \overline{B}} \sqcap \overline{F\overline{A}^\mathsf{T} B}$$
$$= \overline{(AF^\mathsf{T})^\mathsf{T} \overline{B}} \sqcap \overline{\overline{AF^\mathsf{T}}^\mathsf{T} B} = \mathbf{syq}(AF^\mathsf{T}, B). \qquad \square$$

The univalent part from Sect. 4.2 is an example of a symmetric quotient, for it can be written as $\mathbf{unp}(R) = \overline{R\overline{I}} \sqcap R = \overline{R\overline{I}} \sqcap \overline{\overline{R}I} = \mathbf{syq}(R^\mathsf{T}, I)$.

The symmetric quotient can be illustrated as follows: Taken of the two relations $A \subset X \times Y$ and $B \subset X \times Z$, it relates an element $y \in Y$ to an element

$z \in Z$ precisely when $y$ and $z$ have the same set of "inverse images" with respect to $A$ or $B$ respectively. Thus

$$(y, z) \in \mathbf{syq}(A, B) \quad \Longleftrightarrow \quad \forall x : [(x, y) \in A \leftrightarrow (x, z) \in B].$$

In terms of matrices the above condition for $y$ and $z$ means that the corresponding columns of $A$ and $B$ are equal. In Fig. 4.4.1 we have $X = \{\, p, q, r, s, t \,\}$, $Y = \{\, a, b, c, d, e, f \,\}$ and $Z = \{\, 1, \ldots, 9 \,\}$.

A closer examination of $\mathbf{syq}(A, B)$ shows that its matrix falls into constant boxes after suitably rearranging its rows and columns so as to obtain block diagonal form. One has just to arrange equal columns of $A$ and $B$ side by side. We shall later formulate this property algebraically and discuss some consequences of it.

$$
\begin{array}{c}
\begin{array}{c}
\phantom{p}\; a\; b\; c\; d\; e\; f \\
\begin{array}{c}p\\q\\r\\s\\t\end{array}
\left(\begin{array}{cccccc}
0&1&1&0&1&1\\
0&0&0&1&0&0\\
0&1&0&0&1&1\\
0&0&1&0&0&0\\
0&0&0&0&1&0
\end{array}\right)
\end{array}
\quad
\begin{array}{c}
1\;2\;3\;4\;5\;6\;7\;8\;9 \\
\left(\begin{array}{ccccccccc}
0&1&1&0&0&0&1&1&0\\
1&0&0&0&1&0&0&0&0\\
0&1&1&0&0&0&1&1&0\\
0&0&0&0&0&0&0&0&0\\
0&0&1&0&0&0&0&1&0
\end{array}\right)
\end{array}
\quad
\begin{array}{c}
\phantom{a}\;1\;2\;3\;4\;5\;6\;7\;8\;9 \\
\begin{array}{c}a\\b\\c\\d\\e\\f\end{array}
\left(\begin{array}{ccccccccc}
0&0&0&1&0&1&0&0&1\\
0&1&0&0&0&0&1&0&0\\
0&0&0&0&0&0&0&0&0\\
1&0&0&0&1&0&0&0&0\\
0&0&1&0&0&0&0&1&0\\
0&1&0&0&0&0&1&0&0
\end{array}\right)
\end{array}
\quad
\begin{array}{c}
\phantom{a}\;4\;6\;9\;2\;7\;1\;5\;3\;8 \\
\begin{array}{c}a\\b\\f\\c\\d\\e\end{array}
\left(\begin{array}{ccc|cc|cc|cc}
1&1&1&0&0&0&0&0&0\\
\hline
0&0&0&1&1&0&0&0&0\\
0&0&0&1&1&0&0&0&0\\
\hline
0&0&0&0&0&0&0&0&0\\
\hline
0&0&0&0&0&1&1&0&0\\
\hline
0&0&0&0&0&0&0&1&1
\end{array}\right)
\end{array}
\\[2em]
\;\;\;A \qquad\qquad\quad\; B \qquad\qquad\qquad \mathbf{syq}(A, B) \text{ with rearrangement}
\end{array}
$$

**Fig. 4.4.1** Symmetric quotient

Before that we want to analyze in which way $A\,\mathbf{syq}(A, B)$ can differ from $B$. According to our next proposition it is contained in $B$ in a fairly regular fashion: a column of $A\,\mathbf{syq}(A, B)$ is either equal to the column of $B$ or is zero.

**4.4.2 Proposition.** i) $\qquad\qquad A\,\mathbf{syq}(A, B) \;=\; B \sqcap L\,\mathbf{syq}(A, B);$

ii) $\qquad\qquad\qquad \mathbf{syq}(A, B)$ surjective $\implies A\,\mathbf{syq}(A, B) = B.$

**Proof:** i) We have $B \sqcap L\,\mathbf{syq}(A, B) = (B \sqcap A\,\mathbf{syq}(A, B)) \sqcup (B \sqcap \overline{A}\,\mathbf{syq}(A, B)) = A\,\mathbf{syq}(A, B)$, since we already know that $A\,\mathbf{syq}(A, B) \subset B$, and since we have $\overline{A}\,\mathbf{syq}(A, B) = \overline{A}\,\mathbf{syq}(\overline{A}, \overline{B}) \subset \overline{B}$ as a consequence of Prop. 4.4.1.i.

ii) follows from (i). $\qquad\qquad\qquad\qquad\qquad\qquad\qquad\qquad\qquad\qquad\qquad\quad \Box$

Having asked when $A\,\mathbf{syq}(A, B) = B$, a property characteristic of quotients, we now ask whether a cancellation rule holds for these quotients.

**4.4.3 Proposition.** i) For arbitrary relations $A, B, C$ we have

$$\mathbf{syq}(A, B)\,\mathbf{syq}(B, C) = \mathbf{syq}(A, C) \sqcap \mathbf{syq}(A, B)L = \mathbf{syq}(A, C) \sqcap L\,\mathbf{syq}(B, C).$$

ii) If $\mathbf{syq}(A, B)$ is total, or if $\mathbf{syq}(B, C)$ is surjective, then

$$\mathbf{syq}(A, B)\,\mathbf{syq}(B, C) = \mathbf{syq}(A, C).$$

**Proof:** i) Without loss of generality we concentrate on the first equality sign. "$\subset$" follows from

$$(A^{\mathsf{T}}\overline{C} \sqcup \overline{A}^{\mathsf{T}}C)[\mathbf{syq}(B, C)]^{\mathsf{T}} = A^{\mathsf{T}}\overline{C}\,\mathbf{syq}(\overline{C}, \overline{B}) \sqcup \overline{A}^{\mathsf{T}}C\,\mathbf{syq}(C, B) \subset A^{\mathsf{T}}\overline{B} \sqcup \overline{A}^{\mathsf{T}}B$$

using the Schröder equivalences and Prop. 4.4.1.i,ii. "⊃" may be obtained using (i), the Dedekind rule and the partial result just proven:

$$\mathsf{syq}(A,B)\,L \sqcap \mathsf{syq}(A,C)$$
$$\subset \big(\mathsf{syq}(A,B) \sqcap \mathsf{syq}(A,C)L^{\mathsf{T}}\big)\big(L \sqcap [\mathsf{syq}(A,B)]^{\mathsf{T}}\,\mathsf{syq}(A,C)\big)$$
$$\subset \mathsf{syq}(A,B)\,\mathsf{syq}(B,A)\,\mathsf{syq}(A,C) \subset \mathsf{syq}(A,B)\,\mathsf{syq}(B,C).$$

ii) follows from (i). $\qquad\qquad\square$

We list some special cases:

**4.4.4 Corollary.** i)  $\quad \mathsf{syq}(A,B)[\mathsf{syq}(A,B)]^{\mathsf{T}} \subset \mathsf{syq}(A,A);$

ii) $\qquad\qquad\qquad \mathsf{syq}(A,A)\,\mathsf{syq}(A,B) = \mathsf{syq}(A,B);$

iii) $\qquad\qquad\qquad \mathsf{syq}(A,A) \quad$ is an equivalence relation.

**Proof:** For (i) set $C := A$ in Prop. 4.4.3.i and for (ii) $B := A$ and $C := B$. We have already shown the totality of $\mathsf{syq}(A,A)$ in Prop. 4.4.1.iv.

iii) $\mathsf{syq}(A,A)$ is reflexive due to Prop. 4.4.1.iv, symmetric by construction and transitive as a consequence of (ii). $\qquad\qquad\square$

The following two assertions follow immediately from the corollary, the first being a characterization of equivalence relations.

**4.4.5 Proposition.** i)  $A$  equivalence relation  $\iff$  $A = \mathsf{syq}(A,A);$

ii) $\qquad\qquad\quad A$  univalent and surjective  $\implies$  $I = \mathsf{syq}(A,A).$

**Proof:** i) "$\impliedby$" is a consequence of Corollary 4.4.4.iii. "$\implies$": With $I \subset A$ we have $\mathsf{syq}(A,A) \subset A\,\mathsf{syq}(A,A) \subset A$. The opposite containment is obtained from symmetry and transitivity: $AA \subset A \iff A^{\mathsf{T}}\overline{A} \subset \overline{A} \implies A^{\mathsf{T}} = A \subset \overline{A^{\mathsf{T}}\overline{A}}$.
ii) $L = A^{\mathsf{T}}L = A^{\mathsf{T}}\overline{A} \sqcup A^{\mathsf{T}}A$, since $A$ is surjective. Using this together with the univalence of $A$ results in $\mathsf{syq}(A,A) \subset A^{\mathsf{T}}A \subset I$. The construct $\mathsf{syq}(A,A)$ is reflexive because of Prop. 4.4.1.iv. $\qquad\qquad\square$

For symmetric quotients of injective relations, more special results hold.

**4.4.6 Proposition.** For injective relations $A, B$ the following holds

i) $\qquad\qquad\qquad\quad A^{\mathsf{T}}B \subset \mathsf{syq}(A,B);$

ii) $\qquad\qquad\qquad\quad A\,\mathsf{syq}(A,B) = AL \sqcap B;$

iii) $\qquad\qquad\qquad\quad B \subset \mathsf{syq}\left(A^{\mathsf{T}}, \mathsf{syq}(A,B)\right).$

**Proof:** i) $O = A^{\mathsf{T}}(B \sqcap \overline{B}) = A^{\mathsf{T}}B \sqcap A^{\mathsf{T}}\overline{B}$, so that $A^{\mathsf{T}}B \subset \overline{A^{\mathsf{T}}\overline{B}}$. Similarly, $A^{\mathsf{T}}B \subset \overline{\overline{A}^{\mathsf{T}}B}$.
ii) "$\subset$" is trivial and "$\supset$" results from (i), using Dedekind's rule $AL \sqcap B \subset (A \sqcap BL^{\mathsf{T}})(L \sqcap A^{\mathsf{T}}B) \subset AA^{\mathsf{T}}B \subset A\,\mathsf{syq}(A,B).$

iii) Following (i), $A\,\overline{\mathrm{syq}(A,B)} \subset \overline{B}$. Furthermore $\overline{A}\,\mathrm{syq}(A,B) \subset \overline{B}$ because of Prop. 4.4.1.i; together we have

$$B \subset \overline{A\,\overline{\mathrm{syq}(A,B)}} \sqcap \overline{\overline{A}\,\mathrm{syq}(A,B)} = \mathrm{syq}\,(A^{\mathsf{T}}, \mathrm{syq}(A,B))\,. \qquad \square$$

The next proposition deals with the relationship between vectors and symmetric quotients.

**4.4.7 Proposition.** i)   $\mathrm{syq}(A,L)$ is always a vector.

ii)   $\mathrm{syq}(A,B) = L \iff A$ and $B$ are vectors and $AL = BL$;

iii)   $\mathrm{syq}(A,A) = L \iff A = AL$.

**Proof:** i) Obviously, $\mathrm{syq}(L,L) = L$ is surjective; therefore, in conjunction with Prop. 4.4.3.ii, $\mathrm{syq}(A,L)L = \mathrm{syq}(A,L)\,\mathrm{syq}(L,L) = \mathrm{syq}(A,L)$.

ii) For "$\Longrightarrow$" we have

$$L \subset \mathrm{syq}(A,B) \Longrightarrow \{A^{\mathsf{T}}\overline{B} \subset O,\ \overline{A}^{\mathsf{T}}B \subset O\} \iff \{AL \subset B,\ LB^{\mathsf{T}} \subset A^{\mathsf{T}}\}.$$

So we have $AL \subset B \subset BL$ and $BL \subset A \subset AL$. Using $LL = L$ we obtain $AL = BL$. (Note: It *does not* follow that $A = AL = BL = B$, since $A$ and $B$ may have a "different number of columns".) "$\Longleftarrow$": Since $ALL = AL$ we have $(AL)^{\mathsf{T}}\overline{AL} \subset O$, and consequently $L \subset \overline{(AL)^{\mathsf{T}}\overline{AL}} = \overline{(AL)^{\mathsf{T}}\overline{BL}} = \overline{A^{\mathsf{T}}\overline{B}}$ and $L = L^{\mathsf{T}} = \overline{\overline{AL}^{\mathsf{T}}AL} = \overline{\overline{AL}^{\mathsf{T}}BL} = \overline{\overline{A}^{\mathsf{T}}B}$.   iii) follows from (ii).   $\square$

Our aim is now to find an algebraic criterion for a (heterogeneous) relation $A$ to be representable by a matrix of block diagonal form as in Fig. 4.4.1. We first define when a matrix $A$ is said to have **block diagonal form**: If there are rows $i, k$ and a column $j$ such that $(i,j) \in A$ (meaning the $(i,j)$-th entry is $\mathbf{1}$) and $(k,j) \in A$, then, for all columns $m$, $(k,m) \in A$ implies $(i,m) \in A$ (see Fig. 4.4.2).

So the condition formalized reads

$$\forall i : \forall j, k : \{[(i,j) \in A \wedge (k,j) \in A] \to [\forall m : (k,m) \in A \to (i,m) \in A]\}$$

or, equivalently,

$$\forall i, m : \{[\exists j, k : (i,j) \in A \wedge (j,k) \in A^{\mathsf{T}} \wedge (k,m) \in A] \to (i,m) \in A\}.$$

Note that this property is independent of the order in which rows and columns are arranged; so it refers to the *relation* represented by the matrix. Using this preparation, we give the following definition.

**4.4.8 Definition.** We call a relation

$$A \textbf{ difunctional } :\iff AA^{\mathsf{T}}A \subset A \iff AA^{\mathsf{T}}A = A \iff A\overline{A}^{\mathsf{T}}A \subset \overline{A}$$

$$\iff \quad A \text{ can be written in block diagonal form by suitably rearranging rows and columns.} \qquad \square$$

$$
\begin{array}{c}
\begin{array}{cccccccccccccc}
 & & & & j & & & & & m & & & & \\
 & & & & \downarrow & & & & & \downarrow & & & & \\
1 & 2 & 3 & 4 & 5 & 6 & 7 & 8 & 9 & 10 & 11 & 12 & 13 & 14
\end{array}
\end{array}
\qquad
\begin{array}{c}
\begin{array}{cccccccccccccc}
 & & & & & & & j & & m & & & & \\
 & & & & & & & \downarrow & & \downarrow & & & & \\
4 & 6 & 8 & 13 & 1 & 2 & 3 & 5 & 9 & 10 & 11 & 12 & 7 & 14
\end{array}
\end{array}
$$

Fig. 4.4.2 where the left matrix, with row labels $1,2,k{\to}3,4,5,i{\to}6,7,8,9,10,11,12$, reads:

| | 1 | 2 | 3 | 4 | 5 | 6 | 7 | 8 | 9 | 10 | 11 | 12 | 13 | 14 |
|---|---|---|---|---|---|---|---|---|---|---|---|---|---|---|
| 1 | 0 | 0 | 0 | 1 | 0 | 1 | 0 | 1 | 0 | 0 | 0 | 0 | 1 | 0 |
| 2 | 0 | 0 | 0 | 0 | 0 | 0 | 1 | 0 | 0 | 0 | 0 | 0 | 0 | 1 |
| $k{\to}3$ | 1 | 1 | 1 | 0 | ① | 0 | 0 | 0 | 1 | ① | 1 | 1 | 0 | 0 |
| 4 | 1 | 1 | 1 | 0 | 1 | 0 | 0 | 0 | 1 | 1 | 1 | 1 | 0 | 0 |
| 5 | 0 | 0 | 0 | 0 | 0 | 0 | 1 | 0 | 0 | 0 | 0 | 0 | 0 | 1 |
| $i{\to}6$ | 1 | 1 | 1 | 0 | ① | 0 | 0 | 0 | 1 | ☐1 | 1 | 1 | 0 | 0 |
| 7 | 0 | 0 | 0 | 1 | 0 | 1 | 0 | 1 | 0 | 0 | 0 | 0 | 1 | 0 |
| 8 | 1 | 1 | 1 | 0 | 1 | 0 | 0 | 0 | 1 | 1 | 1 | 1 | 0 | 0 |
| 9 | 0 | 0 | 0 | 0 | 0 | 0 | 1 | 0 | 0 | 0 | 0 | 0 | 0 | 1 |
| 10 | 1 | 1 | 1 | 0 | 1 | 0 | 0 | 0 | 1 | 1 | 1 | 1 | 0 | 0 |
| 11 | 0 | 0 | 0 | 1 | 0 | 1 | 0 | 1 | 0 | 0 | 0 | 0 | 1 | 0 |
| 12 | 0 | 0 | 0 | 0 | 0 | 0 | 0 | 0 | 0 | 0 | 0 | 0 | 0 | 0 |

and the right matrix, in block diagonal form, reads:

| | 4 | 6 | 8 | 13 | 1 | 2 | 3 | 5 | 9 | 10 | 11 | 12 | 7 | 14 |
|---|---|---|---|---|---|---|---|---|---|---|---|---|---|---|
| 1 | 1 | 1 | 1 | 1 | 0 | 0 | 0 | 0 | 0 | 0 | 0 | 0 | 0 | 0 |
| 7 | 1 | 1 | 1 | 1 | 0 | 0 | 0 | 0 | 0 | 0 | 0 | 0 | 0 | 0 |
| 11 | 1 | 1 | 1 | 1 | 0 | 0 | 0 | 0 | 0 | 0 | 0 | 0 | 0 | 0 |
| $k{\to}3$ | 0 | 0 | 0 | 0 | 1 | 1 | 1 | ① | 1 | ① | 1 | 1 | 0 | 0 |
| 4 | 0 | 0 | 0 | 0 | 1 | 1 | 1 | 1 | 1 | 1 | 1 | 1 | 0 | 0 |
| $i{\to}6$ | 0 | 0 | 0 | 0 | 1 | 1 | 1 | ① | 1 | ☐1 | 1 | 1 | 0 | 0 |
| 8 | 0 | 0 | 0 | 0 | 1 | 1 | 1 | 1 | 1 | 1 | 1 | 1 | 0 | 0 |
| 10 | 0 | 0 | 0 | 0 | 1 | 1 | 1 | 1 | 1 | 1 | 1 | 1 | 0 | 0 |
| 12 | 0 | 0 | 0 | 0 | 0 | 0 | 0 | 0 | 0 | 0 | 0 | 0 | 0 | 0 |
| 2 | 0 | 0 | 0 | 0 | 0 | 0 | 0 | 0 | 0 | 0 | 0 | 0 | 1 | 1 |
| 5 | 0 | 0 | 0 | 0 | 0 | 0 | 0 | 0 | 0 | 0 | 0 | 0 | 1 | 1 |
| 9 | 0 | 0 | 0 | 0 | 0 | 0 | 0 | 0 | 0 | 0 | 0 | 0 | 1 | 1 |

Fig. 4.4.2 Block diagonal form of a matrix

Again, the equivalence of the definition variants needs proving. Firstly, it must be shown that, for arbitrary $A$, we always have that $A \subset AA^{\mathsf{T}}A$:

$$A = IA = (AL \sqcap I)A \sqcup (\overline{AL} \sqcap I)A \subset AA^{\mathsf{T}}A \sqcup (\overline{AL}^{\mathsf{T}} \sqcap I)A = AA^{\mathsf{T}}A.$$

In the course of the proof, $\overline{AL}^{\mathsf{T}}A \subset O$ and $AL \sqcap I \subset (A \sqcap IL^{\mathsf{T}})(L \sqcap A^{\mathsf{T}}I) \subset AA^{\mathsf{T}}$ have been used. Secondly, we show:

$$(AA^{\mathsf{T}})A \subset A \iff A(A^{\mathsf{T}}\overline{A}) \subset \overline{A} \iff A\overline{A}^{\mathsf{T}}A \subset \overline{A}.$$

We now prove that every symmetric quotient has block diagonal form.


**4.4.9 Proposition.**  $\mathrm{syq}(A, B)$ is difunctional.

**Proof:**  Using Prop. 4.4.3.i and Corollary 4.4.4.ii we obtain

$$\mathrm{syq}(A, B)[\mathrm{syq}(A, B)]^{\mathsf{T}}\,\mathrm{syq}(A, B) \subset \mathrm{syq}(A, A)\,\mathrm{syq}(A, B) = \mathrm{syq}(A, B). \qquad \square$$


We now want to explain why the property discussed above is called "difunctional". One can indeed associate a pair of functions with a difunctional relation. Consider the matrix in Fig. 4.4.3. In practical applications one would like to combine rows and columns with equivalence classes, so as to be able to work with a considerably smaller matrix.

With this formation of equivalence classes of rows or columns are associated natural projections $F$ and $G$, respectively; their domains are uniquely determined "up to isomorphism" when $A$ is difunctional.


**4.4.10 Proposition.**  i)  A product $FG^{\mathsf{T}}$ is always difunctional, if $F$ and $G$ are univalent relations.

ii)  The decomposition $A = FG^{\mathsf{T}}$ of a (necessarily difunctional) relation using two surjective univalent relations $F$ and $G$ may be achieved *in essentially one fashion*.

$$
\begin{array}{c}
\begin{array}{cccccccccccccc} 4 & 6 & 8 & 13 & 1 & 2 & 3 & 5 & 9 & 10 & 11 & 12 & 7 & 14 \end{array}
\end{array}
$$

$$
\begin{array}{c}
1 \\ 7 \\ 11 \\ 3 \\ 4 \\ 6 \\ 8 \\ 10 \\ 12 \\ 2 \\ 5 \\ 9
\end{array}
\left(\begin{array}{cccc|cccccccc|cc}
1 & 1 & 1 & 1 & 0 & 0 & 0 & 0 & 0 & 0 & 0 & 0 & 0 & 0 \\
1 & 1 & 1 & 1 & 0 & 0 & 0 & 0 & 0 & 0 & 0 & 0 & 0 & 0 \\
1 & 1 & 1 & 1 & 0 & 0 & 0 & 0 & 0 & 0 & 0 & 0 & 0 & 0 \\
\hline
0 & 0 & 0 & 0 & 1 & 1 & 1 & 1 & 1 & 1 & 1 & 1 & 0 & 0 \\
0 & 0 & 0 & 0 & 1 & 1 & 1 & 1 & 1 & 1 & 1 & 1 & 0 & 0 \\
0 & 0 & 0 & 0 & 1 & 1 & 1 & 1 & 1 & 1 & 1 & 1 & 0 & 0 \\
0 & 0 & 0 & 0 & 1 & 1 & 1 & 1 & 1 & 1 & 1 & 1 & 0 & 0 \\
0 & 0 & 0 & 0 & 1 & 1 & 1 & 1 & 1 & 1 & 1 & 1 & 0 & 0 \\
\hline
0 & 0 & 0 & 0 & 0 & 0 & 0 & 0 & 0 & 0 & 0 & 0 & 0 & 0 \\
\hline
0 & 0 & 0 & 0 & 0 & 0 & 0 & 0 & 0 & 0 & 0 & 0 & 1 & 1 \\
0 & 0 & 0 & 0 & 0 & 0 & 0 & 0 & 0 & 0 & 0 & 0 & 1 & 1 \\
0 & 0 & 0 & 0 & 0 & 0 & 0 & 0 & 0 & 0 & 0 & 0 & 1 & 1
\end{array}\right)
\left(\begin{array}{ccc}
1 & 0 & 0 \\
1 & 0 & 0 \\
1 & 0 & 0 \\
0 & 1 & 0 \\
0 & 1 & 0 \\
0 & 1 & 0 \\
0 & 1 & 0 \\
0 & 1 & 0 \\
0 & 0 & 0 \\
0 & 0 & 1 \\
0 & 0 & 1 \\
0 & 0 & 1
\end{array}\right)
$$

$$
\left(\begin{array}{cccccccccccccc}
1 & 1 & 1 & 1 & 0 & 0 & 0 & 0 & 0 & 0 & 0 & 0 & 0 & 0 \\
0 & 0 & 0 & 0 & 1 & 1 & 1 & 1 & 1 & 1 & 1 & 1 & 0 & 0 \\
0 & 0 & 0 & 0 & 0 & 0 & 0 & 0 & 0 & 0 & 0 & 0 & 1 & 1
\end{array}\right) G^{\mathsf{T}}
$$

with $F$ the second tall matrix, and

$$
\begin{array}{cc}
& \begin{array}{ccc} l & m & n \end{array} \\
\begin{array}{r} \{1,7,11\} \\ \{3,4,6,8,10\} \\ \{12\} \\ \{2,5,9\} \end{array} &
\left(\begin{array}{ccc}
1 & 0 & 0 \\
0 & 1 & 0 \\
0 & 0 & 0 \\
0 & 0 & 1
\end{array}\right)
\end{array}
$$

where:

$$
\begin{aligned}
l &= \{4,6,8,13\} \\
m &= \{1,2,3,5,9,10,11,12\} \\
n &= \{7,14\}
\end{aligned}
$$

**Fig. 4.4.3** Difunctionality and block diagonal form

**Proof:** i) $FG^{\mathsf{T}}GF^{\mathsf{T}}FG^{\mathsf{T}} \subset FG^{\mathsf{T}}$  ii) Suppose, $A = FG^{\mathsf{T}} = F'G'^{\mathsf{T}}$ with $F^{\mathsf{T}}F = I$ and the corresponding properties for $G, F'$, and $G'$. Then $AG = FG^{\mathsf{T}}G = F$, $F^{\mathsf{T}}A = F^{\mathsf{T}}FG^{\mathsf{T}} = G^{\mathsf{T}}$, and analogously $AG' = F'$, $F'^{\mathsf{T}}A = G'^{\mathsf{T}}$. For $\Phi := F^{\mathsf{T}}F'$ we have therefore $\Phi\Phi^{\mathsf{T}} = F^{\mathsf{T}}F'F'^{\mathsf{T}}F = F^{\mathsf{T}}F'G'^{\mathsf{T}}G'F'^{\mathsf{T}}F = F^{\mathsf{T}}AA^{\mathsf{T}}F = G^{\mathsf{T}}G = I$, and by symmetry $\Phi^{\mathsf{T}}\Phi = I$. So $\Phi$ is a bijective mapping satisfying $F\Phi = FF^{\mathsf{T}}F' = FG^{\mathsf{T}}GF^{\mathsf{T}}F' = AA^{\mathsf{T}}F' = AG' = F'$ and $G\Phi = GF^{\mathsf{T}}F' = A^{\mathsf{T}}F' = G'$. Compare this with the remarks concerning homomorphism in Sects. 7.1 and 7.2.    $\square$

## Relations of Ferrers Type

There is another interesting type of relations, those that may after suitable re-arrangement of rows and columns be decomposed in rectangular blocks of either 0's or 1's. Consider for example the relation $E$ of a linear ordering, that may be represented as an upper triangular matrix. From linearity $L \subset E \sqcup E^{\mathsf{T}}$ and transitivity of $E$ it is easily deduced that $E\overline{E}^{\mathsf{T}}E \subset E$.

This is a formula of a kind similar to that we have just investigated. What we are going to do now is to show that it is characteristic for blockwise staircase shaped relations. They need not even be homogeneous.

Considering Fig. 4.4.4, we again start with a predicate logic formulation. Characteristic for a relation to be of staircase form is that whenever there are rows $i, k$ and columns $j, m$ such that $(i,j) \in A$, $(k,j) \notin A$, and $(k,m) \in A$ then $(i,m) \in A$. Note that this has been described as a property of a relation since the ordering of the matrix has never been used. More formally, this reads

$$\forall i,j,k,m : \{(i,j) \in A \wedge (j,k) \notin A^{\mathsf{T}} \wedge (k,m) \in A \to (i,m) \in A\} \quad \text{or}$$

$$\forall i,m : \{[\exists j,k : (i,j) \in A \wedge (j,k) \notin \overline{A}^{\mathsf{T}} \wedge (k,m) \in A] \to (i,m) \in A\}.$$

The following definition presents this in a relation-algebraic form.

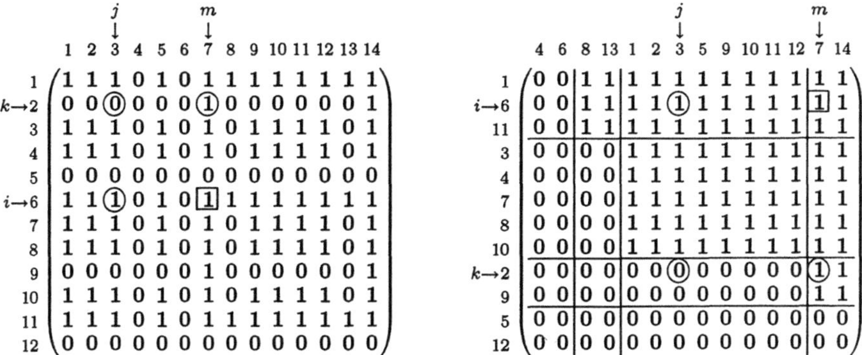

**Fig. 4.4.4** Staircase form of a relation of Ferrers type

**4.4.11 Definition.** We say that a relation

$$A \text{ is of } \textbf{Ferrers type} :\Longleftrightarrow A\overline{A}^{\mathsf{T}}A \subset A \iff \overline{A}A^{\mathsf{T}}\overline{A} \subset \overline{A} \iff A^{\mathsf{T}}\overline{A}A^{\mathsf{T}} \subset A^{\mathsf{T}}$$

$$\iff A \text{ can be written in staircase block form}$$
$$\text{by suitably rearranging rows and columns.} \qquad \square$$

What we have included as definitional variants can also be summed up more formally in part (i) of the following proposition.

**4.4.12 Proposition.** i) Either non or all of $A, \overline{A}, A^{\mathsf{T}}$ are of Ferrers type.

ii)       $A$ is of Ferrers type $\iff$    $A^{\mathsf{T}}\overline{A}$ is of Ferrers type.

iii)     $A$ is of Ferrers type $\iff$    $\overline{A}A^{\mathsf{T}}$ is of Ferrers type.

iv)     $A$ is of Ferrers type $\iff$    $\overline{A}A^{\mathsf{T}}$ is a strict-ordering.

**Proof:** ii) Use that $A = \overline{A\overline{A}^{\mathsf{T}}\overline{A}}$ for every $A$, since Schröder gives both, "$\supset$" and $I \subset \overline{A^{\mathsf{T}}\overline{A}}$. With this, "$\Longrightarrow$" is obtained directly. We show "$\Longleftarrow$" by $A\overline{A}^{\mathsf{T}}A = (\overline{A\overline{A^{\mathsf{T}}\overline{A}}})\overline{A}^{\mathsf{T}}(\overline{A\overline{A^{\mathsf{T}}\overline{A}}}) \subset \overline{A\overline{A^{\mathsf{T}}\overline{A}}} = A$, having in mind that the relation $A^{\mathsf{T}}\overline{A}$ is of Ferrers type precisely when its complement is.

iii) is shown similarly and (iv) is simple. $\qquad \square$

There is a close relationship to the example mentioned above concerning a linear ordering that satisfies $E\overline{E}^{\mathsf{T}}E \subset E$. Consider classes of rows and columns arising from the block structure together with the partial projections $F$ and $G$ on those classes where at least one block of **1**'s occurs. Then *these classes* should be linearly ordered.

**4.4.13 Proposition.** i)   A product $FEG^{\mathsf{T}}$ is always of Ferrers type, if $F$ and $G$ are univalent and surjective relations and $E$ is a linear ordering.

ii)   The decomposition $A = FEG^{\mathsf{T}}$ of a (necessarily Ferrers-type) relation using two surjective univalent relations $F$ and $G$ and a linear ordering $E$ may be achieved *in essentially one* fashion.

**Proof:** Remember that a univalent and surjective relation $F$ satisfies $F^\mathsf{T}F = I$ as well as $F^\mathsf{T}\overline{FX} = \overline{X}$ for arbitrary $X$ (cf. Def. 4.2.1 and Prop. 4.2.4). i) We start with the characteristic property of the linear ordering and continue with two uses of the identity just mentioned:

$$E \supset E\overline{\overline{E}^\mathsf{T}E} = EG^\mathsf{T}\overline{GE^\mathsf{T}}E = EG^\mathsf{T}\overline{GE^\mathsf{T}F^\mathsf{T}}FE.$$

Multiplying $F$ from the left and $G^\mathsf{T}$ from the right, we obtain the desired result $FEG^\mathsf{T}\overline{GE^\mathsf{T}F^\mathsf{T}}FEG^\mathsf{T} \subset FEG^\mathsf{T}$.

ii) Suppose there are two decompositions $A = FEG^\mathsf{T} = F'E'G'^\mathsf{T}$ and define $\Phi := F^\mathsf{T}F'$.

We first prove several identities to be used in the sequel:

$$F^\mathsf{T}\overline{A}G = F^\mathsf{T}\overline{FEG^\mathsf{T}}G = \overline{EG^\mathsf{T}}G = \overline{E};$$
$$F\overline{E}^\mathsf{T}G^\mathsf{T} = FE\overline{\overline{E}^\mathsf{T}EG^\mathsf{T}} = FEG^\mathsf{T}\overline{A}^\mathsf{T}FEG^\mathsf{T} = A\overline{A}^\mathsf{T}A \quad \text{since } \overline{E}^\mathsf{T} = E\overline{E}^\mathsf{T}E.$$

For reasons of symmetry, this will hold for the other decomposition likewise. It implies $F\overline{E}^\mathsf{T} = F\overline{E}^\mathsf{T}G^\mathsf{T}G = A\overline{A}^\mathsf{T}AG$ so that

$$F = FL \sqcap \overline{\overline{FI}} = FL \sqcap \overline{F(\overline{E} \sqcup \overline{E}^\mathsf{T})} = FL \sqcap \overline{F\overline{E}} \sqcap \overline{F\overline{E}^\mathsf{T}} = FE \sqcap \overline{F\overline{E}^\mathsf{T}} = FE \sqcap \overline{A\overline{A}^\mathsf{T}AG}.$$

Now $FG^\mathsf{T}$ has a representation independent from $F, E, G$:

$$FG^\mathsf{T} = (FE \sqcap \overline{A\overline{A}^\mathsf{T}AG})G^\mathsf{T} = FEG^\mathsf{T} \sqcap \overline{A\overline{A}^\mathsf{T}AGG^\mathsf{T}} = A \sqcap \overline{A\overline{A}^\mathsf{T}A}$$
$$= F'G'^\mathsf{T} \quad \text{by symmetry.}$$

Using this, it is possible to prove that $\Phi$ is univalent and surjective:

$$\Phi^\mathsf{T}\Phi = F'^\mathsf{T}FF^\mathsf{T}F' = F'^\mathsf{T}FG^\mathsf{T}GF^\mathsf{T}F' = F'^\mathsf{T}F'G'^\mathsf{T}G'F'^\mathsf{T}F' = I.$$

Totality and injectivity is proved similarly. Furthermore $\Phi$ is isotonic, i.e.,

$$E\Phi = EF^\mathsf{T}F' = F^\mathsf{T}FEG^\mathsf{T}GF^\mathsf{T}F' = F^\mathsf{T}AG'F'^\mathsf{T}F'$$
$$= F^\mathsf{T}AG' = F^\mathsf{T}F'E'G'^\mathsf{T}G' = \Phi E'.$$

In addition $\Phi = G^\mathsf{T}G'$, $F\Phi = FF^\mathsf{T}F' = FG^\mathsf{T}G' = F'G'^\mathsf{T}G' = F'$ and $G\Phi = GF^\mathsf{T}F' = G'F'^\mathsf{T}F' = G'$ can be proved. $\qquad\square$


Finally, we present a relationship between difunctional and Ferrers relations.


**4.4.14 Proposition.** For an arbitrary relation $A$ the construct $A \sqcap \overline{A\overline{A}^\mathsf{T}A}$ is difunctional. If $A$ is of Ferrers type, then $A \sqcap \overline{A\overline{A}^\mathsf{T}A}$ is of Ferrers type.

**Proof:** Define $S := A \sqcap \overline{A\overline{A}^\mathsf{T}A}$ and $W := \overline{A\overline{A}^\mathsf{T}A}$ as abbreviations. Then $AW^\mathsf{T} \subset \overline{A}\overline{A}^\mathsf{T}$ by the Schröder equivalence. Therefore, $SS^\mathsf{T}S \subset AW^\mathsf{T}A \subset \overline{A}\overline{A}^\mathsf{T}A \subset A$ and $SS^\mathsf{T}S \subset WA^\mathsf{T}W \subset \overline{A\overline{A}^\mathsf{T}}W \subset W$, where the last $\subset$ follows from $\overline{A}\overline{A}^\mathsf{T} \subset \overline{A}\overline{A}^\mathsf{T}$, $\overline{A}\overline{A}^\mathsf{T}A \subset A$, and $\overline{A}\overline{A}^\mathsf{T}A\overline{A}^\mathsf{T}A \subset A\overline{A}^\mathsf{T}A$. So $S$ is difunctional.

If $A$ is of Ferrers type, we have $A \sqcap \overline{W} = \overline{W}$. Therefore, we prove $\overline{W}W^\mathsf{T}\overline{W} \subset AW^\mathsf{T}\overline{W} \subset \overline{A}\overline{A}^\mathsf{T}\overline{W} \subset \overline{W}$. Here, the first inclusion holds since $A$ is of Ferrers type, the second follows from a previous statement and the third is proved by

$$\overline{W} \subset \overline{W}, \ \overline{A\overline{A}^\mathsf{T}} AA^\mathsf{T}\overline{A} \subset \overline{A}, \ W\overline{W}^\mathsf{T} = \overline{A\overline{A}^\mathsf{T}} AA^\mathsf{T}\overline{A}A^\mathsf{T} \subset \overline{A}A^\mathsf{T}.$$      □

## Exercises

**4.4.1** Prove that $RLR = R \iff R\overline{R}^\mathsf{T}R = O$. In this case $R$ is called a **rectangular** relation (which is *relation rectangle* in French according to J. Riguet). If $R$ is also symmetric, it is called a **square-shaped** relation or *relation carré* in French.

**4.4.2** A relation $R$ is called right-invertible if there exists a relation $X$ with $RX = I$, and left-invertible if there exists a relation $Y$ with $YR = I$. $X$ and $Y$ are then called the right and left inverse of $R$, respectively. Right- and left-invertible relations are called **invertible**.

i)   For invertible homogeneous relations all right and left inverses coincide; the notion **inverse** $R^{-1}$ is used. Then $R^{-1} = R^\mathsf{T}$ holds.

ii)  Show that for Boolean $n \times n$-matrices

$$R \text{ invertible} \iff R^m = I \quad \text{for some } m \geq 1.$$

($R$ is then called a permutation matrix.)

**4.4.3** If $f$ is univalent and $\Xi$ is an equivalence relation, then $f\Xi$ is difunctional.

**4.4.4** For equivalences $\Xi, \Theta$ and arbitrary $R$ the following implication holds:

$$R^\mathsf{T}\Xi R \subset \Theta \implies \Xi R\Theta \text{ difunctional.}$$

**4.4.5** Define the **rectangular closure** of a relation $R$ by

$$h_{\text{rect}}(R) := \inf \{ H \mid R \subset H, H \text{ is rectangular} \}.$$

Prove that this is indeed a closure operation und show that

$$h_{\text{rect}}(R) = RLR = RL \sqcap LR.$$

**4.4.6** Define the **difunctional closure** of a relation $R$ by

$$h_{\text{dif}}(R) := \inf \{ H \mid R \subset H, H \text{ is difunctional} \}.$$

Prove that this is indeed a closure operation und show that

$$h_{\text{dif}}(R) = R(R^\mathsf{T}R)^+ = (RR^\mathsf{T})^+R = (RR^\mathsf{T})^+R(R^\mathsf{T}R)^+.$$

**4.4.7** Let a symmetric difunctional relation $A$ be partitioned $A = H \sqcap T$, where $H = A \sqcap \overline{A^2}$ is the nontransitive part and $T = A \sqcap A^2$ is what might be called the transitive part (cf. Sect. 6.5) then $HT = TH = O$ and $T$ is a transitive relation.

**4.4.8** Let two disjoint orthogonal relations $S$ and $U$ be given, i.e., $SU = US = O$, where $S$ is symmetric difunctional and satisfies $S \subset \overline{S^2}$ while $U = U^\mathsf{T} = U^2$, i.e., $U$ is a so-called **quasi-equivalence**. Then $A := S \sqcup U$ is a symmetric difunctional relation with disjoint decomposition $S = A \sqcap \overline{A^2}$ and $U = A \sqcap A^2$.

## 4.5 References

See also the references in the appendix.


CHIN LH, TARSKI A: *Distributive and modular laws in the arithmetic of relation algebras.* Univ. California Publ. Math. **1** (1951) 341–384.

ORE O: *Theory of graphs.* Amer. Math. Soc. Colloq. Publ. Vol. XXXVIII, Providence, R. I., 1962.

PÖSCHEL R, KALUŽNIN LA: *Funktionen- und Relationenalgebren.* Birkhäuser, Basel, 1979.

RIGUET J: *Quelques propriétés des relations difonctionelles.* C. R. Acad. Sci. Paris **230** (1950) 1999–2000.

# 5. Graphs:<br>Associated Relation, Incidence, Adjacency

Graphs are models for relations. We have already seen various types of graphs, albeit from a phenomenological point of view: 1-graphs, simple graphs, bipartite graphs. We introduce directed graphs in Sect. 5.1. There is a number of simple concepts, applicable to several types of graphs provided suitable modifications are made. In the graph-theoretical literature these variants are usually specified at the beginning.

These related concepts differ according to whether or not there exist neighbors or successors, whether or not these are unique, etc.; we compare them and study their interdependence. We work with the calculus of relations in order to clarify the differences. The leading theme of Sect. 5.2 is the associated relation; Sect. 5.3 deals with incidence, Sect. 5.4, with adjacency. Finally, in Sect. 5.4 we investigate which incidences can belong to the same adjacency, and how a "canonical form" can be distinguished among them.

## 5.1 Directed Graphs

In contrast to the definition of 1-graphs, we now allow arcs as independent mathematical objects; in general, they may even fail to be "initially incident" with any point—in Fig. 5.1.1 arrow $p_8$ is of this kind. Also, arcs need not be "terminally incident" with any point, such as $p_7$ in Fig. 5.1.1. Arrow $p_{10}$, at last, is neither initially nor terminally incident with any point at all. By requiring that the relations $A$ and $E$ below be univalent, i.e., be partially defined functions, we associate with every arc either a pair consisting of an "initial" point and an "end-"point, or just an initial point or an endpoint, or no point at all (for the "empty" arrow).

**5.1.1 Definition.** We call $G = (P, V, A, E)$ a **directed graph** (graph for short, sometimes digraph), when:

i)  $P$ is a set of **arcs** (or arrows).

ii)  $V$ is a set of **points**.

iii)  $A, E \subset P \times V$ are two univalent relations, called the **outgoing incidence** and the **ingoing incidence**, respectively.

The relation $M := A \sqcup E$ is called the **incidence** of the graph. □

We define some elementary concepts for directed graphs and illustrate them by an example.

**5.1.2 Definition.** For a point $x$ and an arc $p$ in a graph with outgoing incidence $A$ and ingoing incidence $E$ we call

$$Ax \quad \text{the set of arcs starting in } x,$$
$$Ex \quad \text{the set of arcs ending in } x,$$
$$A^\mathsf{T}p \quad \text{the (empty or one-element) set of \textbf{starting points} of } p,$$
$$E^\mathsf{T}p \quad \text{the (empty or one-element) set of \textbf{endpoints} of } p.$$

Furthermore, we define the

$$\textbf{outdegree } d_A(x) := |Ax| = \text{number of arcs with starting point } x,$$
$$\textbf{indegree } \quad d_E(x) := |Ex| = \text{number of arcs with endpoint } x. \qquad \square$$

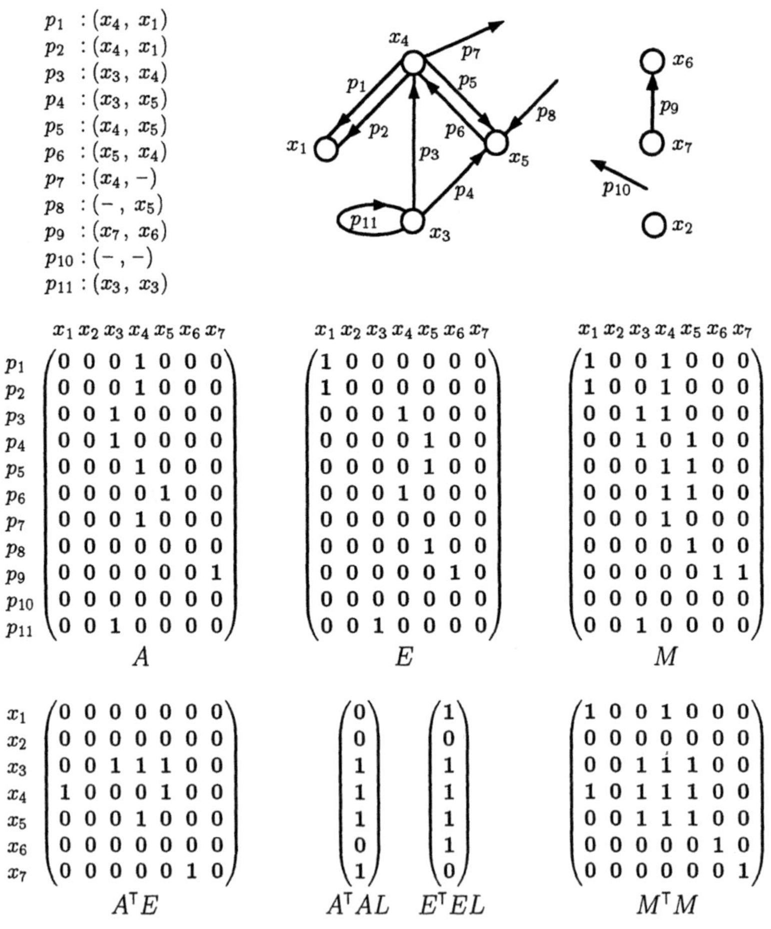

$$
\begin{aligned}
p_1 &: (x_4,\, x_1)\\
p_2 &: (x_4,\, x_1)\\
p_3 &: (x_3,\, x_4)\\
p_4 &: (x_3,\, x_5)\\
p_5 &: (x_4,\, x_5)\\
p_6 &: (x_5,\, x_4)\\
p_7 &: (x_4,\, -)\\
p_8 &: (-,\, x_5)\\
p_9 &: (x_7,\, x_6)\\
p_{10} &: (-,\, -)\\
p_{11} &: (x_3,\, x_3)
\end{aligned}
$$

$$
A=\begin{pmatrix}
0&0&0&1&0&0&0\\
0&0&0&1&0&0&0\\
0&0&1&0&0&0&0\\
0&0&1&0&0&0&0\\
0&0&0&1&0&0&0\\
0&0&0&0&1&0&0\\
0&0&0&1&0&0&0\\
0&0&0&0&0&0&0\\
0&0&0&0&0&0&1\\
0&0&0&0&0&0&0\\
0&0&1&0&0&0&0
\end{pmatrix}
\quad
E=\begin{pmatrix}
1&0&0&0&0&0&0\\
1&0&0&0&0&0&0\\
0&0&0&1&0&0&0\\
0&0&0&0&1&0&0\\
0&0&0&0&1&0&0\\
0&0&0&1&0&0&0\\
0&0&0&0&0&0&0\\
0&0&0&0&1&0&0\\
0&0&0&0&0&1&0\\
0&0&0&0&0&0&0\\
0&0&1&0&0&0&0
\end{pmatrix}
\quad
M=\begin{pmatrix}
1&0&0&1&0&0&0\\
1&0&0&1&0&0&0\\
0&0&1&1&0&0&0\\
0&0&1&0&1&0&0\\
0&0&0&1&1&0&0\\
0&0&0&1&1&0&0\\
0&0&0&1&0&0&0\\
0&0&0&0&1&0&0\\
0&0&0&0&0&1&1\\
0&0&0&0&0&0&0\\
0&0&1&0&0&0&0
\end{pmatrix}
$$

(rows $p_1,\dots,p_{11}$; columns $x_1\,x_2\,x_3\,x_4\,x_5\,x_6\,x_7$)

$$
A^\mathsf{T}E=\begin{pmatrix}
0&0&0&0&0&0&0\\
0&0&0&0&0&0&0\\
0&0&1&1&1&0&0\\
1&0&0&0&1&0&0\\
0&0&0&1&0&0&0\\
0&0&0&0&0&0&0\\
0&0&0&0&0&1&0
\end{pmatrix}
\qquad
A^\mathsf{T}AL=\begin{pmatrix}0\\0\\1\\1\\1\\0\\1\end{pmatrix}
\quad
E^\mathsf{T}EL=\begin{pmatrix}1\\0\\1\\1\\1\\1\\0\end{pmatrix}
\qquad
M^\mathsf{T}M=\begin{pmatrix}
1&0&0&1&0&0&0\\
0&0&0&0&0&0&0\\
0&0&1&1&1&0&0\\
1&0&1&1&1&0&0\\
0&0&1&1&1&0&0\\
0&0&0&0&0&1&0\\
0&0&0&0&0&0&1
\end{pmatrix}
$$

(rows $x_1,\dots,x_7$; columns $x_1\,x_2\,x_3\,x_4\,x_5\,x_6\,x_7$)

**Fig. 5.1.1** Directed graph

So we do not require the relations $A$ and $E$ to be total. When we want to emphasize this, we call $G$ a partial graph, and an arc $p$ such that $A^\mathsf{T}p = O$ or $E^\mathsf{T}p = O$ (or both), **partial**.

We now take up the theme of Definition 2.4.11. There, we defined "loop-carrying" points in 1-graphs, i.e., in graphs which are characterized by their associated relation. Now we introduce a concept for directed graphs which is closely related.

**5.1.3 Definition.** Given a graph with outgoing incidence $A$ and ingoing incidence $E$ we term an arc

$$p \ \textbf{loop} \quad :\Longleftrightarrow \quad A^\mathsf{T}p = E^\mathsf{T}p \neq O. \qquad \square$$

In Fig. 5.1.1 point $x_4$ has indegree 2 and outdegree 4. Arc $p_{11}$ is a loop. The set of starting points of $p_8$ is empty, and $\{\, p_1, p_2, p_5, p_7 \,\}$ is the set of arcs emerging from $x_4$.

There is a broad spectrum of possibilities for representing graphs. In Definition 5.1.1 we defined a graph by means of two relations $A$ and $E$. Graph theory owes its name to the method of representing its objects graphically, as in Fig. 5.1.1, for instance.

The problem of representing arcs without starting points or endpoints is only graphic in nature; whereas the listing of arcs as tuples, implemented as an array of variables, as in Fig. 5.1.1, presents nothing unusual. Arc $p_7$ having no endpoint is partial, although the representing line segment ends in some point of the plane.

The nonvanishing diagonal elements of the products $A^\mathsf{T}A$ and $E^\mathsf{T}E$ determine those points which are starting points or endpoints, respectively, of at least one arc. Analogously, the diagonal elements of $AA^\mathsf{T}$ and $EE^\mathsf{T}$ determine those arcs that possess starting points or endpoints. These characterizations can also be expressed in terms of $M$. We formulate this in the following proposition which is easy to prove.

**5.1.4 Proposition.** For the incidence $M$ of a graph with outgoing incidence $A$ and ingoing incidence $E$ the following holds:

i)
$$I \sqcap M^\mathsf{T}M = A^\mathsf{T}A \sqcup E^\mathsf{T}E;$$

ii)
$$I \sqcap MM^\mathsf{T} = I \sqcap (AA^\mathsf{T} \sqcup EE^\mathsf{T}). \qquad \square$$

## Adjacency and Line Graph

In Sect. 2.2 we studied adjacency for simple graphs and for 1-graphs. Now we are going to see that there is a notion of adjacency for directed graphs as well. We define two *distinct* points to be adjacent if there is an arc incident with both of them. Similarly, two distinct arcs are called adjacent if there is a point that is incident with both. So adjacency can be expressed by the product of the incidence and its transpose.

**5.1.5 Definition.** If $M = A \sqcup E$ is the incidence of the graph $G = (P, V, A, E)$, we call

$$\Gamma := \overline{I} \sqcap M^{\mathsf{T}} M \quad \text{the (point) \textbf{adjacency} and}$$
$$\mathrm{K} := \overline{I} \sqcap M M^{\mathsf{T}} \quad \text{the \textbf{edge-adjacency} of } G.$$

In addition, we call $G := (V, \Gamma)$ the **associated simple graph**. $L(G) := (P, \mathrm{K})$ is called the **line graph** or the **representative graph** of $G$. □

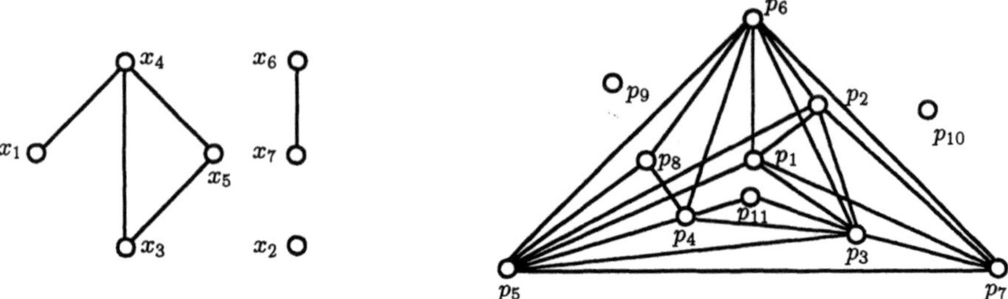

Fig. **5.1.2** Simple graph and line graph of the graph of Fig. 5.1.1

Figure 5.1.2 shows the associated simple graph and the line graph for the directed graph from Fig. 5.1.1.

## Associated Relation and Associated 1-Graph

Directed graphs and 1-graphs look very similar when drawn as pictures. The picture in Fig. 5.1.3 below can be interpreted as either a 1-graph or a directed graph, depending on whether one considers the relation $B$ or the relations $A$ and $E$ that can be obtained from it. So 1-graphs and certain directed graphs have the same graphic representation; their formal definitions, however, differ considerably.

We shall see how to pass from one of these concepts to the other. The following definition shows the straightforward transition from directed graphs to 1-graphs which allows us to apply results about 1-graphs to directed graphs.

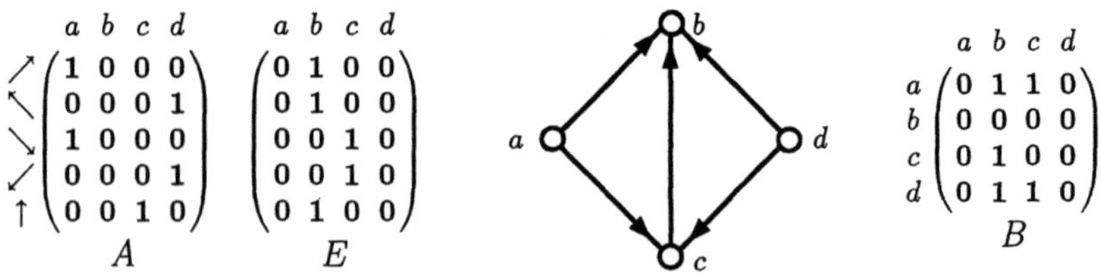

Fig. **5.1.3** Directed graph $G = (P, V, A, E)$ and 1-graph $G = (V, B)$

**5.1.6 Definition.** In a directed graph $G$ with point set $V$, outgoing incidence $A$, and ingoing incidence $E$ we call

$$B := A^{\mathsf{T}} E \quad \text{the \textbf{associated relation}}$$

of $G$. Furthermore, we say that $(V, B)$ is the **associated 1-graph** of the directed graph $G$. $\qquad\square$

Of course, different directed graphs may yield the same 1-graph, so there is no one-to-one correspondence between the two types of graphs. In order to establish a reverse correspondence, additional conditions must be imposed on the corresponding directed graph so that the Factorization Theorem 7.2.1 can be used to distinguish a unique corresponding directed graph.

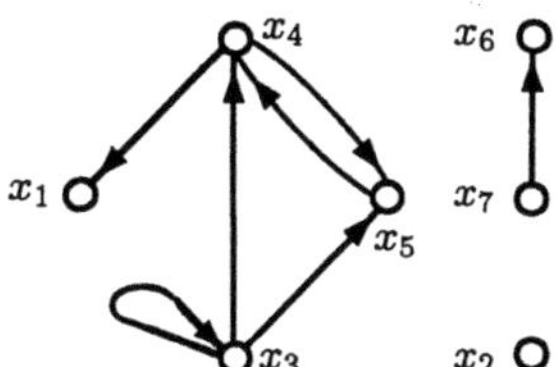


**Fig. 5.1.4** Associated 1-graph of the graph of Fig. 5.1.1

With a directed graph $G = (P, V, A, E)$ one can associate an adjacency relation in two ways: by means of $M := A \sqcup E$ and $\Gamma := \overline{I} \sqcap M^\mathsf{T} M$ (Definitions 5.1.1, 5.1.5) on the one hand, and via $B := A^\mathsf{T} E$ and $\Gamma := \overline{I} \sqcap (B \sqcup B^\mathsf{T})$ (Definitions 5.1.6, 2.2.2) on the other. We show that these two procedures yield the same result.

**5.1.7 Proposition.** In a graph $G = (P, V, A, E)$ with incidence $M = A \sqcup E$ and associated relation $B = A^\mathsf{T} E$ the following holds

$$\overline{I} \sqcap M^\mathsf{T} M = \Gamma = \overline{I} \sqcap (B \sqcup B^\mathsf{T}).$$

**Proof:** From the univalence of $A$ and $E$ we deduce

$$\overline{I} \sqcap M^\mathsf{T} M = \overline{I} \sqcap (A^\mathsf{T} A \sqcup A^\mathsf{T} E \sqcup E^\mathsf{T} A \sqcup E^\mathsf{T} E) = \overline{I} \sqcap (B \sqcup B^\mathsf{T}). \qquad\square$$

Properties of graphs which are not related to the splitting of the incidence into the ingoing and the outgoing part, and which are therefore independent of the "direction" of the arrows, are studied by means of the incidence $M = A \sqcup E$, i.e., in the corresponding hypergraph (cf. Def. 5.3.1). Properties not involving partial or parallel arrows are studied by means of the associated relation $B = A^\mathsf{T} E$ in the associated 1-graph.

One has to be careful, however, when changing the point of view. For example, consider the relationship between "loop" and "loop-carrying point": If $p$ is a loop, $x := A^\mathsf{T} p = E^\mathsf{T} p$ is a loop-carrying point, since $x x^\mathsf{T} = A^\mathsf{T} p p^\mathsf{T} E \subset A^\mathsf{T} E = B$. (Note that $p$ is being counted for the indegree of $x$ as well as for the outdegree.) Conversely, given a loop-carrying point $x$, there need not exist a loop attached to it unless one assumes the point axiom. However, if $O \neq x \subset A^\mathsf{T} E x$ holds, then by the Intermediate Point Theorem 2.4.8 it follows that there exists a $p$ such that $x \subset A^\mathsf{T}p$, which can also be expressed as $x = A^\mathsf{T}p$ and $E^\mathsf{T}p = x$.

We now give a condition on $A$ and $E$ equivalent to loop-freeness (see Definitions 2.1.3, 2.4.11) *without* assuming the point axiom.

**5.1.8 Proposition.** In a graph with associated relation $B = A^\mathsf{T}E$ the following holds:
$$A \sqcap E = O \quad \Longleftrightarrow \quad B \subset \overline{I}.$$

If $A$ and $E$ are total, then, in addition, this is equivalent to $M\overline{I} = L$.

**Proof:** $A \sqcap E = O \iff \overline{A} \sqcup \overline{E} = L \iff AI \subset \overline{E} \iff A^\mathsf{T}E \subset \overline{I}$. If $A, E$ are mappings, we may use Definition 4.2.1 to obtain:
$$A \sqcap E = O \iff \overline{A} \sqcup \overline{E} = L \iff A\overline{I} \sqcup E\overline{I} = L \iff M\overline{I} = L. \qquad \square$$

# 5.2 Graphs via the Associated Relation

In Sect. 2.4 we introduced, for a point $x$, the set $Bx$ of predecessors and the predecessor degree $d_p(x)$, as well as the set $B^\mathsf{T}x$ of successors and the successor degree $d_s(x)$. Now we want to classify the points of a graph according to their degrees; we characterize the points that have no, precisely one, and more than one predecessor.

**5.2.1 Definition.** In a graph with associated relation $B$ we consider a point $x$ (or a point set $x$) and call

$$
\begin{aligned}
x \textbf{ initial (or source)} \quad &:\!\!\iff \quad Bx = O \quad \iff \quad d_p(x) = 0 \\
&\iff \quad x \text{ has no predecessor.} \\[2mm]
x \textbf{ terminal (or sink)} \quad &:\!\!\iff \quad B^\mathsf{T}x = O \quad \iff \quad d_s(x) = 0 \\
&\iff \quad x \text{ has no successor.}
\end{aligned}
$$

$\overline{B^\mathsf{T}L}$ is the set of the initial[1] points and $\overline{BL}$ that of the terminal points. $\qquad \square$

So the points of $\overline{BL}$ can be described as those from which it is *not* possible to proceed *any* further along $B$. In Fig. 5.2.1 examples are given of initial and of terminal points. There are points such as $g$ which are both initial and terminal; we do not call them "isolated", however, because this term will be used in connection with adjacency (Def. 5.4.1) and will then also apply to point $h$.

We have just dealt with points that have *no* predecessor or *no* successor, respectively. Next come points that have *at most* one predecessor or successor. For that purpose we decompose the point set of the graph by means of the univalent part $\mathrm{unp}(B)$ and the multivalent part $\mathrm{mup}(B)$ of the associated relation. The

---

[1] A point $x$ is initial if $Bx \subset O$ which, by the Schröder rule, is equivalent to $x \subset \overline{B^\mathsf{T}L}$. However, in a relation algebra which does not satisfy the point axiom 2.4.6 the odd situation is possible that $\overline{B^\mathsf{T}L} \neq O$, but that there is no initial point.

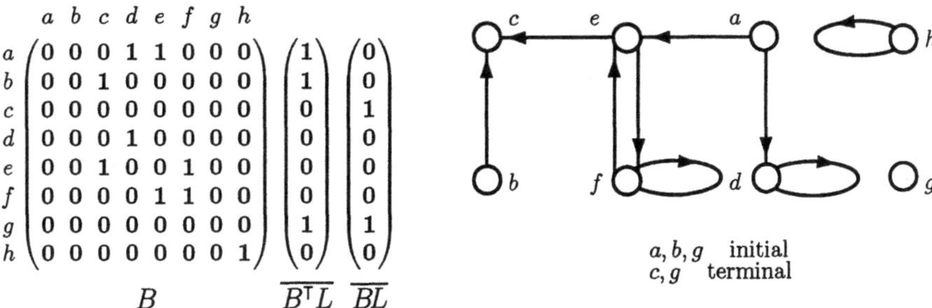



**Fig. 5.2.1** Initial and terminal points

next definition contains several equivalent variants expressing algebraic formulas in ordinary language.

**5.2.2 Definition.** In a graph with associated relation $B$ we call a point

$$
\begin{aligned}
x \text{ functional} \quad :&\Longleftrightarrow\quad x \subset \overline{BL} \sqcup \operatorname{unp}(B)L \\
&\Longleftrightarrow\quad [\operatorname{mup}(B)]^{\mathsf{T}}x = O \qquad \Longleftrightarrow\quad d_s(x) \le 1 \\
&\Longleftrightarrow\quad x \text{ has at most one successor.}
\end{aligned}
$$

$$
\begin{aligned}
x \text{ cofunctional} \quad :&\Longleftrightarrow\quad x \subset \overline{B^{\mathsf{T}}L} \sqcup \operatorname{unp}(B^{\mathsf{T}})L \\
&\Longleftrightarrow\quad [\operatorname{mup}(B^{\mathsf{T}})]^{\mathsf{T}}x = O \qquad \Longleftrightarrow\quad d_p(x) \le 1 \\
&\Longleftrightarrow\quad x \text{ has at most one predecessor.}
\end{aligned}
$$

$$
\begin{aligned}
x \text{ branching point} \quad :&\Longleftrightarrow\quad x \subset \operatorname{mup}(B)L \\
&\Longleftrightarrow\quad x \text{ not functional} \qquad \Longleftrightarrow\quad d_s(x) \ge 2 \\
&\Longleftrightarrow\quad x \text{ has at least two successors.}
\end{aligned}
$$

$$
\begin{aligned}
x \text{ joining point} \quad :&\Longleftrightarrow\quad x \subset \operatorname{mup}(B^{\mathsf{T}})L \\
&\Longleftrightarrow\quad x \text{ not cofunctional} \quad \Longleftrightarrow\quad d_p(x) \ge 2 \\
&\Longleftrightarrow\quad x \text{ has at least two predecessors.} \qquad \square
\end{aligned}
$$

The first two versions of "functional" are equivalent by Prop. 4.2.11. It remains to show that the set $B^{\mathsf{T}}x$ of successors of $x$ is injective. Indeed, since $x$ is injective, $O = [\operatorname{mup}(B)]^{\mathsf{T}}x = (B^{\mathsf{T}} \sqcap \overline{I}B^{\mathsf{T}})x$ is equivalent to $O = B^{\mathsf{T}}x \sqcap \overline{I}B^{\mathsf{T}}x$, hence to $\overline{I}B^{\mathsf{T}}x \subset \overline{B^{\mathsf{T}}x}$, and finally to $(B^{\mathsf{T}}x)(B^{\mathsf{T}}x)^{\mathsf{T}} \subset I$. Figure 5.2.2 illustrates this situation. Every terminal point is also functional.

As an extreme case, a graph may have *only* functional points, i.e., every point has at most one successor. Such a graph is then called a **functional graph**, and this property is characterized by $\operatorname{mup}(B)L = O$. By Prop. 4.2.12.v therefore, a graph is functional if and only if its associated relation is univalent.

Another extreme case arises when the graph has *no* functional points at all, so that it contains *only* branching points. Since $\operatorname{mup}(B)L = L$, the multivalent part of the associated relation is total which, by Prop. 4.2.12.vi, is equivalent to $B\overline{I} = L$.

A directed graph is called an **injective graph** if it contains only cofunctional points. The graph $(V, B)$ is injective if and only if $(V, B^{\mathsf{T}})$ is functional. Thus a

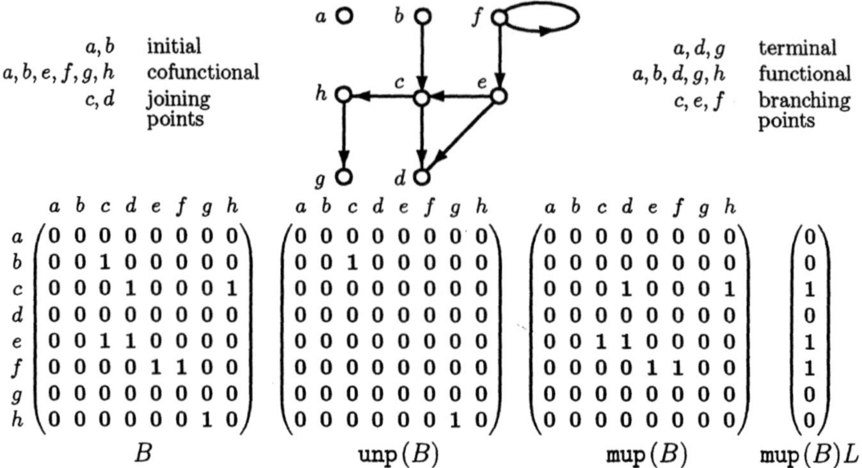

Left labels:

$a, b$    initial
$a, b, e, f, g, h$    cofunctional
$c, d$    joining points

Right labels:

$a, d, g$    terminal
$a, b, d, g, h$    functional
$c, e, f$    branching points

$$
B=\begin{pmatrix}
 & a & b & c & d & e & f & g & h \\
a & 0 & 0 & 0 & 0 & 0 & 0 & 0 & 0 \\
b & 0 & 0 & 1 & 0 & 0 & 0 & 0 & 0 \\
c & 0 & 0 & 0 & 1 & 0 & 0 & 0 & 1 \\
d & 0 & 0 & 0 & 0 & 0 & 0 & 0 & 0 \\
e & 0 & 0 & 1 & 1 & 0 & 0 & 0 & 0 \\
f & 0 & 0 & 0 & 0 & 1 & 1 & 0 & 0 \\
g & 0 & 0 & 0 & 0 & 0 & 0 & 0 & 0 \\
h & 0 & 0 & 0 & 0 & 0 & 0 & 1 & 0
\end{pmatrix}
\qquad
\operatorname{unp}(B)=\begin{pmatrix}
 & a & b & c & d & e & f & g & h \\
a & 0 & 0 & 0 & 0 & 0 & 0 & 0 & 0 \\
b & 0 & 0 & 1 & 0 & 0 & 0 & 0 & 0 \\
c & 0 & 0 & 0 & 0 & 0 & 0 & 0 & 0 \\
d & 0 & 0 & 0 & 0 & 0 & 0 & 0 & 0 \\
e & 0 & 0 & 0 & 0 & 0 & 0 & 0 & 0 \\
f & 0 & 0 & 0 & 0 & 0 & 0 & 0 & 0 \\
g & 0 & 0 & 0 & 0 & 0 & 0 & 0 & 0 \\
h & 0 & 0 & 0 & 0 & 0 & 0 & 1 & 0
\end{pmatrix}
$$

$$
\operatorname{mup}(B)=\begin{pmatrix}
 & a & b & c & d & e & f & g & h \\
a & 0 & 0 & 0 & 0 & 0 & 0 & 0 & 0 \\
b & 0 & 0 & 0 & 0 & 0 & 0 & 0 & 0 \\
c & 0 & 0 & 0 & 1 & 0 & 0 & 0 & 1 \\
d & 0 & 0 & 0 & 0 & 0 & 0 & 0 & 0 \\
e & 0 & 0 & 1 & 1 & 0 & 0 & 0 & 0 \\
f & 0 & 0 & 0 & 0 & 1 & 1 & 0 & 0 \\
g & 0 & 0 & 0 & 0 & 0 & 0 & 0 & 0 \\
h & 0 & 0 & 0 & 0 & 0 & 0 & 0 & 0
\end{pmatrix}
\qquad
\operatorname{mup}(B)L=\begin{pmatrix}
0 \\ 0 \\ 1 \\ 0 \\ 1 \\ 1 \\ 0 \\ 0
\end{pmatrix}
$$

**Fig. 5.2.2** Functional points and branching points

graph is injective if its associated relation is injective, i.e., if every point has at most one predecessor.

## The Associated Relation of a Bipartitioned Graph

In Sect. 4.1 we represented a bipartitioned graph $(X, Y, R, S)$ as a 1-graph on the whole point set $V := X \cup Y$. There, the point sets $X$ and $Y$ were distinguished in the whole set $V$ by the vectors $g$ and $d$ which are *subsets given by a predicate* as in Definition 2.4.1. This, however, was not sufficient for converting $B$ and $R, S$ into one another. We now assume that the two subsets are given by two *injections* $\gamma: X \longrightarrow V$ and $\delta: Y \longrightarrow V$. The relations $\gamma$ and $\delta$ then satisfy

$$\gamma\gamma^{\mathsf{T}} = I, \quad \delta\delta^{\mathsf{T}} = I, \quad \gamma^{\mathsf{T}}\gamma \sqcup \delta^{\mathsf{T}}\delta = I, \quad \gamma\delta^{\mathsf{T}} = O.$$

These conditions mean that both $\gamma$ and $\delta$ are univalent, total, and injective relations, and that the set $V$ is the disjoint union of $X$ and $Y$. In Fig. 5.2.3 these relations are given for the example from Fig. 4.1.3.

$$
\begin{pmatrix} O & R \\ S & O \end{pmatrix}
\qquad
B=\begin{pmatrix}
 & x & y & z \\
a & 1 & 0 & 1 \\
b & 0 & 0 & 0 \\
c & 0 & 1 & 0 \\
d & 0 & 1 & 1
\end{pmatrix}
\qquad
R=\begin{pmatrix}
 & a & b & c & d \\
x & 0 & 0 & 0 & 0 \\
y & 1 & 1 & 0 & 0 \\
z & 1 & 0 & 0 & 1
\end{pmatrix}
$$

$$
\gamma=\begin{pmatrix}
 & a & b & c & d & x & y & z \\
 & 1 & 0 & 0 & 0 & 0 & 0 & 0 \\
 & 0 & 1 & 0 & 0 & 0 & 0 & 0 \\
 & 0 & 0 & 1 & 0 & 0 & 0 & 0 \\
 & 0 & 0 & 0 & 1 & 0 & 0 & 0
\end{pmatrix}
\qquad
\delta=\begin{pmatrix}
 & a & b & c & d & x & y & z \\
 & 0 & 0 & 0 & 0 & 1 & 0 & 0 \\
 & 0 & 0 & 0 & 0 & 0 & 1 & 0 \\
 & 0 & 0 & 0 & 0 & 0 & 0 & 1
\end{pmatrix}
$$

where $S$ has row labels $x, y, z$ and column labels $a, b, c, d$.

**Fig. 5.2.3** Injections for the bipartitioned graph of Fig. 4.1.3

For these injections we have $\gamma^{\mathsf{T}}L = g$, $\delta^{\mathsf{T}}L = d$. Since

$$B = \gamma^{\mathsf{T}}R\delta \sqcup \delta^{\mathsf{T}}S\gamma, \quad R = \gamma B\delta^{\mathsf{T}}, \quad S = \delta B\gamma^{\mathsf{T}},$$

$B$ and $R, S$ can be obtained from one another.

## Directed Graph and 1-Graph

We now formulate properties by which directed graphs can be better compared
to 1-graphs.

**5.2.3 Definition.** Given a directed graph $G = (P, V, A, E)$ and the relation
$H := AA^\mathsf{T} \sqcap EE^\mathsf{T}$, we call

i)   $G$ **total graph**   $:\Longleftrightarrow$   $I \subset H$   $\Longleftrightarrow$   $A$ total and $E$ total

$\qquad\qquad\qquad\qquad\qquad \Longleftrightarrow$   Every arc has precisely one
$\qquad\qquad\qquad\qquad\qquad\qquad$ starting point and one endpoint.

Caution: For a total graph $G$ the incidence $M$ is total, but the converse need
not be true. If only $M$ is total we say that $G$ "has total incidence".

ii)   For a total graph, $H$ is an equivalence relation which may be called "arc
parallelism". We call two arcs

$\qquad p, q$   **parallel**   $:\Longleftrightarrow$   $pq^\mathsf{T} \subset H$   $\Longleftrightarrow$   $A^\mathsf{T}p = A^\mathsf{T}q \neq O,\ E^\mathsf{T}p = E^\mathsf{T}q \neq O$

$\qquad\qquad\qquad\qquad \Longleftrightarrow$   For both $p$ and $q$ there is a starting point and an
$\qquad\qquad\qquad\qquad\qquad$ endpoint, and the respective points coincide.

Given a natural number $s$, the graph $G$ is called an **$s$-graph**, if it is a total
graph, and if there are no $s + 1$ distinct arcs that are pairwise parallel to one
another.                                                                      $\square$

The details of the proof, that the conditions in Definition 5.2.3.ii are equivalent,
are left as Exercise 5.2.4. Usually, a directed graph is understood to be total.
An $s$-graph is, of course, an $(s + 1)$-graph, too. Our terminology is consistent
since for $s = 1$ we do indeed get back our previous 1-graphs.

Next we establish a criterion for a graph to have only total arcs and no distinct parallel arcs. Observe that parallel arcs are understood to have the same
direction.

**5.2.4 Proposition.** Let $G = (P, V, A, E)$ be a graph and put $H := AA^\mathsf{T} \sqcap EE^\mathsf{T}$.
Then

$$H = I \quad \Longleftrightarrow \quad G \text{ is total and has no parallel arcs.}$$

**Proof:** Following Definition 5.2.3.i, $H \supset I$ means that $G$ is total. It remains
to show

$$H \subset I \quad \Longleftrightarrow \quad G \text{ has no parallel arcs.}$$

"$\Longrightarrow$" holds because $pq^\mathsf{T} \subset H \subset I$ and Prop. 2.4.5.ii imply $p = q$. For "$\Longleftarrow$"
assume $H \sqcap \overline{I} \neq O$. Then (using the point axiom 2.4.6) two arcs $p$ and $q$ may be
selected satisfying $pq^\mathsf{T} \subset H \sqcap \overline{I}$. These arcs are then distinct and parallel.     $\square$

The concept of a directed graph is quite intuitive in nature and we have given two definitions of it in terms of relations. Let us recollect: In terms of the associated relation for a 1-graph, practically nothing more is required than a homogeneous relation $B$. In terms of the incidence, a graph is given by two univalent relations $A$ and $E$.

It would be impossible to interpret the graph in Fig. 5.1.1 as a 1-graph, due to the occurrence of parallel and of partial arrows. That picture, therefore, cannot be the graphic representation of a 1-graph. Nevertheless, this graph possesses, by Def. 5.1.6, an associated relation $B = A^\mathsf{T} E$ and therefore an associated 1-graph which looks quite different, though.

When representing graphs in computer science by means of abstract data types, one has to distinguish clearly between 1-graphs, $s$-graphs, and directed graphs, for this has a considerable effect on the implementation ("change of computation structure", "crypt equivalence").

The following problem arises: Given $A$ and $E$, one gets $B := A^\mathsf{T} E$; but how can one regain the relations $A$ and $E$ from $B$? One could introduce separate symbols for the arrows and then determine the outgoing and the ingoing incidence $A$ and $E$ as matrices. This would not be difficult to do, but would not be a mathematically precise construction either. In Proposition 7.2.1 we shall prove that, under the assumption $H = I$, the decomposition of $B$ into $A^\mathsf{T}$ and $E$ is unique to the extent that the graphs arising in this way are all isomorphic.

Only then will it be justified to conceive the theory of 1-graphs as the theory of total graphs without parallel arcs. For example, every 1-graph will then have indegrees and outdegrees. Conversely, every graph will be endowed with predecessor and successor degrees. Exercise 5.2.1 deals with the relationship between these degree concepts.

## Exercises

**5.2.1** Prove the following relationships connecting the predecessor degree and the successor degree in a 1-graph on the one hand, and the indegree and the outdegree on the other. ($d$ is defined in Def. 5.3.2):

   i) $d_s(x) \leq d_A(x)$ and $d_p(x) \leq d_E(x)$

   ii) $d_s(x) = d_A(x)$ and $d_p(x) = d_E(x)$,    if $AA^\mathsf{T} \sqcap EE^\mathsf{T} = I$.

    Therefore, these degrees, coincide in a 1-graph.

   iii) $d(x) \leq d_A(x) + d_E(x)$

   iv) $d(x) = d_A(x) + d_E(x)$,   if $A \sqcap E = O$.

**5.2.2** Show that for a total graph injectivity may be expressed equivalently by the associated relation or the ingoing incidence, i.e.,

$$EE^\mathsf{T} \subset I \quad \Longleftrightarrow \quad BB^\mathsf{T} \subset I.$$

**5.2.3** Prove that $\overline{I} \sqcap B = A^\mathsf{T}(E \sqcap \overline{A})$.

**5.2.4** Let $p$ and $q$ be arcs of a graph. Show that

$$pq^\mathsf{T} \subset AA^\mathsf{T} \quad \Longleftrightarrow \quad A^\mathsf{T} p = A^\mathsf{T} q \neq O.$$

Prove that for a total graph $H := AA^\mathsf{T} \sqcap EE^\mathsf{T}$ is an equivalence relation.

# 5.3 Hypergraphs

Basically, a hypergraph is just a heterogeneous relation disguised as a graph which will be called incidence in this context.

**5.3.1 Definition.** The triple $G = (P, V, M)$ is called a **hypergraph**, if

  i) $P$ is a set of **hyperedges**. (Depending on the situation, these may also be termed edges, arrows, links, or arcs).

  ii) $V$ is a set of **points** (also called vertices or nodes.)

  iii) $M \subset P \times V$ is a relation, called the **incidence**. If $(p, x) \in M$ for a hyperedge $p$ and a point $x$, we say "$p$ and $x$ are incident" or "$x$ lies on $p$".

The hypergraph $G^\mathsf{T} := (V, P, M^\mathsf{T})$ is called the **dual**, or **transposed, hypergraph** of $G$. The hypergraph $\overline{G} := (P, V, \overline{M})$ is called the **complement** of $G$. $G$ is called **finite**, if $P$ and $V$ are finite. Given the directed graph $(P, V, A, E)$, we obtain the **corresponding hypergraph** $(P, V, M)$ by defining $M = A \sqcup E$. Furthermore, $\Gamma := \overline{I} \sqcap M^\mathsf{T} M$ is the (point) **adjacency** and $\mathrm{K} := \overline{I} \sqcap M M^\mathsf{T}$ is the **edge-adjacency** of the hypergraph. $\qquad\qquad\square$

For 1-graphs, which are based on homogeneous relations, we first introduced the notion of "degree". In the heterogeneous case, we now define the "rank".

**5.3.2 Definition.** In a hypergraph $G$ with incidence $M$, a point set $x$, and a set $p$ of hyperedges give rise to the sets

$$Mx \quad \text{of hyperedges that are incident with at least one point of } x,$$
$$M^\mathsf{T}p \quad \text{of points that are incident with at least one hyperedge of } p.$$

For singleton sets $x$ and $p$ we define

$$d(x) := |Mx| \quad \text{the } \textbf{degree} \text{ of the point } x,$$
$$r(p) := |M^\mathsf{T}p| \quad \text{the } \textbf{rank} \text{ of the hyperedge } p.$$

For the whole graph, one usually puts

$$d(G) := \sup_x d(x) \text{ the degree,} \quad r(G) := \sup_p r(p) \text{ the rank of } G \text{ (or } M\text{).} \quad \square$$

Similar concepts of degrees were introduced in Definitions 2.4.10, 2.4.12 and 5.1.2. The example given in Fig. 5.3.1 represents a portion of the class schedule of a school. The points either represent classes $C_i$ or teachers $T_j$. A hyperedge of rank 2 is a regular course, incident with one class and with one teacher; it is denoted by $G$ for German, $Ss$ for social science, etc. Hyperedges of rank 4 typically represent combined courses for two classes and two teachers in one or also in two subjects. A hyperedge of rank 3 may stand, e.g., for physical education taught to the male students of a class by one teacher, and to the female students by another. Hyperedges of rank 1 describe free periods for teachers or for classes. A hyperedge comprising exactly the entire faculty could provide for a faculty meeting. Incidentally, every point of the hypergraph in Fig. 5.3.1 has degree 3, a property which is sometimes termed **regular** of degree 3.

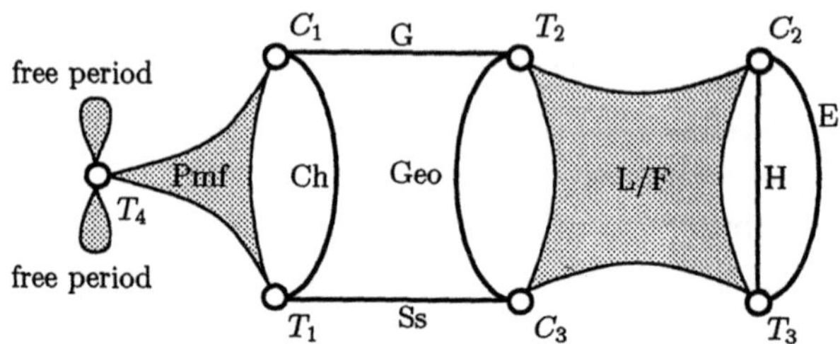


Fig. 5.3.1 Teaching schedule as a hypergraph

There may be points of degree 0, but also hyperedges of rank 0. These conditions can easily be described by means of the incidence matrix, but trying to represent a hyperedge without points graphically would seem unusual. In Fig. 5.1.1 arrow $p_{10}$, viewed as a hyperedge in the associated hypergraph, is empty.

**5.3.3 Definition.** Let a point $x$ (or a point set $x$) and a hyperedge $p$ (or a set $p$ of hyperedges) be given in a hypergraph with incidence $M$. We call

$$x \textbf{ free} \quad :\Longleftrightarrow \quad Mx = O \quad \Longleftrightarrow \quad d(x) = 0$$
$$\Longleftrightarrow \quad x \text{ is not incident with any hyperedge.}$$

$$p \textbf{ empty} \quad :\Longleftrightarrow \quad M^\mathsf{T}p = O \quad \Longleftrightarrow \quad r(p) = 0$$
$$\Longleftrightarrow \quad p \text{ is not incident with any point.}$$

Hypergraphs without empty hyperedges are called **total** and are characterized by $ML = L$. A hypergraph which contains no free point is termed **surjective**; this property is characterized by $M^\mathsf{T}L = L$.  □

Since $Mx \subset O \Longleftrightarrow x \subset \overline{M^\mathsf{T}L}$, the set $\overline{M^\mathsf{T}L}$ consists of all the free points. Similarly, $\overline{ML}$ is the set of all empty hyperedges. The usage of the terms "total" and "surjective" in connection with hypergraphs is in accordance with Definition 4.2.1. Often a hypergraph is assumed total and surjective in the first place, for example when it is given as "a family of nonempty subsets of the point set $V$ which is equal to their union".

We next investigate the interplay between the notions of "free", "initial", and "terminal" points, and then we discuss purely incidence-related properties.

The two properties of a point, being "at the same time initial and terminal" (defined via the associated relation) and being "free" (defined by incidence) are quite similar. However, they differ in nuances which can cause mistakes when not taken into account; also by choosing the less suitable of the two concepts, one may be forced to list a number of exceptional cases.

By definition, a free point is incident with no arrow, therefore it should be at the same time initial and terminal. However, the converse need not hold: a point may well be the starting point of an arrow with undefined endpoint, yet can still be initial and terminal. For the sake of simplicity we confine ourselves to total graphs in the following proposition.

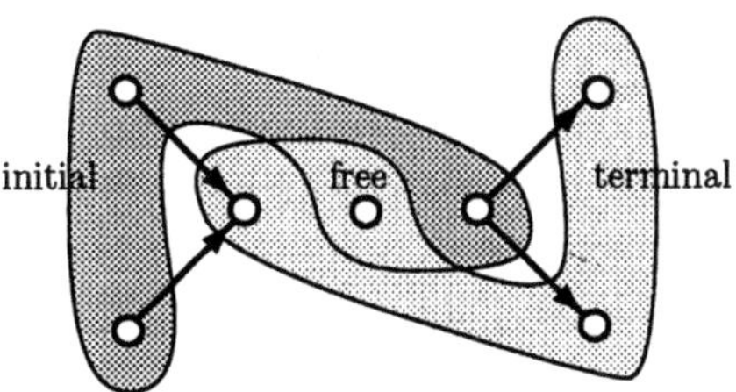


**Fig. 5.3.2** Free, initial, and terminal points

**5.3.4 Proposition.** In a total graph $G$ with incidence $M$ and associated relation $B$ the following inclusions hold:

$$B \sqcup B^{\mathsf{T}} \subset M^{\mathsf{T}}M \quad \text{and} \quad M \subset M(B \sqcup B^{\mathsf{T}}).$$

Thus, for a point $x$ one always has

$$x \text{ initial and terminal} \quad \Longleftrightarrow \quad x \text{ free.}$$

**Proof:** $B \sqcup B^{\mathsf{T}} = A^{\mathsf{T}}E \sqcup E^{\mathsf{T}}A \subset M^{\mathsf{T}}M$. Since $G$ is total, we have $AL = EL = L$, and therefore $M = A \sqcup E = (EL \sqcap A) \sqcup (AL \sqcap E) \subset (E \sqcap AL^{\mathsf{T}})(L \sqcap E^{\mathsf{T}}A) \sqcup (A \sqcap EL^{\mathsf{T}})(L \sqcap A^{\mathsf{T}}E) \subset MB^{\mathsf{T}} \sqcup MB = M(B \sqcup B^{\mathsf{T}})$. Using Definitions 5.3.3 and 5.2.1 leads to the desired equivalence of "free" and "initial and terminal".    □

We now distinguish points of degree 1 and hyperedges of rank 1 from those of degree or rank $\geq 2$. Our procedure is similar to that in Sect. 5.2, and again makes use of the univalent and the multivalent parts, as well as of the decomposition according to Proposition 4.2.11.

**5.3.5 Definition.** Let a hypergraph with incidence $M$ be given together with a point $x$ and a hyperedge $p$. We call

$$x \textbf{ peak} \quad :\Longleftrightarrow \quad x \subset \operatorname{unp}(M^{\mathsf{T}})L$$
$$\Longleftrightarrow \quad O \neq Mx \subset \overline{\overline{I}Mx} \quad \Longleftrightarrow \quad d(x) = 1$$
$$\Longleftrightarrow \quad x \text{ is incident with precisely one hyperedge.}$$

$$x \textbf{ edge-connecting} \quad :\Longleftrightarrow \quad x \subset \operatorname{mup}(M^{\mathsf{T}})L$$
$$\Longleftrightarrow \quad (M^{\mathsf{T}} \sqcap M^{\mathsf{T}}\overline{I})^{\mathsf{T}}x \neq O \quad \Longleftrightarrow \quad d(x) \geq 2$$
$$\Longleftrightarrow \quad x \text{ is incident with at least 2 hyperedges.}$$

$$p \textbf{ drop} \quad :\Longleftrightarrow \quad p \subset \operatorname{unp}(M)L$$
$$\Longleftrightarrow \quad O \neq M^{\mathsf{T}}p \subset \overline{\overline{I}M^{\mathsf{T}}p} \quad \Longleftrightarrow \quad r(p) = 1$$
$$\Longleftrightarrow \quad p \text{ is incident with precisely one point.}$$

$$p \text{ \textbf{point-connecting}} \quad :\Longleftrightarrow \quad p \subset \mathrm{mup}(M)L$$
$$\Longleftrightarrow \quad (M \sqcap M\bar{I})^{\mathsf{T}}p \neq O \quad \Longleftrightarrow \quad r(p) \geq 2$$
$$\Longleftrightarrow \quad p \text{ is incident with at least 2 points.} \qquad \square$$

The proof of the equivalence of the various statements above follows from Definition 4.2.8 and Propositions 4.2.9–4.2.12. Figure 5.3.3 contains examples of the special points and hyperedges just introduced.

We now use Prop. 4.2.12 to characterize hypergraphs all of whose points are edge-connecting, and all of whose edges are point-connecting.

**5.3.6 Proposition.** In a hypergraph with incidence $M$ the following holds:

i)
$$M^{\mathsf{T}}\bar{I} = L \quad \Longleftrightarrow \quad \text{Every point is edge-connecting.}$$
$$M\bar{I} = L \quad \Longleftrightarrow \quad \text{Every edge is point-connecting.}$$

ii)
$$M^{\mathsf{T}} \subset M^{\mathsf{T}}\bar{I} \quad \Longleftrightarrow \quad \text{There are no peaks.}$$
$$M \subset M\bar{I} \quad \Longleftrightarrow \quad \text{There are no drops.}$$

iii)
$$\overline{M^{\mathsf{T}}} = M^{\mathsf{T}}\bar{I} \quad \Longleftrightarrow \quad \text{Every point is a peak.}$$
$$\overline{M} = M\bar{I} \quad \Longleftrightarrow \quad \text{Every edge is a drop.} \qquad \square$$

Furthermore, in a hypergraph the point set decomposes into free points $\overline{M^{\mathsf{T}}L}$, peaks $\mathrm{unp}(M^{\mathsf{T}})L$, and edge-connecting points $\mathrm{mup}(M^{\mathsf{T}})L$. Analogously, the set of hyperedges decomposes into empty hyperedges $\overline{ML}$, drops $\mathrm{unp}(M)L$, and point-connecting edges $\mathrm{mup}(M)L$. In Fig. 5.3.3 these decompositions are visualized.

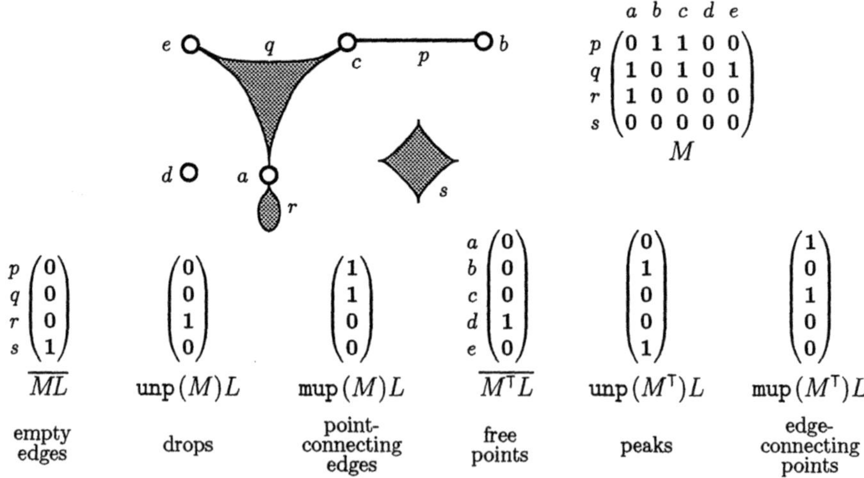


$$
\begin{array}{c}
\begin{array}{ccccc} a & b & c & d & e \end{array} \\
\begin{array}{c} p \\ q \\ r \\ s \end{array}
\begin{pmatrix}
0 & 1 & 1 & 0 & 0 \\
1 & 0 & 1 & 0 & 1 \\
1 & 0 & 0 & 0 & 0 \\
0 & 0 & 0 & 0 & 0
\end{pmatrix} \\
M
\end{array}
$$

$$
\begin{array}{c} p \\ q \\ r \\ s \end{array}
\begin{pmatrix} 0 \\ 0 \\ 0 \\ 1 \end{pmatrix}
\qquad
\begin{pmatrix} 0 \\ 0 \\ 1 \\ 0 \end{pmatrix}
\qquad
\begin{pmatrix} 1 \\ 1 \\ 0 \\ 0 \end{pmatrix}
\qquad
\begin{array}{c} a \\ b \\ c \\ d \\ e \end{array}
\begin{pmatrix} 0 \\ 0 \\ 0 \\ 1 \\ 0 \end{pmatrix}
\qquad
\begin{pmatrix} 0 \\ 1 \\ 0 \\ 0 \\ 1 \end{pmatrix}
\qquad
\begin{pmatrix} 1 \\ 0 \\ 1 \\ 0 \\ 0 \end{pmatrix}
$$

$$\overline{ML} \qquad \mathrm{unp}(M)L \qquad \mathrm{mup}(M)L \qquad \overline{M^{\mathsf{T}}L} \qquad \mathrm{unp}(M^{\mathsf{T}})L \qquad \mathrm{mup}(M^{\mathsf{T}})L$$

| empty edges | drops | point-connecting edges | free points | peaks | edge-connecting points |
|---|---|---|---|---|---|

**Fig. 5.3.3** Peaks, drops, connecting points and edges

We want to carry our analysis one step further and introduce the following notions which will be useful for the treatment of matchings and covering problems.

**5.3.7 Definition.** In a hypergraph with incidence $M$ we call

$$[\mathsf{unp}\,(M^\mathsf{T})]^\mathsf{T}L \quad \text{the } \textbf{set of cusps,}$$
$$[\mathsf{unp}\,(M)]^\mathsf{T}L \quad \text{the } \textbf{set of drop-carrying points.}} \qquad \square$$

$$
\begin{pmatrix} 0 & 1 & 1 & 0 & 0 \\ 1 & 0 & 1 & 0 & 1 \\ 1 & 0 & 0 & 0 & 0 \\ 0 & 0 & 0 & 0 & 0 \end{pmatrix}
\begin{pmatrix} 0 \\ 1 \\ 0 \\ 0 \\ 1 \end{pmatrix}
=
\begin{pmatrix} 1 \\ 1 \\ 0 \\ 0 \end{pmatrix}
\qquad
\begin{pmatrix} 0 & 1 & 1 & 0 \\ 1 & 0 & 0 & 0 \\ 1 & 1 & 0 & 0 \\ 0 & 0 & 0 & 0 \\ 0 & 1 & 0 & 0 \end{pmatrix}
\begin{pmatrix} 0 \\ 0 \\ 1 \\ 0 \end{pmatrix}
=
\begin{pmatrix} 1 \\ 0 \\ 0 \\ 0 \\ 0 \end{pmatrix}
$$

$$M \, \mathsf{unp}\,(M^\mathsf{T})L = [\mathsf{unp}\,(M^\mathsf{T})]^\mathsf{T}L \qquad M^\mathsf{T}\,\mathsf{unp}\,(M)L = [\mathsf{unp}\,(M)]^\mathsf{T}L$$

**Fig. 5.3.4** Cusps and drop-carrying points for Fig. 5.3.3

We shall now show that our terminology is consistent in the sense that an edge is a cusp, if it has at least one peak, and that a point is drop-carrying, if it is incident with at least one drop.

**5.3.8 Proposition.** In a hypergraph with incidence $M$ the following holds:

i) $$M \, \mathsf{unp}\,(M^\mathsf{T})L = [\mathsf{unp}\,(M^\mathsf{T})]^\mathsf{T}L;$$

ii) $$M^\mathsf{T}\,\mathsf{unp}\,(M)L = [\mathsf{unp}\,(M)]^\mathsf{T}L.$$

**Proof:** Apply Prop. 4.2.9.iv using $R^\mathsf{T}RL = R^\mathsf{T}L$ for any relation. $\qquad \square$

In Fig. 5.3.4 we investigate the situation of Fig. 5.3.3 with respect to cusps and drop-carrying points. Edges $p$ and $q$ are cusps by virtue of the peaks $b$ and $e$, respectively. Only point $a$ is drop-carrying in view of drop $r$.

The notions to be introduced next are similar to those in Def. 5.2.3.ii, but are not dependent on the direction of the arcs.

**5.3.9 Definition.** If $M$ is the incidence of a hypergraph and $p, q$ are two hyperedges, we define

i) $\quad p$ **contained** in $q \quad :\Longleftrightarrow \quad M^\mathsf{T}p \subset M^\mathsf{T}q \quad \Longleftrightarrow \quad pq^\mathsf{T} \subset \overline{M\overline{M^\mathsf{T}}},$

ii) $\quad p$ and $q$ **parallel** $\quad :\Longleftrightarrow \quad M^\mathsf{T}p = M^\mathsf{T}q \quad \Longleftrightarrow \quad pq^\mathsf{T} \subset \mathsf{syq}(M^\mathsf{T}, M^\mathsf{T}).$

$H := \mathsf{syq}(M^\mathsf{T}, M^\mathsf{T}) = \overline{M\overline{M^\mathsf{T}}} \sqcap \overline{\overline{M}M^\mathsf{T}}$ is an equivalence relation by Corollary 4.4.4; one might call it "edge-parallelism". $G$ is called **essential** if no hyperedge is contained in another. $G$ is called "parallel-free" or **simple** if no hyperedge is parallel with another, i.e., if no two edges are contained in one another. $\qquad \square$

The versions in (i), hence also those in (ii), are equivalent by Proposition 2.4.4.i. Therefore, we have established the following criterion:

**5.3.10 Proposition.** Let $G$ be a hypergraph with incidence $M$.

i)              $G$ essential $\iff$ $\overline{M\overline{M^\mathsf{T}}} \subset I$;

ii)             $G$ simple $\iff$ $\mathrm{syq}(M^\mathsf{T}, M^\mathsf{T}) \subset I$. □

Indeed, in both cases "$=$" holds, and not just "$\subset$". The containment $\overline{I} \subset M\overline{M}^\mathsf{T}$ has the following concrete meaning: Given any two different hyperedges, there is a point incident with the first, but not with the second. An equivalent form of condition (i) is $\overline{I} \subset \overline{M}M^\mathsf{T}$.

## 5.4 Graphs via the Adjacency Relation

Our present theme arose already in Sect. 2.4 where, for an adjacency $\Gamma$, we introduced the set $\Gamma x$ of neighbors and the simple degree $d_0(x)$ of some point $x$ (cf. Def. 2.4.12). We now distinguish points according to their degree, as far as this can be done by relation-algebraic means: namely, according to whether there is no, one, or more than one neighbor.

**5.4.1 Definition.** In a graph with adjacency $\Gamma$, we call a point

$$x \text{ isolated} \quad :\iff \quad \Gamma x = O$$
$$\iff \quad d_0(x) = 0 \quad \iff \quad x \text{ has no neighbors.} \quad □$$

In Definition 5.3.3 we gave the definition of the incidence-related property of "free". With regard to this, our above choice of terminology is justified by the following sociological analogy: A man is called *isolated*, if he has no relationship with any other person; he is called *free*, if he is not subject to outside rule. We shall see that "isolation" does not necessarily create "freedom"—a man could be bound by an obligation concerning him alone, which could be interpreted as a drop-shaped hyperedge (Fig. 5.4.1).

**Fig. 5.4.1** Two isolated points

After the case of the simple degree $d_0(x) = 0$ we proceed to degrees 1 and $\geq 2$.

**5.4.2 Definition.** In a graph with adjacency $\Gamma$ we call a point

$$x \text{ pendant} \quad :\iff \quad x \subset \mathrm{unp}\,(\Gamma)L \quad \iff \quad d_0(x) = 1$$
$$\iff \quad x \text{ has precisely one neighbor,}$$
$$x \text{ rigid} \quad :\iff \quad x \subset \mathrm{mup}\,(\Gamma)L \quad \iff \quad d_0(x) \geq 2$$
$$\iff \quad x \text{ has more than one neighbor.} \quad □$$

In Fig. 2.2.6 the whole point set breaks down into isolated points (B), pendant points (HB and S), and rigid points (SH, HH, N, NW, RP, H, BW, and BY). In certain situations one would like to exclude isolated or pendant points. The precise formulation refers to Proposition 4.2.12:

**5.4.3 Proposition.** For an adjacency $\Gamma$, the set of pendant points is described by $\mathtt{unp}\,(\Gamma)L$, and $\mathtt{mup}\,(\Gamma)L$ is the set of rigid points. Furthermore,

i)
$$\overline{\Gamma} = \Gamma\overline{I} \quad \Longleftrightarrow \quad \text{The graph has only pendant points.}$$
$$\Gamma \subset \Gamma\overline{I} \quad \Longleftrightarrow \quad \text{The graph has no pendant points.}$$

ii)
$$\Gamma\overline{I} = L \quad \Longleftrightarrow \quad \text{The graph has only rigid points.}$$
$$\Gamma\overline{I} \subset \overline{\Gamma} \quad \Longleftrightarrow \quad \text{The graph has no rigid points.} \qquad \square$$

We have classified points in two ways: With respect to incidence we distinguish free points, peaks, and edge-connecting points; with respect to adjacency there are isolated, pendant, and rigid points. So it is of interest to investigate the relationship between these two sets of properties. We start with "free" and "isolated" (cf. Fig. 5.4.1).

**5.4.4 Proposition.** In a graph with incidence $M$ and adjacency $\Gamma$ we always have the inclusion $\Gamma \subset M^{\mathsf{T}}M$. If, in addition, the incidence has no drops, i.e., if $M \subset M\overline{I}$, we have $M\Gamma \supset M$. Thus, for a point $x$ we always have that

$$x \text{ free} \quad \Longrightarrow \quad x \text{ isolated,}$$
$$x \text{ free} \quad \Longleftarrow \quad x \text{ isolated and } M \text{ drop-free.}$$

The incidence of graphs in which no hyperedge has rank 1, in particular of simple graphs and loop-free 1-graphs, has no drops.

**Proof:** We have $\Gamma = \overline{I} \sqcap M^{\mathsf{T}}M \subset M^{\mathsf{T}}M$; so $Mx \subset O$ always implies $\Gamma x \subset O$. From $M \subset M\overline{I}$ it follows that $M = M\overline{I} \sqcap M \subset (M \sqcap M\overline{I})(\overline{I} \sqcap M^{\mathsf{T}}M) \subset M\Gamma$; Hence $\Gamma x \subset O \implies Mx \subset O$. $\qquad \square$

The set $\overline{M^{\mathsf{T}}L}$ of all free points is contained in the set of isolated points. More precisely (cf. Exercise 5.4.1)

$$\overline{M^{\mathsf{T}}L} \subset \overline{[\mathtt{mup}\,(M)]^{\mathsf{T}}L} = \overline{\Gamma L},$$

so that $M \subset M\overline{I}$ has $\overline{M^{\mathsf{T}}L} = \overline{\Gamma L}$ as an immediate consequence. For isolated points, therefore, we have both, a characterization by adjacency ($\Gamma x = O$ or $x \subset \overline{\Gamma L}$) and a characterization by incidence ($\mathtt{mup}\,(M)x = O$ or $x \subset \overline{[\mathtt{mup}\,(M)]^{\mathsf{T}}L}$).

By contrast, from the point of view of adjacency one can make precise statements on free points only under the general assumption $M \subset M\overline{I}$ under which, however, free and isolated points coincide. Also, some kinship between rigid and edge-connecting points is to be expected, the former being introduced by adjacency, the latter by incidence. Figure 5.4.2 gives an overview of the phenomena that can occur. The notions diverge markedly and can correspond to each other more closely only when restricting conditions are met. The example also shows that the hypotheses in Proposition 5.4.5 cannot be weakened.

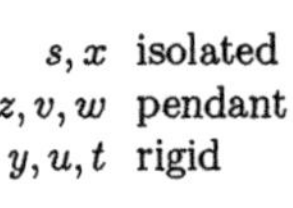


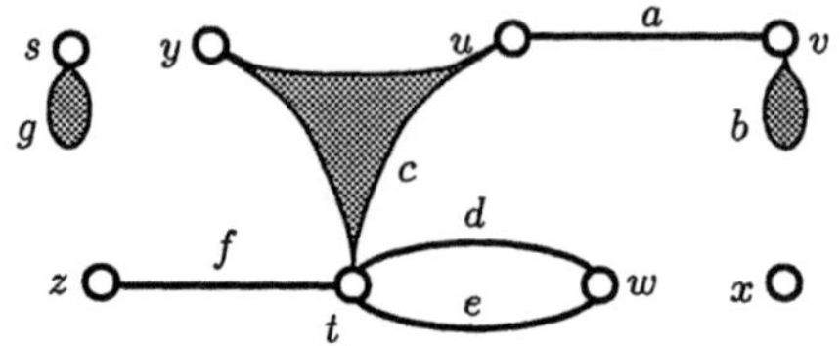


**Fig. 5.4.2** Pendant points and peaks

**5.4.5 Proposition.** An incidence $M$ of rank 2 and its corresponding adjacency $\Gamma$ satisfy $\mathrm{mup}(\Gamma) \subset \mathrm{mup}(M^\mathsf{T})M$. For an essential incidence, i.e., if $\overline{I} \subset M\overline{M^\mathsf{T}}$, in addition $\mathrm{mup}(M^\mathsf{T}) \subset \mathrm{mup}(\Gamma)M^\mathsf{T}$ holds. Therefore, the following implications hold for an incidence $M$ of rank 2:

$$x \text{ rigid} \implies x \text{ edge-connecting;}$$
$$x \text{ rigid} \impliedby x \text{ edge-connecting and } M \text{ essential.}$$

**Proof:** Introducing an orientation for the moment, we express the fact that the rank is less or equal to 2 by writing $M$ as the union of two univalent relations $A$ and $E$. Following Prop. 5.1.7 we have $\mathrm{mup}(\Gamma) = \Gamma \sqcap \Gamma\overline{I}$ where $\Gamma = \overline{I} \sqcap (A^\mathsf{T}E \sqcup E^\mathsf{T}A)$. Without loss of generality we confine ourselves to presenting two of the necessary estimates, employing Prop. 4.2.2.iii and the results on the univalent and the multivalent part from Sect. 4.2:

$$A^\mathsf{T}E \sqcap A^\mathsf{T}E\overline{I} = (A^\mathsf{T} \sqcap A^\mathsf{T}E\overline{I}E^\mathsf{T})E \subset (A^\mathsf{T} \sqcap A^\mathsf{T}E\overline{E}^\mathsf{T})E$$
$$\subset (A^\mathsf{T} \sqcap A^\mathsf{T}\overline{I})E \subset \mathrm{mup}(M^\mathsf{T})M$$
$$\overline{I} \sqcap A^\mathsf{T}E \sqcap E^\mathsf{T}A\overline{I} = A^\mathsf{T}E \sqcap E^\mathsf{T}(E\overline{I} \sqcap A\overline{I}) \subset A^\mathsf{T}E \sqcap E^\mathsf{T}E\overline{I}$$
$$= (A^\mathsf{T} \sqcap E^\mathsf{T}E\overline{I}E^\mathsf{T})E \subset (A^\mathsf{T} \sqcap E^\mathsf{T}\overline{E}E^\mathsf{T})E$$
$$\subset (A^\mathsf{T} \sqcap E^\mathsf{T}\overline{I})E \subset \mathrm{mup}(M^\mathsf{T})M.$$

Therefore, $\mathrm{mup}(\Gamma)L \subset \mathrm{mup}(M^\mathsf{T})L$, so that the set of rigid points is contained in the set of edge-connecting points. From Prop. 5.3.10 we know that an incidence is essential precisely when $\overline{I} \subset \overline{M}M^\mathsf{T}$. This yields $M^\mathsf{T}\overline{I} \subset M^\mathsf{T}\overline{M}M^\mathsf{T} \subset \overline{I}M^\mathsf{T}$, as well as $M\overline{I}M^\mathsf{T} \subset I$ which will be used later on. We can now proceed with the estimation

$$\mathrm{mup}(M^\mathsf{T}) = M^\mathsf{T}\overline{I} \sqcap M^\mathsf{T} \subset M^\mathsf{T}\overline{M}M^\mathsf{T} \sqcap M^\mathsf{T}$$
$$\subset (M^\mathsf{T}\overline{M} \sqcap M^\mathsf{T}M)(M^\mathsf{T} \sqcap \overline{M}^\mathsf{T}MM^\mathsf{T}) \subset (M^\mathsf{T}\overline{M} \sqcap M^\mathsf{T}M)M^\mathsf{T}.$$

The first $M^\mathsf{T}$ in the bracket expression is now decomposed with respect to $\overline{I}M^\mathsf{T}$ and the complement $\overline{\overline{I}M^\mathsf{T}}$. On the one hand this results in

$$(M^\mathsf{T} \sqcap \overline{\overline{I}M^\mathsf{T}})\overline{M} \sqcap M^\mathsf{T}M \subset \overline{\overline{I}M^\mathsf{T}}\,\overline{M} \sqcap M^\mathsf{T}M$$
$$\subset (\overline{\overline{I}M^\mathsf{T}} \sqcap M^\mathsf{T}M\overline{M}^\mathsf{T})(\overline{M} \sqcap M\overline{I}M^\mathsf{T}M) \subset (\,...\,)(\overline{M} \sqcap IM) = O$$

and on the other in

$$(M^\mathsf{T} \sqcap \overline{I}M^\mathsf{T})\overline{M} \sqcap M^\mathsf{T}M = M^\mathsf{T}\overline{M} \sqcap M^\mathsf{T}M \sqcap (\overline{I}M^\mathsf{T} \sqcap M^\mathsf{T})\overline{M}$$
$$\subset \overline{I} \sqcap M^\mathsf{T}M \sqcap (\overline{I} \sqcap M^\mathsf{T}M)(M^\mathsf{T} \sqcap \overline{I}M^\mathsf{T})\overline{M}$$
$$\subset \Gamma \sqcap \Gamma M^\mathsf{T}\overline{M} \subset \Gamma \sqcap \Gamma\overline{I} = \mathtt{mup}(\Gamma).$$

For an essential $M$, the set $\mathtt{mup}(M^\mathsf{T})L$ of edge-connecting points is contained in the set $\mathtt{mup}(\Gamma)L$ of rigid points. $\qquad\qquad\qquad\qquad\qquad\qquad\square$

Point $v$ in Fig. 5.4.2 is edge-connecting, but not rigid because edge $b$ is contained in edge $a$, and therefore the graph is not essential. On the other hand, $y$ is rigid because edge $c$ has rank 3, but $y$ is not edge-connecting.

Propositions 5.4.4 and 5.4.5 yield consequences for the relationship between peaks and pendant points, if we take into account the point set decomposition into free points, peaks, and edge-connecting points on the one hand, and isolated, pendant, and rigid points on the other.

**5.4.6 Corollary.** In a graph of rank 2 the following holds:

$$\begin{array}{llll} x \text{ peak} & \Longrightarrow & x \text{ pendant}, & \text{if there are no drops;} \\ x \text{ peak} & \Longleftarrow & x \text{ pendant}, & \text{if the incidence is essential.} \end{array}$$

In a graph of rank 2 with drop-free essential incidence, e.g., in a simple graph, we have therefore

$$x \text{ peak} \quad \Longleftrightarrow \quad x \text{ pendant}. \qquad\qquad\qquad\qquad\qquad\square$$

There are dual forms of Proposition 5.4.4 and Corollary 5.4.6, concerning the relationship between incidence and edge-adjacency, which we will not elaborate on here.

**Exercise**

**5.4.1** Prove that $[\mathtt{mup}(M)]^\mathsf{T}L = \Gamma L$.

# 5.5 Incidence and Adjacency

Often a graph or a hypergraph is equipped with both incidence *and* adjacency. When only an incidence $M$ is given in the first place, then one can define an adjacency in a natural way by setting $\Gamma := \overline{I} \sqcap M^\mathsf{T}M$. However, the situation is more difficult when one has only an adjacency $\Gamma$ to start with, subject to the condition $\Gamma = \Gamma^\mathsf{T} \subset \overline{I}$. Then a relation $M$ is to be found that satisfies $\Gamma = \overline{I} \sqcap M^\mathsf{T}M$. The upper part of Fig. 5.5.1 shows that this cannot be achieved in a unique way. There, three hypergraphs with different incidences $M$ are given, all having the same adjacency $\Gamma$.

The situation is comparable to that in Sect. 5.2: Given the outgoing and the ingoing incidence $A$ and $E$, one immediately gets the associated relation $B := A^\mathsf{T}E$; but $A$ and $E$ are not uniquely determined by $B$.

$H = (P, V, M)$   with point-adjacency $\Gamma$:

resp.

$H^{\mathsf{T}} = (V, P, M^{\mathsf{T}})$   with edge-adjacency K:

$$\Gamma = \begin{array}{c@{\;}c} & \begin{array}{cccccc} p & q & r & s & t & u \end{array} \\ \begin{array}{c} p \\ q \\ r \\ s \\ t \\ u \end{array} & \left(\begin{array}{cccccc} 0 & 1 & 1 & 1 & 1 & 0 \\ 1 & 0 & 1 & 0 & 1 & 1 \\ 1 & 1 & 0 & 1 & 0 & 1 \\ 1 & 0 & 1 & 0 & 1 & 1 \\ 1 & 1 & 0 & 1 & 0 & 1 \\ 0 & 1 & 1 & 1 & 1 & 0 \end{array}\right) \end{array}$$

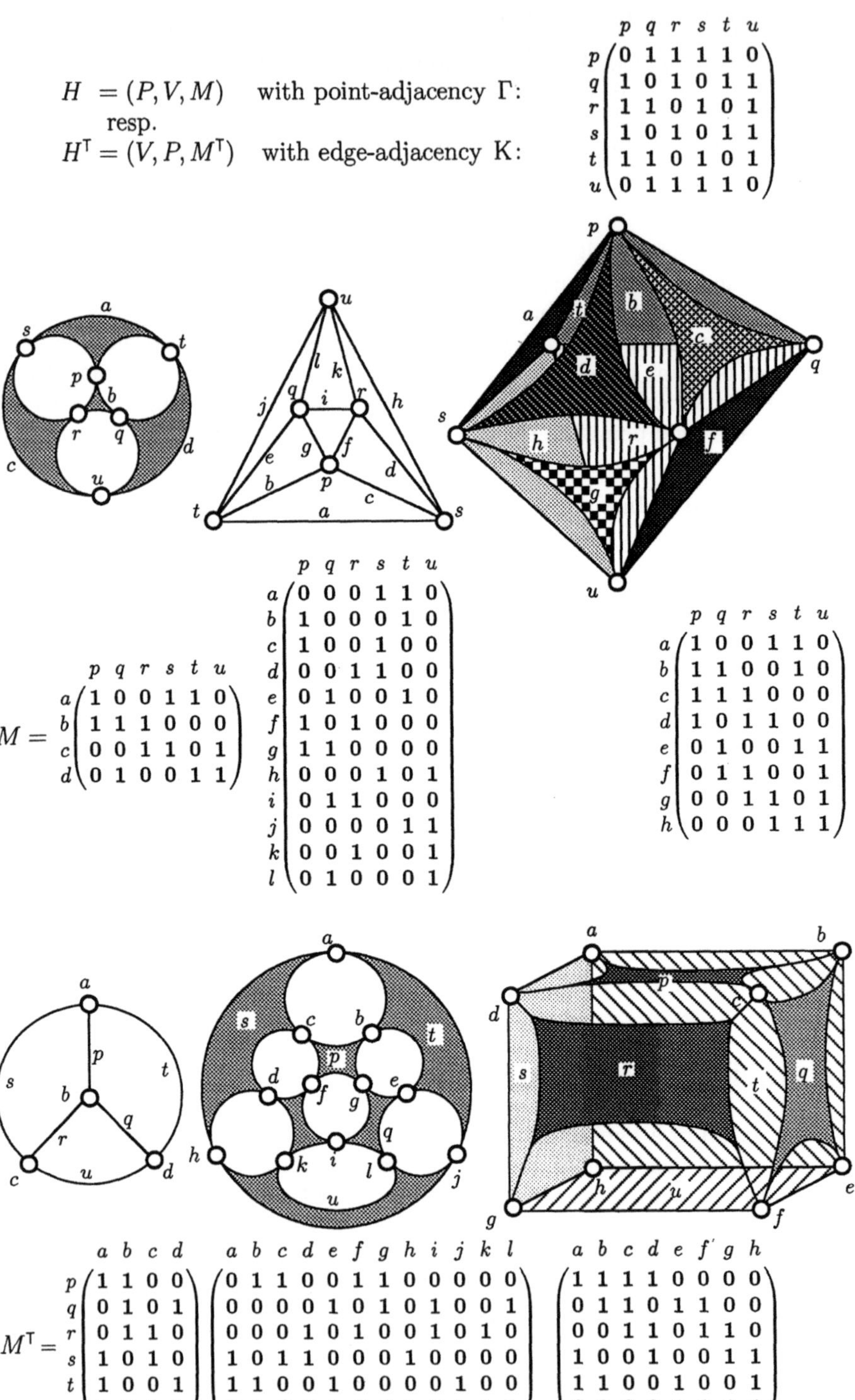

$$M = \begin{array}{c@{\;}c} & \begin{array}{cccccc} p & q & r & s & t & u \end{array} \\ \begin{array}{c} a \\ b \\ c \\ d \end{array} & \left(\begin{array}{cccccc} 1 & 0 & 0 & 1 & 1 & 0 \\ 1 & 1 & 1 & 0 & 0 & 0 \\ 0 & 0 & 1 & 1 & 0 & 1 \\ 0 & 1 & 0 & 0 & 1 & 1 \end{array}\right) \end{array}$$

$$\begin{array}{c@{\;}c} & \begin{array}{cccccc} p & q & r & s & t & u \end{array} \\ \begin{array}{c} a \\ b \\ c \\ d \\ e \\ f \\ g \\ h \\ i \\ j \\ k \\ l \end{array} & \left(\begin{array}{cccccc} 0 & 0 & 0 & 1 & 1 & 0 \\ 1 & 0 & 0 & 0 & 1 & 0 \\ 1 & 0 & 0 & 1 & 0 & 0 \\ 0 & 0 & 1 & 1 & 0 & 0 \\ 0 & 1 & 0 & 0 & 1 & 0 \\ 1 & 0 & 1 & 0 & 0 & 0 \\ 1 & 1 & 0 & 0 & 0 & 0 \\ 0 & 0 & 0 & 1 & 0 & 1 \\ 0 & 1 & 1 & 0 & 0 & 0 \\ 0 & 0 & 0 & 0 & 1 & 1 \\ 0 & 0 & 1 & 0 & 0 & 1 \\ 0 & 1 & 0 & 0 & 0 & 1 \end{array}\right) \end{array}$$

$$\begin{array}{c@{\;}c} & \begin{array}{cccccc} p & q & r & s & t & u \end{array} \\ \begin{array}{c} a \\ b \\ c \\ d \\ e \\ f \\ g \\ h \end{array} & \left(\begin{array}{cccccc} 1 & 0 & 0 & 1 & 1 & 0 \\ 1 & 1 & 0 & 0 & 1 & 0 \\ 1 & 1 & 1 & 0 & 0 & 0 \\ 1 & 0 & 1 & 1 & 0 & 0 \\ 0 & 1 & 0 & 0 & 1 & 1 \\ 0 & 1 & 1 & 0 & 0 & 1 \\ 0 & 0 & 1 & 1 & 0 & 1 \\ 0 & 0 & 0 & 1 & 1 & 1 \end{array}\right) \end{array}$$

$$M^{\mathsf{T}} = \begin{array}{c@{\;}c} & \begin{array}{cccc} a & b & c & d \end{array} \\ \begin{array}{c} p \\ q \\ r \\ s \\ t \\ u \end{array} & \left(\begin{array}{cccc} 1 & 1 & 0 & 0 \\ 0 & 1 & 0 & 1 \\ 0 & 1 & 1 & 0 \\ 1 & 0 & 1 & 0 \\ 1 & 0 & 0 & 1 \\ 0 & 0 & 1 & 1 \end{array}\right) \end{array} \quad \begin{array}{c@{\;}c} \begin{array}{cccccccccccc} a & b & c & d & e & f & g & h & i & j & k & l \end{array} \\ \left(\begin{array}{cccccccccccc} 0 & 1 & 1 & 0 & 0 & 1 & 1 & 0 & 0 & 0 & 0 & 0 \\ 0 & 0 & 0 & 0 & 1 & 0 & 1 & 0 & 1 & 0 & 0 & 1 \\ 0 & 0 & 0 & 1 & 0 & 1 & 0 & 0 & 1 & 0 & 1 & 0 \\ 1 & 0 & 1 & 1 & 0 & 0 & 0 & 1 & 0 & 0 & 0 & 0 \\ 1 & 1 & 0 & 0 & 1 & 0 & 0 & 0 & 0 & 1 & 0 & 0 \\ 0 & 0 & 0 & 0 & 0 & 0 & 0 & 1 & 0 & 1 & 1 & 1 \end{array}\right) \end{array} \quad \begin{array}{c@{\;}c} \begin{array}{cccccccc} a & b & c & d & e & f' & g & h \end{array} \\ \left(\begin{array}{cccccccc} 1 & 1 & 1 & 1 & 0 & 0 & 0 & 0 \\ 0 & 1 & 1 & 0 & 1 & 1 & 0 & 0 \\ 0 & 0 & 1 & 1 & 0 & 1 & 1 & 0 \\ 1 & 0 & 0 & 1 & 0 & 0 & 1 & 1 \\ 1 & 1 & 0 & 0 & 1 & 0 & 0 & 1 \\ 0 & 0 & 0 & 0 & 1 & 1 & 1 & 1 \end{array}\right) \end{array}$$

**Fig. 5.5.1** Different hypergraphs with the same adjacency or edge-adjacency

There is a dual version of our problem. For every incidence $M$ there is the edge-adjacency $\mathsf{K} := \overline{I} \sqcap MM^{\mathsf{T}}$. But given an edge-adjacency $\mathsf{K} = \mathsf{K}^{\mathsf{T}} \subset \overline{I}$, we cannot define offhand an incidence. Underneath the three different hypergraphs in Fig. 5.5.1 we have drawn their dual hypergraphs whose edge-adjacencies $\mathsf{K}$ are the same.

When trying to factorize $B$ we found that the condition $AA^{\mathsf{T}} \sqcap EE^{\mathsf{T}} = I$ enabled us to determine $A$ and $E$ uniquely up to isomorphism, because it has the effect of excluding parallel or partial arcs. We now look for conditions which will determine a hypergraph uniquely. The middle hypergraph in Fig. 5.5.1 suggests itself as the candidate which, by virtue of its additional properties, should be uniquely determined by the adjacency $\Gamma$. (The left-hand hypergraph seems to be "easier", but that is coincidental.)

These additional properties are that, firstly, every edge is incident with precisely two points and, secondly, that no two edges are incident with the same pair of points. We shall prove in Props. 7.2.2 and 7.2.3 that, up to isomorphism, there is at most one factorization of the respective kind. Clearly, one cannot expect an existence theorem: Consider, e.g., the element $\Gamma = \begin{pmatrix} 0 & 1 & 0 & 1 \\ 1 & 0 & 1 & 1 \\ 0 & 1 & 0 & 1 \\ 1 & 1 & 1 & 0 \end{pmatrix}$ in the relation algebra of Boolean $4\times 4$-matrices. Although $\Gamma$ is irreflexive and symmetric, no factorization of the form $\Gamma = \overline{I} \sqcap M^{\mathsf{T}}M$ holds in that relation algebra. Five edges are necessary, and there does indeed exist a factorization in the relation algebra formed by the Boolean matrices of sizes $(4 \times 4)$, $(4 \times 5)$, $(5 \times 4)$, and $(5 \times 5)$. We confine ourselves to the uniqueness problem and will not study possible factorizations in an extended relation algebra, because we want to perform all our calculations in some fixed algebra.

When drawing a simple graph with its given adjacency, one automatically produces a factorization, while assuming some sufficiently large relation algebra in which any two points can be joined by an edge.

There are other possibilities, too, for associating with a given adjacency a hypergraph, which will in essence be unique, provided certain additional conditions are met. One of these possibilities will involve the following notions.


**5.5.1 Definition.** Let an incidence $M$ be given together with the corresponding adjacency $\Gamma := \overline{I} \sqcap M^{\mathsf{T}}M$ and edge-adjacency $\mathsf{K} := \overline{I} \sqcap MM^{\mathsf{T}}$. We call

| | | |
|---|---|---|
| $M$ **conformal** | $:\Longleftrightarrow$ | All vectors $y$ with $yy^{\mathsf{T}} \sqcap \overline{I} \subset \Gamma$ satisfy $\overline{M}y \neq L$. |
| | $\Longleftrightarrow$ | Every set of pairwise adjacent points is contained in an edge. |
| $M$ **Helly-type** | $:\Longleftrightarrow$ | All vectors $p$ with $pp^{\mathsf{T}} \sqcap \overline{I} \subset \mathsf{K}$ satisfy $\overline{M}^{\mathsf{T}}p \neq L$. |
| | $\Longleftrightarrow$ | Every set of pairwise adjacent edges is incident with a common point. $\qquad\qquad\square$ |

So one has conformity when every clique is contained in a hyperedge. Clearly, $M$ has the Helly property if and only if $M^{\mathsf{T}}$ is conformal. In Fig. 5.5.2 the incidences on the left are conformal, those on the right are of Helly type.

There are two important examples for Helly-type incidences: systems of intervals of the real number line, and the ordering between partial functions.

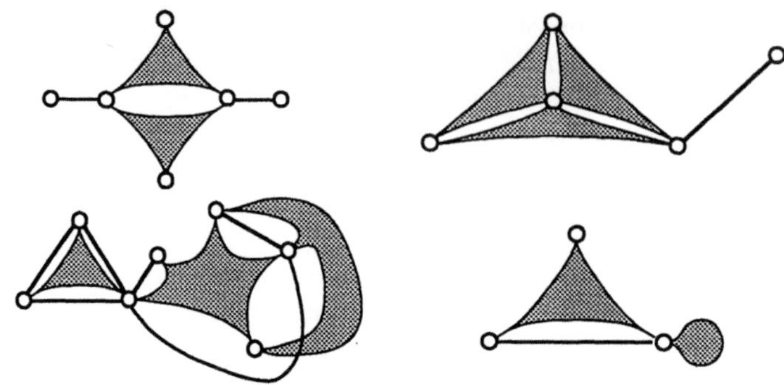

**Fig. 5.5.2** Conformal and Helly-type relations

**5.5.2 Example.** Let $V := \mathbb{R}$ be the point set of a hypergraph, and let $P$ be a system of (open or closed) intervals of $\mathbb{R}$, considered as hyperedges. Each interval is defined to be *incident* with all of its points. Two intervals are then *edge-adjacent* if their intersection is nonvoid. Clearly, any set of intervals, such that any two of them have a nonvoid intersection, has a nonvoid intersection as well. Therefore, this incidence is of Helly type. For higher dimensions this is no longer true, as can be seen from the plane sets $U_1, U_2, U_3$ in Fig. 5.5.3. Occasionally, **interval graphs** are mentioned in the literature—these are simple graphs whose points are intervals and where adjacency is the above edge-adjacency.

$\square$

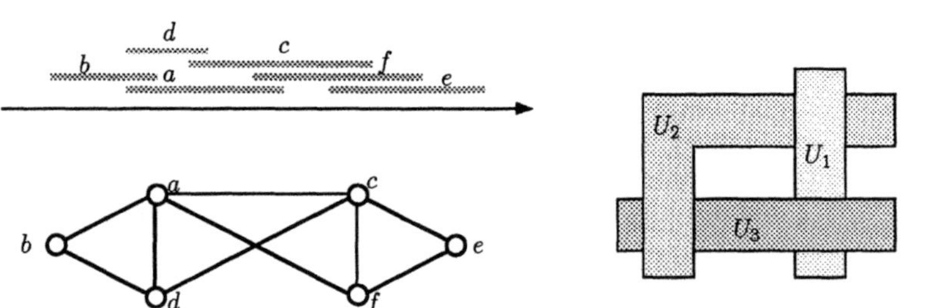

**Fig. 5.5.3** Interval system and interval graph; system of open sets

The Helly property is also of interest in the particular case when $M$ is a (homogeneous) order relation. Two elements are then edge-adjacent if there exists a common upper bound. We give an example:

**5.5.3 Example.** Consider the set of all univalent relations (i.e., partially defined functions) from the set $X$ into the set $Y$. This set will here again serve as both the point set and the set of edges. It is ordered by inclusion, and we take the ordering for the incidence $M$. So one univalent relation $E$ is "incident" with another univalent relation $F$, if $E \subset F$. Moreover, $E$ is "edge-adjacent" to $F$, if there is a univalent $G$ such that $E \subset G$ and $F \subset G$. In other words, two univalent relations are edge-adjacent if they take the same values as functions wherever they are both defined. The following holds:

The ordering of partial functions by inclusion is of Helly type. Indeed, let $E_\iota, \iota \in J$, be a system of pairwise edge-adjacent partial functions. Then $G :=$ $\sup\{E_\iota \mid \iota \in J\}$ is univalent because

$$G^\mathsf{T} G = \sup\{E_\iota^\mathsf{T} E_\chi \mid \iota, \chi \in J\} \subset \sup\{G_{\iota\chi}^\mathsf{T} G_{\iota\chi} \mid \iota, \chi \in J\} \subset I,$$

where $G_{\iota\chi}$ is a partial function containing $E_\iota$ and $E_\chi$, and which exists by edge-adjacency. Clearly, one also has $E_\iota \subset G$ for all $\iota \in J$. Hence every $E_\iota$ is incident with $G$. $\qquad\square$

Next we apply the point axiom in order to prove two identities for conformal incidences.

**5.5.4 Proposition.** For an essential and conformal incidence $M$ the following identities hold:

$$M\overline{M^\mathsf{T} M} = \overline{M} \quad \text{and} \quad M\overline{\Gamma} = L.$$

**Proof:** Only "$\supset$" of the first equation is nontrivial. Let a point $x$ and an edge $p$ with $px^\mathsf{T} \subset \overline{M} \sqcap \overline{M\overline{M^\mathsf{T} M}}$ be given. Then $Mx \subset \overline{p}$. On the other hand, $L = \overline{Mx} \sqcup \overline{p} = \overline{Mx} \sqcup \overline{Ip} \subset \overline{Mx} \sqcup \overline{MM^\mathsf{T} p} = \overline{M}(x \sqcup M^\mathsf{T} p)$, since $M$ is essential and $x^\mathsf{T}$, as well as $p^\mathsf{T}$, are mappings. This results in a contradiction, which we show by defining $y := x \sqcup M^\mathsf{T} p$ and by proving

$$\begin{aligned}
yy^\mathsf{T} \sqcap \overline{I} &= (xx^\mathsf{T} \sqcup xp^\mathsf{T} M \sqcup M^\mathsf{T} px^\mathsf{T} \sqcup M^\mathsf{T} pp^\mathsf{T} M) \sqcap \overline{I} \\
&\subset (I \sqcup \overline{\overline{M^\mathsf{T} M} M^\mathsf{T} M} \sqcup M^\mathsf{T} \overline{M\overline{M^\mathsf{T} M}} \sqcup M^\mathsf{T} M) \sqcap \overline{I} \\
&\subset (I \sqcup M^\mathsf{T} M \sqcup M^\mathsf{T} M \sqcup M^\mathsf{T} M) \sqcap \overline{I} = \Gamma.
\end{aligned}$$

In the case of a conformal $M$, it follows that $\overline{M}y \neq L$. The second equation is proved by $M\overline{\Gamma} = M\overline{\overline{I} \sqcap M^\mathsf{T} M} = MI \sqcup M\overline{M^\mathsf{T} M} = M \sqcup \overline{M} = L$. $\qquad\square$

We shall prove in Proposition 7.2.3 that, for every irreflexive and symmetric relation $\Gamma$, there is up to isomorphism at most one essential, conformal hypergraph having $\Gamma$ as its adjacency. It follows that for an essential incidence there can be either no void edge, or just one such void edge and no other edges; for a conformal incidence there can be no free points. So each clique, in particular a maximal one, is contained in some hyperedge, but no hyperedge is contained in another. Thus there is a one-to-one correspondence between maximal cliques and hyperedges.

**Exercise**

5.5.1 Prove that for $M := A \sqcup E$ and $M' := A' \sqcup E'$ with disjoint functions $A, E$ or $A', E'$, respectively, the relation $\mathsf{syq}(M^\mathsf{T}, M'^\mathsf{T})$ is total, provided $\overline{I} \sqcap M^\mathsf{T} M = \Gamma = \overline{I} \sqcap M'^\mathsf{T} M'$.

## 5.6 References

HELLY E: *Über Mengen konvexer Körper mit gemeinschaftlichen Punkten.* Jahresber. Deutsch. Math.–Verein. **32** (1923) 175–176.

# 6. Reachability

In this chapter we focus once more on the transitive closure of relations. In view of the applications we employ graph-theoretical language and speak of the reachability relation. In Sect. 6.1 we define paths and distinguish between paths in 1-graphs and paths in directed graphs. In the first case one considers points, in the second, points and arcs. We define reachability in terms of paths, and discuss rooted graphs and strong connectedness. In Sect. 6.2 we discuss a similar close relationship between joinability, expressed in terms of chains, and connectedness.

In Sect. 6.3 we investigate terminal points and the reachability of these, which will lead to relation-algebraic forms of deeper concepts, such as transfinite induction. We shall return to this topic in Sect. 10.1 where the correctness of programs is investigated, and the difference between determinism and nondeterminism is analyzed. Section 6.4 deals with confluence and Church-Rosser theorems. In Sect. 6.5 refinability of paths is discussed which is the starting point for distinguishing discrete and nondiscrete relations in terms of some fixedpoint property.

## 6.1 Paths and Circuits

We now investigate the possibility of getting from one point to some other while successively tracing related points. A sequence of points traversed in this fashion is customarily called a path. There is an obvious path concept for 1-graphs; the corresponding concept for directed graphs is based on the sequence of arcs traversed and will be discussed afterwards.

**6.1.1 Definition** (*Path as a sequence of points*). Given the associated relation $B$ of a graph and a sequence $w = (x_0, \ldots, x_h)$, $h \geq 0$, of points, we call

i)  $\quad w$ a **path** from $x_0$ to $x_h$ $\quad :\Longleftrightarrow\quad x_{i-1} \subset Bx_i$ for $i = 1, \ldots, h$,

ii) $\quad w$ a **circuit** in $x_0$ $\qquad\quad :\Longleftrightarrow\quad w$ is a path from $x_0$ to $x_h$, where $x_h = x_0$ and $h \geq 1$.

The **length** of the path $w$ is defined as $|w| := h$. A sequence $(x_0)$ is called the **void path** in $x_0$. A sequence $(x_0, x_1, \ldots)$ of infinite length with $x_{i-1} \subset Bx_i$ for $i \geq 1$ will be termed a **path of infinite length starting from** $x_0$. $\qquad\square$

Point $x_0$ is also called the **starting point** of path $w$, and $x_h$ the **endpoint**. Paths of length 0 are the void paths. Paths of length 1 are sequences consisting of the starting and the endpoint of one arc. Circuits are nonvoid paths whose starting point and endpoint coincide. If $w = (x_0, \ldots, x_h)$ is a circuit passing through $x_0$, then $(x_i, \ldots, x_h = x_0, \ldots, x_i)$, $0 < i < h$, is a *different* circuit passing through $x_i$. These circuits, however, are often identified, i.e., circuits are then considered without their "suspension point" $x_i$.

Through any point $x$ there is the void path $(x)$ of length 0. Through loop-carrying points $x$ there are circuits $(x,x)$, $(x,x,x)$, ... of lengths 1, 2, .... If these trivial circuits through some loop-carrying point are excluded, then one speaks of *proper* circuits. Figure 6.1.1 contains a number of paths and circuits in some graph.

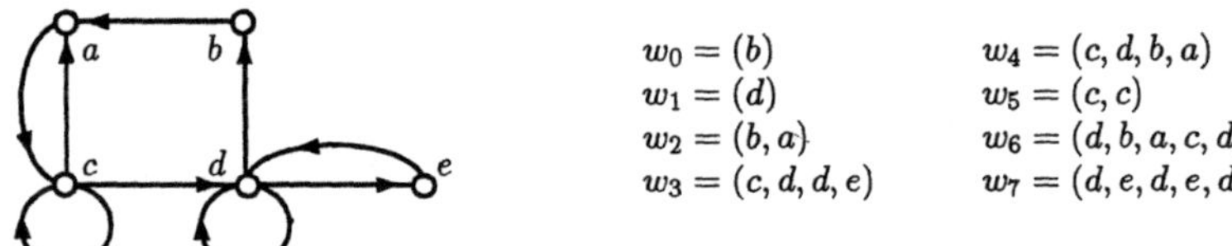

$$w_0 = (b) \qquad\qquad w_4 = (c, d, b, a)$$
$$w_1 = (d) \qquad\qquad w_5 = (c, c)$$
$$w_2 = (b, a) \qquad\qquad w_6 = (d, b, a, c, d)$$
$$w_3 = (c, d, d, e) \qquad\qquad w_7 = (d, e, d, e, d)$$

**Fig. 6.1.1** Paths and circuits

Occasionally, one is interested in paths tracing no point more than once.

**6.1.2 Definition.** A path or a circuit of a graph is called **elementary**, if no point occurs more than once in the sequence. (Of course, in the case of paths that are circuits we do *not* count the starting point as being traversed twice.)

□

Paths $w_0, w_1, w_2, w_4$ and circuits $w_5, w_6$ in Fig. 6.1.1 are elementary; however, path $w_3$ and circuit $w_7$ are not.

The transitive closure of the associated relation $B$ of a graph is closely related to the existence of paths between points. If $(x_0, \ldots, x_h)$ is a path from $x = x_0$ to $y = x_h$, we have $x \subset B^h y$. In the following proposition we consider the converse.

**6.1.3 Proposition.** Let $x$ and $y$ be two arbitrary points of a graph with associated relation $B$, and let $h \geq 0$ be a natural number. Then

i) $\qquad x \subset B^h y \qquad \Longleftrightarrow \qquad$ There exists a path from $x$ to $y$ of length $h$.

ii) $\qquad x \subset B^* y \qquad \Longleftrightarrow \qquad$ There exists a path from $x$ to $y$.

iii) $\qquad x \subset B^+ x \qquad \Longleftrightarrow \qquad$ There exists a circuit through $x$.

iv) $\qquad B^+ \subset \overline{I} \qquad \Longleftrightarrow \qquad$ There are no circuits.

(For a finite graph we have in addition:

$\qquad B^+ \subset \overline{I} \qquad \Longleftrightarrow \qquad$ There is a natural number $m$ with $B^m = O$.)

**Proof:**[1] i) From Prop. 2.4.5.ii we have $x \subset B^0 y = y \Longleftrightarrow x = y$, and $(x)$ is, of course, a path of length 0 from $x$ to $y$. We now proceed by induction from $h - 1$ to $h$: From $x \subset B^h y = B^{h-1} B y$, using Prop. 2.4.8, follows the existence of a point $z$ with $x \subset B^{h-1} z$ and $z \subset B y$. By the induction hypothesis there

---

[1] In order to prove (i) and (iv) we use Proposition 2.4.8 which yields intermediate points of the path. So the point axiom is being used here.

exists a path $(x=x_0, x_1, \ldots, x_{h-1}=z)$ of length $h-1$ from $x$ to $z$. Therefore, $(x=x_0, \ldots, x_{h-1}=z, y)$ is a path of length $h$ from $x$ to $y$.

ii) $x \subset B^* y \iff xy^\mathsf{T} \subset B^*$ as a consequence of Prop. 2.4.4. In Exercise 2.4.2 it is shown that $xy^\mathsf{T}$ is an atom. If an atom is contained in the union $B^* = I \sqcup B \sqcup B^2 \sqcup \ldots$, then it is contained in one of the parts: $xy^\mathsf{T} \subset B^f$. This is equivalent to $x \subset B^f y$, so that a path (of length $f$) exists from $x$ to $y$. Conversely, we obtain from (i) that $x \subset B^h y$ for some $h$, so that $x \subset B^* y$.

iii) In $B^+$ only circuits of length $\geq 1$ are represented; everything else follows from (ii).

iv) Suppose we have $B^+ \subset \bar{I}$, but a circuit passes through point $x$. Then $xx^\mathsf{T} \subset B^+ \subset \bar{I}$ which contradicts injectivity of $x$. In order to prove "$\Longleftarrow$", assume $B^+ \not\subset \bar{I}$, and choose a pair of points $x, y$ with $xy^\mathsf{T} \subset B^+ \sqcap I$, which is possible by Prop. 2.4.6. Then, on the one hand, $x \subset (B^+ \sqcap I)y \subset y$ which by Prop. 2.4.5 implies $x = y$. On the other hand, $x \subset B^+ y$, i.e., there exists a path of length $\geq 1$ from $x$ to $x$, which is a circuit. The result for finite graphs is left as Exercise 6.1.1. $\qquad\square$

Because of (iv), we call a graph **circuit-free** (or **acyclic**), if $B^+ \subset \bar{I}$; there are no proper circuits when $B^+$ is antisymmetric, i.e., when $B^+ \sqcap B^{+\mathsf{T}} \subset I$. If $B$ is transitive, then circuit-free means irreflexive.

After having defined the set $Bx$ of predecessors of $x$ in Definition 2.4.10, we now introduce some "genealogical" terminology.

**6.1.4 Definition.** If an associated relation $B$ of a graph $G$ is given together with two points $x, y$, we call

| | | |
|---|---|---|
| i) | $y$ **reachable** from $x$ | $:\iff xy^\mathsf{T} \subset B^*$, |
| ii) | $B^*$ the **reachability** within the graph $G$, | |
| iii) | $B^+ x$ the set of **ancestors** of $x$, | |
| iv) | $B^{+\mathsf{T}} x$ the set of **descendants** of $x$. | $\square$ |

By Prop. 6.1.3.ii, relation $B^*$ holds between two points $x$ and $y$ if and only if there exists a path from $x$ to $y$ of finite length, i.e. if $y$ is reachable from $x$ after finitely many steps.

So far, we have considered 1-graphs which are, in essence, just homogeneous relations. The situation becomes more involved with directed graphs since there is the set of arcs beside the point set. So one has to deal with heterogeneous relations. If there are two parallel arcs $p, q$ from $x$ to $y$, then one has to distinguish between paths leading from $x$ to $y$ via $p$ or via $q$. So the arcs traced by the path are to be included in the sequence as additional information.

**6.1.5 Definition** (*Path as a point-arc-sequence*). Let $A, E$ be given as the outgoing incidence and the ingoing incidence of a directed graph. The sequence $(x_0)$ is the **void path** at point $x_0$. An alternating sequence of points and arcs

$w = (x_0, p_1, \ldots, p_h, x_h)$, $h > 0$, with $x_0 = A^\mathsf{T} p_1$ (starting point) and $x_h = E^\mathsf{T} p_h$ (endpoint) is called a **path** from $x_0$ to $x_h$, provided $x_i = A^\mathsf{T} p_{i+1} = E^\mathsf{T} p_i$ for $i = 1, \ldots, h-1$. A path is called **simple**, if no arc occurs more than once in the sequence.                                          $\square$

Thus, the arcs of a path necessarily have starting points and endpoints and cannot be partial. Whether a path is meant to be a sequence of points, or of points and arcs, should be clear from the context. The familiar notions from the homogeneous case apply accordingly. The number $h$ is called the **length** of $w$. A sequence $(x_0, p_1, x_1, p_2, \ldots)$ with $x_{i-1} = A^\mathsf{T} p_i$ and $E^\mathsf{T} p_i = x_i$ for $i = 1, 2, \ldots$ is called a **path of infinite length** from $x_0$. A **circuit** through $x_0$ is now a nonvoid path, in the sense of a point-arc-sequence, from $x_0$ to $x_0$.

We next ask whether the entire point-arc-sequence has to be given, or whether less information suffices. First we ask whether the subsequence of arcs is sufficient. If the void path is excluded, then the sequence $(p_1, \ldots, p_h)$, $h > 0$, already contains the whole information because it determines the intermediate points $x_i$ $(1 \le i \le h-1)$, as well as the starting point $x_0$, and the endpoint $x_h$. The void arc sequence $()$, however, does not determine the void path completely because it does not indicate the starting point or the endpoint.

Let us also ask whether the subsequence of points retains all the information. The sequence $(x_0, \ldots, x_h)$ is a path or circuit in the sense of Definition 6.1.1. So the property "elementary" is defined for a path as a point-arc-sequence, too. A path in the sense of Definition 6.1.5 is completely determined by a point sequence only when no parallel arcs occur. An elementary path or circuit is, of course, also simple.

## Strong Connectedness

Of particular interest are pairs of points in a graph that are mutually reachable from one another.

**6.1.6 Definition.** Given a graph $G$ with associated relation $B$, we call

i)  $G$ **strongly connected** $\quad :\Longleftrightarrow \quad B^* = L \quad \Longleftrightarrow \quad I \subset B^+$
$\Longleftrightarrow$ Every point is reachable from any other point.

ii)  The equivalence classes of points with respect to the equivalence $B^* \sqcap B^{*^\mathsf{T}}$ are called the **strongly connected components** of $G$.                    $\square$

Thus, a strongly connected component of $G$ contains a maximum number of points that can all be reached from one another. If there is just one such component, then $G$ itself is called strongly connected. By Prop. 3.1.3, $B^* \sqcap B^{*\mathsf{T}}$ is the greatest equivalence relation contained in $B^*$. It serves for the following characterization:

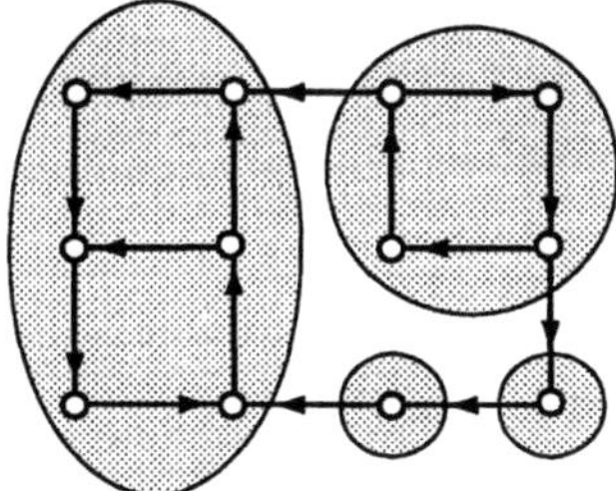

**Fig. 6.1.2** Strongly connected components

**6.1.7 Proposition.** Let $B$ be the associated relation, and let $x$, $y$ be two points of a graph $G$.

i)   $G$ is strongly connected $\iff \overline{I} \subset B^+ \sqcap B^{+\mathsf{T}}$
$\iff$ Any two distinct points lie on a circuit.

ii)  $x, y$ lie in different strongly connected components
$\iff x \neq y$, and there is no circuit through $x, y$.

**Proof:** i) $L = B^* = I \cup B^+ \iff \overline{I} \subset B^+ \iff \overline{I} \subset B^{+\mathsf{T}}$. A formal version of (ii) is $xy^\mathsf{T} \subset \overline{B^* \sqcap B^{*\mathsf{T}}} \iff [\{ xy^\mathsf{T} \not\subset B^+ \text{ or } xy^\mathsf{T} \not\subset B^{+\mathsf{T}} \} \text{ and } xy^\mathsf{T} \subset \overline{I}]$. This is true since $xy^\mathsf{T}$ is an atom.   $\square$

## Rooted Graphs and Rooted Trees

A central role, especially in computer science, is played by graphs in which a particular point is distinguished as an "entry point" from which any other point can be reached. Such graphs occur in connection with flowcharts, finite automata, and heuristic search algorithms, they can be found in referential structures (binary trees, data bases, networks), and they come up as coding trees and syntax diagrams. Here, we investigate the resulting implications for the connectedness of graphs.

**6.1.8 Definition.** Let a graph $G$ be given together with the associated relation $B$, and let $a$ be a point of its point set. We call

$$a \text{ **root** of } G \quad :\iff \quad aL \subset B^* \quad \iff \quad L \subset a^\mathsf{T} B^* \quad \iff \quad a \subset \overline{\overline{B^* L}}$$
$$\iff \quad \text{All points can be reached from } a.$$

A root with respect to $B^\mathsf{T}$ is sometimes called a **final point** of $B$.   $\square$

The first statement is equivalent to the second by Proposition 2.4.4.i, and to the third by the Schröder equivalence. The third statement says that there is *no* point that can*not* be reached from $a$. So roots can be described by lower, as well as upper, bounds.

According to Fig. 6.1.5, a graph may well have several roots; in a strongly connected graph even every point is a root. But in many applications, typically one such root is distinguished as an entry point. This is the reason for the following basic definition.

**6.1.9 Definition.** A pair $W = (G, a)$ consisting of a graph $G$ and a selected root $a$ of $G$ is called a **rooted graph**.                              □

When we later speak of homomorphisms $\Theta: G \longrightarrow G'$ between rooted graphs, then it is understood that they map the distinguished root into the distinguished root, i.e. that $\Theta^{\mathsf{T}} a = a'$.

In Fig. 6.1.3 four rooted graphs are drawn in ways customary to the various fields of application, namely as a finite automaton, a flowchart, a syntax diagram, and a list structure or tree-like structure.

Among the rooted graphs in Fig. 6.1.3, those of type (d) are particularly simple: There are no parallel arcs and no circuits, and each point, except for the root, has precisely one predecessor. We sum up these characteristics in the following definition.

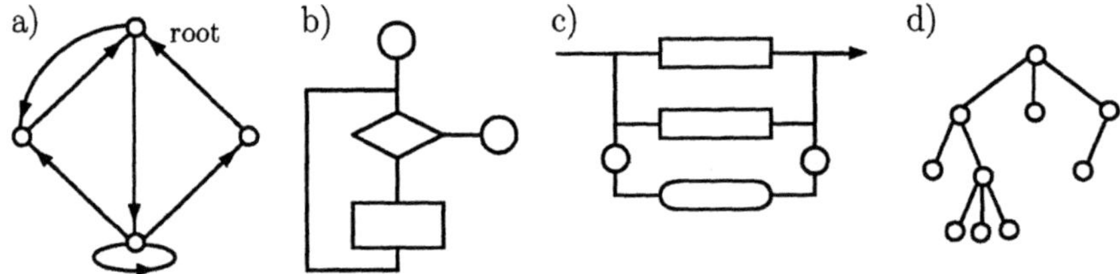



**Fig. 6.1.3** Instances of rooted graphs

**6.1.10 Definition.** An injective and circuit-free rooted graph is called a **rooted tree** (or **arborescence**). Thus, a rooted graph with associated relation $B$ is a rooted tree, provided $BB^{\mathsf{T}} \subset I$ and $B^+ \subset \bar{I}$.                              □

Clearly, a rooted tree has just one root since a second would give rise to a circuit.

An endpoint in a rooted tree is called a **leaf**. The other points are called **inner nodes**. Proposition 6.1.11 contains a number of characterizations of rooted trees which, in part, hold for infinite order, too; the proof is not relation-algebraic.

**6.1.11 Proposition.** For a rooted graph $W = (G, a)$ with associated relation $B$, the following are equivalent:

i)     $W$ is a rooted tree.

ii)    $d_p(x) \leq 1$ for all points of $x \in W$, and $d_p(a) = 0$.

iii)   For each point $x$ of $W$, there is precisely one path from root $a$ to $x$.

Furthermore, if $G$ has finite order $n$, then these are equivalent to:

iv)    There are precisely $n - 1$ arcs in $G$.

**Proof:** i) $\Longrightarrow$ ii) Since $BB^\mathsf{T} \subset I$, a point has at most one predecessor. Of course, the root $a$ cannot have a predecessor since all the points are reachable from the root and such a predecessor would yield a circuit contradicting our assumption $B^+ \subset \overline{I}$.

ii) $\Longrightarrow$ iii) In a rooted graph, there is by definition at least one path from $a$ to any other point. Assume some point $y$ may be reached along two different paths from $a$. These paths meet in $y$ at the latest, possibly earlier; let this occur in point $z$. If $z \neq a$, then this means that $z$ has predecessor degree $\geq 2$. If $z = a$, the difference of the two paths consists in a circuit through $a$. So $a$ would have predecessor degree $\geq 1$.

iii) $\Longrightarrow$ i) Suppose there are two predecessors of some point. Since in a rooted graph there exist paths from the root to both of these predecessors, there would be two different paths to that point. Should there exist a circuit somewhere, one could find infinitely many paths to any of the points of the circuit, contradicting our hypothesis (iii).

i), ii), iii) $\Longrightarrow$ iv) The proposition is valid for $n = 1$. A rooted tree with $n = 2$ has, indeed, precisely one arc. Now we proceed by induction: Let a rooted tree with $(n + 1)$ points be given. Since it is finite and circuit-free, proceeding from $a$, one will finally reach a leaf $x$. This leaf $x$ and the arc leading to it are now cut from its predecessor $y$ that is uniquely determined. The rest of the graph is a rooted tree with $n$ points that has $n - 1$ arcs by induction hypothesis. Now the leaf $x$ and the arc leading to it are reattached to $y$. So we obtain $n + 1$ points and $n$ arcs.

iv) $\Longrightarrow$ i) As any of the $n$ points is reachable from $a$, the set of endpoints of the $n-1$ arcs consists precisely of all points $\neq a$. So predecessors are determined uniquely, and there is no room left for a circuit through $a$. Nor can there exist a circuit outside $a$. For it would be possible to choose a point $x$ on such a circuit and return to the root $a$. Going back along this path, there would be two arcs ending in the last point before leaving the circuit. $\qquad\square$

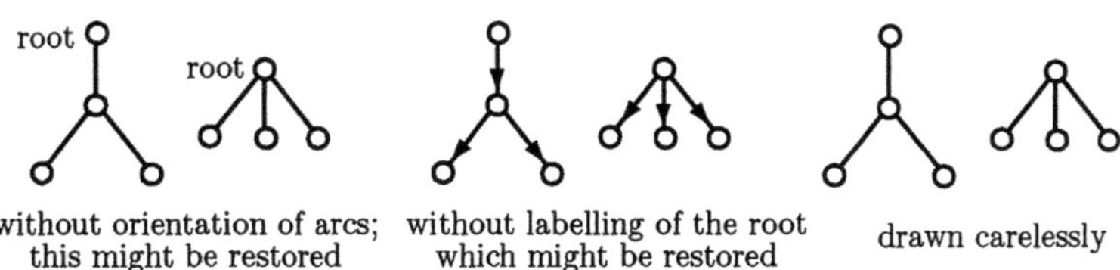


**Fig. 6.1.4** Two different rooted trees

When drawing rooted trees, one usually puts the root on top, in contrast to nature, so as not having to estimate the height of the tree in advance. The orientation of the arcs is often omitted (unfortunately); it can, of course, be restored easily, if the root is given. But consider, for example, the two rooted trees in Fig. 6.1.4: they can easily be confused, if one ignores their orientations as well as their roots, and regards them as undirected trees.

There is a special kind of connectedness closely resembling the situation found in rooted graphs:

**6.1.12 Definition.** We call a graph

$$G \text{ \textbf{quasi-strongly connected}} \quad :\Longleftrightarrow \quad L \subset B^{\mathsf{T}}{}^{*}B^{*}$$

$$\Longleftrightarrow \quad \text{For any two points } x \text{ and } y \text{ there exists a point } z$$
from which $x$ as well as $y$ may be reached.    □

By Definition 6.1.8, $\overline{\overline{B^{*}L}}$ gives the set of all roots of the graph. The graph $G_2$ in Fig. 6.1.5 has no root since $\overline{\overline{B^{*}L}} = O$. On the other hand, graph $G_1$ has the roots $x_1, x_2, x_3$; hence it is clearly quasi-strongly connected because any point can be reached from a root, and the same applies to any pair of points $x, y$. In $G_2$, points $y_3$ and $y_5$ have no common ancestors.

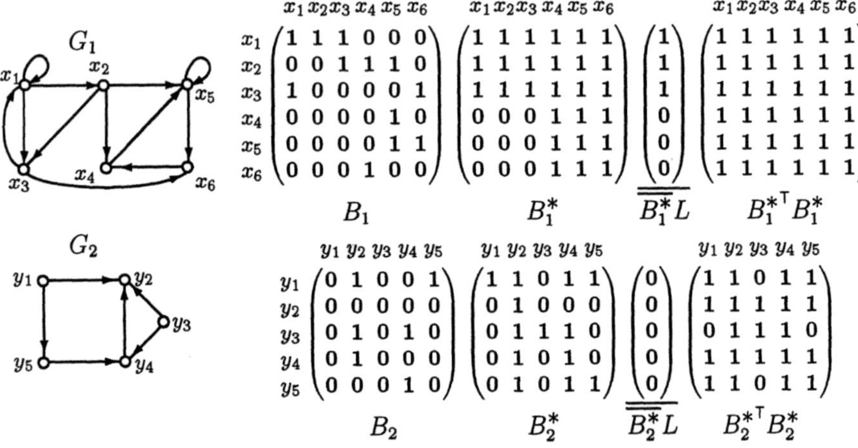

$$
\begin{array}{c}
B_1 \\[4pt]
\begin{array}{c|cccccc}
 & x_1 & x_2 & x_3 & x_4 & x_5 & x_6 \\
x_1 & 1 & 1 & 1 & 0 & 0 & 0 \\
x_2 & 0 & 0 & 1 & 1 & 1 & 0 \\
x_3 & 1 & 0 & 0 & 0 & 0 & 1 \\
x_4 & 0 & 0 & 0 & 0 & 1 & 0 \\
x_5 & 0 & 0 & 0 & 0 & 1 & 1 \\
x_6 & 0 & 0 & 0 & 1 & 0 & 0
\end{array}
\end{array}
\qquad
\begin{array}{c}
B_1^{*} \\[4pt]
\begin{array}{cccccc}
1 & 1 & 1 & 1 & 1 & 1 \\
1 & 1 & 1 & 1 & 1 & 1 \\
1 & 1 & 1 & 1 & 1 & 1 \\
0 & 0 & 0 & 1 & 1 & 1 \\
0 & 0 & 0 & 1 & 1 & 1 \\
0 & 0 & 0 & 1 & 1 & 1
\end{array}
\end{array}
\qquad
\begin{array}{c}
\overline{\overline{B_1^{*}L}} \\[4pt]
\begin{array}{c}
1 \\ 1 \\ 1 \\ 0 \\ 0 \\ 0
\end{array}
\end{array}
\qquad
\begin{array}{c}
B_1^{*\mathsf{T}}B_1^{*} \\[4pt]
\begin{array}{cccccc}
1 & 1 & 1 & 1 & 1 & 1 \\
1 & 1 & 1 & 1 & 1 & 1 \\
1 & 1 & 1 & 1 & 1 & 1 \\
1 & 1 & 1 & 1 & 1 & 1 \\
1 & 1 & 1 & 1 & 1 & 1 \\
1 & 1 & 1 & 1 & 1 & 1
\end{array}
\end{array}
$$

$$
\begin{array}{c}
B_2 \\[4pt]
\begin{array}{c|ccccc}
 & y_1 & y_2 & y_3 & y_4 & y_5 \\
y_1 & 0 & 1 & 0 & 0 & 1 \\
y_2 & 0 & 0 & 0 & 0 & 0 \\
y_3 & 0 & 1 & 0 & 1 & 0 \\
y_4 & 0 & 1 & 0 & 0 & 0 \\
y_5 & 0 & 0 & 0 & 1 & 0
\end{array}
\end{array}
\qquad
\begin{array}{c}
B_2^{*} \\[4pt]
\begin{array}{ccccc}
1 & 1 & 0 & 1 & 1 \\
0 & 1 & 0 & 0 & 0 \\
0 & 1 & 1 & 1 & 0 \\
0 & 1 & 0 & 1 & 0 \\
0 & 1 & 0 & 1 & 1
\end{array}
\end{array}
\qquad
\begin{array}{c}
\overline{\overline{B_2^{*}L}} \\[4pt]
\begin{array}{c}
0 \\ 0 \\ 0 \\ 0 \\ 0
\end{array}
\end{array}
\qquad
\begin{array}{c}
B_2^{*\mathsf{T}}B_2^{*} \\[4pt]
\begin{array}{ccccc}
1 & 1 & 0 & 1 & 1 \\
1 & 1 & 1 & 1 & 1 \\
0 & 1 & 1 & 1 & 0 \\
1 & 1 & 1 & 1 & 1 \\
1 & 1 & 0 & 1 & 1
\end{array}
\end{array}
$$

**Fig. 6.1.5** Roots and quasi-strong connectedness

It might seem that a graph is quasi-strongly connected, if and only if it has a root; things are not that simple, however, as can be seen from Fig. 6.1.6.

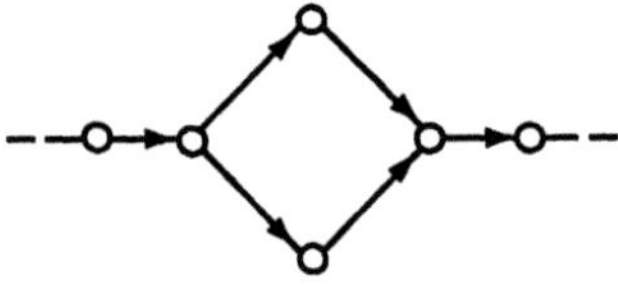

**Fig. 6.1.6** Quasi-strongly connected graph without root

Next we prove that roots imply quasi-strong connectedness, and that the converse holds for finite graphs. A similar result will be given in Proposition 6.3.7.

**6.1.13 Proposition.** i)  The following is valid for any graph:

$$\text{quasi-strongly connected} \quad \Longleftarrow \quad \text{There exists a root.}$$

ii)   For finite graphs we have in addition:

$$\begin{matrix}\text{quasi-strongly} \\ \text{connected}\end{matrix} \quad \Longleftrightarrow \quad \text{There exists a root.}$$

**Proof:** i) $B^{\mathsf{T}*}B^* \supset (aL)^{\mathsf{T}}aL = La^{\mathsf{T}}aL = L$, provided $a$ is the root. ii) It remains to show "$\Longrightarrow$". Let $V = \{\,x_1,\ldots,x_n\,\}$ be the point set. For $x_1$ and $x_2$ find a $z_1 \in V$ from which both $x_1$ and $x_2$ may be reached; then do the same for $z_1$ and $x_3$ so as to obtain $z_2 \in V$, etc. This process eventually terminates with $z_{n-1} \in V$ which is a root by construction. One should be aware, however, that this proof is by no means relation-algebraic.     □

## Terminating Reachability

For a number of questions it is of importance to know whether and when some point is last reachable, i.e., it is reachable and there are no more points reachable beyond it. According to Definition 5.2.1, the terminal points are described by $\overline{BL}$. We are interested only in reached points that are terminal, so we have to intersect $B^*$ with the transpose $\overline{BL}^{\mathsf{T}}$.

**6.1.14 Definition.** The relation $C := B^* \sqcap \overline{BL}^{\mathsf{T}}$ is called the **terminating reachability** of the relation $B$.     □

A simple example is sketched in Fig. 6.1.7. Note that $b$ and $c$ are not related by relation $C$ because there is a loop at $c$.

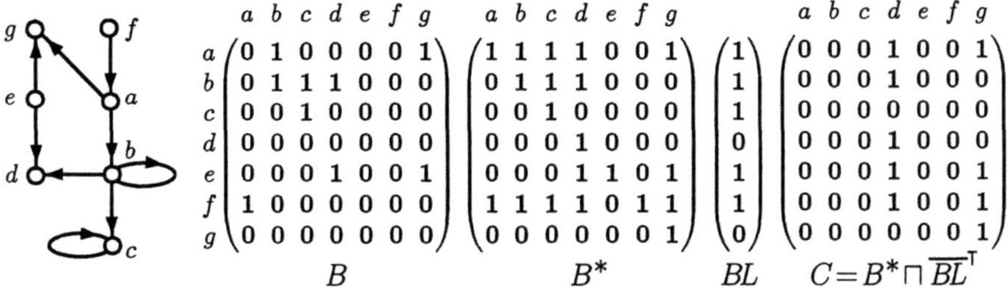

$$
B = \bordermatrix{
 & a & b & c & d & e & f & g \cr
a & 0 & 1 & 0 & 0 & 0 & 0 & 1 \cr
b & 0 & 1 & 1 & 1 & 0 & 0 & 0 \cr
c & 0 & 0 & 1 & 0 & 0 & 0 & 0 \cr
d & 0 & 0 & 0 & 0 & 0 & 0 & 0 \cr
e & 0 & 0 & 0 & 1 & 0 & 0 & 1 \cr
f & 1 & 0 & 0 & 0 & 0 & 0 & 0 \cr
g & 0 & 0 & 0 & 0 & 0 & 0 & 0 \cr}
\qquad
B^* = \bordermatrix{
 & a & b & c & d & e & f & g \cr
 & 1 & 1 & 1 & 1 & 0 & 0 & 1 \cr
 & 0 & 1 & 1 & 1 & 0 & 0 & 0 \cr
 & 0 & 0 & 1 & 0 & 0 & 0 & 0 \cr
 & 0 & 0 & 0 & 1 & 0 & 0 & 0 \cr
 & 0 & 0 & 0 & 1 & 1 & 0 & 1 \cr
 & 1 & 1 & 1 & 1 & 0 & 1 & 1 \cr
 & 0 & 0 & 0 & 0 & 0 & 0 & 1 \cr}
\qquad
BL = \begin{pmatrix} 1 \\ 1 \\ 1 \\ 0 \\ 1 \\ 1 \\ 0 \end{pmatrix}
\qquad
C = B^* \sqcap \overline{BL}^{\mathsf{T}} = \bordermatrix{
 & a & b & c & d & e & f & g \cr
 & 0 & 0 & 0 & 1 & 0 & 0 & 1 \cr
 & 0 & 0 & 0 & 1 & 0 & 0 & 0 \cr
 & 0 & 0 & 0 & 0 & 0 & 0 & 0 \cr
 & 0 & 0 & 0 & 1 & 0 & 0 & 0 \cr
 & 0 & 0 & 0 & 1 & 0 & 0 & 1 \cr
 & 0 & 0 & 0 & 1 & 0 & 0 & 1 \cr
 & 0 & 0 & 0 & 0 & 0 & 0 & 1 \cr}
$$

**Fig. 6.1.7** Terminating reachability

We now prove some identities for the terminating reachability $C$ and interpret these afterwards in more concrete terms.

**6.1.15 Proposition.** For a graph $G$ or a relation $B$ with reachability $B^*$ and terminating reachability $C := B^* \sqcap \overline{BL}^{\mathsf{T}}$ the following identities hold:

i) $$B^*C = C, \qquad\qquad C^2 = C;$$

ii) $$BC = C \sqcap \overline{I}, \qquad \mathrm{mup}\,(BC) = \mathrm{mup}\,(C).$$

**Proof:** i) $B^*C = B^*(B^* \sqcap \overline{BL}^\mathsf{T}) = B^*B^* \sqcap \overline{BL}^\mathsf{T} = B^* \sqcap \overline{BL}^\mathsf{T} = C$, as a consequence of Prop. 2.4.2. This yields $C^2 \subset B^*C = C$ and also the reverse containment $C = CI = C(I \sqcap C) \sqcup C(I \sqcap \overline{C}) \subset C^2 \sqcup O$, since

$$C(I \sqcap \overline{C}) = C(I \sqcap \overline{B^* \sqcap \overline{BL}^\mathsf{T}}) \subset C\left(I \sqcap (\overline{I} \sqcup (BL)^\mathsf{T})\right) \subset C\left(I \sqcap (BL)^\mathsf{T}\right)$$
$$= C(I \sqcap BL) \subset CBL \subset (B^* \sqcap \overline{BL}^\mathsf{T})BL = B^*(BL \sqcap \overline{BL}) = O.$$

ii) $C \sqcap \overline{I} = B^* \sqcap \overline{BL}^\mathsf{T} \sqcap \overline{I} \subset B^+ \sqcap \overline{BL}^\mathsf{T} = BB^* \sqcap \overline{BL}^\mathsf{T} = B(B^* \sqcap \overline{BL}^\mathsf{T}) = BC$ because of Prop. 2.4.2. For the opposite direction, (i) implies $BC \subset C$, and $BC \subset \overline{I}$ is valid since $B^\mathsf{T}I \subset \overline{B^*} \sqcup (BL)^\mathsf{T} = \overline{C}$. Using $C = (C \sqcap I) \sqcup (C \sqcap \overline{I})$ together with $C \sqcap I \subset \overline{BL}$ and $C \sqcap \overline{I} \subset BL$, we prove by elementary reasoning that $\mathrm{mup}\,(C) = C \sqcap C\overline{I} = (C \sqcap \overline{I}) \sqcap (C \sqcap \overline{I})\overline{I} = \mathrm{mup}\,(C \sqcap \overline{I})$. $\qquad\square$

In (i) the reachability $B^*$ acts as a neutral element when multiplying the terminal reachability by it. The terminating reachability, in turn, is transitive. Part (ii) can be interpreted as follows: If *more* than one terminal point can be reached from point $x$, then $x$ itself can*not* be terminal. The terminal points in question can also be reached by first traveling one step along $B$ and then continuing along the terminating reachability $C$.

Our results on the terminating reachability have so far been quite general, inasmuch as they are independent of the number of terminal points that can be reached from any given point. In Sect. 6.4 we shall consider the situation where from any point *precisely* one, or *at most* one, terminal point can be reached. For example, $B$ can describe the steps of some reduction process towards a canonical form (the corresponding terminal point is uniquely determined). When $C$ is total, there are further identities.

**6.1.16 Proposition.** If the terminating reachability $C$ of $B$ is total, then

i) $\qquad\qquad\qquad\qquad L = B^*\overline{BL}, \qquad B^*B^{*\mathsf{T}} = CC^\mathsf{T};$

ii) $\qquad\qquad\qquad\qquad C \text{ univalent} \quad \Longleftrightarrow \quad B^{\mathsf{T}*}C = C.$

**Proof:** i) $L = CL = (B^* \sqcap \overline{BL}^\mathsf{T})L = B^*\overline{BL}$. According to Prop. 6.1.15, for $C$ total we have the containment

$$B^*B^{*\mathsf{T}} = B^*IB^{*\mathsf{T}} \subset B^*CC^\mathsf{T}B^{*\mathsf{T}} = CC^\mathsf{T}$$

which is an equality because $C \subset B^*$.

ii) For "$\Longrightarrow$" we infer $B^\mathsf{T}C \subset CC^\mathsf{T}B^\mathsf{T}C \subset C(B^*C)^\mathsf{T}C = CC^\mathsf{T}C \subset C$, using totality and univalence of $C$, as well as $B^*C = C$. Therefore, $B^{\mathsf{T}i}C \subset C$ for all $i \geq 0$, and hence $B^{\mathsf{T}*}C \subset C$. Direction "$\Longleftarrow$" follows from $C^\mathsf{T}C = (B^{\mathsf{T}*} \sqcap \overline{BL})C = B^{\mathsf{T}*}C \sqcap \overline{BL} = C \sqcap \overline{BL} \subset B^* \sqcap \overline{BL} = (I \sqcap \overline{BL}) \sqcup (B^+ \sqcap \overline{BL}) \subset I \sqcup O$. $\qquad\square$

Clearly, $L = B^*\overline{BL}$ implies that $C$ is total. There is a genealogical interpretation of the second statement in (i): Suppose both $x$ and $y$ are ancestors of $z$. Then there exists a childless person $z'$ of whom both $x$ and $y$ are ancestors. Of course, here one has to make the realistic assumption that everybody has a childless descendant, possibly himself as descendant of degree zero.

Part (i) does not apply, for example, to the relation $\leq$ between natural numbers because there are no terminal elements. If $B$ represents a reduction step towards a uniquely determined canonical form, then statement (ii) says that *the same* canonical form is associated with a point $x$ and all the other points from which $x$ can be reached by reduction steps.

## Exercise

**6.1.1** Prove the finite case of Prop. 6.1.3.iv using the pigeon-hole principle.

## 6.2 Chains and Cycles

A close companion to reachability is the concept of joinability where one is also allowed to proceed against the direction of $B$. This is achieved by the symmetrized relation $B \sqcup B^{\mathsf{T}}$ of $B$. But loops are usually suppressed, and one therefore works with the adjacency $\Gamma$.

**6.2.1 Definition** (*Chain as sequence of points*). Let $\Gamma$ be the adjacency of a simple graph and let $k = (x_0, \ldots, x_h)$, $h \geq 0$, be a sequence of points with $x_{i-1} \subset \Gamma x_i$, $i = 1, \ldots, h$. Then we call $k$ a **chain** of **length** $h$ from $x_0$ to $x_h$. The chain $(x_0)$ of length $0$ is called a **void chain** in $x_0$. A chain is called **elementary**, if no point occurs more than once in the sequence—with the exception of $x_0 = x_h$. $\qquad\square$

**Fig. 6.2.1** Chains in a simple graph

Figure 6.2.1 shows some chains. Two consecutive points in a chain are necessarily distinct.

In Proposition 6.1.3 we could read off the existence of paths from the transitive closure of the associated relation. We indicate without proof the corresponding result on the existence of chains with respect to the transitive closure of the adjacency.

**6.2.2 Proposition.** Given the adjacency $\Gamma$, two arbitrary points $x$ and $y$, and a natural number $h \geq 0$, the following statements hold:

i)   $x \subset \Gamma^h y \iff$ There exists a chain of length $h$ from $x$ to $y$.

ii)  $x \subset \Gamma^* y \iff$ There exists a chain from $x$ to $y$. $\qquad\square$

The notion of circuit can*not* be adapted to the present situation. The (symmetric) adjacency $\Gamma$ is not suited for that because too many "circuits" would arise merely by traversing an edge in both directions.

The analog of reachability $B^*$ along paths is joinability $\Gamma^*$ by means of chains.

**6.2.3 Definition.** Given an adjacency $\Gamma$, we call $\Gamma^*$ the corresponding **joinability**. If $x$, $y$ are two points of the graph $G$, we call

$$x \text{ and } y \text{ \textbf{joinable}} \quad :\Longleftrightarrow \quad xy^{\mathsf{T}} \subset \Gamma^*. \qquad \square$$

In Definition 6.2.1 we defined chains as sequences of points where any two consecutive points are related by the adjacency $\Gamma$. Since graphs and hypergraphs are endowed with an adjacency, chains as point sequences and the notion of "joinable" are therefore available, too. But for many problems a stronger chain concept is required which is related to the incidence $M$. The following is modeled on Definition 6.1.5.

**6.2.4 Definition** (*Chain and cycle as a point-edge-sequence*). Let $M$ be the incidence of a graph or hypergraph. An alternating sequence $(x_0, p_1, \ldots, p_h, x_h)$, $h > 0$, of points and edges, with $x_0 \subset M^{\mathsf{T}}p_1$ (starting point) and $x_h \subset M^{\mathsf{T}}p_h$ (endpoint), as well as $x_i \subset M^{\mathsf{T}}p_i \sqcap M^{\mathsf{T}}p_{i+1}$ for $i = 1, \ldots, h-1$ and $x_i \neq x_{i+1}$, $i = 0, \ldots, h-1$ is called a **chain** of length $h$ from $x_0$ to $x_h$. The chain $(x_0)$ is called the **void chain** in $x_0$. A chain is called **simple**, if no edge occurs more than once in the sequence.

A simple chain of finite, positive length with coinciding endpoints is called a **cycle** or a **closed chain**. To any cycle an **orientation** can be assigned by numbering its vertices. $\qquad \square$

If a chain is given as an alternating point-edge-sequence, then clearly the subsequence of points is a chain in the sense of Definition 6.2.1. If only the edge subsequence of such a chain is given, then the points cannot, in general, be got back, because two consecutive (hyper-)edges of rank 3 could, for instance, be incident with two common points. In a graph in which all arcs have rank 2 the points of a chain are uniquely determined by its edge subsequence, provided it is nonvoid. Because of the condition $x_i \neq x_{i+1}$, a chain contains no loops or drops.

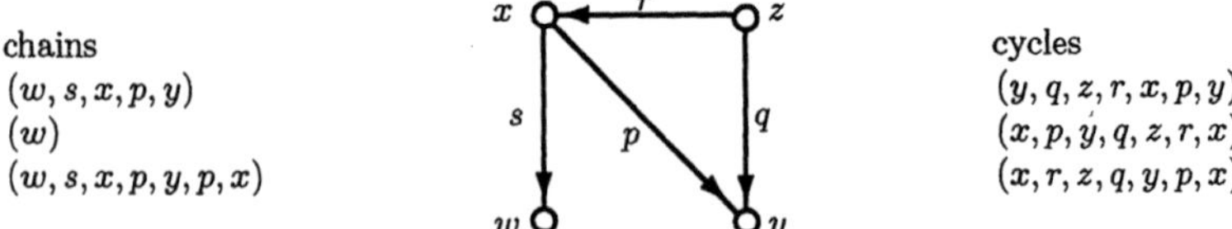


chains

$(w, s, x, p, y)$

$(w)$

$(w, s, x, p, y, p, x)$

cycles

$(y, q, z, r, x, p, y)$

$(x, p, y, q, z, r, x)$

$(x, r, z, q, y, p, x)$

**Fig. 6.2.2** Chains and cycles

Cycles and circuits are defined in terms of sequences. So the sequences $(x, p, y, q, z, r, x)$ and $(y, q, z, r, x, p, y)$ in Fig. 6.2.2 represent *distinct* cycles. But their orientation is the same, and they can be "cyclically transformed" into one another while preserving their orientation. Therefore, such sequences are often identified; see also the remarks after Definition 6.1.1. In homology theory even chains with different orientation such as $(x, p, y, q, z, r, x)$ and $(x, r, z, q, y, p, x)$ are identified.

In contrast to circuits, cycles are not just closed but also simple. Therefore, the results on circuits from Sect. 6.1 do *not* apply to cycles.

Using cycles we can give a necessary and sufficient criterion for a graph to be bipartite or 2-colorable: This is the case precisely when the graph contains no cycles of odd length.

We now define connectedness based on joinability and adjacency. It plays a similar role for them as strong connectedness does for reachability and the associated relation.

**6.2.5 Definition.** Let a simple graph be given with adjacency $\Gamma$.

i)  $G$ **connected**   $:\Longleftrightarrow$   $\Gamma^* = L$
   $\Longleftrightarrow$   Any two points are joinable.

ii)  The equivalence classes of points with respect to the equivalence $\Gamma^*$ are called the **connected components** of $G$.   $\square$

As a consequence of $B^* \subset B^{\mathsf{T}}{}^*B^* \subset (B \sqcup B^{\mathsf{T}})^*$, the concepts of connectedness are related as follows:

$$\text{strongly connected} \implies \text{quasi-strongly connected} \implies \text{connected}.$$

Here, $\Gamma^* = (B \sqcup B^{\mathsf{T}})^*$ is the equivalence closure of $B$ according to Proposition 3.2.3. In a special case, the chain of implications may be closed:

**6.2.6 Proposition.** For a total graph $G$ the following holds:

$$G \text{ strongly connected} \quad \Longleftrightarrow \quad \begin{cases} G \text{ connected and} \\ \text{every edge belongs to a circuit.} \end{cases}$$

**Proof:** Exercise 6.2.1.   $\square$

## Articulation

The following concepts serve for determining "how tightly" a graph is connected. Some graphs stay connected when removing a point, others split into several connected components.

**6.2.7 Definition.** Given a point set $x$ of a simple graph $G$, we call $x$ an **articulation set** (or a **point**, if $x$ is a singleton set), if the subgraph generated by $\bar{x}$ contains *more* connected components than $G$.   $\square$

Clearly, the $n$-clique has no articulation sets; conversely, it is straightforward to see that a connected graph without articulation sets is complete. The exceptional behavior of $n$-cliques has therefore to be taken into account in the following definition.

**6.2.8 Definition.** For a connected simple graph $G$ we call

$$\chi := \begin{cases} \min\{\, |x| \mid x \ \text{articulation set} \,\}, & \text{if } G \text{ is not a clique} \\ n - 1, & \text{if } G \text{ is an } n\text{-clique} \end{cases}$$

the **cohesion number**. If $\chi \geq h$, then we call $G$ **$h$-connected**.  □

Thus, $\chi$ is the minimum number of points after the removal of which $G$ splits or shrinks to a single point. For non-cliques we have $\chi \leq |V| - 2$. The well-known theorem of Menger, which we will not elaborate on here, states that $h$-connectedness is equivalent to the existence of at least $h$ noncrossing paths between any two points of the graph.

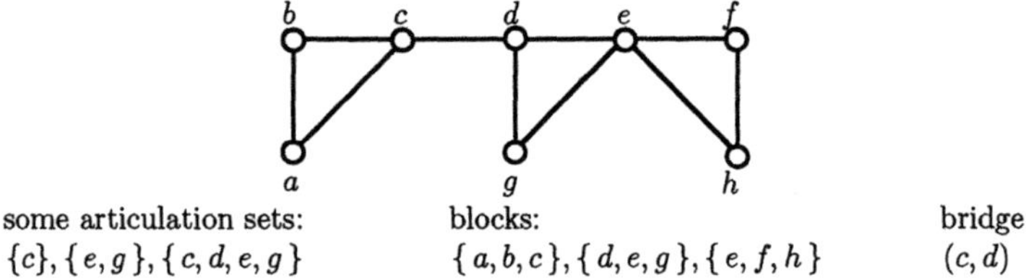

<table>
<tr><td>some articulation sets:</td><td>blocks:</td><td>bridge:</td></tr>
<tr><td>$\{c\}, \{e,g\}, \{c,d,e,g\}$</td><td>$\{a,b,c\}, \{d,e,g\}, \{e,f,h\}$</td><td>$(c,d)$</td></tr>
</table>

**Fig. 6.2.3** Articulation

Of interest are also the pieces of a graph resulting from cutting it at all of its articulation points (which then are supposed to belong to all the respective pieces).

**6.2.9 Definition.** The point set $x$ of a simple graph $G$ is called a **block**, if the subgraph generated by $x$ is a connected subgraph without articulation points, and if there is no larger point set with this property.  □

The graph in Fig. 6.2.3 has cohesion number 1. We have given its block decomposition.

If a graph contains an articulation point $x$, then this restricts the number of related pairs of points because $x$ cannot be "bypassed". Actually, there is the following quantitative result: If the connected, simple graph $G$ has $n \geq 2$ points and $r$ articulation points, then $G$ can have at most $\binom{n-r}{2} + r$ edges, this bound being sharp.

If one removes an edge instead of a point, then the graph may split up, too. In this case the edge is called a **bridge**. Clearly, every edge in a tree is a bridge. But no edge belonging to some cycle of a graph can be a bridge. The relation "no bridge separates $a$ and $b$" is an equivalence relation on the points of the graph. The corresponding equivalence classes are called **bridge components**.

**Exercise**

**6.2.1** Prove Proposition 6.2.6.

## 6.3 Terminality and Foundedness

In Definition 5.2.1 we called points terminal when they have no successor, i.e., when no further points can be reached from them. In the present section we are interested in those points from which paths of any given finite length or even of infinite length start. Both these properties are closely related, and they agree for finite graphs but differ for infinite graphs.

**6.3.1 Definition.** Given a graph $G$ with associated relation $B$ and some point $x$, we call

$x$ **progressively bounded** $\quad :\Longleftrightarrow \quad x \subset \sup_{h \geq 0} \overline{B^h L}$

$\qquad\qquad\qquad\qquad\qquad\quad \Longleftrightarrow \quad$ There is an upper bound for the lengths of all paths starting from $x$.

$B$ **progressively bounded** $\quad :\Longleftrightarrow \quad L = \sup_{h \geq 0} \overline{B^h L}.$

Some authors use the term **regressively bounded** when progressive boundedness holds with respect to $B^\mathsf{T}$. $\qquad\qquad\qquad\qquad\qquad\Box$

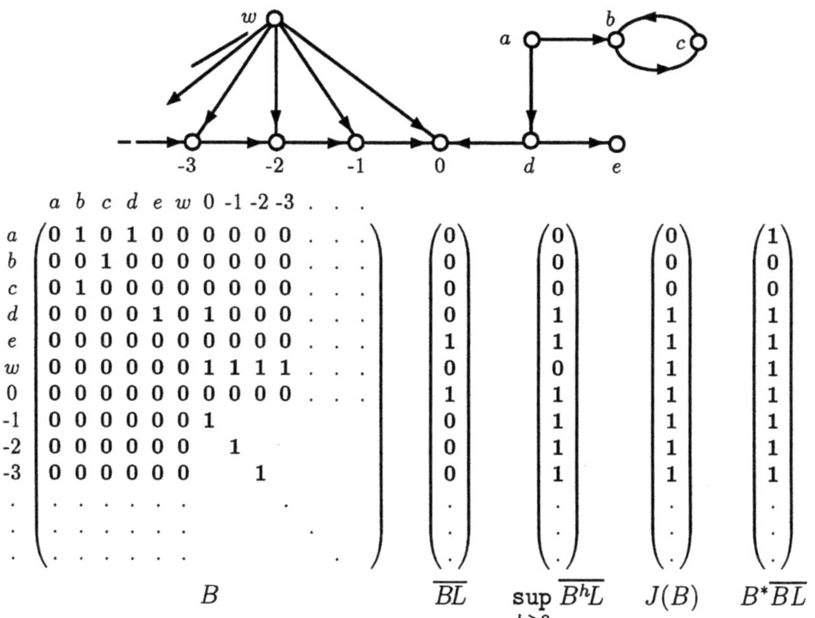

Fig. **6.3.1** Progressively bounded and progressively finite points

The concept of progressively bounded is quite close to that of terminal since $x \subset \overline{BL}$ means that $x$ has no successor, and that hence no path of length 1 starts from $x$. Since points $x$ are atoms in the set of vectors (Prop. 2.4.5.i), it follows from $x \subset \sup_h \overline{B^h L}$ that there is an $h := h_x$ such that $x \subset \overline{B^h L}$. Because of $x = xL$, this is equivalent to $x^{\mathsf{T}} B^h L \subset O$ and hence to $B^{h\mathsf{T}} x = O$. So $h_x - 1$ is an upper bound for the lengths of paths starting from $x$. In particular, there are no paths of infinite length starting from $x$. In Fig. 6.3.1, for example, one has $h_d = 2$, $h_0 = 1$ and $h_{-3} = 4$.

Note, however, the situation at point $w$. Every path starting from $w$ has finite length. But in spite of this, $w$ is not progressively bounded because, given any $n$, $(w, -n, -n+1, \ldots, -1, 0)$ is a path of length $n+1$ from $w$.

Point $w$ shows that one has to distinguish between progressively bounded points and those from which no paths of infinite length start. We now want to characterize this latter kind of points.

If $x$ is the starting point of some path of infinite length, then so are all of its predecessors; hence the set $y$ of all such points $x$ satisfies $By \subset y$. But also $y \subset By$ holds because if a path $w_0$ of infinite length starts from $x$, then $w_0$ contains a successor of $x$ which again is the starting point of such a path. So we have $By = y$ which we interpret as an equation of the form $Bx = \lambda x$ for an eigenvector $x$ corresponding to the eigenvalue $\lambda = 1$. Such a fixedpoint condition[2] was already considered by P. Hitchcock and D. Park in 1972 in connection with the problem of program termination.

The set $y$ of starting points of paths of infinite length is not yet completely determined by the equation $y = By$, as can be seen from the graph in Fig. 6.3.2.

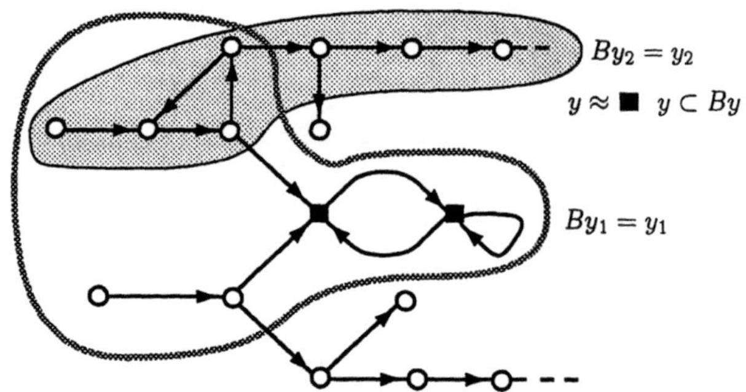


**Fig. 6.3.2** Sets of starting points of infinite paths

In order to obtain the entire set $y$, we must therefore take the largest point set $z$ satisfying $Bz = z$. This largest set indeed exists because $O = BO$, and given $y_i = By_i$ $(i = 1, 2)$ it follows that $y_1 \sqcup y_2 = By_1 \sqcup By_2 = B(y_1 \sqcup y_2)$.

---

[2] In Appendix A.3 the definitions and propositions below will be taken up again in the following more general setting. We define two antitone mappings $\sigma, \pi \colon \mathbf{2}^V \longrightarrow \mathbf{2}^V$ by $\sigma(x) := B\overline{x}$ and $\pi(x) := \overline{x}$, and consider their concatenations $\pi(\sigma(x)) = \overline{B\overline{x}}$ and $\sigma(\pi(x)) = Bx$. What we are doing here can then be achieved by lattice-theoretic arguments using solely that $\sigma$ and $\varphi$ are antitonic. At this stage, however, we prefer to avoid the lattice-theoretic apparatus and give ad hoc arguments.

So the set $y$ of starting points of paths of infinite length is equal to $\sup\{\,y \mid y = By\,\}$. This supremum does not change when extended even over all sets $y$ such that $y \subset By$ because given such a $y$ we can find a larger set $w \supset y$ satisfying $w = Bw$. Just take the set $w := B^+y$ of predecessors of $y$. Then, indeed, we have $y \subset By \subset B^+y = w$, $Bw = BB^+y \subset B^+y = w$ and, in the reverse direction, $w = B^+y \subset B^+By = BB^+y = Bw$. In Fig. 6.3.2 the $w$ corresponding to $y$ is $w := y_1$.

We return to our initial problem and consider the complement of the set $\sup\{\,y \mid y = By\,\}$. First, we express the condition on a relation $B$ to the effect that one can proceed only finitely many steps along it. This topic is of interest in many areas of computer science and mathematics, and is related to recursion and induction principles, well-orderings, the axiom of choice, and also to chain conditions in ideal theory.

**6.3.2 Definition.** Let us consider a relation $B$ and vectors $x$.

i)   The set of all points, from which only paths of finite length emerge,

$$J(B) := \inf\{\,x \mid \overline{x} = B\overline{x}\,\} = \inf\{\,x \mid \overline{x} \subset B\overline{x}\,\},$$

is called the **initial part** of $B$.

ii)   $x$ **progressively finite** $\quad :\Longleftrightarrow\quad x \subset J(B)$

$\qquad\qquad\qquad\qquad\qquad\Longleftrightarrow\quad$ All paths with respect to $B$, starting from $x$, have finite length.

$\quad B$ **progressively finite** $\quad :\Longleftrightarrow\quad J(B) = L.$

A synonym of "progressively finite" is **Noetherian** as well as "satisfying the **ascending chain condition**". Progressive finiteness with respect to the transpose $B^{\mathsf{T}}$ of $B$ is termed **well-founded, Artinian, regressively finite**, or "$B$ satisfies the **descending chain condition**".     $\square$

We have thus defined the properties of Noetherian, Artinian, etc., not just for ordered sets but for relations in general. The induction principles to be discussed later need not be based on orderings, either. A (reflexive) ordering cannot be progressively finite. An Artinian and connex order relation is a well-ordering.

Clearly, $B$ is progressively finite, if and only if, for every point set $y$,

$$y \subset By \quad\Longrightarrow\quad y = O.$$

So a graph is progressively finite precisely when it contains no path of infinite length. The condition above is equivalent to

$$y \neq O \quad\Longrightarrow\quad y \sqcap \overline{By} \neq O,$$

which in turn means the following: Every nonvoid point set $y$ in a progressively finite graph contains some point that has no successor in $y$.

When $B$ is an irreflexive ordering, one could also say that every nonvoid point set contains (at least) one maximal point (as did already VON NEUMANN, MORGENSTERN 44).

We shall now investigate the relationship between the properties of "progressively finite" and "progressively bounded", as well as their impact on the reachability of terminal points. Some of our results require additional assumptions.

**6.3.3 Proposition.** In a graph with associated relation $B$ the following hold:

i) $\qquad\qquad\qquad\qquad\qquad\qquad \sup_{h\geq 0} \overline{B^h L} \subset J(B) \subset B^* \overline{BL}$;

ii) $\quad B$ univalent $\qquad\qquad \Longrightarrow \qquad \sup_{h\geq 0} \overline{B^h L} = J(B) = B^* \overline{BL}$;

iii) $\quad B$ progressively bounded $\;\Longrightarrow\; L = \sup_{h\geq 0} \overline{B^h L} = J(B) = B^* \overline{BL}$;

iv) $\quad B$ progressively finite $\qquad \Longrightarrow \quad \sup_{h\geq 0} \overline{B^h L} \subset J(B) = B^* \overline{BL} = L$;

v) $\quad B$ finite $\qquad\qquad\qquad \Longrightarrow \quad \sup_{h\geq 0} \overline{B^h L} = J(B) \subset B^* \overline{BL}$.

**Proof:** i) The set $x_0 := B^* \overline{BL}$ satisfies $\overline{x}_0 \subset B\,\overline{x_0}$ because $x_0 \sqcup B\,\overline{x_0} = B^*\overline{BL} \sqcup B\,\overline{B^*\overline{BL}} = (\overline{BL} \sqcup BB^*\overline{BL}) \sqcup B\,\overline{B^*\overline{BL}} = \overline{BL} \sqcup BL = L$. Therefore $J(B) \subset x_0$. Furthermore, for any $x$ with $B\overline{x} = \overline{x}$, firstly $\overline{B^0 L} = O \subset x$ and secondly, by induction, $\overline{B^h L} \subset x \Longleftrightarrow \overline{x} \subset B^h L \Longrightarrow B\overline{x} = \overline{x} \subset B^{h+1}L \Longleftrightarrow \overline{B^{h+1}L} \subset x$. This results in $\sup_{h\geq 0} \overline{B^h L} \subset x$.

ii) If $B^{\mathsf T} B \subset I$, we always have

$$(B^h)^{\mathsf T} B^{h+1} L \subset (B^{h-1})^{\mathsf T} B^h L \subset \ldots \subset BL$$

for $h \geq 0$ such that $B^h \overline{BL} \subset \overline{B^{h+1}L}$. This yields $B^* \overline{BL} \subset \sup_{h\geq 0} \overline{B^h L}$.

iii) and iv) follow immediately from the definition of progressive finiteness and boundedness.

v) For a finite graph, the sequence $L \supset BL \supset B^2 L \supset \ldots$ becomes stationary, i.e., there is a $k$ with $B^k L = BB^k L$. Therefore $\inf_{h\geq 0} B^h L = B^k L = BB^k L = B \inf_{h\geq 0} B^h L$. So $x := \sup_{h\geq 0} \overline{B^h L}$ satisfies $\overline{x} = B\overline{x}$ and thus contains $J(B)$.

$\qquad\qquad\qquad\qquad\qquad\qquad\qquad\qquad\qquad\qquad\qquad\qquad\qquad\qquad\qquad\quad \square$

Every progressively bounded point is also progressively finite. Given a progressively finite point, there is always a terminal point that can be reached from it: $J(B) \subset B^* \overline{BL}$. Indeed, no path of infinite length can possibly start from point $x$, if all finite such lengths are bounded above by $h_x$. So in this situation any path must ultimately lead to some point beyond which it cannot be extended any further, that endpoint then being terminal.

On the other hand, there may well be points from which some path leads to a terminal point, whereas some other path has infinite length. Thus the inclusion signs in (i) are all strict in general. When $B$ is univalent, however, there cannot be paths of different sorts as above, and the three terms in (i) are all equal. This is the reason why deterministic programs are so much easier to handle than nondeterministic.

Let us now ask how far apart the properties of progressively bounded and progressively finite can actually be for infinite graphs. The difference can best be understood when one tries to exhaust the point set starting from the terminal points, as is done in Fig. 6.3.3. Beginning with the set $z_0 = \overline{BL}$ of terminal points, successively determine the set of points which have no successor in the complement of the preceding. We are going to see that this exhaustion process yields all those points from which, after a specified, finite number $h_x$ of steps, a terminal point is reached.

**6.3.4 Proposition** (*Exhaustion Theorem*). In a graph with associated relation $B$, all the progressively bounded points are obtained by the iteration

$$z_0 := \overline{BL},$$
$$z_{i+1} := \overline{B\,\overline{z_i}} \quad \text{for } i \geq 0,$$

i.e., the following holds

$$\sup_{h \geq 0} z_h = \sup_{h \geq 0} \overline{B^h L}.$$

**Proof:** We show by induction that $z_h = \overline{B^{h+1}L}$. For $h = 0$ this is true due to the given starting condition. From the assumption $z_{h-1} = \overline{B^h L}$ we deduce $\overline{z_{h-1}} = B^h L$, $B\,\overline{z_{h-1}} = B^{h+1}L$, and $\overline{B\,\overline{z_{h-1}}} = \overline{B^{h+1}L}$. $\qquad\qquad$ $\square$

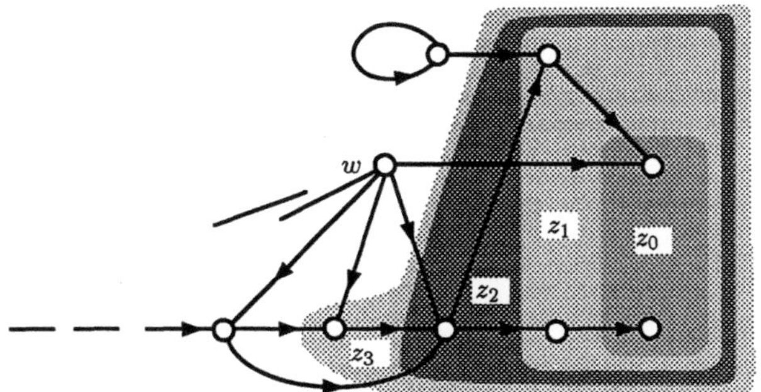

Fig. **6.3.3** Exhausting the progressively bounded points

The exhaustion process could also be started with $z_{-1} := O$ and $z_0 := \overline{B\,\overline{z_{-1}}}$. In Fig. 6.3.3 the process does not, however, yield the progressively finite point $w$ because it has infinitely many successors which cannot be covered by finitely many applications of the process (since we do not extend it transfinitely). As one would expect, for a graph in which every point $x$ has *finite* successor degree $d_s(x)$, the properties of progressively bounded and progressively finite are equivalent.

**6.3.5 Proposition.** The following holds for a graph $G$ with associated relation $B$, where all points have finite successor degree:

i) $$\sup_{h \geq 0} \overline{B^h L} = J(B);$$

ii) $\qquad G$ progressively bounded $\quad \Longleftrightarrow \quad G$ progressively finite.

**Proof:** Given Prop. 6.3.3.i, only "$\supset$" must be shown. Let us assume there is a progressively finite point $x$ that is not progressively bounded. Then for any natural number $h$, there exists a path $w_{f(h)}$ of length $|w_{f(h)}| \geq h$. These infinitely many paths all traverse the *finitely* many successors of $x$. So, among the successors there is at least one point $z$ contained in infinitely many of these paths. This pigeon-hole principle may be repeated for $z$ and so on, yielding a path of infinite length. This contradicts the assumption that $G$ is progressively finite.    □

The " $\Longrightarrow$ " part of Proposition 6.3.5.ii was already established earlier. The " $\Longleftarrow$ " part is known as **König's graph lemma**. Dénes König proved in 1927 that a graph which has finite successor degrees and is not progressively bounded contains a path of infinite length. There are further contributions to this subject by von Neumann and, in connection with game theory and chess in particular, by Zermelo (1912), Kalmár (1928), and Euwe (1929).

For graphs with finite successor degrees and no circuits one has moreover

$$G \ \text{progressively bounded} \quad \Longleftrightarrow \quad G \ \text{progressively finite}$$
$$\Longleftrightarrow \quad \text{For every point } x \text{ the set } B^{*\mathsf{T}}x \text{ of all points reachable from } x \text{ is finite.}$$

The lack of circuits is responsible for " $\Longleftarrow$ " holding true in the second statement. Since a point that has been reached from $x$ can never again be traversed when proceeding along a given path, there can only be finitely many paths from $x$.

In a progressively finite graph there are only paths of finite length. In particular, there are no paths traversing infinitely many distinct points, and no paths that arise from traveling along circuits of finite length infinitely many times. Hence a progressively finite graph contains no circuits. We give the proof as Exercise 6.3.2.

**6.3.6 Proposition.** Every progressively finite graph is circuit-free, i.e., for the associated relation $B$ of any graph, the following implication holds

$$L = J(B) \quad \Longrightarrow \quad B^+ \subset \overline{I}.    □$$

For a finite and circuit-free graph $G = (V, B)$ some power of $B$ vanishes, hence so does $\inf_h B^h L$, and $G$ is progressively bounded. So we have the chain of implications

$$G \ \begin{matrix} \text{progressively} \\ \text{bounded} \end{matrix} \quad \Longrightarrow \quad G \ \begin{matrix} \text{progressively} \\ \text{finite} \end{matrix} \quad \Longrightarrow \quad G \ \text{circuit-free}$$

$$\text{if } G \text{ is finite}$$

For finite graphs, therefore, all three conditions are equivalent.

In the nonfinite but regressively finite case, Proposition 6.1.13 can be sharpened as follows:

**6.3.7 Proposition.** The following holds for regressively finite graphs:

$$\begin{matrix} \text{quasi-strongly} \\ \text{connected} \end{matrix} \quad \Longleftrightarrow \quad \text{There exists } \textit{precisely} \text{ one root.}$$

**Proof:** By quasi-strong connectedness we have

$$L = B^{\mathsf{T}*}B^* = (I \sqcup B^{\mathsf{T}}B^{\mathsf{T}*})B^* = B^* \sqcup B^{\mathsf{T}}B^{\mathsf{T}*}B^* = B^* \sqcup B^{\mathsf{T}}L,$$

so that $\overline{B^*}$ as well as $\overline{B^*L}$ are contained in $B^{\mathsf{T}}L$. From Prop. 6.3.6 we know that a regressively finite graph is circuit-free. So $B^+ = BB^* \subset \overline{I}$, or equivalently, $B^{\mathsf{T}} \subset \overline{B^*}$. Altogether we have $\overline{B^*L} = B^{\mathsf{T}}L$. Using Prop. 6.3.3.iv for $B^{\mathsf{T}}$, from $L = J(B^{\mathsf{T}}) = B^{\mathsf{T}*}\overline{B^{\mathsf{T}}L}$ it follows that $\overline{B^{\mathsf{T}}L} \neq O$. So there exist points without predecessors, and the set $a := \overline{B^*L} = \overline{B^{\mathsf{T}}L}$ of roots is not empty. Finally, there is at most one root, because given two different roots there would exist a circuit containing both. Formally, injectivity of $a$ has to be shown: on the one side $aa^{\mathsf{T}} \subset B^*$, since $\overline{B^*}a \subset \overline{B^*}L = \overline{a}$, and on the other $aa^{\mathsf{T}} \subset \overline{B^{+\mathsf{T}}}$, because $B^{+\mathsf{T}}a \subset \overline{a} = B^{\mathsf{T}}L$. Putting this together yields $aa^{\mathsf{T}} \subset B^* \sqcap \overline{B^+} \subset I$.     $\square$

## Induction Principles

We shall now discuss an important method for proving statements on graphs which is related to the well-known transfinite induction principle. The statements in question will be whether or not some point belongs to a given subset $x$. Interpreting $x$ as a predicate, the question is then whether it is satisfied by a given point. First we give a name to the sets over which the infimum in Definition 6.3.2 is extended.

**6.3.8 Definition.** Given the associated relation $B$ of a graph and a subset $x$ of points, we call

$$x \ \textbf{\textit{B}-complete} \ :\Longleftrightarrow \ \overline{x} \subset B\overline{x}$$

$\Longleftrightarrow$ Every point with no successor in the complement of $x$ belongs to $x$.     $\square$

Let now a $B$-complete set $x$ be given. We interpret subsets $x$ as the property of points to belong to $x$. Thus, if $x$ is $B$-complete, then a point $a$ has property $x$ if and only if *all* of its successors have. So property $x$ is enjoyed by all points of the initial part, hence by the entire point set in case the graph is progressively finite. In particular, property $x$ holds for terminal points (for lack of successors). The following is an immediate consequence of Definition 6.3.2.

**6.3.9 Proposition** (*Generalized Induction Principle*). Given a point set $x$ of a graph with associated relation $B$, the following hold:

i)     $x$ $B$-complete     $\Longrightarrow$ $J(B) \subset x$;

ii)     $x$ $B$-complete and $B$ progressively finite     $\Longrightarrow$ $L = x$.     $\square$

As a reformulation and specialization for the regressively finite direction, we obtain the following important corollary.

**6.3.10 Corollary** (*Transfinite Induction*). Let $B$ be a well-founded strict-ordering, and let $x$ be a property of points of the underlying set of $B$. Suppose that a point has property $x$ whenever all of its predecessors have. Then every point has property $x$. $\qquad\square$


Orderings can often be represented most economically by their Hasse diagrams. We now investigate what completeness with respect to some relation $B$ has to do with completeness with respect to its transitive closure $B^+$. Obviously,

$$x \ \ B\text{-complete} \quad \Longrightarrow \quad x \ \ B^+\text{-complete.}$$

But the converse does *not* hold generally as can be seen from the relation $B = $ "is immediate predecessor of", defined on the set of natural numbers, and the subset $x$ of even numbers. So $B$-completeness is stronger than $B^+$-completeness, and one may ask whether Proposition 6.3.9 still holds for the latter. Since $BL = B^+L$, $B$-terminal points and $B^+$-terminal points agree, and indeed we have:


**6.3.11 Proposition.** Given a relation $B$ and a point set $x$, the following holds:

i) $$J(B) = J(B^+);$$

ii) $\qquad x \ \ B^+\text{-complete} \qquad\qquad\qquad\qquad \Longrightarrow \quad J(B) \subset x;$

iii) $\qquad x \ \ B^+\text{-complete and } B \text{ progressively finite} \quad \Longrightarrow \quad L = x.$

**Proof:** i) From $\overline{x} \subset B\,\overline{x}$ it follows that $\overline{x} \subset B^+\overline{x}$, yielding $J(B^+) \subset J(B)$ for the initial parts since these are defined as infima. On the other hand, every $y$ with $\overline{y} = B^+\overline{y}$ satisfies

$$\overline{y} = B^+\overline{y} = BB^*\overline{y} = B(I \sqcup B^+)\overline{y} = B\,\overline{y} \sqcup BB^+\overline{y} = B\,\overline{y};$$

therefore $J(B) \subset J(B^+)$. (ii) and (iii) are immediate consequences of (i). $\qquad\square$


## Exercises

**6.3.1** In Definition 6.3.2, the least fixedpoint of the functional $\tau(x) := \overline{B\,\overline{x}}$ has been called the initial part of a graph; see also Appendix A.3. Show that $\tau$ is isotonic, i.e., that $x_1 \subset x_2 \Longrightarrow \tau(x_1) \subset \tau(x_2)$. Give an example of a noncontinuous $\tau$, i.e., show that in general $x_1 \subset x_2 \subset \ldots$ does not imply $\tau(\sup_{h\geq1} x_h) = \sup_{h\geq1} \tau(x_h)$.

**6.3.2** Prove that the associated relation $B$ of any graph satisfies

$$L = J(B) \quad \Longrightarrow \quad B^+ \subset \overline{I},$$

i.e., that any progressively finite graph is circuit-free.

**6.3.3** If there exists a homomorphism of a graph $G$ into a progressively bounded or progressively finite graph, then $G$ itself is progressively bounded or progressively finite, respectively.

**6.3.4** Prove the following identities for the initial part $J(B)$:
$$B^{*\mathsf{T}}J(B) = J(B), \qquad B^*\overline{J(B)} = \overline{J(B)}.$$

**6.3.5** Show that in general $J(B) \subset \inf\{\, x \mid \overline{BL} \sqcup Bx \subset x \,\}$ and that for a univalent relation $B$ this may be sharpened to $J(B) = \inf\{\, x \mid \overline{BL} \sqcup Bx \subset x \,\}$.

# 6.4 Confluence and Church-Rosser Theorems

The graph in Fig. 6.4.1 represents a deduction process of the type that occurs in syntax analysis and in problems related to term rewriting or to conversion in $\lambda$-calculus. The point set here is part of the (infinite) set of words over the set of letters $\{\, a, b, c, S \,\}$. Arrows mark reduction steps according to the rule

$$S ::= aSbS \mid aS \mid c.$$

$S$ stands for the axiom.

For the construction of compilers as well as for automatic theorem proving it is of great importance to know whether the given set of deduction rules allows an analysis without dead ends or not. Such a dead end occurs in the left hand part of Fig. 6.4.1 because the given set of rules provides no path from $SbS$ to $S$. If one wants to reach axiom $S$ in this graph, one has to return to some earlier ramification point and then proceed differently. A set of deduction rules will be particularly convenient when it allows to find a path to the axiom irrespective of the decision made at any ramification point.

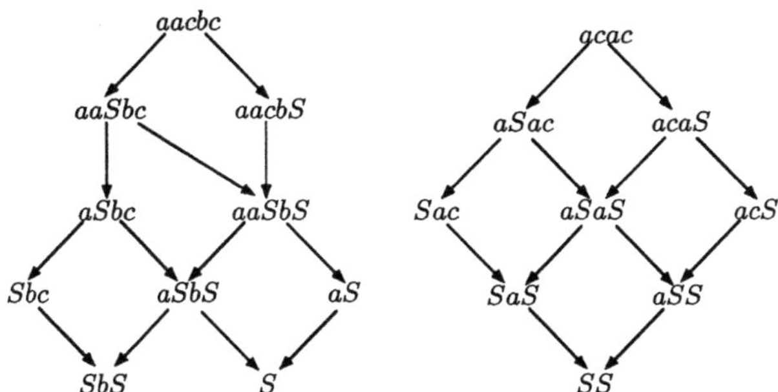

**Fig. 6.4.1** Graph representing a deduction structure

We now formulate such a property in terms of relations. The concept and some of the results below are essentially due to A. Church and B. Rosser (1936), and also to M. H. A. Newman (1942).

**6.4.1 Definition.** We call a relation

$$B \textbf{ confluent} \;\; :\Longleftrightarrow\;\; B^{\mathsf{T}*}B^* \subset B^*B^{\mathsf{T}*} \;\;\Longleftrightarrow\;\; B^{\mathsf{T}}B^* \subset B^*B^{\mathsf{T}*}$$
$$\Longleftrightarrow\;\; (B^{\mathsf{T}} \sqcup B)^* \subset B^*B^{\mathsf{T}*}$$
$$\Longleftrightarrow\;\; \text{Any two points with a common ancestor}$$
$$\text{have common descendants.}$$

The property of confluence, especially its last formulation, is known as the
**Church-Rosser property**.                                                    □


Before proving that the three properties above are equivalent we interpret them
in Fig. 6.4.2 in terms of kinship relations. Assume that any two points $x, y$ have
a descendant $z$ in common whenever they have a common ancestor $v$. It may
then seem a weaker requirement, if the common descendant need only exist when
the common ancestor $u$ is an (immediate) antecedent of $x$. On the other hand,
it may seem stronger a condition to require a common descendant for $x$ and $y$
if they are merely akin, which may have to be verified by repeatedly going up
and down in the genealogical tree.

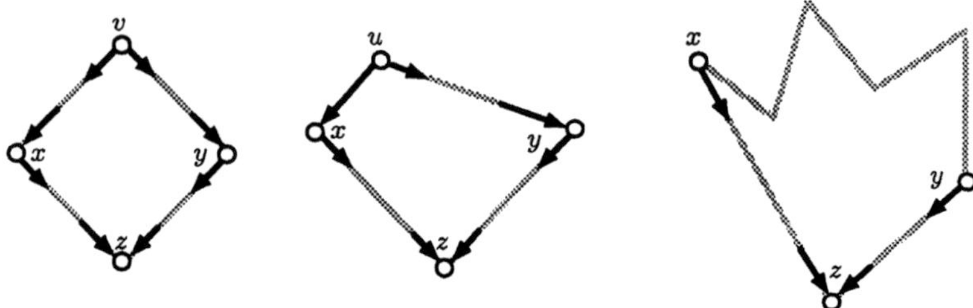

**Fig. 6.4.2** Variants of the confluence condition


Following Proposition 3.2.6.iii, $(B^\mathsf{T} \sqcup B)^* = B^* B^{\mathsf{T}*}$ is a condition equivalent to the
third definitional variant without equality. Transposing the second variant results
in $B^{\mathsf{T}*} B \subset B^* B^{\mathsf{T}*}$. Indeed, the definitional variants are equivalent: Only "$\Longleftarrow$"
is nontrivial between the first two. It may be obtained from $B^\mathsf{T} B^* \subset B^* B^{\mathsf{T}*}$
resulting in $B^\mathsf{T} B^\mathsf{T} B^* \subset B^\mathsf{T} B^* B^{\mathsf{T}*} \subset B^* B^{\mathsf{T}*} B^{\mathsf{T}*} = B^* B^{\mathsf{T}*}$. Formulating this as
an induction $B^{\mathsf{T}i} B^* \subset B^* B^{\mathsf{T}*}$ for $i \geq 0$ gives $B^{\mathsf{T}*} B^* \subset B^* B^{\mathsf{T}*}$. Direction "$\Longleftarrow$"
between the first and the third variant is a special case of Proposition 3.2.6.iii. It
remains to prove "$\Longrightarrow$" for this case: Due to $B^{\mathsf{T}*} B^* \subset B^* B^{\mathsf{T}*}$ the relation $B^* B^{\mathsf{T}*}$
is transitive and, therefore, an equivalence, since reflexivity and symmetry are
trivially satisfied. As $(B^\mathsf{T} \sqcup B)^*$ is the least equivalence containing $B^\mathsf{T}$ and $B$,
we have $(B^\mathsf{T} \sqcup B)^* \subset B^* B^{\mathsf{T}*}$.

The following result should be proved as Exercise 6.4.1.


### 6.4.2 Proposition.

$$B \text{ confluent} \quad \Longleftrightarrow \quad B^* B^{*\mathsf{T}} \text{ difunctional} \quad \Longleftrightarrow \quad B^* B^{*\mathsf{T}} \text{ equivalence.} \quad \square$$


If confluence holds, then points related by "kinship" always have common descendants. If there are several common descendants, then these are again akin
and have in turn, by confluence, common descendants. It is then conceivable
that this goes on forever, or that the family "dies out", meaning that a point
without successor, i.e., a terminal point, is ultimately reached. Next we prove
that a point can have at most one "childless" descendant.

**6.4.3 Proposition.** Let $C := B^* \sqcap \overline{BL}^\mathsf{T}$ denote the terminating reachability of the associated relation $B$.

i)   In a confluent graph, from any point at most one terminal point may be reached:

$$B \text{ confluent} \quad \Longrightarrow \quad C \text{ univalent.}$$

ii)   A graph is confluent, if from any point precisely one terminal point may be reached:

$$B \text{ confluent} \quad \Longleftarrow \quad C \text{ mapping.}$$

**Proof:** i) Firstly, $C^\mathsf{T}C \subset (B^* \sqcap \overline{BL}^\mathsf{T})^\mathsf{T} L \subset \overline{BL}L = \overline{BL}$ and analogously $C^\mathsf{T}C \subset \overline{BL}^\mathsf{T}$; secondly, from confluence we deduce $C^\mathsf{T}C \subset B^{\mathsf{T}*}B^* \subset B^*B^{\mathsf{T}*} = (I \sqcup B^+)(I \sqcup B^{+\mathsf{T}}) \subset I \sqcup BL \sqcup (BL)^\mathsf{T}$. Putting this together, we obtain $C^\mathsf{T}C \subset I$.

ii) Totality and univalence of $C$ are used together with Prop. 6.1.16.ii to prove that $B^{\mathsf{T}*}B^* \subset B^{\mathsf{T}*}CC^\mathsf{T}B^* = CC^\mathsf{T} \subset B^*B^{\mathsf{T}*}$.   □

Note that we did not assume the graph to be circuit-free, or progressively bounded, or progressively finite.

The above result is intuitively clear: If from some point two terminal points could be reached, then these would be akin, but no successor could be reached from them, let alone a common descendant. This would contradict the confluence condition.

Figure 6.4.3 contains four confluent graphs. It shows that Proposition 6.4.3.i does not imply that terminal points can actually be reached; nor that the graph need be circuit-free, or progressively bounded, or progressively finite in case a terminal point can be reached. Although points $a$ and $b$ are terminal in their respective confluent graphs, one of these graphs contains a circuit and the other a path of infinite length. On the other hand, if one assumes that from each point precisely one terminal point can be reached (Prop. 6.4.3.ii), then it is clear that all possible ramifications cannot ultimately diverge, and this means confluence.

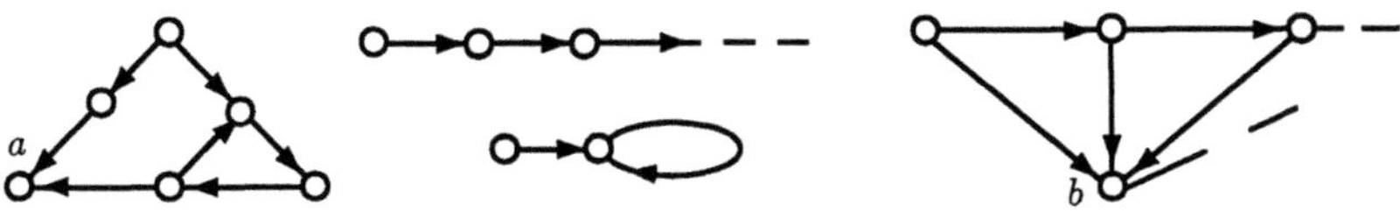

**Fig. 6.4.3** Confluence and terminating reachability

Confluence has been defined using reachability $B^*$. We now consider a similar condition involving $B$ instead of $B^*$.

**6.4.4 Proposition.** Any "rhombic" relation is confluent:

$$B^\mathsf{T}B \subset BB^\mathsf{T} \quad \Longrightarrow \quad B \text{ confluent.}$$

**Proof** by induction: $B^{\mathsf{T}i}B^j \subset B^j B^{\mathsf{T}i}$ yields $B^{\mathsf{T}*}B^* \subset B^*B^{\mathsf{T}*}$.   □

This proposition immediately yields a simple criterion for confluence based on a Hasse diagram of rhombic shape directed downwards as in Fig. 6.4.4. However, the situation can well be more complicated as we shall see in examples to come. We leave the proof to the reader.

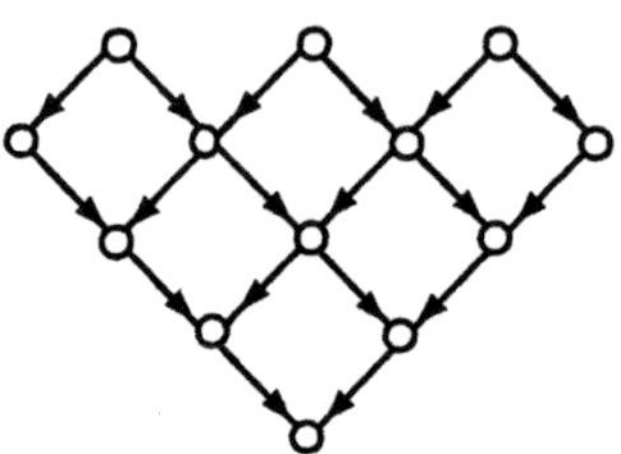

**Fig. 6.4.4** Rhombic confluence

**6.4.5 Corollary.** If for the relation $B$ there exists a relation $S$ satisfying $S^* = B^*$ and $S^\mathsf{T}S \subset SS^\mathsf{T}$, then $B$ is confluent. $\qquad\square$

By virtue of Definition 6.4.1 it is equivalent to consider relatives in general (version 1) or kinship with descendants of the father (version 2). We want to pursue this further and consider only kinship between brothers and sisters. As in ancient Egypt, we now postulate descendants only from them.

**6.4.6 Definition.** We call a relation

$B$ **locally confluent**   $:\Longleftrightarrow$   $B^\mathsf{T}B \subset B^*B^{\mathsf{T}*}$

$\qquad\qquad\qquad\qquad\Longleftrightarrow$   Any two points with a common immediate
$\qquad\qquad\qquad\qquad\qquad$ predecessor have a common descendant.   $\square$

Local confluence is easier to check than confluence proper. Obviously,

$$\text{confluent} \quad \Longrightarrow \quad \text{locally confluent.}$$

The converse need not hold. There is a simple counterexample due to R. Hindley (1974). Points $x$ and $y$ in Fig. 6.4.5 are in both cases related by $B^{\mathsf{T}*}B^*$, but not by $B^\mathsf{T}B$ nor by $B^*B^{\mathsf{T}*}$. So for the example on the left, local confluence holds but confluence does not. The example on the right is due to M. H. A. Newman and shows that the converse continues to fail even when there are no circuits. (This graph is a covering of the graph on the left, cf. Sect. 7.3.)

Newman showed already that "locally confluent" implies "confluent" for progressively finite graphs. The proof can be found in G. Huet; it is usually done by Noetherian induction.

**6.4.7 Proposition.** If $C$ is the terminating reachability of a progressively finite relation $B$, then the following holds:

$$B \text{ confluent} \quad \Longleftrightarrow \quad B \text{ locally confluent} \quad \Longleftrightarrow \quad C \text{ mapping.}$$

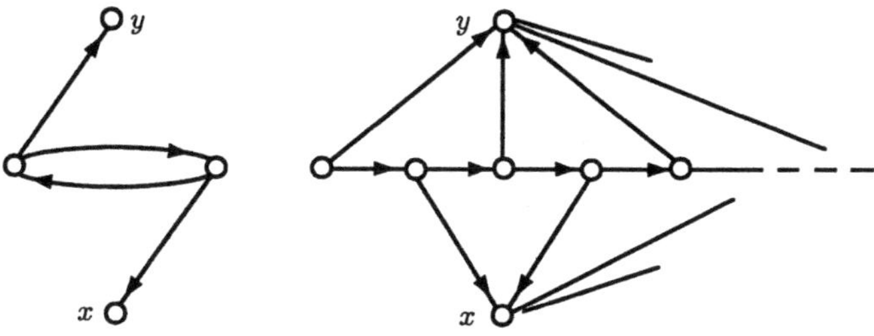

**Fig. 6.4.5** Local confluence without confluence

**Proof:** Direction "$\Longrightarrow$" of the first equivalence is trivial. If we manage to prove "$\Longrightarrow$" for the second equivalence, then by Prop. 6.4.3.ii, the proof may be closed cyclically.

Totality of $C$ is obtained from progressive finiteness by Prop. 6.3.3.iv:

$$L = J(B) = B^*\overline{BL} = (B^* \sqcap \overline{BL}^{\mathsf{T}})L = CL.$$

For the multivalent part $\mathrm{mup}\,(C) = C \sqcap C\overline{I} =: X$, we now show that $X \subset BX$. This is done in order to be able to conclude from progressive finiteness that $X = O$, i.e., that $C$ is univalent. From Prop. 6.1.15.ii we have $X = \mathrm{mup}\,(BC) = BC \sqcap BC\overline{I}$. There, the product $C\overline{I}$ is to be intersected with $C$ and with $\overline{C}$. The intersection with $C$ is obviously contained in $BX$ as required. For the intersection with $\overline{C}$ we obtain

$$BC \sqcap B(\overline{C} \sqcap C\overline{I}) \subset BC \sqcap B\overline{C} \subset (B \sqcap B\overline{C}C^{\mathsf{T}})(C \sqcap B^{\mathsf{T}}B\overline{C})$$

$$\subset B(C \sqcap B^*B^{\mathsf{T}*}\overline{C}), \qquad \text{due to local confluence}$$

$$= B(C \sqcap CC^{\mathsf{T}}\overline{C}), \qquad \text{according to Prop. 6.1.16.i}$$

$$\subset B(C \sqcap C\overline{I}) = BX. \qquad \qquad \square$$

Remembering how we arrived at the condition for local confluence $B^{\mathsf{T}}B \subset B^*B^{\mathsf{T}*}$, we may ask what happens if we replace $B^*$ by $B$ not only in the left hand side: The so modified condition is again a sufficient criterion for confluence. It is visualized in Fig. 6.4.6.

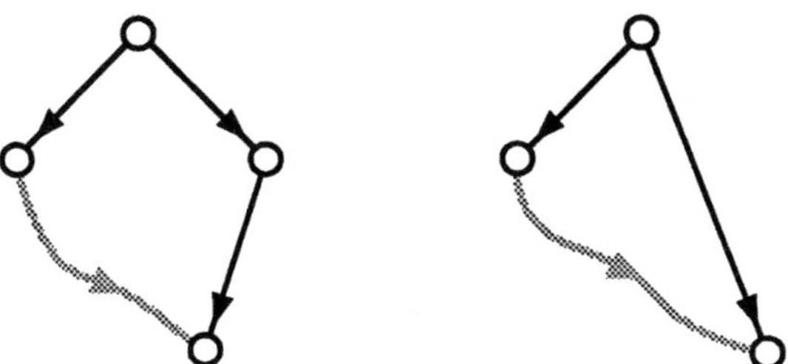

**Fig. 6.4.6** Sufficient criterion for confluence

**6.4.8 Proposition.**        $B^{\mathsf{T}}B \subset B^*(B \sqcup I)^{\mathsf{T}} \quad \Longrightarrow \quad B$ confluent.

**Proof:** The premise may be modified to $(B \sqcup I)^\mathsf{T} B \subset B^*(B \sqcup I)^\mathsf{T}$. In this form, we obtain

$$(B \sqcup I)^\mathsf{T} B^2 \subset B^*(B \sqcup I)^\mathsf{T} B \subset B^* B^*(B \sqcup I)^\mathsf{T} = B^*(B \sqcup I)^\mathsf{T}$$

and, again by induction, $(B \sqcup I)^\mathsf{T} B^* \subset B^*(B \sqcup I)^\mathsf{T}$. This gives $(B \sqcup I)^\mathsf{T} B^* \subset B^*(B \sqcup I)^{\mathsf{T}*}$ and, finally, $B^\mathsf{T} B^* \subset B^* B^{\mathsf{T}*}$.    □

In anticipation of such considerations we mention: The confluence property for a relation $B$ can be generalized to a condition on two relations $R$ and $S$,

$$R^{\mathsf{T}*} S^* \subset S^* R^{\mathsf{T}*},$$

it can also be formulated "modulo a relation" $Q$ (mostly an equivalence relation),

$$R^{\mathsf{T}*} Q^\mathsf{T} S^* \subset S^* Q R^{\mathsf{T}*}.$$

The techniques developed above mostly apply to these generalizations, too.


## Braid Relations

A realistic example for confluence is the plaiting of braids. Suppose the parents have combed their daughter's hair, have divided it into several strands, and start independently to make a pigtail, one parent working on the left, the other on the right. After numbering the spaces between the strands the resulting braid can be represented as the sequence of these numbers with plus or minus signs attached to them according as to whether the strand on the right was laid over the strand on the left, or conversely (Fig. 6.4.7).

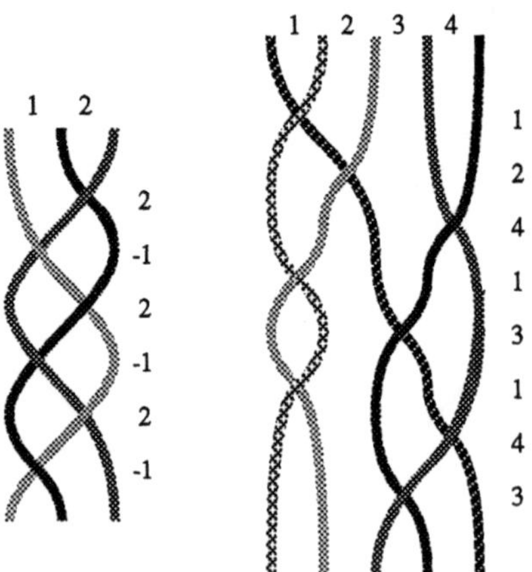


**Fig. 6.4.7** Braids as sequences of crossings

For the sake of simplicity we insist that the respective right-hand strands be laid *over* their neighboring left-hand strands. We see that the crossing sequences drawn opposite each other in Fig. 6.4.8 represent the same braid (after being fastened). So we obtain the defining relations for the semigroup of one-sidedly plaited braids

$$ik = ki, \quad \text{for} \quad |k - i| \geq 2$$
$$i(i+1)i = (i+1)i(i+1).$$

The two parents have meanwhile continued plaiting independently of one another, and now they see that the crossing sequences they have produced so far are distinct. But they want their kid to get two identical pigtails and are reluctant to start all over again. Can they now *continue* in such a way that, by virtue of the braid relations, they will eventually end up with the same result?

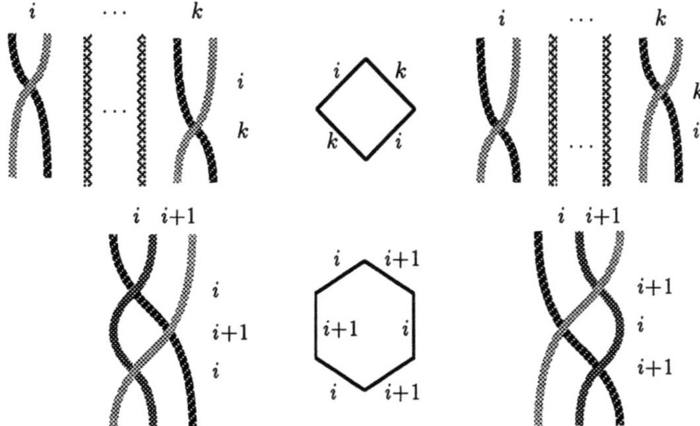

**Fig. 6.4.8** Braid relations

In the example in Fig. 6.4.9 it is shown explicitly, by paving with quadrangles and hexagons, how the father's braid 3142 and the mother's braid 215 can be continued to 3142134235432 or 2153213453423, respectively, both of which represent the same braid. The braids represented by both crossing sequences are once

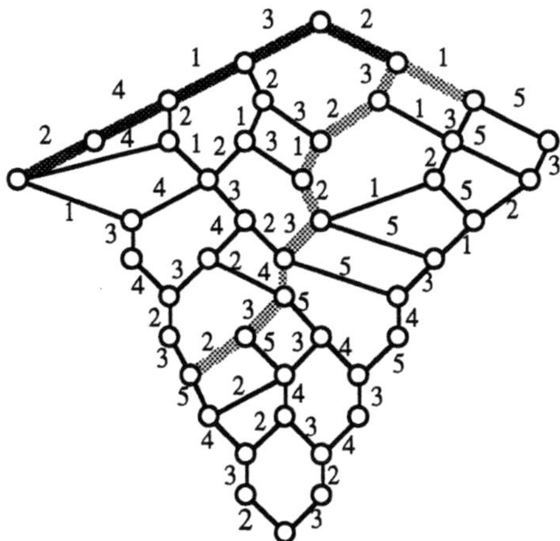

**Fig. 6.4.9** Paving by braid relations

more drawn in Fig. 6.4.10. The question arises whether such a procedure is always possible, irrespective of how both parents began. Mathematically speaking the question is whether the semigroup of braids plaited in the same direction is "right-reversible". Think of the edges in Fig. 6.4.9 as downward pointing arrows: The question is then whether the confluence condition holds. The leftmost and the rightmost point have the top point as a common predecessor. We are asking for a common successor (the bottom point). Clearly, local confluence holds by virtue of the basic building stones $\diamond$ and $\bigcirc$. Also, it is clear, in principle, how to continue the plaiting process: wherever there is a downward opening angle left over, pave it with quadrangles and hexagons. The only question that remains is whether this process terminates, i.e., whether the paving graph is progressively finite.

As one can see from the example in Fig. 6.4.9, attempts of a simple induction proof will fail. The shaded line seems to present a paving problem more difficult than the dark line. Would the father at position 3142 manage to catch up with the mother's first crossing 2, he would be faced with the problem of synchronizing the *longer* braid 32123432 with the crossing 1.

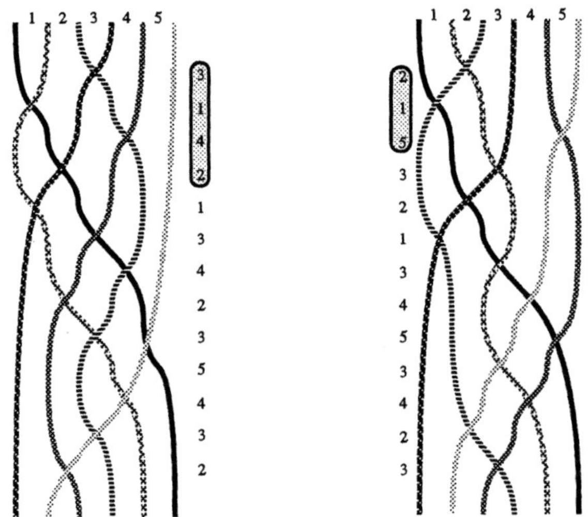

**Fig. 6.4.10** Two representations of a braid
by crossing sequences

Braids have been studied since the early investigations by E. Artin (1926) and K. Reidemeister (1932). The paving problem mentioned above is related to the word problem and the conjugacy problem for the braid group which has been treated by F. A. Garside (1969), G. S. Makanin (1968), and G. Schmidt (1969). There is a connection with classes of inessentially distinct deduction sequences in programming languages, established by H. Langmaack and G. Schmidt (1969).

## Exercises

**6.4.1** Prove Proposition 6.4.2.

**6.4.2** If $R_1$ and $R_2$ are confluent relations with $R_1^{\mathsf{T}*} R_2^* \subset R_2^* R_1^{\mathsf{T}*}$, then $R_1 \sqcup R_2$ is confluent. (Remark: The premise may be weakened; see STAPLES 75.)

# 6.5 Hasse Diagrams and Discreteness

In Sect. 6.1 we discussed paths of infinite length starting *from* one point in a graph, but we did not systematically study paths of infinite length *between* two given points, as we are going to do now. The results will be mainly applied to order relations.

H. Hermes terms a lattice **length-finite** if "every chain connecting any two elements is of finite length". The lattice in Fig. 6.5.3 is thus length-finite, the one in Fig. 6.5.2 is not. But for the time being we do not assume the relation under consideration to be a lattice.

For the purpose of this section we choose the following terminology: By saying that there is **a path of infinite length between two points** we mean that there is a path of finite length between them that can be refined indefinitely (in a proper fashion). By (properly) refining a path we mean that at least one of its arcs is replaced by a proper path, i.e., a path of length $> 1$ from the tail to the head of the arc. In Fig. 6.5.1 various ways of refining a path are illustrated. Refinements of paths have been investigated in BERGHAMMER, SCHMIDT 83.

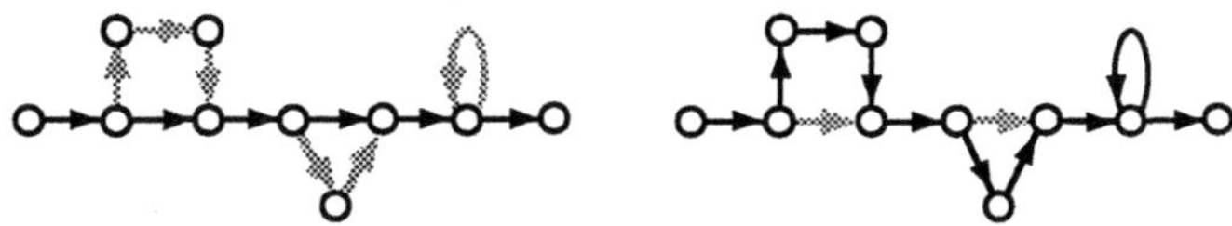

**Fig. 6.5.1** Path and refinement

Points $x$ and $0$ in Fig. 6.5.2 are connected by a path of infinite length because the paths

$$(x, 0), \ (x, 1, 0), \ (x, 2, 1, 0), \ (x, 3, 2, 1, 0), \ \ldots$$

are proper refinements of one another, the respective first arcs being systematically replaced by paths of length 2.

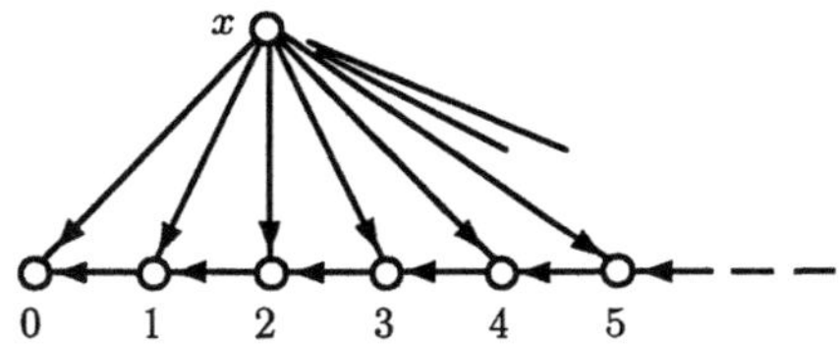


**Fig. 6.5.2** Non path-discrete and non discrete relation

In the sequel paths of infinite length are understood to be what we defined above.

**6.5.1 Definition.** We call a graph or its associated relation **path-discrete**, if there is no path of infinite length between any two of its points.                    □

So the relation in Fig. 6.5.2 is not path-discrete. Figure 6.5.3 represents a path-discrete relation where the lengths of paths from $x$ to $y$ are not bounded, however.

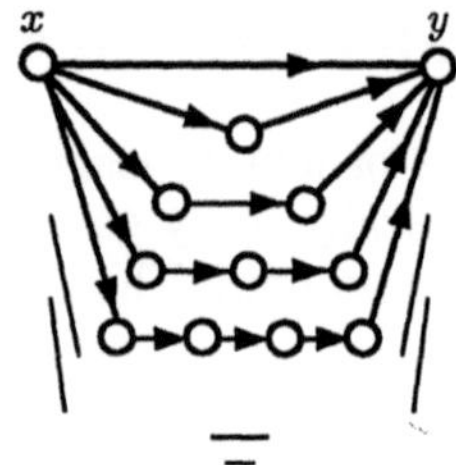

**Fig. 6.5.3** Path-discrete relation

Our definition of relations with bounded path lengths is based on the predicate logic formulation:

$$\forall x, y : \exists n_{xy} \in \mathbb{N} : \forall m \geq n_{xy} : xy^{\mathsf{T}} \subset \overline{B^m}.$$

**6.5.2 Definition.** We call a relation

$$B \text{ \textbf{path-length bounded}} \quad :\Longleftrightarrow \quad L = \sup_{n \geq 1} \inf_{m \geq n} \overline{B^m}. \qquad \square$$

For a transitive relation the condition is much simpler. Because a transitive relation satisfies $\inf_{m \geq n} \overline{B^m} = \overline{B^n}$, it is path-length bounded if $L = \sup_{n \geq 1} \overline{B^n}$.

A path-length bounded relation is always irreflexive; this is easily proved since $R \subset I \implies R^2 = R$, from which we obtain:

$$(I \sqcap B) = (I \sqcap B)^2 = (I \sqcap B)^3 = \ldots \subset \inf_{n \geq 1} \sup_{m \geq n} B^m = O.$$

There is a connection with progressive boundedness, $\sup_{n \geq 1} \overline{B^n L} = L$: A path-length bounded relation need not be progressively bounded (take the ordering of the natural numbers), but the converse does hold:

**6.5.3 Proposition.** For an arbitrary relation $B$ the following holds:

$$B \text{ progressively bounded} \quad \implies \quad B \text{ path-length bounded}.$$

**Proof:** Since $\overline{B^n L} \subset \overline{B^m}$ for all $m \geq n$, we obtain $\overline{B^n L} \subset \inf_{m \geq n} \overline{B^m}$. $\qquad \square$

We want to find a relation-algebraic formulation for path discreteness. For that purpose it is helpful to remember what we did in Sect. 6.3 in connection with progressive finiteness. There, we collected starting points of paths of infinite length in point sets $x$ satisfying the condition $x \subset Bx$. For a fixed relation $B$ we now consider the functional

$$\tau_B(X) := B \sqcap (BX \sqcup XB).$$

Clearly, $\tau_B$ is isotonic. By the Tarski-Knaster fixedpoint theorem (see Proposition A.3.1), therefore, $\tau_B$ has fixedpoints, and there is even a greatest such which is the supremum

$$N_B := \sup \mathcal{M}_B$$

of the set $\mathcal{M}_B := \{ X \mid X \subset \tau_B(X) \}$ consisting of the relations expanded by $\tau_B$. Belonging to $\mathcal{M}_B$ is inherited by joins ($X_i \in \mathcal{M}_B$, $i = 1, 2$ implies $X_1 \sqcup X_2 \in \mathcal{M}_B$) but not necessarily by meets (see Exercise 6.5.1). So the greatest fixedpoint $N_B$ cannot, in general, be approximated from above $L \supset \tau_B(L) \supset \tau_B^2(L) \dots$.

The following two concepts will be needed for our next results.

**6.5.4 Definition.** For an arbitrary relation $B$, we define $N_B$ to be the **nondiscrete part** of $B$. We call a relation

$$B \ \textbf{discrete} \quad :\Longleftrightarrow \quad N_B = O. \qquad\qquad \square$$

By part (i) of the following, a relation that is contained in some discrete relation is again discrete. By part (ii) all path-length bounded relations are discrete, hence, by Prop. 6.5.3, so are all progressively bounded relations. Finite, discrete, and transitive relations are easy to characterize: they are just the finite strict-orderings (Exercise 6.5.4). Compare a related concept in FELSCHER, SCHMIDT 58.

**6.5.5 Proposition.** Let $B, D$ be relations.

i) $\qquad\qquad B$ discrete and $D \subset B \quad \Longrightarrow \quad D$ discrete.

ii) $\qquad\qquad B$ path-length bounded $\quad \Longrightarrow \quad B$ discrete.

**Proof:** i) Every $Y \in \mathcal{M}_D$ satisfies $Y \subset D \sqcap (DY \sqcup YD) \subset B \sqcap (BY \sqcup YB)$, therefore, $Y \in \mathcal{M}_B$. So we have $\sup \mathcal{M}_D \subset \sup \mathcal{M}_B = O$.

ii) It is easy to prove by induction that

$$X \subset B^m \quad \text{for all } m \geq 1 \text{ and } X \in \mathcal{M}_B,$$

yielding $X \subset \sup_{m \geq n} B^m$ for all $n \geq 1$. So,

$$\sup \mathcal{M}_B \subset \inf_{n \geq 1} \sup_{m \geq n} B^m = O,$$

proving discreteness as desired. $\qquad\qquad \square$

With regard to discreteness there is a line separating progressive *boundedness* and progressive *finiteness*. While the former implies discreteness by Propositions 6.5.3 and 6.5.5, the latter does not, as can be seen from the example in Fig. 6.5.2 where $N_B$ is the set of arrows pointing downward from $x$. Conversely, discrete relations need not be progressively finite, not even when they are transitive, as can be seen from the strictly ordered natural numbers which are path-length bounded.

The transitive closure $B^+$ of a discrete relation $B$ need *not* be discrete as well, as is indicated in Fig. 6.5.4. (This fact can cause misunderstanding if an *ordering relation* is meant, but only its *Hasse diagram* is being considered.)

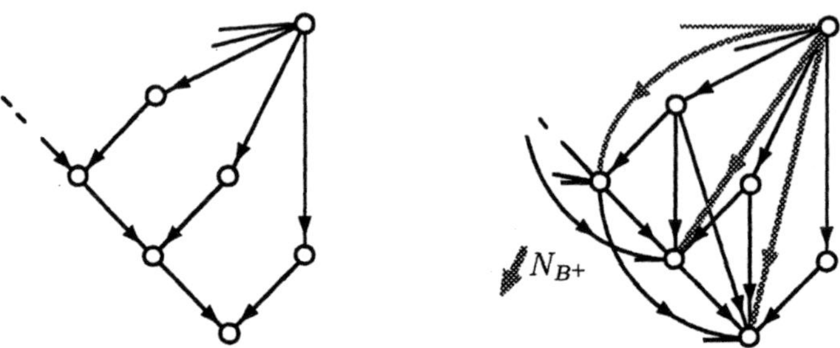

Fig. 6.5.4 Discrete relation with nondiscrete transitive closure

We now want to make clear that the abstract concept of discreteness, defined by means of fixedpoints, is essentially the same as path-discreteness.

**6.5.6.i Proposition.** For an arbitrary relation $B$ the following holds:

$$B \text{ path-discrete} \implies B \text{ discrete.}$$

**Proof:** Assume $B$ to be not discrete. Then there exists a relation $N \neq O$ with $N = B \sqcap (BN \sqcup NB)$. Using the point axiom this means that there are points $x$ and $y$ with $xy^\mathsf{T} \subset N$. Employing the intermediate point theorem 2.4.8, we obtain that there is a point $z$ with $xz^\mathsf{T} \subset B$ and $zy^\mathsf{T} \subset N$. Continuing this process, one obviously gets an infinite sequence of refinements $(x,y)$, $(x,z,y)$, ...    $\square$

For the converse statement one needs that $B$ is transitive.

**6.5.6.ii Proposition.** For a transitive relation $B$ the following holds:

$$B \text{ discrete} \implies B \text{ path-discrete.}$$

**Proof:** Let the relation $R$ describe the existence of paths of infinite length between two points. We are going to prove that $R \in \mathcal{M}_B$. Since $B$ is transitive, we have $R \subset B$. In order to prove $R \subset BR \sqcup RB$, we consider a pair of points satisfying $uv^\mathsf{T} \subset R$ and a path $(u = x_0, x_1, \ldots, x_n = v)$ with $n \geq 2$ which is infinitely refinable. On this path, we choose an arbitrary intermediate point $w = x_k$, $0 < k < n$. Then $uw^\mathsf{T} \subset R$ or $wv^\mathsf{T} \subset R$, because if there were no path of infinite length, neither between $u$ and $w$ nor between $w$ and $v$, then the path from $u$ to $v$ could not be infinitely refined. In the first case we have, by transitivity of $B$, that $uv^\mathsf{T} = uw^\mathsf{T} wv^\mathsf{T} \subset RB$ while in the second case $uv^\mathsf{T} \subset BR$. So, indeed, $R \in \mathcal{M}_B$ and $R \subset \sup \mathcal{M}_B = O$, since $B$ was supposed to be discrete.    $\square$

Discreteness is particularly desirable with orderings which one would like to represent economically by indicating their successor relations. We recall Prop. 3.2.5 and consider the set

$$\mathcal{G}_B := \{\, R \mid R \subset B,\ R^+ = B^+ \,\},$$

of elements we called *generators*. The minimal elements of $\mathcal{G}_B$ are the transitively irreducible kernels. If the generator $R$ is "small", then it seldom occurs that $R$ relates $x, y$ and $y, z$, and also $x, z$. We make this precise in the following

**6.5.7 Definition.** For a relation $B$ we call $H_B := B \sqcap \overline{B^2}$ the **nontransitive part** of $B$. If two points $x, y$ are related by $H_B$, then we call $y$ an **immediate successor** of $x$. If $B = H_B^+$, then the relation $H_B$ will be called the **Hasse diagram** of $B$. $\qquad\square$

Returning to the relations $R$ in Fig. 3.2.1 we see that the transitively irreducible kernel $K_1$ is equal to the nontransitive part $R \sqcap \overline{R^2}$, whereas the transitively irreducible kernel $K_2$ is different from $R \sqcap \overline{R^2}$. Transitively irreducible kernels, being minimal elements of $\mathcal{G}_B$, are by no means unique. In Fig. 6.5.2 $B \sqcap \overline{B^2}$ is the subgraph with the horizontal arrows and is contained in every generator.

Alternatively, the nontransitive part of a relation $B$ can be defined by expressions like

$$B \sqcap \overline{BB^+} = B \sqcap \overline{B^+B^+};$$

cf. ERNÉ 82. (Of course, for $B$ transitive one has $H_B = B \sqcap \overline{BB^+}$.) We give it as Exercise 6.5.2 to show that $B \sqcap \overline{BB^+}$, as well as $H_B$, is contained in every generator. In contrast to $H_B$, transitivity is not required here.

For reflexive relations the situation becomes irregular. The nontransitive part of the 3-clique equals $O$, and the two circuits are generators. But also the graph in which one point is joined to the two others by forward and backward pointing arrows is a generator for the 3-clique.

If $B$ has a Hasse diagram $H_B$, then $B$ is transitive by definition and $H_B$ belongs to $\mathcal{G}_B$.

**6.5.8 Proposition.** $B$ has a Hasse diagram $\implies$ $B$ is a strict-ordering.

**Proof:** Transitivity is obvious. In order to show irreflexivity, we subdivide

$$B = H_B^+ = H_B \sqcup H_B^+ H_B = H_B \sqcup B H_B$$

and prove $H_B \subset \overline{I}$ and $BH_B \subset \overline{I}$ separately. The first inclusion is an immediate consequence of $I \sqcap B = (I \sqcap B)^2 \subset B^2$, while $BH_B \subset \overline{I}$ follows from

$$BH_B \sqcap I \subset (B \sqcap IH_B^{\mathsf{T}})(H_B \sqcap B^{\mathsf{T}}I) \quad \text{and}$$
$$H_B \sqcap B^{\mathsf{T}} = B \sqcap \overline{B^2} \sqcap B^{\mathsf{T}} = (B^{\mathsf{T}}I \sqcap B) \sqcap \overline{B^2}$$
$$\subset (B^{\mathsf{T}} \sqcap BI^{\mathsf{T}})(I \sqcap BB) \sqcap \overline{B^2} \subset BB^2 \sqcap \overline{B^2} \subset BB \sqcap \overline{B^2} = O. \qquad\square$$

The converse to what we just proved does not hold in general, as can be seen from the strictly ordered real numbers.

The following lemma serves as a tool for establishing below the criterion for Hasse diagrams to exist.

**6.5.9 Lemma.** $B \sqcap \overline{H_B^+} \in \mathcal{M}_B$ for every $B$.

**Proof:** We have $B \sqcap \overline{H_B^+} \subset B$, and

$$B^2 = [(B \sqcap H_B^+) \sqcup (B \sqcap \overline{H_B^+})]^2 \subset H_B^+ \sqcup B(B \sqcap \overline{H_B^+}) \sqcup (B \sqcap \overline{H_B^+})B,$$

since $H_B^+$ is transitive. Therefore

$$B^2 \sqcap \overline{H_B^+} \subset B(\overline{H_B^+} \sqcap B) \sqcup (\overline{H_B^+} \sqcap B)B,$$

from which the result is deduced as follows

$$\begin{aligned}
B \sqcap \overline{H_B^+} &= B \sqcap \overline{H_B} \sqcap \overline{H_B^+} = B \sqcap (\overline{B} \sqcup B^2) \sqcap \overline{H_B^+} \\
&= [(B \sqcap \overline{B}) \sqcup (B \sqcap B^2)] \sqcap \overline{H_B^+} \subset B^2 \sqcap \overline{H_B^+}.
\end{aligned} \qquad \square$$

The nondiscrete relation $B$ in Fig. 6.5.2 obviously has no Hasse diagram since the transitive closure of the nontransitive part of $B$ fails to give $B$.

**6.5.10 Proposition.** $B$ is a discrete strict-ordering $\implies B$ has a Hasse diagram.

**Proof:** Proposition 6.5.9 and discreteness yield $B \sqcap \overline{H_B^+} = O$, so $B \subset H_B^+$. The containment $H_B^+ \subset B$ is a consequence of $H_B \subset B$ and $B = B^+$. $\qquad \square$

For a better understanding of the criterion, we suggest a direct proof of it in the finite case, without making use of the lemma above (Exercise 6.5.3). The example in Fig. 6.5.4 shows that the converse to Proposition 6.5.10 is false in general. Finite relations with Hasse diagrams can easily be characterized, however, cf. Exercise 6.5.4.

## Exercises

**6.5.1** The functional $\tau_B$ satisfies

$$\tau_B(X \sqcup Y) = \tau_B(X) \sqcup \tau_B(Y), \quad \tau_B(X \sqcap Y) \subset \tau_B(X) \sqcap \tau_B(Y).$$

Give an example with " $\subsetneq$ ".

**6.5.2** Prove for arbitrary $B$:

$B \sqcap \overline{BB^+}$ is a lower bound of $\{ R \mid R \subset B, \ R^+ = B^+ \}$.

**6.5.3** Prove that for $B$ a finite transitive irreflexive relation, $H_B := B \sqcap \overline{B^2}$ is the uniquely determined Hasse diagram. Hint: $B \subset \sup_{1 \leq i < n} H_B^i \sqcup B^n$.

**6.5.4** For finite transitive relations $B$ the following holds:

$B$ is discrete $\iff$ $B$ has a Hasse diagram $\iff$ $B$ is a strict-ordering.

**6.5.5** Prove that $H_{B^+} \subset H_B = H_{H_B}$ .


# 6.6 References


ARTIN E: *Theorie der Zöpfe.* Abh. Math. Sem. Univ. Hamburg **4** (1926) 47–72.

ARTIN E: *The theory of braids.* Ann. of Math. (2) **48** (1947) 101–126.

BERGHAMMER R, SCHMIDT G: *Discrete ordering relations.* Discrete Math. **43** (1983) 1–7.

CHURCH A, ROSSER JB: *Some properties of conversion.* Trans. Amer. Math. Soc. **39** (1936) 472–482.

ERNÉ M: *Einführung in die Ordnungstheorie.* Bibliographisches Institut, Mannheim, 1982.

EUWE M: *Mengentheoretische Betrachtungen über das Schachspiel.* Proc. Section of Sciences. Koninklijke Akad. van Wetensch. **32** (1929) 633–642.

FELSCHER W, SCHMIDT J: *Natürliche Zahlen, Ordnung, Nachfolge.* Arch. Math. Logik Grundlagenforsch. **4** (1958) 81–94.

GARSIDE FA: *The braid group and other groups.* Quart. J. Math. **20** (1969) 235–254.

HERMES H: *Einführung in die Verbandstheorie.* Springer, Berlin, 1955

HINDLEY R: *An abstract Church-Rosser theorem. II: Applications.* J. Symbolic Logic **39** (1974) 1–21.

HITCHCOCK P, PARK D: *Induction rules and termination proofs.* In: Nivat M (ed.): *Automata, languages and programming.* Proc. of a Symp. organized by IRIA, July 3–7 1972, Rocquencourt, North-Holland Publ. Co., Amsterdam, 1973, 225–251.

HUET G: *Confluent reductions: Abstract properties and applications to term rewriting systems.* J. Assoc. Comput. Mach. **27** (1980) 797–821.

KALMÁR L: *Zur Theorie der abstrakten Spiele.* Acta Sci. Math. (Szeged) **4** (1928/29) 65–85.

KÖNIG D: *Über eine Schlußweise aus dem Endlichen ins Unendliche.* Acta Sci. Math. (Szeged) **3** (1927) 121–130.

LANGMAACK H, SCHMIDT G: *Klassen unwesentlich verschiedener Ableitungen als Verbände.* In: Dörr J, Hotz G (eds.): *Automatentheorie und formale Sprachen.* Proc., Math. Forschungsinstitut Oberwolfach, 1969, Bibliographisches Institut, Mannheim, 1970, 169–182.

MAKANIN GS: *The conjugacy problem in the braid group.* Soviet Math. Dokl. **9** (1968) 1156–1157.

MENGER K: *Zur allgemeinen Kurventheorie.* Fund. Math. **10** (1927) 96–115.

NEUMANN J VON, MORGENSTERN O: *Theory of Games and Economic Behavior.* Princeton Univ. Press., Princeton, N. J. (1944).

NEWMAN MHA: *On theories with a combinatorial definition of "equivalence".* Ann. of Math. (2) **43** (1942) 223–243.

REIDEMEISTER K: *Knotentheorie.* Erg. der Math. **1** (1932).

SCHMIDT G: *Der Verband der Zöpfe.* Abteilung Math. der Technischen Universität München, Bericht 6914, 1969.

STAPLES J: *Church-Rosser theorems for replacement systems.* In: Crossley JN (ed.): *Algebra and logic.* Papers from the 1974 Summer Research Institute of the Australian Math. Soc. Monash Univ., Australia, Lecture Notes in Mathematics **450**, Springer, Berlin, 1975, 291–307.

ZERMELO E: *Über eine Anwendung der Mengenlehre auf die Theorie des Schachspiels.* In: Proc. 5th International Congress Mathematics, Vol. II, Cambridge, 1912, 501–504.

# 7. The Category of Graphs

Given any mathematical structure, one is interested in homomorphisms, substructures, and congruences. There is a characteristic difference between algebraic and relational structures.

*Algebraic* structures are defined by composition laws such as a binary multiplication $mult\colon A \times A \longrightarrow A$ or the unary operation of forming the inverse $inv\colon A \longrightarrow A$. These operations can, of course, be interpreted as relations. The first example furnishes a "ternary" relation $R_{mult} \subset (A \times A) \times A$, the second, an ordinary relation $R_{inv} \subset A \times A$, and both are univalent and total.

*Relational* structures are also defined by certain relations, but these need no longer be univalent or total. Purely relational structures are orderings, equivalences, and all types of graphs. Typically, however, mixed structures with both algebraic and relational features occur, such as ordered fields, for example.

In Sect. 7.1 homomorphisms and substructures for algebraic and relational structures are treated by our relation-algebraic method. In connection with substructures we resume our distinction between subsets given by specification or by injection (cf. Sect. 4.2).

For the sake of clarity, we first discuss our concepts in the case of 1-graphs. In Sect. 7.2, then, we define homomorphisms for hypergraphs and simple graphs in an analogous way. With the concept of homomorphy we can complete some results on factorization from Sects. 5.2 and 5.5.

Important special homomorphisms are coverings, to be discussed in Sect. 7.3. There are connections with topology, complex analysis, and with the equivalence problem for flow chart programs.

Section 7.4 deals with the relationship between mappings and equivalence relations. The topics include the so-called substitution property and the forming of quotients.

So far we have dealt only with binary relations (hence with mappings of just one variable). In Sect. 7.5 we show how to lift this restriction by introducing $n$-ary relations. This will also enable us to give the $n$-ary data base relations from Sect. 4.3 a relation-algebraic treatment. However, unforeseen problems will arise. There are relation algebras satisfying all axioms except the point axiom but which do not conform with our understanding of relations. Elementary examples will be given.

## 7.1 Homomorphisms of 1-Graphs

In general, a mathematical structure comprises several basic sets with various composition laws and relations defined between them. Take, for instance, a vector space over an ordered field with its addition of vectors and multiplication by scalars. Another example is furnished by the points and arrows of a graph with ingoing and outgoing incidence.

First we study the simplest case where just *one* set $V$ is given, and *one* relation $B$ defined on it, in other words, we study a 1-graph $G = (V, B)$.

If a 1-graph $G$ is to be mapped into another 1-graph $G'$, so that the structure is preserved, one needs a mapping $\Phi$ from the point set $V$ of $G$ into that of $G'$ which sends couples of points related by the associated relation $B$ into couples related by $B'$. If $x, y$ are two points from $V$ with images $\Phi^\mathsf{T}x$ and $\Phi^\mathsf{T}y$, then $xy^\mathsf{T} \subset B$ shall imply that $\Phi^\mathsf{T}x(\Phi^\mathsf{T}y)^\mathsf{T} \subset B'$, which is equivalent to $xy^\mathsf{T} \subset \Phi B' \Phi^\mathsf{T}$ since $\Phi$ is a mapping.

**7.1.1 Definition.** If $G = (V, B)$ and $G' = (V', B')$ are 1-graphs, we call the relation $\Phi$ a (1-graph-) **homomorphism** from $G$ to $G'$, provided

$$\Phi^\mathsf{T}\Phi \subset I, \quad I \subset \Phi\Phi^\mathsf{T}, \quad B \subset \Phi B' \Phi^\mathsf{T};$$

in other words, if $\Phi$ is a mapping from $V$ to $V'$ satisfying $B \subset \Phi B' \Phi^\mathsf{T}$. We will often write $\Phi: G \longrightarrow G'$ to indicate such a homomorphism.     □

Using $\Phi^\mathsf{T}\Phi \subset I$ and $I \subset \Phi\Phi^\mathsf{T}$, one immediately sees that $B \subset \Phi B' \Phi^\mathsf{T}$ is equivalent to each of the following three inclusions

$$B\Phi \subset \Phi B', \quad \Phi^\mathsf{T}B\Phi \subset B', \quad \Phi^\mathsf{T}B \subset B'\Phi^\mathsf{T}.$$

So the equivalent conditions for homomorphy are produced in a cyclic way by suitably multiplying by $\Phi$.

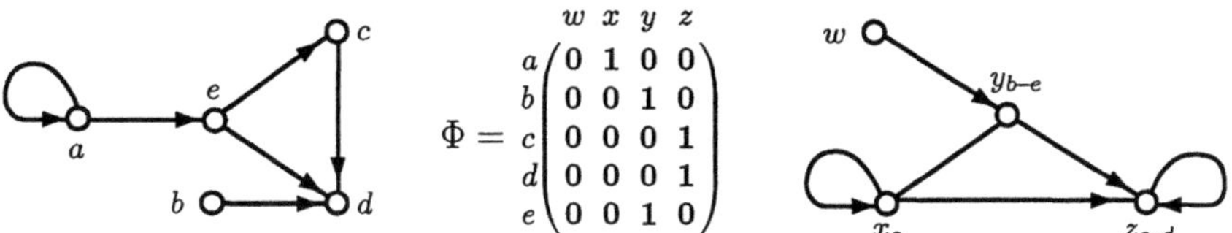



$$\Phi = \begin{array}{c} \\ a \\ b \\ c \\ d \\ e \end{array}\begin{array}{cccc} w & x & y & z \\ \left(\begin{array}{cccc} 0 & 1 & 0 & 0 \\ 0 & 0 & 1 & 0 \\ 0 & 0 & 0 & 1 \\ 0 & 0 & 0 & 1 \\ 0 & 0 & 1 & 0 \end{array}\right)\end{array}$$

**Fig. 7.1.1** A 1-graph homomorphism

Clearly, the composite $\Phi\Phi'$ of homomorphisms $\Phi: G \longrightarrow G'$ and $\Phi': G' \longrightarrow G''$ is again a homomorphism. The formal proof is as follows,

$$B \subset \Phi B' \Phi^\mathsf{T} \subset \Phi\Phi' B'' \Phi'^\mathsf{T}\Phi^\mathsf{T} = (\Phi\Phi')B''(\Phi\Phi')^\mathsf{T}.$$

In Fig. 7.1.1 a homomorphism is given. The action of the mapping $\Phi$ on the points is indicated by the matrix $\Phi$ and also by the subscripts attached to the points of the graph on the right. Looking at this homomorphism in detail we see that certain couples of points are identified, and then other points and arrows are added. If two points connected by an arrow are identified, such as $c$ and $d$, then the resulting point $z_{c-d}$ must carry a loop. If two points without an arrow between them are identified, such as $b$ and $e$ in the example, then the resulting point $y_{b-e}$ need not have a loop. Homomorphisms identifying just one couple of points in the manner described above are sometimes called **elementary homomorphisms**.

## Isomorphisms

It is well-known that isomorphisms of groups or vector spaces (without any further structure such as topology) are just invertible homomorphisms.

**Fig. 7.1.2** Bijective homomorphism
of a relational structure

This is *not* true for relational structures, as can be seen from Fig. 7.1.2 where the mapping $\Phi$ is a bijective homomorphism of two 1-graphs which one would not consider as isomorphic. So one has to require that the inverse mapping be a homomorphism, too.

**7.1.2 Definition.** Let two 1-graphs $G = (V, B)$ and $G' = (V', B')$, and a relation $\Phi$ be given. We call $\Phi$ a (1-graph-) **isomorphism** of $G$ and $G'$, if both $\Phi$ *and* $\Phi^\mathsf{T}$ are homomorphisms. $\qquad\square$

Thus an isomorphism is characterized by the following conditions

$$\Phi\colon V \longrightarrow V' \text{ bijective mapping,} \qquad B\Phi \subset \Phi B', \qquad B'\Phi^\mathsf{T} \subset \Phi^\mathsf{T} B.$$

The two inclusions on the right may equally well be replaced by those cyclically produced, equivalent conditions from above.

If $\Phi$ is an isomorphism, then clearly each of the cyclically produced homomorphy conditions becomes an equality,

$$B = \Phi B' \Phi^\mathsf{T}, \quad B\Phi = \Phi B', \quad \Phi^\mathsf{T} B \Phi = B', \quad \Phi^\mathsf{T} B = B' \Phi^\mathsf{T}.$$

However, these four equations are no longer equivalent to one another if $\Phi$ is not bijective; see Prop. 7.1.6.

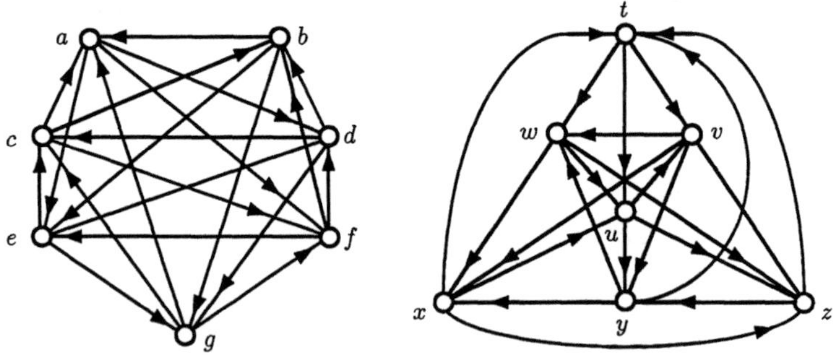

**Fig. 7.1.3** Two isomorphic graphs

It is a frequently occurring task in computer science to decide whether two given graphs are isomorphic[1], i.e., whether they can be identified by means of a suitable

---

bijection of their point sets. In simple enough cases this task can be accomplished in an ad hoc manner, but the computational effort for general algorithms is considerable. As an exercise, establish an isomorphism between the two graphs in Fig. 7.1.3.

## Partial Graphs and Subgraphs

We introduced the concept of isomorphy in order to characterize graphs that are "essentially the same". One would also like to know when a given graph can be considered as a part of some other graph. To this end, a suitable injective mapping $\iota$ of the points has to be found[2].

First of all, given a piece of some graph consisting of certain points and arcs, we want to endow it with a graph structure. We recall Definitions 2.4.1 and 4.2.7, and do not require $V$ to be a subset given by specification on the point set $V''$. Instead, we postulate an explicitly given injective mapping $\iota \subset V \times V''$ which is a homomorphism.

**7.1.3 Definition.** If $G = (V, B)$ and $G'' = (V'', B'')$ are two 1-graphs and if $\iota'' \colon G \longrightarrow G''$ is an injective homomorphism, we call $G$ a **partial subgraph** of $G''$. $\qquad\qquad\square$

Graph $G$ in Fig. 7.1.4 is a partial subgraph of $G''$ comprising only certain of the points of $G''$ and just certain (not all) of the arcs between these selected points.

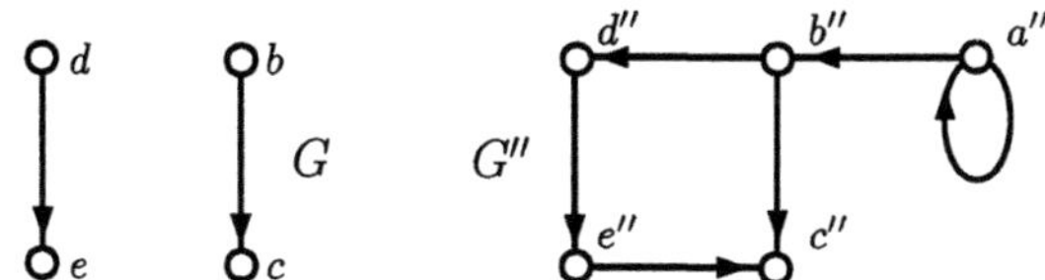

**Fig. 7.1.4** Partial subgraph as a separate graph

Since $\iota''$ maps the point set $V$ injectively into the point set $V''$, one may identify $V$ with a subset of $V''$ and need not draw $G$ separately. As in Fig. 7.1.5 it suffices to mark the points (squares) and arcs (dotted lines) constituting the partial subgraph.

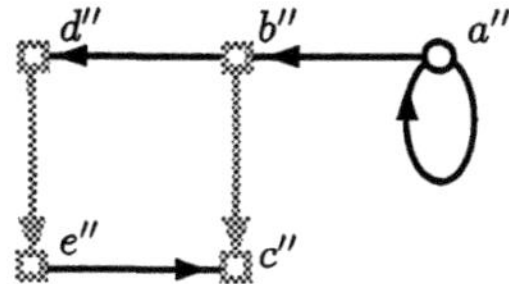

**Fig. 7.1.5** Partial subgraph embedded in the supergraph

---

[2] This problem has been shown to be $\mathcal{NP}$-complete.

There are two distinguished special cases of partial substructures. For one, the whole of the points can be chosen together with a possibly proper subset of arcs. The result is called a partial graph. A subgraph, on the other hand, consists of a subset of points together with *all* the arcs originally joining them.

**7.1.4 Definition.** If $G' = (V', B')$ is a partial subgraph of $G'' = (V'', B'')$ via the injective homomorphism $\iota'$, we call $G'$ the **subgraph of** $G''$, **generated by** $V'$, if $B' = \iota' B'' \iota'^\mathsf{T}$. $\qquad\square$

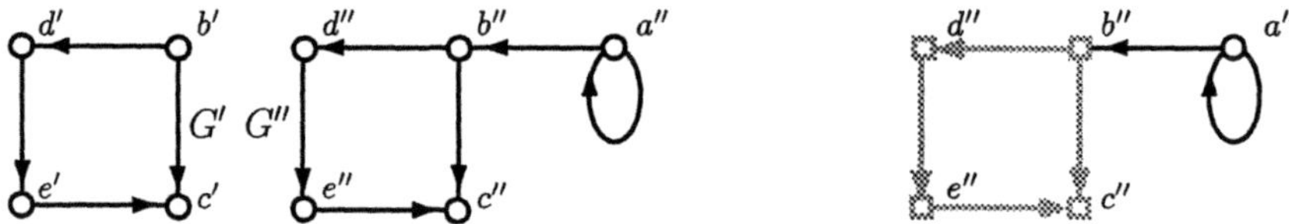

**Fig. 7.1.6** Subgraph, drawn separately, and embedded in the supergraph

Figure 7.1.6 shows the subgraph $G'$ generated by the point subset $V'$. Since it contains all the arrows that exist between its points in $G''$, it is unnecessary to mark its arrows explicitly when defining it *within* the bigger graph. The points of the subgraph are again depicted as squares.

We now come to the second special case.

**7.1.5 Definition.** If $G = (V, B)$ is a partial subgraph of $G' = (V', B')$ by the injective homomorphism $\iota$, we call $G$ **partial graph of** $G'$ **generated by** $B$, if $\iota$ is surjective. $\qquad\square$

Since $\iota$ is injective and maps $V$ onto $V'$, these two sets may be identified. A partial graph contains all the points but not necessarily all the arrows between them. So when representing a partial graph within its supergraph, one need only mark its arrows (dotted in Fig. 7.1.7).

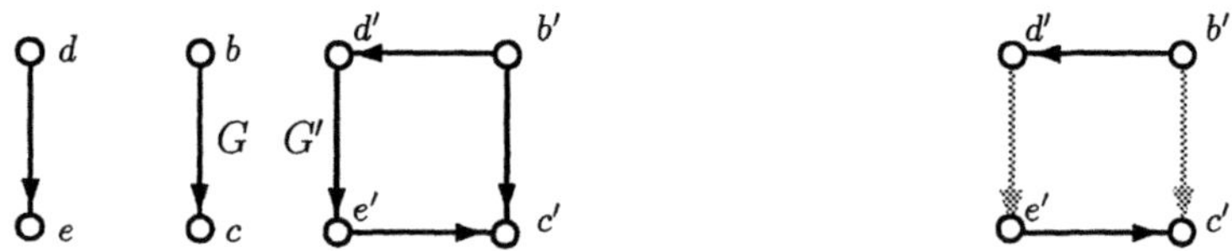

**Fig. 7.1.7** Partial graph, drawn separately, and embedded in the supergraph

Any partial subgraph $G$ of $G''$ can be thought of as a partial graph $G$ of some subgraph $G'$ of $G''$.

To give an example, think of the graph representing the relation "$x$ owes $y$ a sum of money". One may not want to consider all the debtor relations at once, but instead focus on a certain group of persons or on certain amounts owed. In the first case a subgraph is formed, in the second, a partial graph.

## Homomorphisms and Algebraic Structures

As we have seen, homomorphisms of graphs are not characterized by equations (such as $B\Phi = \Phi B'$), in contrast to other mathematical theories such as the theory of groups where the homomorphy condition is given by the well-known equation

$$\varphi(x + y) = \varphi(x) +' \varphi(y).$$

In Fig. 7.1.1 we have $a \not\subset B\Phi z$, but $a \subset \Phi B'z$. Groups, however, constitute an algebraic structure as opposed to the relational structure of graphs. So we ask what the impact on the concept of homomorphy will be when the set $V$ carries both a relational and an algebraic structure. We stick to the simplest case and assume that $B$ is a mapping from $V$ into $V$ (i.e., a unary operation on $V$). Our results remain valid for $n$-ary operations.

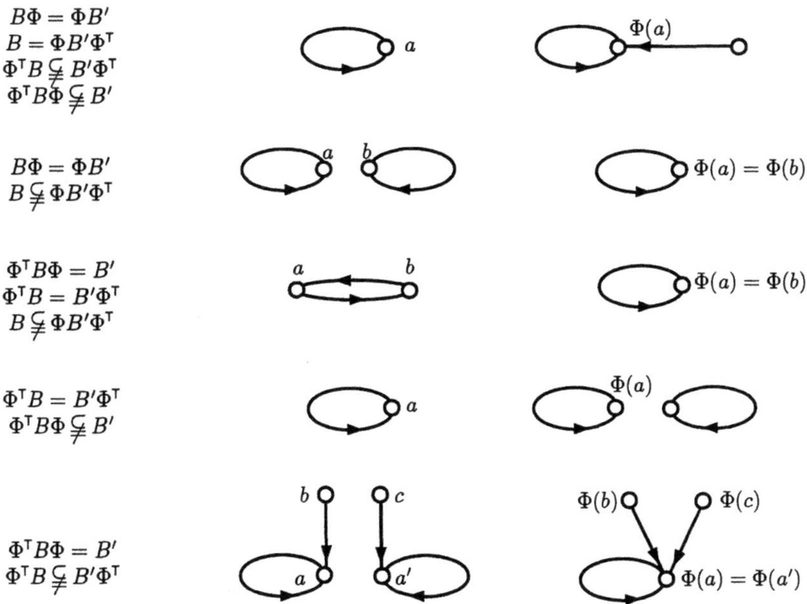

Fig. 7.1.8 Counterexamples related to Prop. 7.1.6.ii

**7.1.6 Proposition.** Let the graphs $G = (V, B)$ and $G' = (V', B')$ be given with $B$ and $B'$ mappings, and let $\Phi \subset V \times V'$ be a mapping.

i) $$B\Phi \subset \Phi B' \quad \Longleftrightarrow \quad B\Phi = \Phi B'.$$

ii) *Each* of the following three equations implies $B\Phi = \Phi B'$:
$$B = \Phi B'\Phi^{\mathsf{T}}, \quad \Phi^{\mathsf{T}} B\Phi = B', \quad \Phi^{\mathsf{T}} B = B'\Phi^{\mathsf{T}}.$$

For any other two of the four equations there exists an example where one is satisfied while the other is not.

**Proof:** i) Only direction "$\Longrightarrow$" needs a proof. In $B\Phi \subset \Phi B'$, the relation $\Phi B'$ is univalent, since $\Phi$ and $B'$ are mappings. The domains $B\Phi L = L = \Phi B'L$ of $B\Phi$ and $\Phi B'$ coincide, because all the relations are mappings and, therefore, are total. So we may use Proposition 4.2.2.iv and obtain $B\Phi = \Phi B'$.

ii) From any of the equations, $B\Phi \subset \Phi B'$ may be obtained by cyclic permutation, so that using (i) we have $B\Phi = \Phi B'$. For the cases of nonexistence, consider the counterexamples of Fig. 7.1.8. In every depicted case, there is a homomorphism from the left graph into the right graph, satisfying certain equations and violating others. $\qquad\square$

The condition for homomorphy, $B\Phi \subset \Phi B'$, holds for algebraic as well as for relational structures. The equation $B\Phi = \Phi B'$ applies to algebraic structures only.

Finally, we want to explain why the notion of isomorphism is simpler for algebraic than for relational structures. Given an algebraic structure, both $B$ and $B'$ are mappings. Therefore, the homomorphy condition $B\Phi \subset \Phi B'$ already implies the corresponding condition $B' \subset \Phi^\mathsf{T} B\Phi$ for the inverse mapping $\Phi^\mathsf{T}$. To see this, recall the equivalence from Proposition 7.1.6.i,

$$B\Phi \subset \Phi B' \quad \Longleftrightarrow \quad B\Phi = \Phi B'.$$

Multiplying the right-hand equation by $\Phi^\mathsf{T}$ from the left and using surjectivity $I \subset \Phi^\mathsf{T}\Phi$ yields that $B' \subset \Phi^\mathsf{T}\Phi B' = \Phi^\mathsf{T} B\Phi$.

## Exercise

**7.1.1** If $\Phi\colon G' \longrightarrow G''$ is a homomorphism, then $\Phi\overline{B''L}\Phi^\mathsf{T} \subset \overline{B'L}$, i.e., a terminal point has only terminal inverse images.

# 7.2 More Graph Homomorphisms

After having introduced homomorphisms for 1-graphs in the previous section we now extend this notion to hypergraphs, directed graphs, and simple graphs. The corresponding concepts of isomorphism and substructure follow the same pattern as before and will not be discussed in detail.

The only new feature of hypergraphs, compared to 1-graphs, is that they relate *several* sorts of objects. If the hypergraph $G$ is to be mapped homomorphically into the hypergraph $G'$, then a mapping $\Phi_V$ of the points *and* a mapping $\Phi_P$ of the hyperedges is to be given. Since $M \subset P \times V$, the homomorphy condition reads $M \subset \Phi_P M'\Phi_V^\mathsf{T}$, and is equivalent to three more such conditions fabricated as before,

$$M \subset \Phi_P M'\Phi_V^\mathsf{T}, \quad M\Phi_V \subset \Phi_P M', \quad \Phi_P^\mathsf{T} M\Phi_V \subset M', \quad \Phi_P^\mathsf{T} M \subset M'\Phi_V^\mathsf{T}.$$

Figure 7.2.1 is an example for a homomorphism of hypergraphs.

A directed graph is formed from a hypergraph if its incidence $M$ is given as the union $M = A \sqcup E$ of the univalent ingoing and outgoing incidences. Hence

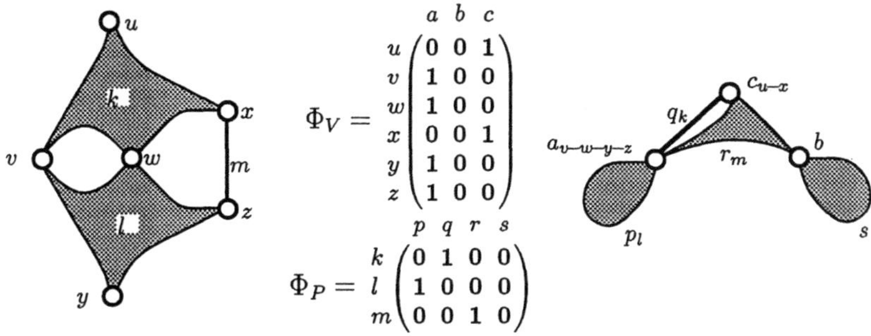

**Fig. 7.2.1** Homomorphism of hypergraphs

a homomorphism of directed graphs is one for hypergraphs which, in addition, is compatible with both $A$ and $E$. So a homomorphism must respect *several* sorts as well as *several* relations. Since $A \subset P \times V$, $E \subset P \times V$, this is expressed by

$$A \subset \Phi_P A' \Phi_V^\mathsf{T}, \quad E \subset \Phi_P E' \Phi_V^\mathsf{T}.$$

In Fig. 7.2.2 a graph homomorphism is shown which also involves partial graphs.

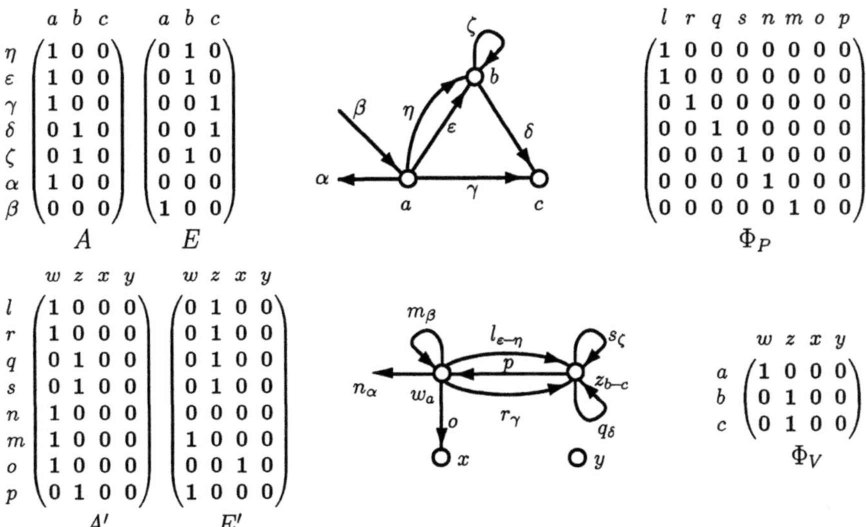

**Fig. 7.2.2** Homomorphism of directed graphs

Here again we distinguish two types of homomorphisms and call them elementary. The first such **elementary homomorphism** is one that sends two given points of $G$ onto one point in $G'$ and maps the remaining points and arrows of $G$ bijectively into $G'$. All the arrows starting from, or ending in, one of the distinguished points have then images that start from, or end in, the image point. An arrow between the two distinguished points becomes a loop at the image point.

The second elementary homomorphism maps two given arrows of $G$ into one in $G'$, while acting bijectively on the remaining arrows and points of $G$. The two distinguished arrows must then be parallel. As a result, surjective homomorphisms between finite graphs without partial arrows can be written as composites of elementary homomorphisms.

For simple graphs one has $\Gamma \subset V \times V$. Hence Definition 7.1.1 applies, and the condition for homomorphy is $\Gamma \subset \Phi \Gamma' \Phi^\mathsf{T}$. But elementary homomorphisms must be defined differently because the adjacency $\Gamma$ is always symmetric and irreflexive. A homomorphism could not possibly identify two points related by $\Gamma$ because the image point would then carry a loop, contradicting simplicity.

Thus we define an **elementary adjacency homomorphism** of simple graphs as one that identifies two *nonadjacent* points. The operation of identifying two adjacent points (and suppressing the resulting loop) is also needed sometimes, although it is not a homomorphism, and is called an **elementary contraction**. Figure 7.2.3 shows an adjacency homomorphism.

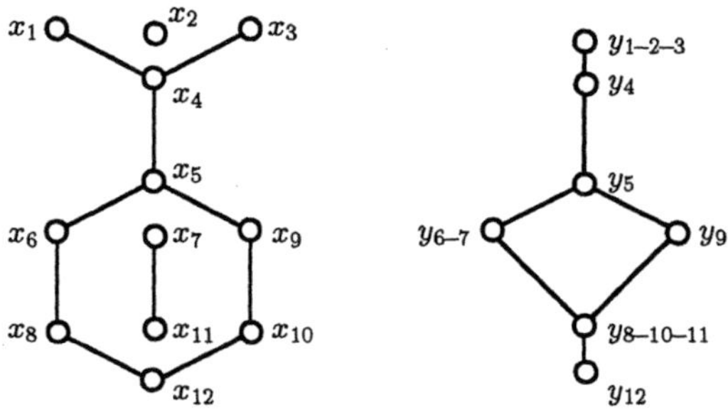


**Fig. 7.2.3** Adjacency homomorphism

## Factorization of the Associated Relation

We recall our considerations at the end of Sect. 5.2. A 1-graph can be given in two ways, either by its associated relation or by its two incidence relations if these satisfy $H = I$. In the first case a homogeneous relation $B$ on the point set $V$ is required, in the second, two univalent relations $A$ and $E$.

If the ingoing and the outgoing incidence relations $E$ and $A$ are given, then the associated relation is simply $B := A^\mathsf{T} E$. But, conversely, for a given $B$ there may exist several pairs of incidence relations $A, E$ producing the same $B = A^\mathsf{T} E$. In applications the following difficulty can arise: One starts with a 1-graph, deemed to be a sufficient model for the problem at the beginning, and is later compelled to augment the graph structure by admitting parallel arrows. Which of the directed graphs should one use?

The following completes Sect. 5.2. We show that under certain assumptions the factorizing of $B$ into $A^\mathsf{T}$ and $E$ is essentially unique.

**7.2.1 Proposition** (*Factorization of an associated relation into an outgoing and an ingoing incidence*). If $G_1 = (V, B)$ is a 1-graph, there exists, up to isomorphism, at most one total graph without parallel arcs $G = (P, V, A, E)$ such that $B = A^\mathsf{T} E$.

**Proof:** Suppose there exists another graph $G' = (P', V, A', E')$ satisfying the properties mentioned in Proposition 5.2.4, i.e., $H' := A'A'^\mathsf{T} \sqcap E'E'^\mathsf{T} = I$ and $B = A'^\mathsf{T} E'$. Then we may write down the isomorphism of $G$ and $G'$ directly as

$$\Phi_V = I, \quad \Phi_P := AA'^\mathsf{T} \sqcap EE'^\mathsf{T}.$$

So the image of an arc in $G'$ must start where its inverse image starts *and* must end where its inverse image ends. Both $\Phi_V$ and $\Phi_P$ are mappings, since

$$\begin{aligned}
\Phi_P \Phi_P^\mathsf{T} &= (AA'^\mathsf{T} \sqcap IEE'^\mathsf{T})(E'E^\mathsf{T} \sqcap A'A^\mathsf{T}I) \\
&\supset AA'^\mathsf{T}E'E^\mathsf{T} \sqcap I = ABE^\mathsf{T} \sqcap I = AA^\mathsf{T}EE^\mathsf{T} \sqcap I \supset II \sqcap I = I, \\
\Phi_P^\mathsf{T} \Phi_P &= (A'A^\mathsf{T} \sqcap E'E^\mathsf{T})(AA'^\mathsf{T} \sqcap EE'^\mathsf{T}) \\
&\subset A'A^\mathsf{T}AA'^\mathsf{T} \sqcap E'E^\mathsf{T}EE'^\mathsf{T} \subset A'A'^\mathsf{T} \sqcap E'E'^\mathsf{T} = H' = I.
\end{aligned}$$

The homomorphism conditions are satisfied because, using Prop. 4.2.2.iii twice, we obtain for $A$:

$$\begin{aligned}
A\Phi_V = A &= (AA^\mathsf{T} \sqcap EE^\mathsf{T})A = A \sqcap EE^\mathsf{T}A \\
&= A \sqcap EB^\mathsf{T} = A \sqcap EE'^\mathsf{T}A' = (AA'^\mathsf{T} \sqcap EE'^\mathsf{T})A' = \Phi_P A'.
\end{aligned}$$

Since the situation is completely symmetric with respect to $G$ and $G'$, the results obtained remain valid when exchanging $G$ and $G'$. Then $\Phi_P$ is replaced by $\Phi_P^\mathsf{T}$ and conversely, showing that $\Phi$ is an isomorphism. $\qquad\square$


## Factorization of the Adjacency Relation

A similar factorization problem came up already in Sect. 5.5. We saw that the adjacency relation $\Gamma$ belonging to the incidence relation $M$ is gotten by the simple formula $\Gamma := \overline{I} \sqcap M^\mathsf{T}M$, whereas given an adjacency relation $\Gamma$ one needs additional assumptions in order to arrive at an incidence relation. Specifically, one has to assume that every edge is incident with precisely two points, and that no two distinct edges are incident with the same pair of points.


**7.2.2 Proposition** (*First factorization of an adjacency into an incidence*). For an irreflexive symmetric relation $\Gamma$ there is, up to isomorphism, at most one essential hypergraph with edges of rank 2 and adjacency relation $\Gamma$.

**Proof:** Assume there exist two incidences $M$ and $M'$ of the type required with $\overline{I} \sqcap M^\mathsf{T}M = \Gamma = \overline{I} \sqcap M'^\mathsf{T}M'$. The obvious candidate for an isomorphism is the following $\Phi = (\Phi_P, \Phi_V)$:

$$\Phi_V := I, \quad \Phi_P := \mathsf{syq}(M^\mathsf{T}, M'^\mathsf{T}) = \overline{\overline{M}M'^\mathsf{T}} \sqcap \overline{M\overline{M'}^\mathsf{T}}.$$

By this relation, the set of points that are $M$-incident with an edge $p$ is considered, and is then related to those edges which are $M'$-incident with this set of points. For reasons of symmetry, only univalence and totality of $\Phi_P$, as well as the homomorphism condition, need to be shown. $\Phi_P$ is univalent, since

$$
\begin{aligned}
\Phi_P^\mathsf{T}\Phi_P &= \mathsf{syq}(M^\mathsf{T}, M'^\mathsf{T})^\mathsf{T}\,\mathsf{syq}(M^\mathsf{T}, M'^\mathsf{T}) \\
&\subset \mathsf{syq}(M'^\mathsf{T}, M'^\mathsf{T}), \qquad\qquad \text{using Props. 4.4.1 and 4.4.4,} \\
&\subset \overline{\overline{M'}M'^\mathsf{T}} \subset I, \qquad\qquad\quad \text{since } M' \text{ is essential.}
\end{aligned}
$$

Here, the homomorphism condition $\Phi_P^\mathsf{T}M\Phi_V \subset M'$ is equivalent to $\overline{M'}M^\mathsf{T} \subset \overline{\Phi_P^\mathsf{T}}$, which is obviously satisfied.

The proof of the totality of $\Phi_P$ has already been given as Exercise 5.5.1 and is relatively complicated. Expressing in terms of relations that every edge has rank 2 can be done by setting $M = A \sqcup E$ as the union of two mappings. $\qquad\square$

Note that we do not claim that there exists a factorization, rather that there is at most one (in a given relation algebra). However, Proposition 7.2.2 is not the only method of factorizing an adjacency into a hypergraph that is uniquely determined, at least to a certain extent.

### 7.2.3 Proposition (*Second factorization of an adjacency into an incidence*).
For an irreflexive symmetric relation $\Gamma$ there exists, up to isomorphism, at most one essential conformal hypergraph having adjacency relation $\Gamma$.

**Proof:** Again, we assume two hypergraphs with essential and conformal incidences $M$ and $M'$ to exist such that $\overline{I} \sqcap M^\mathsf{T}M = \Gamma = \overline{I} \sqcap M'^\mathsf{T}M'$. For a conformal incidence, we have in addition that $M^\mathsf{T}M = M'^\mathsf{T}M'$, since there can be no free point $x$, i.e., $I \subset M^\mathsf{T}M$ and $I \subset M'^\mathsf{T}M'$. If there were a free point $x$, then $Mx = O$, and we would have $xx^\mathsf{T} \sqcap \overline{I} \subset I \sqcap \overline{I} = O$. Employing Def. 5.5.1, this implies by conformity that $L \neq \overline{M}x = \overline{M}x \sqcup O = \overline{M}x \sqcup Mx = Lx = LxL = L$.

There is just one possibility of comparing the two incidences, namely comparing the $M$-incident or the $M'$-incident points of any hyperedge. As in Prop. 7.2.2 we define the mappings

$$
\Phi_V := I, \quad \Phi_P := \mathsf{syq}(M^\mathsf{T}, M'^\mathsf{T}) = \overline{M\overline{M'^\mathsf{T}}} \sqcap \overline{\overline{M}M'^\mathsf{T}}.
$$

Using Proposition 5.5.4 twice, we have

$$
\overline{M}M'^\mathsf{T} = M\overline{M^\mathsf{T}M}M'^\mathsf{T} = M\overline{M'^\mathsf{T}M'}M'^\mathsf{T} = M\overline{M'^\mathsf{T}},
$$

hence $\Phi_P = \overline{M\overline{M'^\mathsf{T}}} = \overline{\overline{M}M'^\mathsf{T}}$. Therefore, as $\Phi_P^\mathsf{T}$ is defined analogously to $\Phi_P$, homomorphy needs to be proved only for $(\Phi_P, \Phi_V)$.

The condition for a homomorphism $\Phi_P^\mathsf{T}M\Phi_V \subset M'$, or equivalently, $\overline{M'}M^\mathsf{T} \subset \overline{\Phi_P^\mathsf{T}}$, is obviously satisfied. Because $M'$ is essential, univalence of $\Phi_P$ results from

$$
\Phi_P\overline{I} = \overline{M\overline{M'^\mathsf{T}}}\,\overline{M'}M'^\mathsf{T} \subset \overline{MM'^\mathsf{T}} = \overline{\Phi_P}.
$$

At last we show that $\Phi_P$ is total: Let $p$ be an arbitrary hyperedge. From $pp^\mathsf{T} \subset I$ we obtain $M^\mathsf{T}pp^\mathsf{T}M \sqcap \overline{I} \subset \Gamma$ and further, using conformity of $M'$, that $L \neq \overline{M'}M^\mathsf{T}p$. Assuming that $\Phi_P$ does not assign any hyperedge to $p$, i.e., $\Phi_P^\mathsf{T}p \subset O$, yields $Lp^\mathsf{T} \subset \overline{\Phi_P^\mathsf{T}} = \overline{M'}M^\mathsf{T}$. This, however, is by Prop. 2.4.4

equivalent to $L = \overline{M'M^{\mathsf{T}}p}$, resulting in a contradiction. In the final steps of the proof we have again used the point axiom.      □

Returning to Fig. 5.5.1 we see that the third hypergraph is uniquely determined by the adjacency relation if, in accordance with the previous proposition, we allow only essential and conformal hypergraphs as candidates.

## 7.3 Covering of Graphs and Path Equivalence

Computers can perform only finite operations. *Infinite* objects therefore have to be described in some *finite* way. A language for example can be described by a grammar, a regular set by a regular expression or a finite automaton, and the variety of intended program executions by a program. In these examples the grammar, the automaton, and the program are not uniquely determined. The language can be described by several grammars, different programs can yield the same sequence of executed basic operations. So one may want to know which grammars or programs are related in this fashion. We shall first study this question in a graph-theoretical setting by running through rooted graphs and rooted trees. In Sect. 10.5 the results on path equivalence will be applied to flow chart equivalence and to program transformation.

The effect of a flow chart diagram is particularly easy to understand when the underlying rooted graph is also a rooted *tree*, so that the view is not blocked by nested circuits which can be traveled several times. When dealing with a rooted graph that does contain such circuits one has to keep track of the possibly infinite number of ways in which the graph can be traversed from its root. To this end we develop a theory of coverings which is based on the notion of homomorphy.

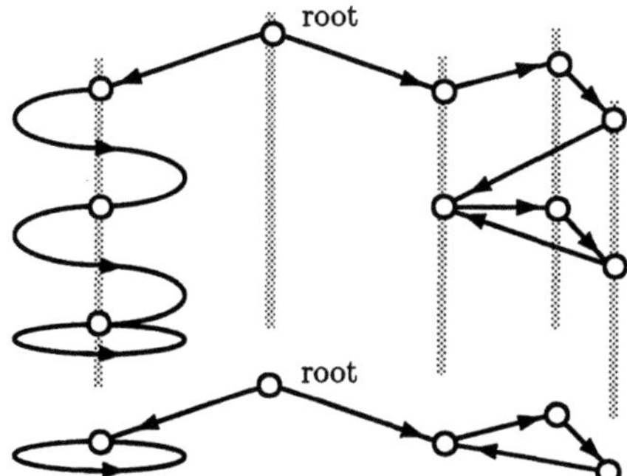


**Fig. 7.3.1** Partially unfolding circuits of a rooted graph

The idea is to unfold circuits as is done in Fig. 7.3.1. We want to characterize those homomorphisms of a graph that preserve to a certain extent the possibilities of traversal. We shall see that such a homomorphism is surjective and that it carries the successor relation at any point onto that at the image point.

**7.3.1 Definition.** A surjective homomorphism $\Phi: G \longrightarrow G'$ is called a **covering**, if

$$\Phi B' \subset B\Phi, \quad B^\mathsf{T} B \sqcap \Phi\Phi^\mathsf{T} \subset I.$$

We will also say that "$G$ is a covering of $G'$ by $\Phi$".                □

Clearly the composite of two coverings is again a covering. The first inclusion is actually an equality, $B\Phi = \Phi B'$, because "⊃" holds by homomorphy; it compares two relations between the points of $G$ and of $G'$ and ensures that for any inverse image point $x$ of some point $x'$ and successor $y'$ of $x'$ there is *at least* one successor $y$ of $x$ which is mapped onto $y'$. The second condition guarantees that there is *at most* one such $y$ since it requires that the relation "have a common predecessor according to $B$, and have a common image under $\Phi$" is contained in the identity. There is an elegant way of representing a covering: it is usually drawn by arranging the inverse images of a point as the fiber above it and by interpreting $\Phi$ as the projection onto the base point.

After these graph-theoretic considerations we give a more concrete interpretation of coverings. In Fig. 7.3.2 we consider the relationship between two grammars producing the same language, the arrows being labeled with symbols from a given set.

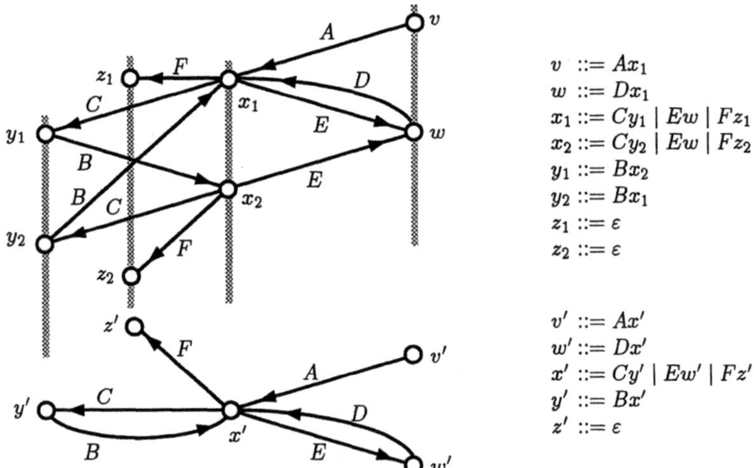


**Fig. 7.3.2** Covering of the rooted graphs related to two grammars yielding the same language

The first condition of Def. 7.3.1 immediately yields that $\Phi B'^* = B^*\Phi$. So a covering also carries the reachability relations into one another. Moreover, we shall show that for every path in $G'$ starting from $x'$ there is precisely one inverse image path starting from any inverse image point $x$ of $x'$. This closely resembles what is known for Riemann surfaces.

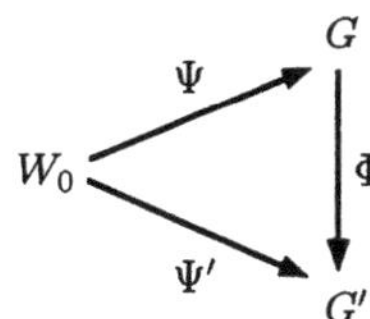

**Fig. 7.3.3** Lifting
to the covering graph

The following proposition is slightly more general, in as much as it treats a whole bundle of paths which, for technical reasons, is represented by the homomorphic image of a rooted tree. The point to be shown is that the homomorphism $\Psi'$ of the rooted tree $W_0$ in Fig. 7.3.3 can be lifted to a homomorphism $\Psi$. Theorems of this kind are familiar from topological covering theory. A relation-algebraic proof for rooted trees is given in SCHMIDT 76.

**7.3.2 Proposition** (*Lifting of paths*). Let $\Phi$ be a covering of the graph $G$ onto the graph $G'$, and let a point $a'$ of $G'$ be given together with an inverse image $a$ of $a'$ in $G$. Then for any given homomorphism $\Psi'$ of a rooted tree $W_0$ into the (covered) graph $G'$, sending the root $a_0$ into the given point $a'$, there exists a uniquely determined homomorphism $\Psi$ from $W_0$ into the covering $G$, sending $a_0$ into $a$ and satisfying $\Psi\Phi = \Psi'$.

**Proof:** Let $B, B'$, and $B_0$ denote the associated relations. From Def. 6.1.8 we have $L = B_0^{\mathsf{T}*}a_0$, so we may exhaust the point set $W_0$ by $L = \sup_{i \geq 0} V_i$ where $V_i := B_0^{\mathsf{T}i}a_0$ is the set of all points having distance $i$ from the root $a_0$. Denoting by $\Psi_i$ the restriction to $V_i$ of the mapping to be constructed, we define recursively:

$$\Psi_0 := a_0 a^{\mathsf{T}}$$
$$\Psi_i := B_0^{\mathsf{T}}\Psi_{i-1}B \sqcap \Psi'\Phi^{\mathsf{T}}$$
$$\Psi := \sup_{i \geq 0} \Psi_i.$$

Obviously, for $i = 0$ the root $a_0$ is mapped into the given point $a$. We shall prove that the relation $\Psi$ is univalent and total, and furthermore satisfies the homomorphism condition $B_0\Psi \subset \Psi B$ together with the "factorization condition" $\Psi\Phi = \Psi'$.

For $i > 0$, each point $x_0 \in V_i$ has a uniquely determined predecessor in $V_{i-1}$, the image of which under $\Psi_{i-1}$ is already known. The $\Psi_i$-image of $x_0$ in $G$ is chosen from the successors, so that $\Psi$ will satisfy the homomorphism condition. In addition, among the successors, that one belonging to the $\Phi$-fiber over the $\Psi'$-image of $x_0$ is taken, so $\Psi\Phi = \Psi'$ will hold. The formal proofs are as follows: Using Prop. 2.4.5 we have by definition $\Psi_0 = a_0 a^{\mathsf{T}} = a_0 a'^{\mathsf{T}}(aa'^{\mathsf{T}})^{\mathsf{T}} \subset \Psi'\Phi^{\mathsf{T}}$ as well as $\Psi_i \subset \Psi'\Phi^{\mathsf{T}}$, hence $\Psi \subset \Psi'\Phi^{\mathsf{T}}$. Therefore, $\Psi$ satisfies the first part $\Psi\Phi \subset \Psi'$ of the factorization condition. We obtain easily that $\Psi$ is a homomorphism:

$$B_0\Psi = \sup_{i \geq 0} B_0\Psi_i \subset B_0 a_0 a^{\mathsf{T}} \sqcup \sup_{i \geq 1} B_0 B_0^{\mathsf{T}}\Psi_{i-1}B \subset O \sqcup \sup_{i \geq 1} \Psi_{i-1}B \subset \Psi B,$$

remembering that the root $a_0$ of a rooted tree has no predecessor, $(B_0 a_0 = O)$, and that predecessors—if they exist—are uniquely determined, $(B_0 B_0^{\mathsf{T}} \subset I)$.

We manage to prove univalence of $\Psi$ by $\Psi_0^{\mathsf{T}}\Psi_0 \subset I$, $\Psi_0^{\mathsf{T}}\Psi_j = aa_0^{\mathsf{T}}\Psi_j = a(a_0^{\mathsf{T}}B_0^{\mathsf{T}}\Psi_{j-1}B \sqcap a_0^{\mathsf{T}}\Psi'\Phi^{\mathsf{T}}) = a(O \sqcap \ldots) = O$ if $j > 1$, and

$$\Psi_i^{\mathsf{T}}\Psi_j \subset B^{\mathsf{T}}\Psi_{i-1}^{\mathsf{T}}B_0\Psi_j \sqcap \Phi\Psi'^{\mathsf{T}}\Psi_j \subset B^{\mathsf{T}}\Psi_{i-1}^{\mathsf{T}}\Psi_{j-1}B \sqcap \Phi\Psi'^{\mathsf{T}}\Psi'\Phi^{\mathsf{T}},$$

using that $B_0\Psi_j \subset \Psi_{j-1}B$ is a further consequence of the preceding arguments. Now we proceed by induction on $i$ and $j$. Assuming $\Psi_{i-1}^\mathsf{T}\Psi_{j-1} \subset I$, we get $\Psi_i^\mathsf{T}\Psi_j \subset B^\mathsf{T}B \sqcap \Phi\Phi^\mathsf{T}$ and, by Def. 7.3.1, $\Psi_i^\mathsf{T}\Psi_j \subset I$ for all $i,j$. Therefore we have $\Psi^\mathsf{T}\Psi = \sup_{i,j\geq 0}\Psi_i^\mathsf{T}\Psi_j \subset I$, meaning that $\Psi$ is univalent.

In order to prove that $\Psi$ is total, we relate the domains of the $\Psi_i$:

$$
\begin{aligned}
\Psi_i L &= (B_0^\mathsf{T}\Psi_{i-1}B \sqcap \Psi'\Phi^\mathsf{T})L \\
&\supset (\Psi'\Phi^\mathsf{T} \sqcap B_0^\mathsf{T}\Psi_{i-1}B)(B^\mathsf{T} \sqcap \Phi\Psi'^\mathsf{T}B_0^\mathsf{T}\Psi_{i-1})L \\
&\supset (\Psi'\Phi^\mathsf{T}B^\mathsf{T} \sqcap B_0^\mathsf{T}\Psi_{i-1})L && \text{applying the Dedekind rule} \\
&= (\Psi'B'^\mathsf{T}\Phi^\mathsf{T} \sqcap B_0^\mathsf{T}\Psi_{i-1})L && \text{because of Def. 7.3.1 for } \Phi \\
&\supset (B_0^\mathsf{T}\Psi'\Phi^\mathsf{T} \sqcap B_0^\mathsf{T}\Psi_{i-1})L && \text{since } \Psi' \text{ is a homomorphism} \\
&= B_0^\mathsf{T}(\Psi'\Phi^\mathsf{T} \sqcap \Psi_{i-1})L && \text{using Prop. 4.2.2.ii for univalent } B_0^\mathsf{T} \\
&= B_0^\mathsf{T}\Psi_{i-1}L && \text{since } \Psi_{i-1} \subset \Psi'\Phi^\mathsf{T}.
\end{aligned}
$$

Iterating this, we get $\Psi_i L \supset B_0^{\mathsf{T}i}\Psi_0 L$, whence and by
$$
\Psi L = \sup_{i\geq 0}\Psi_i L \supset \sup_{i\geq 0} B_0^{\mathsf{T}i}\Psi_0 L = B_0^{\mathsf{T}*}\Psi_0 L = B_0^{\mathsf{T}*}a_0 = L
$$

we obtain that $\Psi$ is total. If $\Psi$ is total, so is $\Psi\Phi$; from $\Psi\Phi \subset \Psi'$, using Prop. 4.2.2.iv, we may now conclude that $\Psi\Phi = \Psi'$ holds.　　□

The inductive proof we have just given can be replaced by an argument more lattice-theoretic in nature where the definition by "stepwise exhaustion", $\Psi := \sup_{i\geq 0}\Psi_i$, is to be replaced by the "descriptive" form

$$\Psi := \inf\{\, X \mid a_0 a^\mathsf{T} \sqcup (B_0^\mathsf{T}XB \sqcap \Psi'\Phi^\mathsf{T}) \subset X \,\}.$$

The details are the same as in the proof given, with suitable modifications due to the definition of $\Psi$ as an infimum. We give this alternative proof as Exercise 7.3.1; it is shorter and more elegant, and avoids indices.

Proposition 7.3.2 yields the following corollary the proof of which is given as Exercise 7.3.2.

**7.3.3 Corollary.** If in the setting of Prop. 7.3.2 the homomorphism $\Psi'$ is in addition a covering, then $\Psi$ is a covering as well.　　□

By virtue of this corollary the coverings of a rooted graph $(G', a')$ by a rooted *tree* $(G, a)$ are distinguished, for they are at the same time coverings for any other covering of $(G', a')$. As a consequence, it can be shown that there is up to isomorphism at most one[3] covering rooted tree for any rooted graph.

With the aid of coverings we can now give a survey of *all* the paths in a rooted graph $W$ that start from the root $a$. The collection of all these paths carries the structure of a tree (infinite in general) which is a covering of $W$.

---

[3] We did not discuss the question whether covering rooted trees exist. Indeed, in a given relation algebra a rooted graph does not necessarily have such a covering tree. But if the relation algebra is "sufficiently large", then covering rooted trees can be constructed according to Proposition 7.3.4.

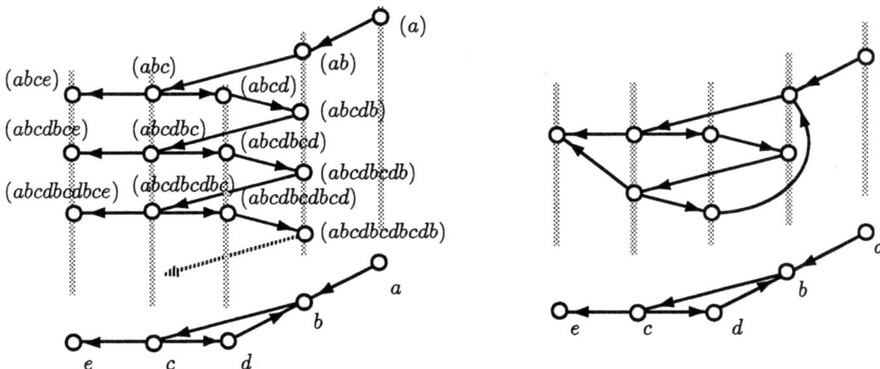

**Fig. 7.3.4** Universal covering and covering of a graph

**7.3.4 Proposition.** Let $(V, B)$ be a rooted graph with root $a$. We define the set $\widetilde{V}$ of all paths of finite length starting from $a$ as a set of points (of a new type). A point $\widetilde{v} \in \widetilde{V}$ will be connected by an arc with a point $\widetilde{w} \in \widetilde{V}$ precisely when the path $\widetilde{w}$ is longer by one point (of the old type) $y$ than the path $\widetilde{v}$:

$$\widetilde{v} \subset \widetilde{B}\widetilde{w} \quad :\Longleftrightarrow \quad x \subset By, \quad \text{where } \widetilde{v} = (a, \ldots, x) \text{ and } \widetilde{w} = (a, \ldots, x, y).$$

The graph $(\widetilde{V}, \widetilde{B})$ obtained in this way is a rooted tree whose root $\widetilde{a}$ corresponds to the path $(a)$ in point $a$ of the old graph.

**Proof:** $\widetilde{B}$ is indeed a relation on $\widetilde{V}$, hence $(\widetilde{V}, \widetilde{B})$ is a 1-graph. Since every finite path in the graph $(V, B)$ starting from $a$ may be obtained by successively extending the void path $(a)$ by an arc, $\widetilde{a} = (a)$ is a root of the graph $(\widetilde{V}, \widetilde{B})$. Finally, to every path of the old graph there corresponds at most (and to all paths unequal to $(a)$ precisely) one path that is shortened by one arc, so by Proposition 6.1.11.ii this constitutes a rooted tree. $\qquad\square$

In Fig. 7.3.4 two coverings of a graph are shown. By the covering on the right the graph is only partially unfolded; on the left, the beginning of the universal covering is indicated.

Our next result is that $(\widetilde{V}, \widetilde{B})$ is even a covering of $(V, B)$. We show that up to isomorphism this is the only rooted tree that covers $(V, B)$.

**7.3.5 Proposition.** i) Given the assumptions of Prop. 7.3.4, the mapping $\widetilde{\Phi}$ of the rooted tree $(\widetilde{V}, \widetilde{B})$ into the rooted graph $(V, B)$ which assigns to every $\widetilde{v} \in \widetilde{V}$ the endpoint of $\widetilde{v}$ (as a path in $V$) is a covering of $(\widetilde{V}, \widetilde{B})$ onto $(V, B)$.

ii) There is (up to isomorphism) at most one rooted *tree* covering $(V, B)$.

**Proof:** i) $\widetilde{\Phi}$ is a homomorphism: The successors of a point $\widetilde{v} \in \widetilde{V}$ are those paths obtained by attaching one more arc to $\widetilde{v}$. The images under $\widetilde{\Phi}$ of these paths are indeed the successors of the image of $\widetilde{v}$ in $V$. Considered modulo $\widetilde{\Phi}$, the fan of successors, in $\widetilde{v}$ and in the image of $\widetilde{v}$ coincide.

ii) The proposition does not say anything as to existence, so it may be proved relation-algebraically. Assume there are two rooted *trees* $(V_0, B_0)$ and $(V_0', B_0')$, both coverings of $(V, B)$ via the mappings $\Phi_0$ and $\Phi_0'$, respectively. Then the situation looks like Fig. 7.3.5.

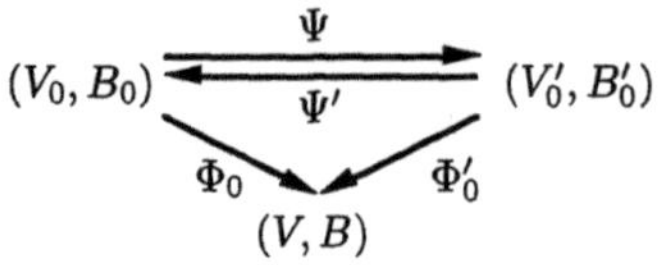

$$(V_0, B_0) \underset{\Psi'}{\overset{\Psi}{\rightleftarrows}} (V_0', B_0')$$
$$\Phi_0 \searrow \quad \swarrow \Phi_0'$$
$$(V, B)$$

**Fig. 7.3.5** Isomorphism of mutually covering rooted trees

Proposition 7.3.2 guarantees the existence of mappings $\Psi$ and $\Psi'$ factorizing $\Phi_0'$ and $\Phi_0$ to $\Phi_0' = \Psi'\Phi_0$ and $\Phi_0 = \Psi\Phi_0'$, respectively; by the remark following Proposition 7.3.2 they may be obtained as

$$\Psi := \inf \{\, Y \mid a_0 a_0'^{\mathsf{T}} \sqcup (B_0^{\mathsf{T}} Y B_0' \sqcap \Phi_0 \Phi_0'^{\mathsf{T}}) \subset Y \,\}$$
$$\Psi' := \inf \{\, X \mid a_0' a_0^{\mathsf{T}} \sqcup (B_0'^{\mathsf{T}} X B_0 \sqcap \Phi_0' \Phi_0^{\mathsf{T}}) \subset X \,\}.$$

Here, one is the transpose of the other. $\qquad\qquad$ □

We give a name to what we have established in the previous two propositions.

**7.3.6 Definition.** The rooted tree $(\widetilde{V}, \widetilde{B})$ defined in Proposition 7.3.4 is called *the* **covering rooted tree** or *the* **universal covering** of $(V, B)$. $\qquad$ □

The covering rooted tree is also termed the "initial element" among all coverings.


### Exercises

**7.3.1** Prove Prop. 7.3.2 with $\Psi := \inf \{\, X \mid a_0 a^{\mathsf{T}} \sqcup (B_0^{\mathsf{T}} X B \sqcap \Psi' \Phi^{\mathsf{T}}) \subset X \,\}$.

**7.3.2** Prove Corollary 7.3.3.

**7.3.3** Let rooted graphs $G$ and $G'$ be given together with a (not necessarily surjective) rooted graph homomorphism $\Phi$. Show that $\Phi$ is a covering, provided $\Phi B' \subset B\Phi$ and $B^{\mathsf{T}} B \sqcap \Phi \Phi^{\mathsf{T}} \subset I$.

**7.3.4** Prove that the composition of two coverings is again a covering.


# 7.4 Congruences

A mathematical structure on a supporting set often comes with an equivalence. If that equivalence relation is compatible with the structure, then one may want to divide it out, a common procedure in mathematics. Also the point set of a graph may carry an equivalence relation besides the associated relation $B$.

**7.4.1 Definition.** Given the 1-graph $G = (V, B)$ and the equivalence $\Xi$ on its point set $V$, we call

$$\Xi \;\; \boldsymbol{B}\text{-\textbf{congruence}} \;\; :\Longleftrightarrow \;\; \Xi B \subset B\Xi. \qquad\qquad □$$

Within a given relation algebra it is in general not possible to pass on to equivalence classes as one would normally do by dividing out a congruence. So we can only show that there is at most one quotient set up to isomorphism.

**7.4.2 Proposition.** If a 1-graph $(V, B)$ is given together with an equivalence $\Xi$, there is (up to isomorphism) at most one surjective homomorphism $\Phi$ onto a 1-graph $(V', B')$, satisfying

$$\Xi = \Phi\Phi^{\mathsf{T}} \quad \text{and} \quad B' = \Phi^{\mathsf{T}} B \Phi.$$

**Proof:** Assume, there is a second homomorphism $\Phi'$ onto $(V'', B'')$ satisfying $\Xi = \Phi'\Phi'^{\mathsf{T}}$ and $B'' = \Phi'^{\mathsf{T}} B \Phi'$. We are going to prove that $\Psi := \Phi^{\mathsf{T}}\Phi'$ is an isomorphism between $(V', B')$ and $(V'', B'')$. $\Psi$ is univalent and surjective:

$$\Psi^{\mathsf{T}}\Psi = \Phi'^{\mathsf{T}}\Phi\Phi^{\mathsf{T}}\Phi' = \Phi'^{\mathsf{T}}\Xi\Phi' = \Phi'^{\mathsf{T}}\Phi'\Phi'^{\mathsf{T}}\Phi' = II = I.$$

Injectivity and totality are shown analogously. We now prove one direction of the homomorphism condition:

$$B'\Psi = \Phi^{\mathsf{T}} B \Phi\Phi^{\mathsf{T}}\Phi' = \Phi^{\mathsf{T}} B\Xi\Phi' = \Phi^{\mathsf{T}} B\Phi'\Phi'^{\mathsf{T}}\Phi' \subset \Phi^{\mathsf{T}}\Phi' B''I = \Psi B''. \qquad \square$$

Particular congruences are generated by coverings discussed in Sect. 7.3.

**7.4.3 Proposition.** Let $\Phi$ be a homomorphism of the 1-graph $(V, B)$ into the 1-graph $(V', B')$ satisfying $B' = \Phi^{\mathsf{T}} B \Phi$, and define $\Xi := \Phi\Phi^{\mathsf{T}}$. Then $\Xi$ is a $B$-congruence if and only if $\Phi$ satisfies $\Phi B' \subset B\Phi$.

**Proof:** First, let $\Phi$ satisfy the condition. Clearly, $\Xi$ is reflexive, symmetric and transitive. The congruence condition is proved using $\Phi B' \subset B\Phi$ in the third position below

$$\Xi B = \Phi\Phi^{\mathsf{T}} B \subset \Phi B'\Phi^{\mathsf{T}} \subset B\Phi\Phi^{\mathsf{T}} = B\Xi.$$

Assume, on the other hand $\Xi$ to be a $B$-congruence. Then

$$\Phi B' = \Phi\Phi^{\mathsf{T}} B\Phi = \Xi B\Phi \subset B\Xi\Phi = B\Phi\Phi^{\mathsf{T}}\Phi \subset B\Phi. \qquad \square$$

In addition to the associated relation $B$, a homomorphism may preserve a congruence. There is some terminology for this situation.

**7.4.4 Definition.** Given equivalences $\Xi$, $\Xi'$ on the sets $V$, $V'$, respectively, and a mapping $\Phi : V \longrightarrow V'$ for which $\Xi\Phi \subset \Phi\Xi'$ holds true, we say that "$\Phi$ has the **substitution property** with respect to $\Xi$ and $\Xi'$". $\qquad \square$

This, of course, just means that $\Phi$ is a homomorphism from the 1-graph $(V, \Xi)$ into the 1-graph $(V', \Xi')$. A similar situation arises with $n$-ary mappings. When $n = 2$ and the equivalence relations $\Xi$ on $X$, $\Theta$ on $Y$, and $\Omega$ on the image set

$Z$ are given in infix notation, then the substitution property for $\varphi$ reads

$$\forall x_1, x_2 \in X \;\; \forall y_1, y_2 \in Y \; : \; x_1 \,\Xi\, x_2, \; y_1 \,\Theta\, y_2 \;\rightarrow\; \varphi(x_1, y_1) \,\Omega\, \varphi(x_2, y_2).$$

The general $n$-ary case will be treated in Sect. 7.5.

## Reduced Graphs

The following is an example of a congruence relation. We identify the points in each strongly connected component and investigate the resulting object (cf. Section 3.1 on equivalence relations and quotient sets).

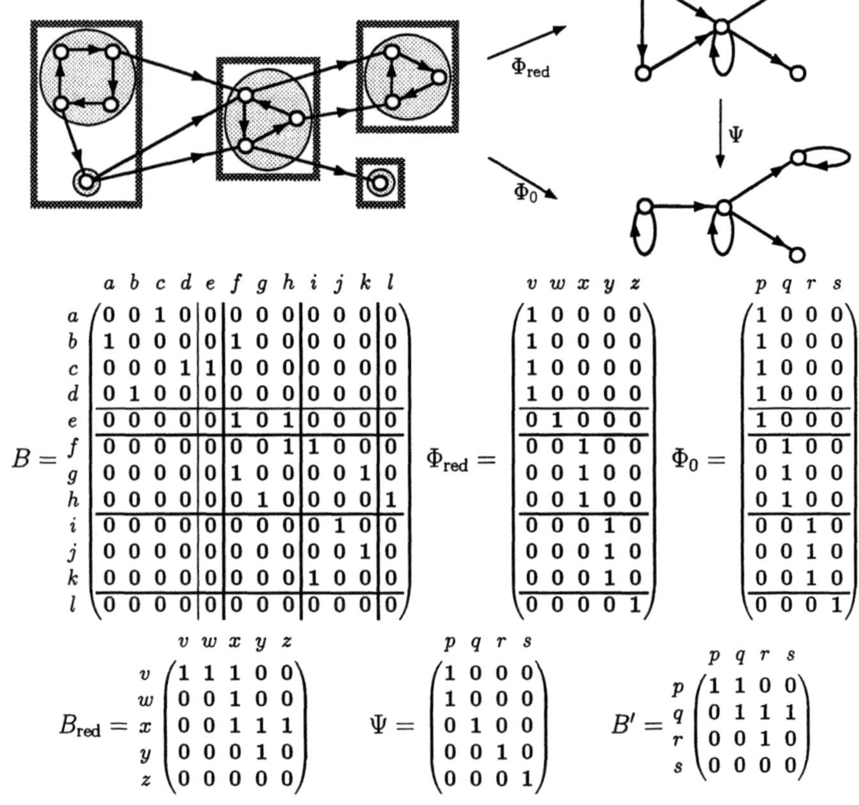

$$B = \begin{array}{c|cccc|c|ccc|ccc|c}
 & a & b & c & d & e & f & g & h & i & j & k & l \\
\hline
a & 0 & 0 & 1 & 0 & 0 & 0 & 0 & 0 & 0 & 0 & 0 & 0 \\
b & 1 & 0 & 0 & 0 & 0 & 1 & 0 & 0 & 0 & 0 & 0 & 0 \\
c & 0 & 0 & 0 & 1 & 1 & 0 & 0 & 0 & 0 & 0 & 0 & 0 \\
d & 0 & 1 & 0 & 0 & 0 & 0 & 0 & 0 & 0 & 0 & 0 & 0 \\
e & 0 & 0 & 0 & 0 & 0 & 1 & 0 & 1 & 0 & 0 & 0 & 0 \\
f & 0 & 0 & 0 & 0 & 0 & 0 & 0 & 1 & 1 & 0 & 0 & 0 \\
g & 0 & 0 & 0 & 0 & 0 & 1 & 0 & 0 & 0 & 0 & 1 & 0 \\
h & 0 & 0 & 0 & 0 & 0 & 0 & 1 & 0 & 0 & 0 & 0 & 1 \\
i & 0 & 0 & 0 & 0 & 0 & 0 & 0 & 0 & 0 & 1 & 0 & 0 \\
j & 0 & 0 & 0 & 0 & 0 & 0 & 0 & 0 & 0 & 0 & 1 & 0 \\
k & 0 & 0 & 0 & 0 & 0 & 0 & 0 & 0 & 1 & 0 & 0 & 0 \\
l & 0 & 0 & 0 & 0 & 0 & 0 & 0 & 0 & 0 & 0 & 0 & 0
\end{array}$$

$$\Phi_{\text{red}} = \begin{array}{c|ccccc}
 & v & w & x & y & z \\
\hline
a & 1 & 0 & 0 & 0 & 0 \\
b & 1 & 0 & 0 & 0 & 0 \\
c & 1 & 0 & 0 & 0 & 0 \\
d & 1 & 0 & 0 & 0 & 0 \\
e & 0 & 1 & 0 & 0 & 0 \\
f & 0 & 0 & 1 & 0 & 0 \\
g & 0 & 0 & 1 & 0 & 0 \\
h & 0 & 0 & 1 & 0 & 0 \\
i & 0 & 0 & 0 & 1 & 0 \\
j & 0 & 0 & 0 & 1 & 0 \\
k & 0 & 0 & 0 & 1 & 0 \\
l & 0 & 0 & 0 & 0 & 1
\end{array}
\qquad
\Phi_0 = \begin{array}{c|cccc}
 & p & q & r & s \\
\hline
a & 1 & 0 & 0 & 0 \\
b & 1 & 0 & 0 & 0 \\
c & 1 & 0 & 0 & 0 \\
d & 1 & 0 & 0 & 0 \\
e & 1 & 0 & 0 & 0 \\
f & 0 & 1 & 0 & 0 \\
g & 0 & 1 & 0 & 0 \\
h & 0 & 1 & 0 & 0 \\
i & 0 & 0 & 1 & 0 \\
j & 0 & 0 & 1 & 0 \\
k & 0 & 0 & 1 & 0 \\
l & 0 & 0 & 0 & 1
\end{array}$$

$$B_{\text{red}} = \begin{array}{c|ccccc}
 & v & w & x & y & z \\
\hline
v & 1 & 1 & 1 & 0 & 0 \\
w & 0 & 0 & 1 & 0 & 0 \\
x & 0 & 0 & 1 & 1 & 1 \\
y & 0 & 0 & 0 & 1 & 0 \\
z & 0 & 0 & 0 & 0 & 0
\end{array}
\qquad
\Psi = \begin{array}{ccccc}
 v & w & x & y & z \\
\begin{pmatrix} 1 & 0 & 0 & 0 \\ 1 & 0 & 0 & 0 \\ 0 & 1 & 0 & 0 \\ 0 & 0 & 1 & 0 \\ 0 & 0 & 0 & 1 \end{pmatrix}
\end{array}
\qquad
B' = \begin{array}{c|cccc}
 & p & q & r & s \\
\hline
p & 1 & 1 & 0 & 0 \\
q & 0 & 1 & 1 & 1 \\
r & 0 & 0 & 1 & 0 \\
s & 0 & 0 & 0 & 0
\end{array}$$

**Fig. 7.4.1** Factorization of a homomorphism
into a graph without proper circuits

**7.4.5 Definition.** Let $G$ be a graph with point set $V$, associated relation $B$, and strongly connected components $y_1, y_2, \ldots \subset V$. These components are now considered to be "points" and the point set

$$V_{\text{red}} := \{y_1, y_2, \ldots\}$$

is formed. Now, a point of $G$ is mapped by $\Phi_{\text{red}} \colon V \longrightarrow V_{\text{red}}$ onto that new type point, to which, considered as a strongly connected component, the former belongs. On $V_{\mathrm{red}}$ an associated relation is defined by

$$B_{\mathrm{red}} := \Phi_{\mathrm{red}}^{\mathsf{T}} B \Phi_{\mathrm{red}}.$$

Then $G_{\mathrm{red}} := (V_{\mathrm{red}}, B_{\mathrm{red}})$ is called the **reduced graph** of $G$.                  □


Thus the mapping $\Phi_{\mathrm{red}} : G \longrightarrow G_{\mathrm{red}}$ is a surjective homomorphism whose image contains no proper circuits, since if $G_{\mathrm{red}}$ did contain a proper circuit $(y_0, y_1, \ldots)$ through $y_0$, then every sequence $(x_0, x_1, \ldots)$ of inverse images could be completed to a circuit that passed through several connected components, contradicting Prop. 6.1.7. The strongly connected components of the reduced graph are one-element. $B_{\mathrm{red}}^*$ is by construction an order relation, for it is reflexive and transitive, and also antisymmetric since $\Phi_{\mathrm{red}}$ identifies mutually reachable points.

Our definition above can also be given in more formal terms. The equivalence relation $B^* \sqcap B^{\mathsf{T}*}$ gives rise to strongly connected components. It is difunctional since it is symmetric and transitive, or by Props. 4.4.5.i and 4.4.9. By virtue of Prop. 4.4.10 there is (up to isomorphism) *at most* one decomposition $B^* \sqcap B^{\mathsf{T}*} = FG^{\mathsf{T}}$ with surjective univalent relations $F$ and $G$. They are mappings since $I \subset B^* \sqcap B^{\mathsf{T}*}$. If this construction is possible in the given relation algebra, then for reasons of symmetry one has that $F = G = \Phi_{\mathrm{red}}$ and hence that $B^* \sqcap B^{\mathsf{T}*} = \Phi_{\mathrm{red}}\Phi_{\mathrm{red}}^{\mathsf{T}}$. Note that $B^* \sqcap B^{\mathsf{T}*}$ is a $B^*$-congruence rather than a $B$-congruence. One finds that $B_{\mathrm{red}}^* = \Phi_{\mathrm{red}}^{\mathsf{T}} B^* \Phi_{\mathrm{red}}$. Any other homomorphism $\Phi_0$ from $G$ into a graph $G'$ without proper circuits is necessarily "coarser" in the sense that it identifies at least all those points identified by $\Phi_{\mathrm{red}}$. This can be expressed in a formal way by saying that $\Phi_0$ breaks down into $\Phi_{\mathrm{red}}$ followed by some homomorphism $\Psi$, so that $\Phi_0 = \Phi_{\mathrm{red}}\Psi$. This homomorphism is given by $\Psi := \Phi_{\mathrm{red}}^{\mathsf{T}} \Phi_0$, cf. Exercise 7.4.1. In Fig. 7.4.1, $\Phi_{\mathrm{red}}$ maps the circles onto one point, and $\Phi_0$ acts in the same way on the quadrangles.


## Exercise

**7.4.1**  Prove that every homomorphism $\Phi_0$ of a graph $G$ into a graph $G'$ without proper circuits $(B'^* \sqcap B'^{\mathsf{T}*} \subset I)$ may be factorized into a homomorphism $\Phi_{\mathrm{red}} : G \longrightarrow G_{\mathrm{red}}$ and a homomorphism $\Psi : G_{\mathrm{red}} \longrightarrow G'$.


# 7.5 Direct Product and $n$-ary Relations


We have so far studied homomorphisms, substructures, and congruences for binary relations only. So we did not even deal with functions of two variables since they are ternary relations. Most operations, however, that occur in real life involve several arguments and sorts. In order to treat these we need the Cartesian product and the projection mappings which associate with each pair its first and second component, respectively.

Let us first look at a simple example. Let two sets, $X := \{a, b, c\}$ and $Y := \{1, 2\}$, be given and form their Cartesian product

$$X \times Y = \{ (a,1), (a,2), (b,1), (b,2), (c,1), (c,2) \}.$$

Then there are the two natural projections $\pi$ and $\rho$ as drawn in Fig. 7.5.1.

$$
\pi = \begin{array}{c} \\ (a,1) \\ (a,2) \\ (b,1) \\ (b,2) \\ (c,1) \\ (c,2) \end{array}
\begin{pmatrix} a & b & c \\ 1 & 0 & 0 \\ 1 & 0 & 0 \\ 0 & 1 & 0 \\ 0 & 1 & 0 \\ 0 & 0 & 1 \\ 0 & 0 & 1 \end{pmatrix}
\qquad
\rho = \begin{pmatrix} 1 & 0 \\ 0 & 1 \\ 1 & 0 \\ 0 & 1 \\ 1 & 0 \\ 0 & 1 \end{pmatrix}
\qquad
t = \begin{pmatrix} 1 \\ 0 \\ 1 \\ 1 \\ 0 \\ 1 \end{pmatrix}
\qquad
R = \begin{array}{c} \\ a \\ b \\ c \end{array}\begin{pmatrix} 1 & 0 \\ 1 & 1 \\ 0 & 1 \end{pmatrix}
$$

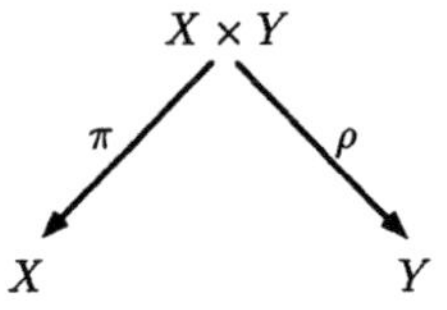

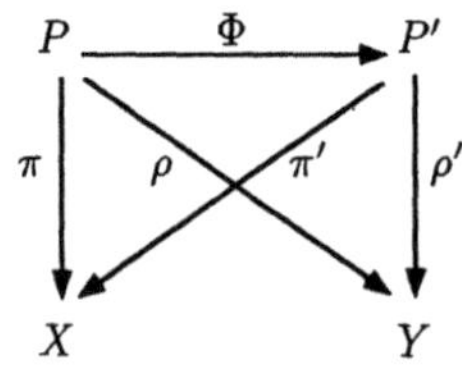


**Fig. 7.5.1** Natural projections

We give some properties of natural projections $\pi, \rho$ as a definition and show below to what extent these are characterizing.

**7.5.1 Definition.** Given two relations $\pi$ and $\rho$, we call the pair

$$(\pi, \rho) \ \textbf{direct product} \quad :\Longleftrightarrow \quad \begin{cases} \pi^{\mathsf{T}}\pi = I, & \rho^{\mathsf{T}}\rho = I, \\ \pi\pi^{\mathsf{T}} \sqcap \rho\rho^{\mathsf{T}} = I, & \pi^{\mathsf{T}}\rho = L. \end{cases}$$

In this setting, $\pi$ and $\rho$ are called the **natural projections.**    □

By this definition the natural projections are clearly surjective mappings. It even suffices to require that $\pi^{\mathsf{T}}\pi \subset I$, for from Prop. 4.2.2.iii it then follows that $\pi^{\mathsf{T}}\pi = \pi^{\mathsf{T}}(\pi\pi^{\mathsf{T}} \sqcap \rho\rho^{\mathsf{T}})\pi = I \sqcap \pi^{\mathsf{T}}\rho\rho^{\mathsf{T}}\pi = I \sqcap LL = I$. The condition $\pi\pi^{\mathsf{T}} \sqcap \rho\rho^{\mathsf{T}} \subset I$ ensures that there is *at most* one pair with given images in $X$ and $Y$.

Finally, the condition $\pi^{\mathsf{T}}\rho = L$ implies that for *every* element in $X$ and *every* element in $Y$ there is indeed a pair in $X \times Y$. If one were to omit all the rows belonging to $(c,2)$ in Fig. 7.5.1, then the resulting relations would satisfy all the conditions of the definition with the exception of $\pi^{\mathsf{T}}\rho = L$.

**Fig. 7.5.2** Monomorphism of the direct product

For the Cartesian product to be uniquely determined we need to show that given two sets $X$ and $Y$ there can be no two essentially distinct pairs of natural projections $\pi, \rho$ and $\pi', \rho'$. It is an obvious question whether (a model for) a set $X \times Y$ with projections $\pi, \rho$ does at all exist—in a given relation algebra it need not. Assuming that two different direct products are given, we now establish an isomorphism between them which is easily recognized in Fig. 7.5.2.

**7.5.2 Proposition** (*Monomorphic characterization of direct products*). If direct products $(\pi,\rho)$ and $(\pi',\rho')$ are given for which the constructs $\pi\pi'^{\mathsf{T}}$, $\rho\rho'^{\mathsf{T}}$ are defined, then

$$\Phi := \pi\pi'^{\mathsf{T}} \sqcap \rho\rho'^{\mathsf{T}}$$

is a bijective mapping satisfying

$$\pi = \Phi\pi', \quad \rho = \Phi\rho' \qquad (\pi' = \Phi^{\mathsf{T}}\pi, \quad \rho' = \Phi^{\mathsf{T}}\rho).$$

Therefore, the relation $\Phi$ is an isomorphism with respect to the direct product structure.

**Proof:** Using Prop. 4.2.2.iii and univalence of $\pi'$, we obtain

$$\Phi\pi' = (\pi\pi'^{\mathsf{T}} \sqcap \rho\rho'^{\mathsf{T}})\pi' = \pi \sqcap \rho\rho'^{\mathsf{T}}\pi' = \pi \sqcap \rho L = \pi.$$

Other isomorphism formulae for $\pi$ and $\rho$ may be derived similarly. We now show that the relation $\Phi$ is a mapping. Since $\pi'$ and $\rho'$ are mappings, we can use Prop. 4.2.4.iii:

$$\Phi\overline{I} = \Phi\overline{\pi'\pi'^{\mathsf{T}} \sqcap \rho'\rho'^{\mathsf{T}}} = \Phi(\overline{\pi'\pi'^{\mathsf{T}}} \sqcup \overline{\rho'\rho'^{\mathsf{T}}}) = \Phi\overline{\pi'\pi'^{\mathsf{T}}} \sqcup \Phi\overline{\rho'\rho'^{\mathsf{T}}}$$
$$= \overline{\pi\pi'^{\mathsf{T}}} \sqcup \overline{\rho\rho'^{\mathsf{T}}} = \overline{\pi\pi'^{\mathsf{T}} \sqcap \rho\rho'^{\mathsf{T}}} = \overline{\Phi}.$$

For reasons of symmetry, $\Phi^{\mathsf{T}}$ is a function and $\Phi$ is a bijective mapping.    □

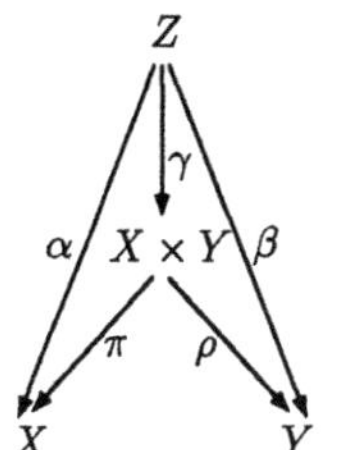


**Fig. 7.5.3** Universal characterization of the direct product

The direct product has the following universal property (Fig. 7.5.3): Given two mappings $\alpha, \beta$ from a set $Z$ into $X, Y$, then there exists a mapping $\gamma$ from $Z$ into $X \times Y$ yielding the factorizations $\alpha = \gamma\pi$ and $\beta = \gamma\rho$. This mapping is given explicitly by $\gamma := \alpha\pi^{\mathsf{T}} \sqcap \beta\rho^{\mathsf{T}}$. The factorization conditions may be proved using Prop. 4.2.2.iii.

For direct products formed from several factors we now give the relation-algebraic formulation for the associative property.

**7.5.3 Proposition** (*Associativity of the direct product*). Let the direct products $(\pi,\rho)$, $(\lambda,\nu)$, $(\xi,\eta)$ and $(\alpha,\beta)$ be given, as indicated in Fig. 7.5.4. Then there exists a bijective mapping $\Phi$ satisfying

$$\lambda\pi = \Phi\alpha, \quad \lambda\rho = \Phi\beta\xi \quad \text{and} \quad \nu = \Phi\beta\eta. \qquad \square$$

This mapping is actually $\Phi := \lambda\pi\alpha^{\mathsf{T}} \sqcap (\lambda\rho\xi^{\mathsf{T}} \sqcap \nu\eta^{\mathsf{T}})\beta^{\mathsf{T}}$, but we omit the proof.

To an arbitrary relation $R$ between $X$ and $Y$ there corresponds a subset $t$ of $X \times Y$, and vice versa (Fig. 7.5.1). With the next proposition we investigate this correspondence further.

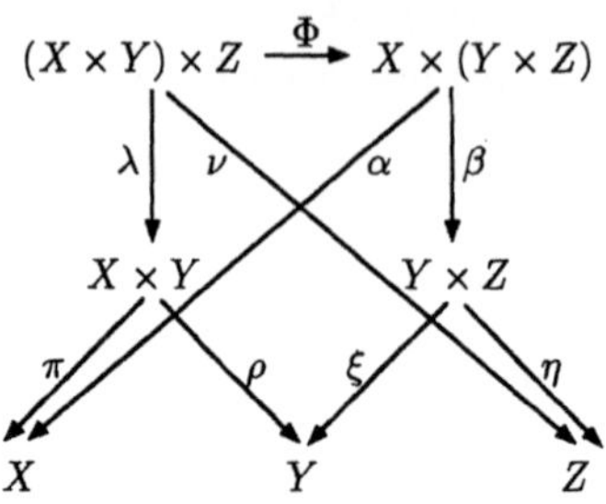

**Fig. 7.5.4** Associativity of the direct product

**7.5.4 Proposition.** Let $(\pi, \rho)$ be a direct product as before. If for a relation $R$ the subset (or vector) $\tau(R)$ is formed, and for the subset (or vector) $t$, the relation $\sigma(t)$

$$\tau(R) := (\pi R \sqcap \rho)L, \qquad \sigma(t) := \pi^{\mathsf{T}}(\rho \sqcap tL),$$

then

$$\tau\big(\sigma(t)\big) = t \quad \text{and} \quad \sigma\big(\tau(R)\big) = R.$$

**Proof:** The following estimates are immediate:

$$\begin{aligned}
\tau\big(\sigma(t)\big) = \big(\pi\sigma(t) \sqcap \rho\big)L &= \big(\pi\pi^{\mathsf{T}}(\rho \sqcap tL) \sqcap \rho\big)L \\
&\subset \big(\pi\pi^{\mathsf{T}} \sqcap \rho(\rho \sqcap tL)^{\mathsf{T}}\big)(\rho \sqcap tL \sqcap \pi\pi^{\mathsf{T}}\rho)L \\
&\subset (\pi\pi^{\mathsf{T}} \sqcap \rho\rho^{\mathsf{T}})tL = ItL = tL = t.
\end{aligned}$$

In the opposite direction we have

$$\begin{aligned}
\big(\pi\pi^{\mathsf{T}}(\rho \sqcap tL) \sqcap \rho\big)L &\supset \big((\rho \sqcap tL) \sqcap \rho\big)L = (\rho \sqcap tL)L \\
&\supset (\rho \sqcap tL)(L \sqcap \rho^{\mathsf{T}}t)L \supset (\rho L \sqcap t)L = (L \sqcap t)L = tL = t.
\end{aligned}$$

In order to prove $\sigma\big(\tau(R)\big) = R$, we show $\pi^{\mathsf{T}}\big((\pi R \sqcap \rho)L \sqcap \rho\big) = R$. This is obtained from the Dedekind rule

$$\pi^{\mathsf{T}}(\pi R \sqcap \rho \sqcap \rho L^{\mathsf{T}})\big(L \sqcap (\pi R \sqcap \rho)^{\mathsf{T}}\rho\big) \subset \pi^{\mathsf{T}}\pi R \rho^{\mathsf{T}}\rho = R$$

and from

$$\begin{aligned}
R = \pi^{\mathsf{T}}\rho I \sqcap R &\subset (\pi^{\mathsf{T}}\rho \sqcap RI)(I \sqcap (\pi^{\mathsf{T}}\rho)^{\mathsf{T}}R) \\
&= \pi^{\mathsf{T}}(\rho \sqcap \pi R)(I \sqcap LR) \subset \pi^{\mathsf{T}}(\rho \sqcap (\pi R \sqcap \rho)L). \qquad \square
\end{aligned}$$

The functionals $\sigma$ and $\tau$ are clearly isotonic, i.e.,

$$R \subset R' \implies \tau(R) \subset \tau(R'), \qquad t \subset t' \implies \sigma(t) \subset \sigma(t').$$

We want to know to what extent the structure of the relation algebra given by the operations $O, I, L, {}^{-}, {}^{\mathsf{T}}, \sqcap, \sqcup, \circ$ is preserved by the functionals $\sigma$ and $\tau$. The following proposition is a partial answer as far as the Boolean operations are concerned.

**7.5.5 Proposition.**

i) $\qquad\qquad \tau(O) = O, \qquad\qquad\qquad\qquad \sigma(O) = O;$

ii) $\qquad\qquad \tau(L) = L, \qquad\qquad\qquad\qquad \sigma(L) = L;$

iii) $\qquad\qquad \tau\left(\overline{R}\right) = \overline{\tau(R)}, \qquad\qquad\qquad \sigma\left(\overline{t}\right) = \overline{\sigma(t)};$

iv) $\qquad \tau(R \sqcup S) = \tau(R) \sqcup \tau(S), \qquad \sigma(t \sqcup s) = \sigma(t) \sqcup \sigma(s);$

v) $\qquad \tau(R \sqcap S) = \tau(R) \sqcap \tau(S), \qquad \sigma(t \sqcap s) = \sigma(t) \sqcap \sigma(s).$

**Proof:** (i, ii, iv) are straightforward. One direction of (iii) is obtained by

$$L = \rho L = (\pi L \sqcap \rho)L = \left((\pi R \sqcup \pi \overline{R}) \sqcap \rho\right)L = (\pi R \sqcap \rho)L \sqcup (\pi\overline{R} \sqcap \rho)L,$$
$$\overline{\tau(R)} = \overline{(\pi R \sqcap \rho)L} \subset (\pi\overline{R} \sqcap \rho)L = \tau\left(\overline{R}\right),$$

while for the other the Dedekind rule is used in

$$(\pi R \sqcap \rho)L \sqcap \rho \subset \left((\pi R \sqcap \rho) \sqcap \rho L\right)\left(L \sqcap (\pi R \sqcap \rho)^{\mathsf{T}}\rho\right) \subset \pi R \rho^{\mathsf{T}}\rho \subset \pi R.$$

Therefore, $L = \overline{(\pi R \sqcap \rho)L} \sqcup \overline{\rho} \sqcup \pi R$ and, since $\pi$ is a mapping,

$$\tau\left(\overline{R}\right) = (\pi\overline{R} \sqcap \rho)L = \overline{(\pi R \sqcap \rho)}L \subset \overline{(\pi R \sqcap \rho)L L} = \overline{\tau(R)}.$$

The right-hand equation is easily obtained using the first:

$$\sigma\left(\overline{t}\right) = \sigma\left(\overline{\tau(\sigma(t))}\right) = \sigma\left(\tau\left(\overline{\sigma(t)}\right)\right) = \overline{\sigma(t)}.$$

Now (v) holds by the de Morgan rules. $\qquad\qquad\qquad\qquad\qquad\qquad\square$

We now consider the above situation in the special case of a product $X \times X$ of two *equal* factors. The additional operations are then $I$, $^{\mathsf{T}}, \circ$. In Prop. 7.5.5 all the operations were to be replaced by similarly denoted operations on the subsets. Now completely different operations on the subset side are to be investigated, which we denote in a similar way:

$$\mathtt{ident} := (\pi \sqcap \rho)L;$$
$$\mathtt{transp}(t) := (\pi\rho^{\mathsf{T}} \sqcap \rho\pi^{\mathsf{T}})t;$$
$$\mathtt{mult}(t, s) := \left(\pi\pi^{\mathsf{T}}(\rho \sqcap tL) \sqcap \rho\rho^{\mathsf{T}}(\pi \sqcap sL)\right)L.$$

In $\mathtt{ident}$, pairs with two equal components are singled out by $\pi \sqcap \rho$, in $\mathtt{transp}$ the components of pairs are interchanged by $\pi\rho^{\mathsf{T}} \sqcap \rho\pi^{\mathsf{T}}$. The third case cannot be described in such a simple way. We need some technical results.

**7.5.6 Proposition.** Let $(\pi, \rho)$ be a direct product for which $\pi\rho^{\mathsf{T}}$ is defined. Then:

i) The relation $\zeta := \pi\rho^{\mathsf{T}} \sqcap \rho\pi^{\mathsf{T}}$ is a bijective mapping satisfying

$$\zeta\pi = \rho, \quad \zeta\rho = \pi, \quad \zeta^{\mathsf{T}} = \zeta.$$

ii) $\qquad\qquad\qquad (\pi R^{\mathsf{T}} \sqcap \rho)L = (\rho R \sqcap \pi)L.$

iii) $\qquad\qquad (\pi RS \sqcap \rho)L = (\pi R \sqcap \rho S^{\mathsf{T}})L = (\rho S^{\mathsf{T}}R^{\mathsf{T}} \sqcap \pi)L.$

**Proof:** i) We restrict ourselves to showing two of the equations using Prop. 4.2.2.iii:

$$\zeta\pi = (\pi\rho^\mathsf{T} \sqcap \rho\pi^\mathsf{T})\pi = \pi\rho^\mathsf{T}\pi \sqcap \rho = L \sqcap \rho = \rho;$$
$$\zeta\bar{I} = \zeta\overline{\pi\pi^\mathsf{T} \sqcap \rho\rho^\mathsf{T}} = \zeta(\overline{\pi\pi^\mathsf{T}} \sqcup \overline{\rho\rho^\mathsf{T}}) = \zeta\overline{\pi\pi^\mathsf{T}} \sqcup \zeta\overline{\rho\rho^\mathsf{T}} = \rho\overline{\pi^\mathsf{T}} \sqcup \pi\overline{\rho^\mathsf{T}}$$
$$= \overline{\rho\pi^\mathsf{T}} \sqcup \overline{\pi\rho^\mathsf{T}} = \overline{\rho\pi^\mathsf{T} \sqcap \pi\rho^\mathsf{T}} = \bar{\zeta}.$$

The proofs for (ii) and (iii) are completely analogous; they may be proved by cyclically permuted estimates using the Dedekind rule:

$$(\pi R^\mathsf{T} \sqcap \rho)L \subset (\pi \sqcap \rho R)(R^\mathsf{T} \sqcap \pi^\mathsf{T}\rho)L \subset (\rho R \sqcap \pi)L$$
$$\subset (\rho \sqcap \pi R^\mathsf{T})(R \sqcap \rho^\mathsf{T}\pi)L \subset (\pi R^\mathsf{T} \sqcap \rho)L. \qquad \square$$

We are now in a position to describe transposition and multiplication of relations by means of the corresponding operations on subsets defined above. Thus, we complete the answer to our question for the remaining operations.

**7.5.7 Proposition.** If $(\pi, \rho)$ is a direct product for which $\pi\rho^\mathsf{T}$ is defined, then:

$$\tau(I) = \texttt{ident}, \qquad\qquad \sigma(\texttt{ident}) = I;$$
$$\tau(R^\mathsf{T}) = \texttt{transp}\big(\tau(R)\big), \qquad \sigma\big(\texttt{transp}(t)\big) = [\sigma(t)]^\mathsf{T};$$
$$\tau(RS) = \texttt{mult}\big(\tau(R), \tau(S)\big), \qquad \sigma\big(\texttt{mult}(t, s)\big) = \sigma(t)\sigma(s).$$

**Proof:** We consider the left-hand column of equations: The first is trivial.

$$\texttt{transp}\big(\tau(R)\big) = \zeta\tau(R) = \zeta(\pi R \sqcap \rho)L = (\zeta\pi R \sqcap \zeta\rho)L$$
$$= (\rho R \sqcap \pi)L = (\pi R^\mathsf{T} \sqcap \rho)L = \tau(R^\mathsf{T}),$$
$$\texttt{mult}\big(\tau(R), \tau(S)\big) = [\pi\pi^\mathsf{T}\big(\rho \sqcap \tau(R)\big) \sqcap \rho\rho^\mathsf{T}\big(\pi \sqcap \tau(S)\big)]L$$
$$= [\pi\sigma\big(\tau(R)\big) \sqcap \rho[\sigma\big(\tau(S)\big)]^\mathsf{T}]L$$
$$= (\pi R \sqcap \rho S^\mathsf{T})L = (\pi RS \sqcap \rho)L = \tau(RS).$$

The second column of these equations is now easily obtained, e.g., $\sigma(\texttt{ident}) = \sigma\big(\tau(I)\big) = I.$ \qquad $\square$

## Relation Algebras Without Points

We have established the Cartesian product and have shown that it is essentially unique. This enables us to consider relations as *elements* of some relation algebra where they can be concatenated with others, on the one hand, and as *subsets* of some Cartesian product on the other. Moreover, the transition between these two aspects can be performed in a rigorous way. The usual set-theoretic point of view and notation does not lend itself particularly well to distinguishing between these two aspects. However, it is of great importance in computer science to be able to make this distinction since *functions* can appear *as arguments of higher order functionals* in addition to their traditional role of *operating on arguments*.

The need for making that distinction becomes even more evident when one considers relation algebras to which all of our relational calculus applies, but which cannot be represented as the set of relations between the elements of some set. We next have a first look at such unusual models.

To this end we let $\Theta$ be an equivalence relation on $X \times X$, and we consider pairs that "are known modulo $\Theta$ only", or "cannot be observed with complete precision". We ask what the effect on a relation is if its underlying set is given only modulo $\Theta$. This effect can indeed be surprising as the following example shows.

**7.5.8 Example.** Consider the set $X = \{a, b\}$, its set of pairs $X \times X$ with the projections $\pi, \rho$, and the equivalence relation $\Theta$ on $X \times X$ as indicated in Fig. 7.5.5.

$$\Theta = \begin{array}{c} \\ aa \\ ab \\ ba \\ bb \end{array}\!\!\begin{array}{c} aa\ ab\ ba\ bb \\ \begin{pmatrix} 1 & 0 & 0 & 1 \\ 0 & 1 & 1 & 0 \\ 0 & 1 & 1 & 0 \\ 1 & 0 & 0 & 1 \end{pmatrix} \end{array} \quad \pi = \begin{array}{c} a\ \ b \\ \begin{pmatrix} 1 & 0 \\ 1 & 0 \\ 0 & 1 \\ 0 & 1 \end{pmatrix} \end{array} \quad \rho = \begin{array}{c} a\ \ b \\ \begin{pmatrix} 1 & 0 \\ 0 & 1 \\ 1 & 0 \\ 0 & 1 \end{pmatrix} \end{array} \quad t_1 = \begin{pmatrix} 0 \\ 0 \\ 0 \\ 0 \end{pmatrix} \quad t_2 = \begin{pmatrix} 1 \\ 0 \\ 0 \\ 1 \end{pmatrix} \quad t_3 = \begin{pmatrix} 0 \\ 1 \\ 1 \\ 0 \end{pmatrix} \quad t_4 = \begin{pmatrix} 1 \\ 1 \\ 1 \\ 1 \end{pmatrix}$$

$$R_1 = \begin{array}{c} \\ a \\ b \end{array}\!\!\begin{array}{c} a\ \ b \\ \begin{pmatrix} 0 & 0 \\ 0 & 0 \end{pmatrix} \end{array} \quad R_2 = \begin{array}{c} a\ \ b \\ \begin{pmatrix} 1 & 0 \\ 0 & 1 \end{pmatrix} \end{array} \quad R_3 = \begin{array}{c} a\ \ b \\ \begin{pmatrix} 0 & 1 \\ 1 & 0 \end{pmatrix} \end{array} \quad R_4 = \begin{array}{c} a\ \ b \\ \begin{pmatrix} 1 & 1 \\ 1 & 1 \end{pmatrix} \end{array}$$

**Fig. 7.5.5** A relation algebra without points

Certain subsets $t$ of $X \times X$ are closed with respect to multiplication by $\Theta$. There are exactly four such subsets, denoted by $t_1, t_2, t_3, t_4$ in Fig. 7.5.5. To these subsets correspond four relations $R_i := \sigma(t_i)$ which constitute a relation algebra. Indeed, the set $\mathcal{R} = \{R_1, R_2, R_3, R_4\}$ is closed with respect to all the operations of a relation algebra. Equally, all the rules of a relation algebra hold true since they did already before restricting to the subset under consideration. There is, however, a little blemish to this relation algebra—it has no points because its only atoms are $R_2, R_3$ which are not row-constant and hence are not points. So this relation algebra does not necessarily satisfy conditions that are dependent on the point axiom or the intermediate point theorem which is a consequence of it. Specifically, Propositions 5.2.4, 6.1.3, 8.1.3, and 9.3.12 do not apply. $\qquad\square$

We expand that example into a general method for constructing new relation algebras. To this end, we generalize the construct $\sigma\big(\tau(R)\big)$ slightly by considering $\sigma\big(\Theta\tau(R)\big)$.

**7.5.9 Definition.** Let $\Theta$ be an equivalence relation on $X \times X$. Then for every relation $R$ on $X$ the $\Theta$-**hull**

$$h_\Theta(R) := \sigma\big(\Theta\tau(R)\big) = \pi^{\mathsf{T}}\big(\rho \sqcap \Theta(\pi R \sqcap \rho)L\big)$$
$$= \inf\{X \mid R \subset X,\ X \subset \sigma(\Theta\tau(X))\}$$

can be defined. We call a relation $\Theta$-**saturated** if it satisfies $R = h_\Theta(R)$. By

$$\mathcal{S} := \{\, R \subset X \times X \mid R = h_\Theta(R)\,\}$$

we denote the set of all $\Theta$-saturated relations on $X$.    □

Let us show that $h_\Theta$ is indeed a closure operator on the set of relations on $X$. Clearly, $h_\Theta$ is *isotonic*, i.e.,

$$R \subset S \quad\Longrightarrow\quad h_\Theta(R) \subset h_\Theta(S),$$

for $\tau$, multiplication with $\Theta$, and $\sigma$ are isotonic. Because of $\tau(R) \subset \Theta\tau(R)$, and since $\sigma$ is isotonic, $h_\Theta$ is *extensive*:

$$R \subset h_\Theta(R).$$

Finally we have *idempotence* of $h_\Theta$, since

$$h_\Theta\big(h_\Theta(R)\big) = \sigma\left(\Theta\tau\left(\sigma\left(\Theta\tau(R)\right)\right)\right) = \sigma\big(\Theta\Theta\tau(R)\big) = \sigma\big(\Theta\tau(R)\big) = h_\Theta(R)$$

using transitivity of $\Theta$.

The properties of belonging to $\mathcal{S}$ and of being closed with respect to multiplication by $\Theta$, respectively, are closely related.

**7.5.10 Proposition.**    $R \in \mathcal{S} \quad\Longleftrightarrow\quad \tau(R) = \Theta\tau(R).$

**Proof:** "$\Longrightarrow$": $\Theta\tau(R) = \tau\left(\sigma\left(\Theta\tau(R)\right)\right) = \tau\big(h_\Theta(R)\big) = \tau(R).$
"$\Longleftarrow$": $h_\Theta(R) = \sigma\big(\Theta\tau(R)\big) = \sigma\big(\tau(R)\big) = R.$    □

We now want to find which conditions are necessary for $\mathcal{S}$ to be a relation algebra, i.e., for $\mathcal{S}$ to be closed with respect to the operations $O, I, L, ^-, \ ^\mathsf{T}, \sqcap, \sqcup, \circ$. Some of these operations require no additional assumptions, others do. We proceed in a general way and investigate $n$-ary operations and relations together with their subset counterparts.

**7.5.11 Definition.** We consider an equivalence relation $\Theta$ on $X \times X$ together with an $n$-ary operation $\Psi$ on relations, to which an operation on subsets is associated by

$$\psi(t_1,\ldots,t_n) := \tau\big(\Psi\big(\sigma(t_1),\ldots,\sigma(t_n)\big)\big).$$

The equivalence $\Theta$ will be called $\Psi$-**compatible**, if for all $t_1,\ldots,t_n$

$$\Theta\psi(\Theta t_1,\ldots,\Theta t_n) = \psi(\Theta t_1,\ldots,\Theta t_n).$$    □

When $n = 0$, the condition reduces to $\Theta\psi = \psi$. We shall see that compatibility with some operation corresponds to $\mathcal{S}$ being closed with respect to the corresponding operation on subsets (denoted by lowercase letters).

**7.5.12 Proposition.** If the equivalence relation $\Theta$ is compatible with the $n$-ary operation $\Psi$, then $\mathcal{S}$ is closed with respect to the operation $\Psi$.

**Proof:** For convenience, we prove only the unary case:

$$
\begin{aligned}
h_\Theta\big(\Psi(R)\big) &= \sigma\left(\Theta\tau\left(\Psi(R)\right)\right) && \text{by the definition of } h_\Theta \\
&= \sigma\left(\Theta\psi\left(\tau(R)\right)\right) && \text{by the definition of } \psi \\
&= \sigma\left(\Theta\psi\left(\Theta\tau(R)\right)\right) && \text{since } R \in \mathcal{S} \text{ by Prop. 7.5.10} \\
&= \sigma\left(\psi\left(\Theta\tau(R)\right)\right) && \text{since } \Theta \text{ is } \Psi\text{-compatible} \\
&= \sigma\left(\psi\left(\tau(R)\right)\right) && \text{since } R \in \mathcal{S} \text{ by Prop. 7.5.10} \\
&= \sigma\left(\tau\left(\Psi(R)\right)\right) && \text{by the definition of } \psi \\
&= \Psi(R) && \text{using Prop. 7.5.2.} \qquad \square
\end{aligned}
$$

With the aid of the condition of Def. 7.5.11 we check compatibility with respect to the operations $O, L, {}^{\boldsymbol{-}}, {}^{\mathsf{T}}, \sqcap, \sqcup, \circ$.

**7.5.13 Proposition.**  i)  Every equivalence $\Theta$ is $O$-, $L$-, ${}^{\boldsymbol{-}}$-, $\sqcup$- and $\sqcap$-compatible.

ii)   $\Theta$ is $I$-compatible, provided $\Theta(\pi \sqcap \rho)L = (\pi \sqcap \rho)L$.

iii)   $\Theta$ is ${}^{\mathsf{T}}$-compatible, provided $\zeta\Theta = \Theta\zeta$.

iv)   $\Theta$ is $\circ$-compatible, provided $\Theta\,\mathtt{mult}(\Theta t, \Theta s) = \mathtt{mult}(\Theta t, \Theta s)$ for all $t, s$.

**Proof:** Concerning (i), all the cases except for complement and intersection are straightforward. $\Theta\overline{\Theta R} = \overline{\Theta R}$ for arbitrary $R$ and an equivalence $\Theta$ follows from Prop. 3.1.6.i. Due to Prop. 3.1.6.ii, $\Theta(\Theta t \sqcap \Theta s) = \Theta t \sqcap \Theta s$ for arbitrary $t, s$ and an equivalence $\Theta$. (iii) follows from

$$
\Theta\,\mathtt{transp}(\Theta t) = \Theta\zeta\Theta t = \zeta\Theta\Theta t = \zeta\Theta t = \mathtt{transp}(\Theta t).
$$

All the other cases are obtained directly from the definition. $\qquad\square$

Thus all the conditions have been checked. It is remarkable that the relation algebra $\mathcal{S}$ is completely determined by the equivalence relation $\Theta$.

**7.5.14 Corollary.** The set $\mathcal{S} = \{\, R \subset X \times X \mid R = h_\Theta(R)\,\}$ of all $\Theta$-compatible relations on $X$ constitutes a relation algebra, if the equivalence $\Theta$ is compatible with respect to identity, transposition and multiplication. $\qquad\square$

We can now give an example of a relation algebra for which the intermediate point theorem 2.4.8 fails to hold. This relation algebra consists of those Boolean $3 \times 3$-matrices whose upper left $2 \times 2$ submatrix belongs to the algebra from Fig. 7.5.5 and whose other three submatrices contain either all zeroes or all ones.

This algebra contains the sole point $x$. Clearly, $xx^{\mathsf{T}}$ is contained in $RR$, but there is no intermediate point in the sense of Prop. 2.4.8.

$$
\left(\begin{array}{cc|c}0&0&0\\0&0&0\\\hline0&0&0\end{array}\right)
\left(\begin{array}{cc|c}0&0&1\\0&0&1\\\hline0&0&0\end{array}\right)
\left(\begin{array}{cc|c}0&0&0\\0&0&0\\\hline1&1&0\end{array}\right)
\left(\begin{array}{cc|c}0&0&1\\0&0&1\\\hline1&1&0\end{array}\right)
\left(\begin{array}{cc|c}0&0&0\\0&0&0\\\hline0&0&1\end{array}\right)
\left(\begin{array}{cc|c}0&0&1\\0&0&1\\\hline0&0&1\end{array}\right)
\left(\begin{array}{cc|c}0&0&0\\0&0&0\\\hline1&1&1\end{array}\right)
\left(\begin{array}{cc|c}0&0&1\\0&0&1\\\hline1&1&1\end{array}\right)
$$

$$
\qquad\qquad\qquad\qquad\quad R\qquad\quad xx^{\mathsf{T}}\qquad\qquad\qquad\quad x
$$

$$
\left(\begin{array}{cc|c}1&0&0\\0&1&0\\\hline0&0&0\end{array}\right)
\left(\begin{array}{cc|c}1&0&1\\0&1&1\\\hline0&0&0\end{array}\right)
\left(\begin{array}{cc|c}1&0&0\\0&1&0\\\hline1&1&0\end{array}\right)
\left(\begin{array}{cc|c}1&0&1\\0&1&1\\\hline1&1&0\end{array}\right)
\left(\begin{array}{cc|c}1&0&0\\0&1&0\\\hline0&0&1\end{array}\right)
\left(\begin{array}{cc|c}1&0&1\\0&1&1\\\hline0&0&1\end{array}\right)
\left(\begin{array}{cc|c}1&0&0\\0&1&0\\\hline1&1&1\end{array}\right)
\left(\begin{array}{cc|c}1&0&1\\0&1&1\\\hline1&1&1\end{array}\right)
$$

$$
\left(\begin{array}{cc|c}0&1&0\\1&0&0\\\hline0&0&0\end{array}\right)
\left(\begin{array}{cc|c}0&1&1\\1&0&1\\\hline0&0&0\end{array}\right)
\left(\begin{array}{cc|c}0&1&0\\1&0&0\\\hline1&1&0\end{array}\right)
\left(\begin{array}{cc|c}0&1&1\\1&0&1\\\hline1&1&0\end{array}\right)
\left(\begin{array}{cc|c}0&1&0\\1&0&0\\\hline0&0&1\end{array}\right)
\left(\begin{array}{cc|c}0&1&1\\1&0&1\\\hline0&0&1\end{array}\right)
\left(\begin{array}{cc|c}0&1&0\\1&0&0\\\hline1&1&1\end{array}\right)
\left(\begin{array}{cc|c}0&1&1\\1&0&1\\\hline1&1&1\end{array}\right)
$$

$$
\left(\begin{array}{cc|c}1&1&0\\1&1&0\\\hline0&0&0\end{array}\right)
\left(\begin{array}{cc|c}1&1&1\\1&1&1\\\hline0&0&0\end{array}\right)
\left(\begin{array}{cc|c}1&1&0\\1&1&0\\\hline1&1&0\end{array}\right)
\left(\begin{array}{cc|c}1&1&1\\1&1&1\\\hline1&1&0\end{array}\right)
\left(\begin{array}{cc|c}1&1&0\\1&1&0\\\hline0&0&1\end{array}\right)
\left(\begin{array}{cc|c}1&1&1\\1&1&1\\\hline0&0&1\end{array}\right)
\left(\begin{array}{cc|c}1&1&0\\1&1&0\\\hline1&1&1\end{array}\right)
\left(\begin{array}{cc|c}1&1&1\\1&1&1\\\hline1&1&1\end{array}\right)
$$

$$
RR
$$

**Fig. 7.5.6** A relation algebra where the intermediate point theorem fails to hold

It may be of interest to find a useful interpretation of that effect. One could think of several parallel processes which may be observed in the Cartesian product of their respective sets of states, under the condition that no "simultaneous observation" of the states of all those processes is possible. The non-existence of a point lends itself to this interpretation.

## Congruences in the Presence of Several Sorts

With the aid of the direct product, functions of several variables[4] can be represented by binary relations. Instead of interpreting a given binary function $\varphi\colon X \times Y \longrightarrow Z$ as a ternary relation $R_\varphi \subset X \times Y \times Z$, it is convenient to first form the direct product $X \times Y$ as in Fig. 7.5.7, and then apply the unary function $F$.

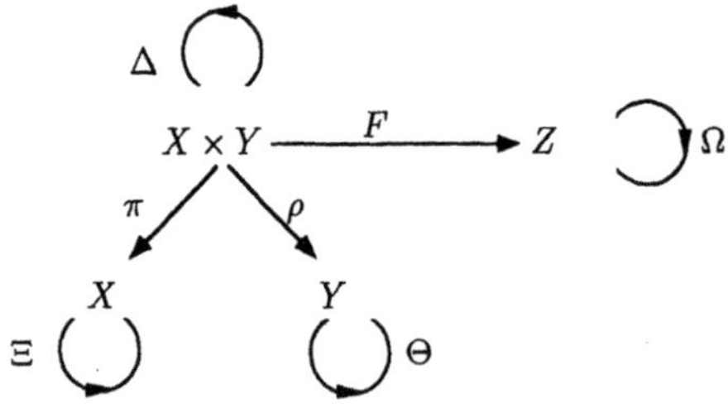


**Fig. 7.5.7** $n$-arity

The situation becomes more challenging when there are additional congruences

---

[4] Concerning functions of several variables the following aspects are of particular relevance in computer science: strictness on the one hand and the possibility of partial evaluation when only one argument is given, on the other. The composition of partially evaluable functions poses serious algebraic problems.

to be taken into account. The substitution property

$$\forall x_1, x_2 \in X \; \forall y_1, y_2 \in Y \; : \; x_1 \,\Xi\, x_2, \; y_1 \,\Theta\, y_2 \;\rightarrow\; \varphi(x_1, y_1) \,\Omega\, \varphi(x_2, y_2)$$

leads to using the product congruence $\Delta := \pi\Xi\pi^\mathsf{T} \sqcap \rho\Theta\rho^\mathsf{T}$ on $X \times Y$. The unary function $F$ then satisfies the substitution property $\Delta F \subset F\Omega$ (cf. Definition 7.4.4). But there are also congruences on $X \times Y$ which are not induced by a couple of congruences $\Xi$ on $X$, $\Theta$ on $Y$, i.e., which are not product congruences. They may nevertheless be respected by $\pi$, $\rho$ and $F$

$$\Delta\pi \subset \pi\Xi, \qquad \Delta\rho \subset \rho\Theta, \qquad \Delta F \subset F\Omega.$$

We return to the case of a relation $B_0$ on the direct product $X \times Y$ which is the relation product

$$B_0 := \pi B\pi^\mathsf{T} \sqcap \rho B'\rho^\mathsf{T}$$

of the two relations $B$ on $X$ and $B'$ on $Y$. We think of $B_0$ as representing the two independently running parallel processes $B$ and $B'$. $B_0$ is then the product process whose states are pairs consisting of the states of the two simultaneous processes. The transitions according to $B_0$ are then defined componentwise, i.e., by each individual process separately performing a transition. It is surprising that the identity

$$B_0^2 = \pi B^2\pi^\mathsf{T} \sqcap \rho B'^2\rho^\mathsf{T},$$

cannot be inferred by relation-algebraic arguments, although one would expect it to represent the composition of the two simultaneous processes. In fact, only "$\subset$" can be proved. This seems to indicate that there are models for which the above identity fails to hold.

A closer investigation seems to be worthwhile, not least for learning more about nonstandard relation algebras. One would hope that the relation-algebraic method could provide more insight into the possibility of observing, and the logic of, processes that are characterized by terms such as "parallel", "concurrent", "cooperating", or "simultaneous". Results in this direction would certainly be of interest for computer science as well as for physics.

## 7.6 References


COHN PM: *Universal algebra*. Harper and Row, New York, 1965.

GRÄTZER G: *Universal algebra*. Springer, New York, 2nd ed. 1979.

SCHMIDT G: *Eine Überlagerungstheorie für Wurzelgraphen*. In: Noltemeier H (ed.): Graphen, Algorithmen, Datenstrukturen. Proc. 2nd Workshop Graph-Theoret. Concepts in Computer Science, June 16-18, 1976, Hanser, München, 1976, 65–76.

# 8. Kernels and Games

The present chapter is almost independent of the preceding material. Given a homogeneous, binary relation $B$, we shall look for solutions $x$ of the equation $\overline{x} = Bx$. These Boolean eigenvectors have nice applications in combinatorial games. We proceed step by step and investigate in Sect. 8.1 the cases $\overline{x} \subset Bx$ ($x$ absorbant or externally stable) and $Bx \subset \overline{x}$ ($x$ stable) separately. Subsets $x$ that have both properties are kernels of $B$. In Sect. 8.2 we discuss when kernels exist. Here again, the powerful tool comes to bear that we used already in the first part of Sect. 6.3; its lattice-theoretic nature will be clarified in Appendix A.3. In Sect. 8.3 the results on kernels will be applied to game theory, in particular to nim-type games and to chess endings.

## 8.1 Absorptiveness and Stability

*Absorption*[1] is a relation between points which need not be symmetric; it is described by the associated relation $B$.

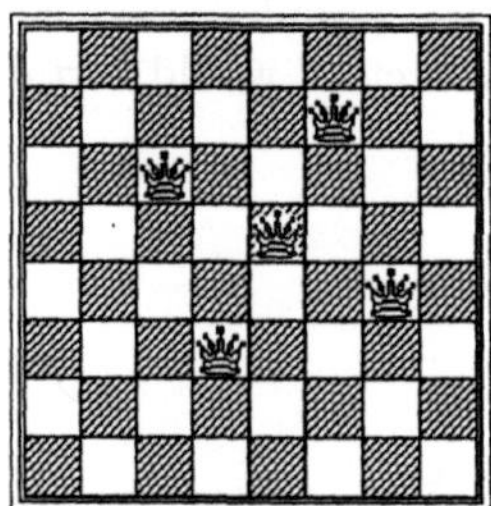

Fig. 8.1.1 5-queens problem

We begin with a well-known example. Given a chess board with a number of queens which can be moved according to the usual rules, how can the entire board be checked by as few queens as possible, positioned in a suitable way? Figure 8.1.1 gives a solution to this problem with five queens.

A similar problem is how to place observation posts in such a way that all information one is interested in is being "absorbed" by them. Following C. Berge, we let Fig. 8.1.2 represent the corridors of a prison whose doors (the points of the graph) are to be watched by as few warders as possible.

Since the corridors (the edges of the graph) can be watched from both of their ends, the graph has a symmetric $B$, and each corridor stands for two opposing arrows.

---

[1] *Stability* is defined by means of the associated relation, too, and is quite similar to the adjacency-related notion of independence, to be introduced later. The occurrence of both absorptiveness and stability is characteristic of kernels.

**8.1.1 Definition.** Given the associated relation $B$ of a graph and a point set $x$, we call

i)   $x$ **absorbant** $\quad :\Longleftrightarrow\quad \overline{x} \subset Bx \quad \Longleftrightarrow \quad \overline{Bx} \subset x$

$\qquad\qquad\qquad\qquad\qquad\Longleftrightarrow\quad$ From every point outside $x$ there is (at least) one arc leading into $x$.

ii)   $x$ **stable** $\qquad :\Longleftrightarrow\quad Bx \subset \overline{x} \quad \Longleftrightarrow \quad x \subset \overline{Bx}$

$\qquad\qquad\qquad\qquad\qquad\Longleftrightarrow\quad$ If an arc ends in $x$, it necessarily begins outside of $x$. (No two points of $x$ are related.)

iii)   $\beta := \min\{\,|x| \in \mathbb{N} \mid x \text{ absorbant}\,\}$   the **absorption number** [2].

Instead of absorbant we will often use **externally stable** and instead of stable, **internally stable**. If $\overline{x} \subset B^{\mathsf{T}}x$ then we call $x$ **dominating**.   □

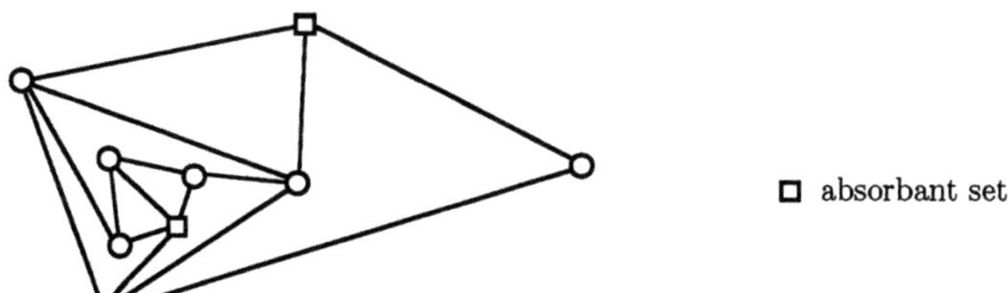


**Fig. 8.1.2** Problem of the warders

Since $Bx \subset \overline{x} \iff B^{\mathsf{T}}x \subset \overline{x}$ stability does *not* depend on the orientation of the arcs; absorption, however, does. Since $\overline{x} \subset Bx \iff \overline{x} \subset (B \sqcup I)x \iff \overline{x} \subset (B \sqcap \overline{I})x$, absorption does *not* depend on the existence of loops; stability, however, does.

Of course, $L$ is always absorbant and $O$ is always stable. Furthermore [3]:

$$x \text{ absorbant}, \quad x \subset y \quad \Longrightarrow \quad y \text{ absorbant};$$
$$x \text{ stable}, \quad z \subset x \quad \Longrightarrow \quad z \text{ stable}.$$

Since sets that contain an absorbant set are again absorbant, the smallest possible absorbant sets are of interest. Since subsets of stable sets are stable as well, one is interested in the largest possible stable sets.

One can now consider the sets whose number of elements is maximum, or which are maximal with respect to inclusion. When an additional point is added to the sets $x, x', x''$ in Fig. 8.1.3, they will no longer be stable. So all these three sets are maximal with respect to inclusion. But only $x$ and $x''$ are stable sets with a maximum number of elements. A largest stable set therefore does not exist.

---

[2] In Definition 9.1.1 below, a notion closely related to stability will be introduced which, however, applies to symmetric and irreflexive relations only. In analogy with $\beta$, the cardinality $\alpha$ of a maximum stable set, which also might be called the "stability number", is then defined to be the point independence number.

[3] Both facts have to do with the mapping $f(x) := \overline{Bx}$ being antitonic. The fixedpoints of $f$ will be studied in greater detail in Sect. 8.2 with the aid of the lattice-theoretic methods from Appendix A.3.

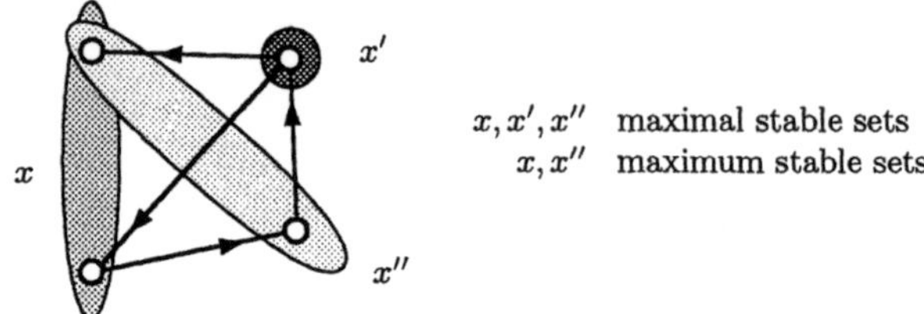

Fig. 8.1.3 Maximum and maximal stable sets

The absorbant sets $x, x', x''$ in Fig. 8.1.4 are minimal with respect to inclusion, for removing a point makes them lose their absorptive property. Thus, there is no absorbant set that is smallest with respect to inclusion.

**8.1.2 Proposition.** For the associated relation $B$ of a graph and a set $x$ of points the following holds:

$$x = \overline{Bx} \quad \Longleftrightarrow \quad \begin{cases} x & \text{is a maximal stable and} \\ & \text{a minimal absorbant set.} \end{cases}$$

**Proof:** Only "$\Longrightarrow$" needs a proof. Assuming $y \supset x$ to be stable, we obtain $y \subset \overline{By} \subset \overline{Bx} = x$. Assuming $z \subset x$ to be absorbant, we obtain $x = \overline{Bx} \subset \overline{Bz} \subset z$. $\qquad\Box$

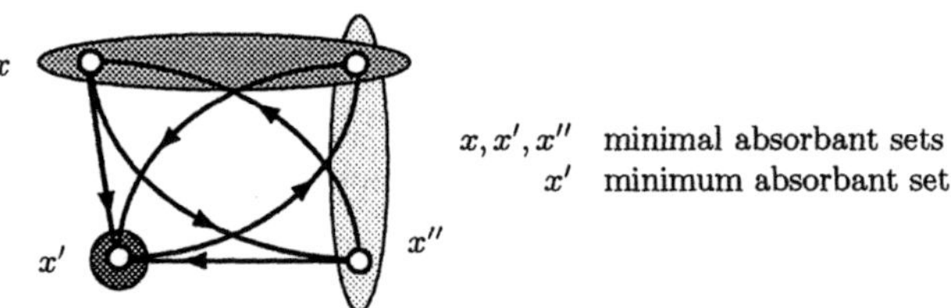

Fig. 8.1.4 Minimum and minimal absorbant sets

Clearly, the statement above cannot be sharpened in the sense that just one of the properties on the right-hand side is retained, i.e., either stable and maximal with respect to inclusion or absorbant and minimal with respect to inclusion. To give a counterexample for the first case, consider the graph on one single point with a loop attached: $O$ is the only stable set, and is therefore maximal with respect to inclusion, but $O \neq \overline{BO}$. The second case is dealt with by the set $x$ in Fig. 8.1.5 which is absorbant and minimal with respect to inclusion, but which satisfies $\overline{x} \subsetneq Bx$ and therefore is not stable.

The implications we have just discussed can, however, be proved under additional hypotheses. The obstacle in the first case is the loop, cf. Proposition 8.1.3. If in the second case the graph were transitive, so that in Fig. 8.1.5 one also had arrows from $c$ and $d$ to $a$, then one could clearly disregard point $b$ when minimizing the absorbant set. This argument remains valid in the general case, provided successively removing points from the absorbant set ultimately stops. The following two results hold.

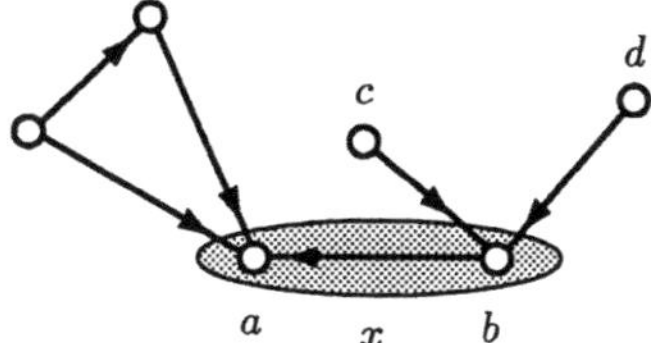



**Fig. 8.1.5** Minimal absorbant point set which is not stable

**8.1.3 Proposition.** In a symmetric graph without loops the associated relation $B$ together with any maximal stable set $x$ satisfies $x = \overline{Bx}$.

**Proof:** Assume $Bx \subsetneqq \overline{x}$ to hold for a maximal stable $x$. Then, choosing a point $v$ from $\overline{x} \sqcap \overline{Bx}$ (point axiom!), $x$ could be properly enlarged to $x \sqcup v$. We prove that the enlarged set is stable:

$$Bx \subset \overline{x}, \qquad\qquad Bv \subset \overline{x} \quad\text{since}\quad B = B^{\mathsf{T}};$$
$$Bx \subset \overline{v} \quad\Longleftrightarrow\quad v \subset \overline{Bx}, \qquad Bv \subset \overline{v} \quad\text{since}\quad vv^{\mathsf{T}} \subset I \subset \overline{B}. \qquad\square$$

The proposition above should be compared to the later considerations on independence in Proposition 9.1.2.i.

**8.1.4 Proposition.** If the associated relation $B$ of a graph is transitive, and if from every point a terminal point may be reached (i.e., $L = B^* \overline{BL}$, a condition satisfied, e.g., for progressively finite graphs, see Prop. 6.3.3.iv), then there exists a smallest absorbant point set $x_0$. Furthermore, $x_0$ is the set of all terminal points, $x_0 = \overline{BL}$, and satisfies $x_0 = \overline{Bx_0}$.

**Proof:** The point set $x_0 := \overline{BL}$ is absorbant, since $L = B^* \overline{BL} = \overline{BL} \sqcup B^+ \overline{BL}$ and $B^+ = B$ imply $\overline{x_0} \subset Bx_0$. The inclusion $Bx_0 \subset BL = \overline{x_0}$ is trivial. Therefore, $\overline{x_0} = Bx_0$, showing by Proposition 8.1.2 that $x_0$ is a minimal absorbant set. An arbitrary absorbant set $y$ necessarily satisfies $\overline{BL} \subset \overline{By} \subset y$; i.e., $\overline{BL}$ is smallest. $\qquad\square$

Similar questions will be taken up again in the next section where no use of transitivity will be made.

More information on the absorption number $\beta$ follows.

**8.1.5 Proposition.** In a 1-graph with $n$ points and $m$ arcs, in which every point has at most (and at least one point has precisely) $h$ proper predecessors, the following inequality holds

$$n - m \le \beta \le n - h.$$

**Proof:** First we show the left inequality: Let $x$ be a minimum absorbant set. Then every point of $\overline{x}$ is the starting point of an arc; therefore $m \ge |\overline{x}| = n - |x| = n - \beta$. In order to prove the right inequality, we choose a point $x_0$ with the maximum number $h$ of proper predecessors. The complement of the set of its proper predecessors contains $n - h$ points, including $x_0$, and is absorbant, hence $\beta \leq n - h$.                                                              □

**Fig. 8.1.6** A graph with $n - m = \beta = n - h$

The bounds above are loose; the lower bound is negative except when the number of arrows is *very* small. For the 5-queens problem one has $-1392 \leq \beta \leq 37$ whereas $\beta = 5$. The graph in Fig. 8.1.6 shows that the bounds cannot be improved, which is of little use, though.


## 8.2 Kernels

A kernel in a graph is a point set $s$ which is at the same time stable and absorbant. With regard to proceeding along an arc, the set $s$ then imposes the following conditions: "When starting from within one *must* leave $s$; when starting from outside one *can* enter $s$". This leads to interesting questions in particular in game theory where kernels were actually first investigated.


**8.2.1 Definition.** Given a graph $G$ with its associated relation $B$ and a point set $s$, we call

$$s \text{ kernel} \quad :\Longleftrightarrow \quad Bs = \bar{s} \quad \Longleftrightarrow \quad s \text{ stable and absorbant}$$

$\Longleftrightarrow$ No arc begins *and* ends in $s$, and from every point outside $s$ there is an arc leading into $s$.            □


The statements on the right in the definition are all equivalent by Def. 8.1.1. In Fig. 8.2.1 an example is given with two kernels in a non-correlated position and of distinct cardinalities. Neither their intersection nor their union is a kernel.

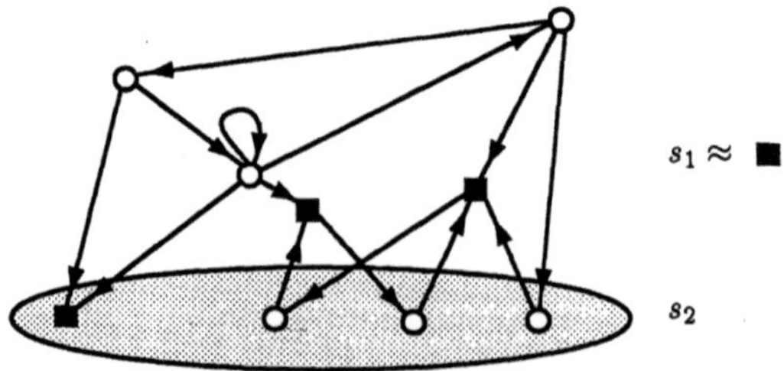



**Fig. 8.2.1** Kernels of a graph

Other examples of particular kernels can be derived from the defining property of a mapping, $R\overline{I} = \overline{R}$. From the form $\overline{I}R^{\mathsf{T}} = \overline{R^{\mathsf{T}}}$ one sees that all "columns" of $R^{\mathsf{T}}$ are kernels with respect to the distinctness relation $\overline{I}$. Thus, each "row" of $R$ must contain precisely one entry of $\mathbf{1}$ which, when multiplying by $\overline{I}$, meets the vanishing diagonal entry once.

**8.2.2 Definition.** Given a graph with associated relation $B$, we call a point set

$$s \ \textbf{basis} \ :\Longleftrightarrow \ (\overline{I} \sqcap B^*)s = \overline{s}. \qquad\qquad \Box$$

In other words, a point set $s$ is a basis[4], if no point in $s$ can be reached from some *other* point in $s$, and if from every point in $\overline{s}$ some point in $s$ can be reached.

The two defining properties of a kernel, *stability* and *absorptiveness*, are in a sense contrary to each other since stability is inherited by subsets and absorptiveness by supersets. So the existence problem may prove difficult. To give an example, the oriented circuit of length 3, considered as a graph, has no kernel because all stable sets maximal with respect to inclusion are of cardinality 1 whereas the minimal absorbant sets all have two elements. As an immediate consequence of Proposition 8.1.2, we state the following without proof.

**8.2.3 Proposition.** If $s$ is the kernel of a graph, then $s$ is at the same time a maximal stable and a minimal absorbant set. $\qquad\qquad \Box$

So having a "stable *and* absorbant set" according to Definition 8.2.1 automatically means that this is a "maximal stable and minimal absorbant set" (both with respect to inclusion).

By Proposition 8.1.3, the inclusion-maximal stable sets in a symmetric and loop-free graph are kernels.

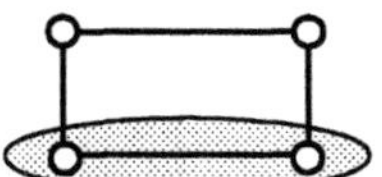

**Fig. 8.2.2** Inclusion-minimal absorbant set which is not a kernel

One can see from Fig. 8.2.2 that in the situation where $B = B^{\mathsf{T}} \subset \overline{I}$ there is no analog regarding absorption, since an absorbant set minimal with respect to inclusion need not be a kernel.

**8.2.4 Proposition.** For a symmetric loop-free graph $G$ the following holds:

i)    $G$ has at least one kernel.

ii)    Kernels are precisely the inclusion-maximal stable sets.

---

[4] Kernels were termed by D. König "point basis of the second kind"; whereas a "point basis of the first kind" in a graph $(V, B)$ would be in our notation a kernel in $(V, \overline{I} \sqcap B^*)$. This terminology is no longer in use.

**Proof:** The empty set $O$ is stable, so stable sets do indeed exist. Every stable set maximal with respect to inclusion satisfies $Bs = \overline{s}$ by Proposition 8.1.3, and is thus a kernel.    □

In the existence theorem above the point axiom is involved: In Proposition 8.1.3 we made a certain choice, as one would automatically do when working with a finite graph drawn on paper. The proof of the following does not require such arguments.

**8.2.5 Proposition.** For a transitive graph $G$ with $B^*\overline{BL} = L$ (e.g., a finite loop-free graph or a progressively finite transitive graph), the following holds:

i)   $G$ has precisely one kernel.

ii)  The kernel is the smallest absorbant set, i.e., the set $\overline{BL}$ of terminal points.

**Proof:** From Proposition 8.1.4 together with Proposition 8.2.3.    □

Examples for these special cases are given in Fig. 8.2.3.

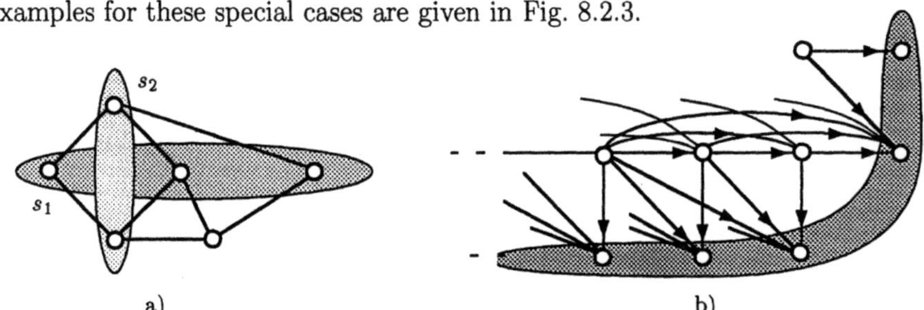

Fig. **8.2.3** Kernels in a loop-free symmetric graph (a)
and in a transitive progressively finite graph (b)

If $B$ is progressively finite, then by Proposition 6.3.11.i so is $B^+$ where $B^+ = \overline{I} \sqcap B^*$. From this one can infer the existence of bases in graphs.

**8.2.6 Corollary.** A finite and loop-free, or progressively finite graph has precisely one basis, the set of all terminal points.    □

We now want to check whether kernels exist in other cases. We shall see in particular that progressively finite graphs always possess kernels, even if they are not transitive. The starting point is the defining condition for a kernel, $s = \overline{Bs}$. Putting $f(x) := \overline{Bx}$, a kernel $s$ is thus a fixedpoint for this mapping $f : 2^V \longrightarrow 2^V$ of the power set into itself. So one would think of applying the well-developed apparatus of fixedpoint theory, summarized in Appendix A.3. Clearly, $f$ is *antitonic*, i.e.,

$$x \subset y \quad \Longrightarrow \quad f(y) \subset f(x).$$

So the fixedpoint theorem A.3.1 cannot be applied readily since it deals with *isotonic* mappings.

The situation can be mended by considering $f^2(x) = \overline{B\overline{Bx}}$, instead of $f(x) = \overline{Bx}$, which is indeed isotonic:

$$x \subset y \quad \Longrightarrow \quad f^2(x) \subset f^2(y).$$

Clearly, every fixedpoint $s$ of $f$ also satisfies $s = \overline{B\overline{Bs}}$ and hence is a fixedpoint for $f^2$ as well. But, conversely, some fixedpoints of $f^2$ may fail to be fixedpoints of $f$; cf. Fig. 8.2.4. In view of Proposition A.3.1, we therefore first study the fixedpoints of $f^2$; they need not be fixedpoints for $f$ but are nevertheless closely related. In this connection the following two types of point sets are relevant which are characterized by $s \subset f^2(s)$ or $f^2(s) \subset s$, respectively.

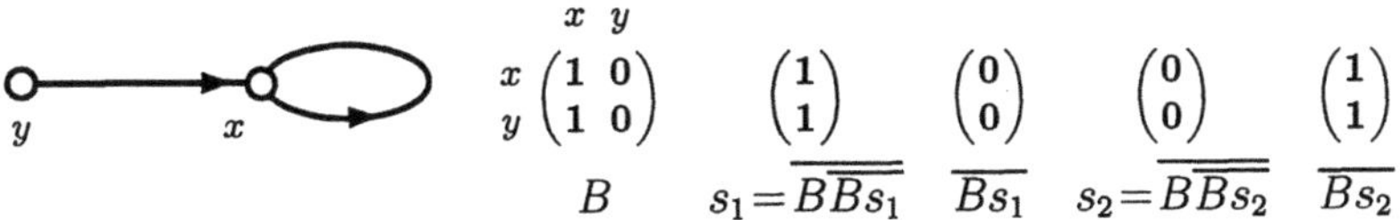

Fig. 8.2.4 Two sets $s$ with $s = \overline{B\overline{Bs}}$, but $s \neq \overline{Bs}$, and which are therefore no kernels

**8.2.7 Definition.** In a graph with associated relation $B$ we call a point set

i) $s$ **retarding** $\quad :\Longleftrightarrow \quad B\overline{Bs} \subset \overline{s}$

$\qquad\qquad\qquad\quad \Longleftrightarrow \quad$ There is no point in $s$ having a successor without a successor in $s$.

ii) $z$ **expansive** $\quad :\Longleftrightarrow \quad \overline{z} \subset B\overline{Bz}$

$\qquad\qquad\qquad\quad \Longleftrightarrow \quad$ Every point whose successors all possess a successor in $z$ belongs to $z$. $\qquad\qquad$ □

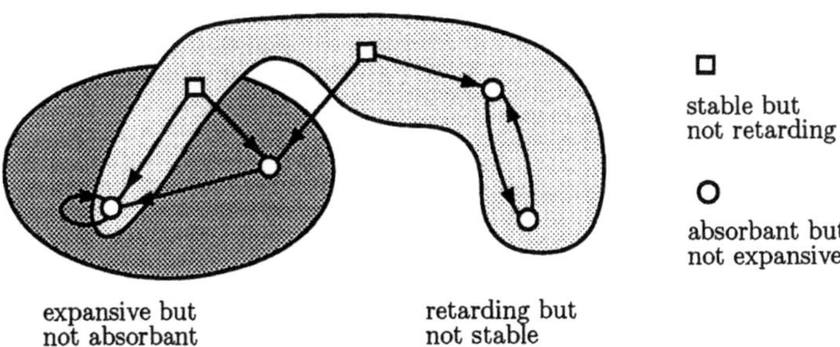


Fig. 8.2.5 Retarding and expansive sets

The points of a retarding set are thus characterized as follows: For any arc leaving $s$ there is another (successor) arc leading into $s$. In Fig. 8.2.5 retarding and expansive sets are compared to stable or absorbant sets. There are four possible cases shown. In the progressively finite case the situation is more rigid:

**8.2.8 Proposition.** In progressively finite graphs a point set $s$ satisfies

$$s \text{ retarding} \quad \Longrightarrow \quad s \text{ stable}.$$

**Proof:** Given the associated relation $B$, we prove for $y := s \sqcap Bs$ that:

$$y = Bs \sqcap s \subset (B \sqcap ss^{\mathsf{T}})(s \sqcap B^{\mathsf{T}}s) \subset B(s \sqcap Bs) = By.$$

We have used that $B^{\mathsf{T}}s \subset Bs$ is equivalent to $B\overline{Bs} \subset \overline{s}$. Since $B$ is progressively finite we obtain $y = O$ according to the remark following Definition 6.3.2. $\quad\square$

By virtue of Propositions A.3.1 and A.3.2, we obtain from the expansive and the retarding sets the least and the greatest fixedpoint of $f^2$, namely

$$a := \inf\{\, z \mid z \text{ expansive}\,\} = \inf\{\, z \mid z = f^2(z)\,\},$$
$$b := \sup\{\, s \mid s \text{ retarding}\,\} = \sup\{\, s \mid s = f^2(s)\,\}.$$

Since kernels are not just stable and absorbant, but also expansive and retarding, one has for every kernel $s$ that

$$\overline{Bb} = a \subset s \subset b = \overline{Ba}.$$

Therefore, we call $a$ and $b$ the **descriptive bounds for kernels**. Here, we have used that $a = \overline{Bb}$ and $b = \overline{Ba}$ by Proposition A.3.8. In particular, $a$ and $b$ are both expansive and retarding. In Fig. 8.2.6 the descriptive bounds for some graph are given, which in that simple case are just the union and the intersection of the sole two kernels, respectively.

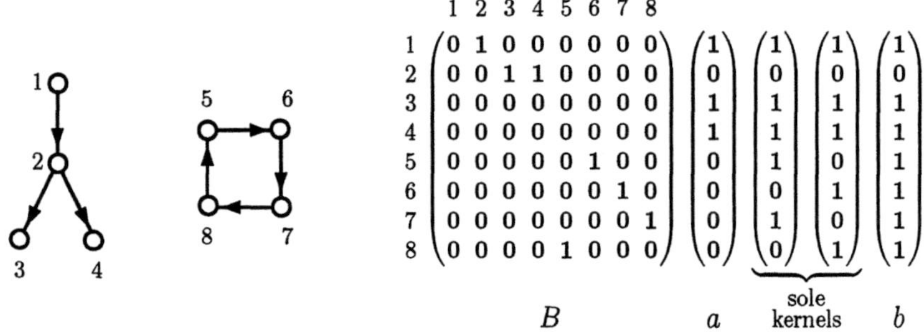

Fig. 8.2.6 Bounds for kernels

Bound $a$ is always retarding and, by Proposition 8.2.8, also stable, $Ba \subset \overline{a}$, provided that the graph is progressively finite. So in this situation every kernel $s$ satisfies $a = s = b$, and we have the following existence theorem.

**8.2.9 Proposition.** For a progressively finite graph $G$ the following holds:

i)     $G$ has precisely one kernel.

ii)    This kernel $s$ satisfies $a = s = b$. $\quad\square$

In order to localize kernels we use the iteration $a_{i+2} := \overline{B\overline{Ba_i}}$ which approximates the least fixedpoint of $f^2$ (cf. Prop. A.3.3):

$$a_0 := O, \quad a_2 := f^2(a_0), \quad a_4 := f^2(a_2), \ \ldots$$

Firstly, all those points, from where no arcs emerge, belong to the kernel; so we start with $a_0 := O$, $a_1 := \overline{BL}$ (see Fig. 8.2.7). The predecessors $Ba_1$ of these points do not belong to the kernel; we define $\overline{b_1} := Ba_0$. Points having all their successors in $\overline{b_1}$ are collected for the kernel, i.e., $a_2 := \overline{B\overline{b_1}} = \overline{B\overline{Ba_0}}$; etc.

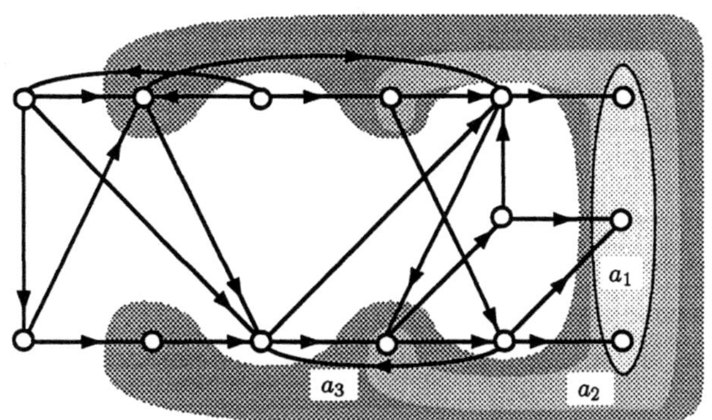

Fig. 8.2.7 Iteration to obtain a kernel,<br>
working also for a graph that is not progressively finite

Splitting this into two iterations (according to Prop. A.3.9), one obtains

$$\begin{aligned}
a_0 &:= O, & b_0 &:= L, \\
a_{i+1} &:= \overline{B\overline{b_i}}, & b_{i+1} &:= \overline{B\overline{a_i}}, & i \geq 0 \\
a_\infty &:= \sup_{i\geq 0} a_i, & b_\infty &:= \inf_{i\geq 0} b_i
\end{aligned}$$

resulting in the **iterative bounds** $a_\infty, b_\infty$ **for kernels**. Altogether:

$$a_\infty \subset \overline{B\overline{b_\infty}} \subset \overline{B\overline{b}} = a \subset s \subset b = \overline{B\overline{a}} \subset \overline{B\overline{a_\infty}} = b_\infty.$$

Notice the lack of symmetry in the above chain of inclusions. It is due to multiplication being $\sqcup$-distributive but only $\sqcap$-*sub*distributive. Bound $a_\infty$ is stable and $b_\infty$ is absorbant, but $a_\infty$ is not necessarily maximal with respect to inclusion, nor is $b_\infty$ minimal, as can be seen from simple counterexamples such as the circuit of length 2. Also, $b_\infty$ need not be stable and $a_\infty$ need not be absorbant, as the graph in Fig. 8.2.8 shows where in addition $a_\infty \subsetneq \overline{B\overline{b_\infty}}$.

We next prove a general relationship between the iterative kernel bounds and the associated relation.

**8.2.10 Proposition.** In a graph with associated relation $B$ the two iterative bounds $a_\infty, b_\infty$ for the kernels satisfy

$$\sup_{i\geq 0} \overline{B^i L} \subset \overline{b_\infty} \sqcup a_\infty.$$

**Proof** by induction: Starting with $L = b_0 \subset a_0 \sqcup B^0 L = L$ we assume $b_i \subset a_i \sqcup B^i L$ to hold true. This immediately leads to

$$\overline{a_{i+1}} = Bb_i \subset B(a_i \sqcup B^i L) = \overline{b_{i+1}} \sqcup B^{i+1} L.$$

Using the equivalent inclusions $\overline{B^i L} \subset \overline{b_i} \sqcup a_i$, we obtain

$$\sup{}_{i \geq 0} \overline{B^i L} \subset \sup{}_{i \geq 0} \overline{b_i} \sqcup \sup{}_{i \geq 0} a_i = \overline{b_\infty} \sqcup a_\infty. \qquad \Box$$

For a progressively bounded graph, therefore, $L \subset \overline{b_\infty} \sqcup a_\infty$ or $b_\infty \subset a_\infty$. Hence $a_\infty = b_\infty$.

As an immediate consequence we have an existence theorem for kernels in progressively bounded graphs.

**8.2.11 Corollary.** If $G$ is a progressively bounded graph with associated relation $B$ (e.g., a finite circuit-free graph), the following holds:

i)    $G$ has precisely one kernel.

ii)   The kernel is the limit $\sup_{i \geq 0} a_i = a_\infty = b_\infty = \inf_{i \geq 0} b_i$ of any of the two iterations

$$a_0 := O, \qquad\qquad b_0 := L,$$
$$a_{i+2} := \overline{B\overline{Ba_i}}, \qquad\qquad b_{i+2} := \overline{B\overline{Bb_i}}, \qquad i \geq 0. \qquad \Box$$

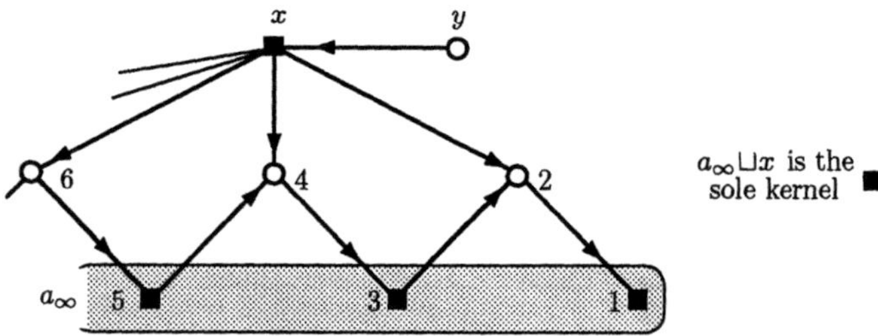


**Fig. 8.2.8** Iteration and kernel of a progressively finite,
but not progressively bounded graph

Algorithm 8.2.11 can be viewed as a two-level variant of the exhaustion theorem 6.3.4 for progressively bounded graphs. The first thorough investigation of this iteration process seems to have been carried out in VON NEUMANN-MORGEN-STERN 44.

Our next aim is to show that kernels exist in graphs which are no longer progressively finite or bounded, and which may, for instance, contain circuits. In view of the simple examples in Fig. 8.2.9, the presence of circuits of odd length is an obstacle which, however, can be overcome.

So, in Proposition 8.2.12, we first have a closer look at graphs without circuits of odd length.

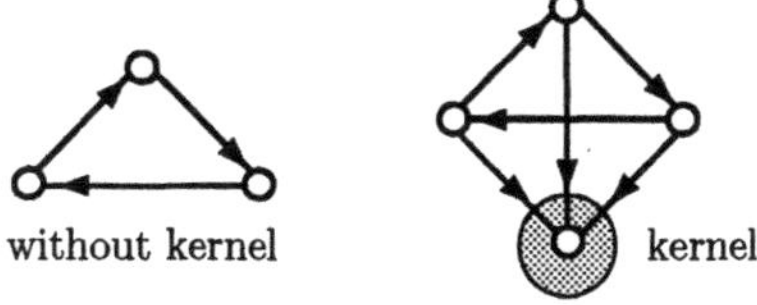



**Fig. 8.2.9** Kernels and circuits of odd length

**8.2.12 Proposition.** The following holds for a strongly connected graph $G$ without circuits of odd length:

i)  $G$ possesses a kernel.

ii)  If $G$ is not just one point, there exist at least two different kernels. Given any point $x$, two complementary kernels may be obtained as those points reachable from $x$ by paths of even or odd length, respectively.

**Proof:** Given the associated relation $B$ of $G$ and putting $D := (B^2)^*$, we write down the assumptions formally:

$$G \text{ strongly connected} \iff L = B^* = D \sqcup BD \iff \overline{D} \subset BD;$$
$$G \text{ without odd circuits} \iff BD = BDD \subset \overline{I} \iff (BD)^\mathsf{T} \subset \overline{D}.$$

Altogether, one obtains $(BD)^\mathsf{T} \subset \overline{D} \subset BD$ which implies $(BD)^\mathsf{T} = BD$ and $BD = \overline{D}$. Every "column" of $D = (B^2)^*$ is a kernel—there are just two types of "columns" representing the sets of points that may be reached along paths of odd or even length. As shown in Fig. 8.2.10, there may exist further kernels.    □

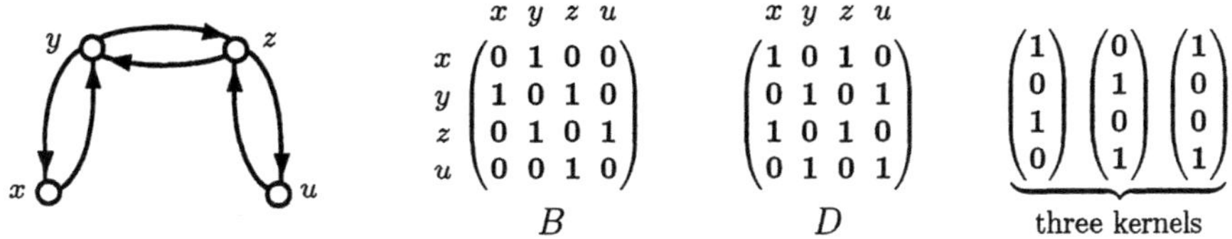

**Fig. 8.2.10** Kernels of a strongly connected graph
without circuits of odd length

The original proof of the following result was based on Corollary 8.2.6; it is quite complicated. Here we give an inductive proof.

**8.2.13 Proposition** (*M. Richardson, 1953*). A finite graph without circuits of odd length has at least one kernel.

**Proof** by induction on the number of points of the graph: The (only) graph without circuits of odd length that has just one point is loop-free, and its sole point constitutes a kernel. Assume the theorem to hold for all graphs with less than $n$ points.

In a graph with $n$ points we consider a terminal strongly connected component which again is without circuits of odd length. The subgraph generated by that component has a kernel $t$ according to Prop. 8.2.12. The (possibly empty) graph generated by $\overline{(I \sqcup B)t}$, i.e., by the complement of the union of $t$ with all predecessors of $t$, has a kernel $t_1$ because of the induction hypothesis. The situation is depicted in Fig. 8.2.11. Obviously, $t \sqcup t_1$ is a kernel of the entire

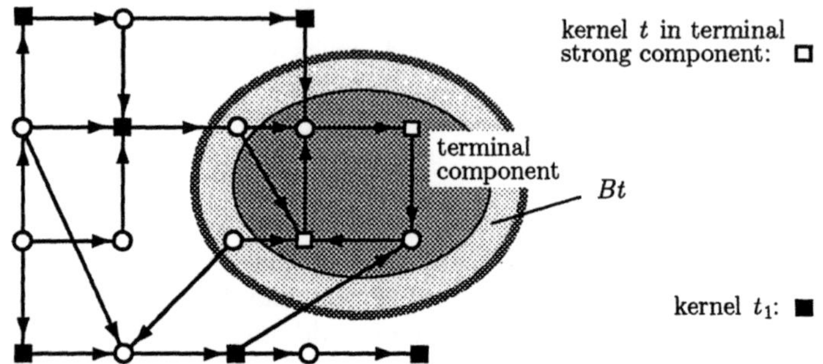


**Fig. 8.2.11** Decomposition according to the proof of the theorem of Richardson

graph. Points inside $t$ or inside $t_1$ are unrelated, since both are kernels. Arcs joining $t$ and $t_1$ cannot exist because $t_1 \subset \overline{(I \sqcup B)t} \implies t_1 t^{\mathsf{T}} \subset \overline{B}$ and because the terminal component $t$ can never be left. The point set $\overline{t \sqcup t_1}$ decomposes into $\overline{t} \sqcap (I \sqcup B)t$ and $\overline{t_1} \sqcap \overline{(I \sqcup B)t}$. In the first term an arc leading into $t$ must exist and in the second, by definition, an arc leading into $t_1$ must exist.     □

The result above can be strengthened by an argument due to T. Ströhlein (1970) who investigated combinatorial games and proved the existence of kernels in bipartitioned game graphs. Bipartite graphs clearly have no circuits of odd length and therefore contain kernels by Prop. 8.2.13. Notice that a connected graph can be bipartitioned in just one way (up to interchanging "left" and "right"). In the following proposition we indicate two kernels that have a distinguished position with respect to all other kernels.

**8.2.14 Proposition.** If the graph $G$ is bipartitioned as $L = g \sqcup d$ we have.

i)  $G$ has (not necessarily distinct) kernels $A'$ and $A''$ whose intersection and union give the lower and upper bound of any kernel $s$ of $G$:
$$A' \sqcap A'' \subset s \subset A' \sqcup A''.$$

ii) The two distinguished kernels $A'$ and $A''$ are determined by the descriptive bounds for kernels $a$ and $b$ as
$$A' := (g \sqcap b) \sqcup (d \sqcap a), \quad A'' := (g \sqcap a) \sqcup (d \sqcap b).$$

**Proof:** We prove only that $A'$ is a kernel:

$$\begin{aligned}
B A' &= B(g \sqcap b) \sqcup B(d \sqcap a) \\
&= (d \sqcap Bb) \sqcup (g \sqcap Ba), &&\text{because of Prop. 4.1.3} \\
&= (d \sqcap \overline{a}) \sqcup (g \sqcap \overline{b}), &&\text{see the remarks prior to Prop. 8.2.9} \\
&= (d \sqcup \overline{b}) \sqcap (g \sqcup \overline{a}), &&\text{by distributivity} \\
&= \overline{A'}.
\end{aligned}$$

The bounding property follows from $a \subset b$ since

$$\begin{aligned}
A' \sqcap A'' &= [(g \sqcap b) \sqcup (d \sqcap a)] \sqcap [(g \sqcap a) \sqcup (d \sqcap b)] \\
&= [(g \sqcap b) \sqcap (g \sqcap a)] \sqcup [(d \sqcap a) \sqcap (d \sqcap b)] \\
&= (g \sqcap b \sqcap a) \sqcup (d \sqcap a \sqcap b) = b \sqcap a = a.
\end{aligned}$$

In essentially the same way $A' \sqcup A'' = b$ is obtained and, indeed, every kernel lies between $a$ and $b$. $\qquad\qquad\square$

If in the proof above one were to utilize the iterative kernel bounds instead,

$$a' := (g \sqcap b_\infty) \sqcup (d \sqcap a_\infty), \quad a'' := (g \sqcap a_\infty) \sqcup (d \sqcap b_\infty),$$

then one would only get $Ba' \subset \overline{a'}$, $\quad Ba'' \subset \overline{a''}$, i.e., two distinguished stable sets would result. (For in the second line of the proof of $BA' = \overline{A'}$ only $a_\infty \subset \overline{Bb_\infty}$ would hold instead of the previous $a = \overline{Bb}$.) But the bounding property would be retained. Taking for $g$ the square points of the graph from Fig. 8.2.8, we obtain an example of a bipartitioned, not progressively bounded graph with $Ba' \subsetneq \overline{a'}$. Note that the iterative and the descriptive bounds coincide for finite or progressively bounded graphs.

In the bipartite case there is an important alternative way of obtaining the iterative bounds $a'$ and $a''$. Above we started with $a_0$, $b_0$, got $a_\infty$ and $b_\infty$ by iteration, and composed $a'$ and $a''$ after intersecting with $g$ and $d$, respectively. The same iteration process can be started with $a'_0 := g$ and $a''_0 := d$, so as to put the composition step at the beginning. The result of the iteration will again be $a'$ and $a''$.

### Exercise

**8.2.1** If $s$ is the basis of a confluent relation $B$, then the relation $B^* \sqcap (sL)^\mathsf{T}$ is a mapping.

## 8.3 Games

The results on kernels from Sect. 8.2 will now be applied to games. We deal with two-player games where both players make their moves alternately, and where the outcome is rated a "win" or a "loss"; we may also allow the game to end in a draw. Thus, we do not consider games involving integer-valued or even real-valued scores and treat only a small section of game theory, which has become an extensive discipline.

We shall only discuss games with "complete information", where questions regarding probability play no role, in contrast to games of cards or of dice. Games of this kind are typically board games such as chess, go, hex, and games of matches such as nim. These games have in common that one player *winning* is equivalent to the opponent *losing*. Moreover, the so-called rule of loss holds: "If some player is to make a move but can no longer make any (terminal situation), then he loses[5]."

A first example of such a game is provided by the bipartitioned graph in Fig. 8.3.1. That game consists of the **situations** $\{l_1, l_2, l_3, l_4, r_1, r_2, r_3\}$ and the possible **moves** represented by arrows. At any given stage of the play one of the situations is the **current** one. In the situations on the left-hand side it is always **player** $\mathcal{L}$'s turn to make the next move, on the right it is player $\mathcal{R}$'s turn. So each situation determines the player who is to make a move which in turn means that he chooses one of the arrows emanating from the current situation. The new situation after making the move is then the one described by the head of the arrow chosen. Because of the bipartition of the graph, the right of making a move is alternating. Situations $l_2$ and $r_1$ are **terminal**. According to the rule of losing when unable to make a move, player $\mathcal{L}$ loses in situation $l_2$, and $\mathcal{R}$ loses in $r_1$, the player being determined by the current situation. The respective other player is then the happy winner.

The wise player $\mathcal{R}$ in the current situation $r_2$ will move to $l_2$ whereupon $\mathcal{L}$ will lose. It would be unwise for $\mathcal{R}$ to move to $l_1$ because there his opponent $\mathcal{L}$ would be given the chance of moving to $r_1$ and hence of winning.

Further, we observe: Situation $l_3$ means victory for player $\mathcal{R}$. Although it is not $\mathcal{R}$'s turn there, it is impossible for $\mathcal{L}$ to move in any way so as to prevent $\mathcal{R}$ from making his next move from $r_2$ to $l_2$ which is terminal and means loss for $\mathcal{L}$.

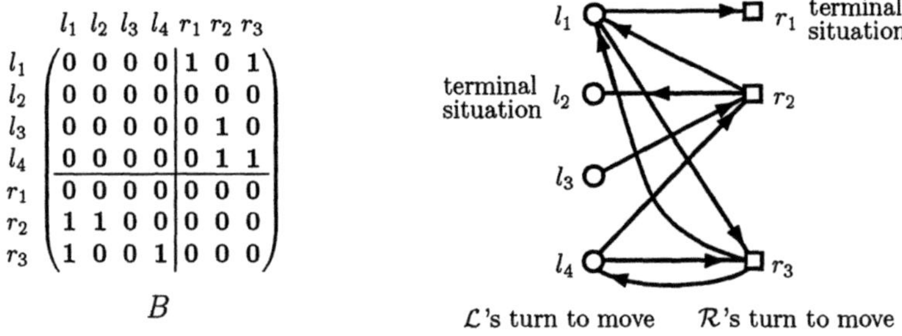


$$B = \begin{pmatrix}
 & l_1 & l_2 & l_3 & l_4 & r_1 & r_2 & r_3 \\
l_1 & 0 & 0 & 0 & 0 & 1 & 0 & 1 \\
l_2 & 0 & 0 & 0 & 0 & 0 & 0 & 0 \\
l_3 & 0 & 0 & 0 & 0 & 0 & 1 & 0 \\
l_4 & 0 & 0 & 0 & 0 & 0 & 1 & 1 \\
r_1 & 0 & 0 & 0 & 0 & 0 & 0 & 0 \\
r_2 & 1 & 1 & 0 & 0 & 0 & 0 & 0 \\
r_3 & 1 & 0 & 0 & 1 & 0 & 0 & 0
\end{pmatrix}$$

**Fig. 8.3.1** A simple game

Situations $l_4$ and $r_3$, finally, both result in a draw (for player $\mathcal{L}$ as well as for $\mathcal{R}$). It would be a bad idea for $\mathcal{L}$ to go from $l_4$ to $r_2$, for then $\mathcal{R}$ could move from $r_2$ to $l_2$ and would win. $\mathcal{L}$ should rather move to $r_3$ where $\mathcal{R}$ would be in similar troubles. Also $\mathcal{R}$ would be ill-advised to move from $r_3$ to $l_1$, for $\mathcal{L}$ could then move from $l_1$ to $r_1$ which is terminal for $\mathcal{R}$. In order to avoid that, $\mathcal{R}$ should instead move from $r_3$ to $l_4$. Thus, the fear of losing leads to repeating certain moves indefinitely which is termed a *draw by infinite repetition*. (In chess

---

[5] This rule is also referred to as "the last player losing". But one might argue as to who is the last player, the one who makes the last move or the first to be unable to do so? It is clearer to follow Žagler and term the rule "losing when unable to make a move". Other rules of loss are also studied, such as "the last player winning". But here completely different problems may arise; see e.g., STRÖHLEIN, ZAGLER 77.

there is also the terminal situation of stalemate which is scored as a draw; such situations are excluded here.)

We have tried to indicate how to assess a given situation of the game with regard to winning, losing, or ending in a draw, on the basis that *terminal* situations entail loss for the player whose turn it is to make a move. A closer analysis of Fig. 8.3.1 yields the following result relative to the player who is about to make a move (and who is uniquely determined by the situation):

Situations implying loss for the player about to move:    $l_2, l_3, r_1,$
Situations implying win for the player about to move:    $l_1, r_2,$
Situations resulting in a draw:    $l_4, r_3.$

Since, by definition, one player wins if and only if the other loses, we have the following "absolute" characterization with respect to $\mathcal{L}$ and $\mathcal{R}$:

|  | for $\mathcal{L}$ | for $\mathcal{R}$ |
|---|---|---|
| losing situations | $l_2, l_3, r_2$ | $l_1, r_1$ |
| winning situations | $l_1, r_1$ | $l_2, l_3, r_2$ |
| draws | $l_4, r_3$ | $l_4, r_3$ |

Indeed,

- a "losing situation for the player to move" is a losing situation for $\mathcal{L}$ with $\mathcal{L}$ to move, or a losing situation for $\mathcal{R}$ with $\mathcal{R}$ to move.

- a "losing situation for $\mathcal{L}$" is a losing situation for the player to move where $\mathcal{L}$ is to move, or a winning situation for the player to move where $\mathcal{R}$ is to move.

It seems difficult to find a more precise terminology for distinguishing between the "relative" and the "absolute" characterizations above. So one should be aware of possible confusion at this point.

We want to study another important fact by means of an example, without giving the general proof: The graph from Fig. 8.3.1 has precisely two kernels, namely

$$s_1 := \{\, l_2, l_3, l_4, r_1 \,\} \quad \text{and} \quad s_2 := \{\, l_2, l_3, r_1, r_3 \,\}.$$

Its kernel bounds (which are both iterative and descriptive) are

$$a = \{\, l_2, l_3, r_1 \,\} \subset s_i \subset \{\, l_2, l_3, l_4, r_1, r_3 \,\} = b.$$

Also, $s_1$ and $s_2$ are the distinguished kernels established in Proposition 8.2.14, whose intersection and union are the kernel bounds $a$ and $b$, respectively.


**8.3.1 Fact.** Given two players and a bipartitioned graph on which they play a game as described above, then

$\quad a$    represents the losing situations for the player to move,

$\quad \overline{a} \sqcap b$    represents the draw situations for the player to move,

$\quad \overline{b}$    represents the winning situations for the player to move.    $\square$

Clearly, given any bipartitioned graph, a game as described above can be played by two players $\mathcal{L}$ and $\mathcal{R}$, and we need not give a formal definition. The game is started with a current situation which we call the **starting situation**. In many games the starting situation is laid down by convention; in chess, for example, by the well-known starting position and the rule that White moves first. On the other hand, in chess one likes to ask whether a player in a given situation (position of the pieces and the right to move) still has a chance to win. Here then, the starting situation would be different. Chess problems are usually normalized in the sense that White moves first.

## Games with Symmetric Rules

The game in Fig. 8.3.1 lacks left-right symmetry. Therefore, the two players must argue quite differently in order to assess their chances of winning. In games such as nim or chess both players will follow rather parallel strategies. (In chess the symmetry is distorted by the rule that White moves first. While this rule gravely affects the initial phase, its influence wanes during the middle and the final part of the play.)

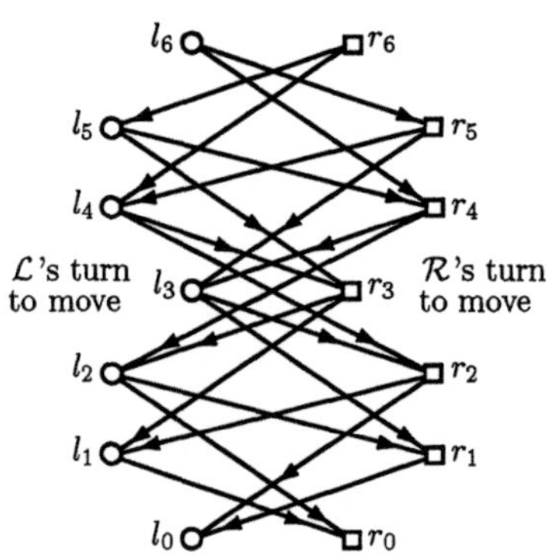


**Fig. 8.3.2** Situations of a game of matches

We now focus on games possessing that kind of symmetry, and consider the bipartitioned graph from Fig. 8.3.2. It can be interpreted as a game of matches: From a pile of 6 matches two players $\mathcal{L}$ and $\mathcal{R}$ may alternately take one or two matches. The starting situation is $l_6$, if $\mathcal{L}$ is to begin, and $r_6$, if $\mathcal{R}$ is.

In contrast to the game from Fig. 8.3.1, both players now have the same options apart from their different starting times. The corresponding graph is therefore "geometrically symmetric" with respect to the middle vertical line. This can also be seen from its associated relation in Fig. 8.3.3, although $B$ is not a symmetric relation. But there is the possibility of switching the point of view[6], i.e., there is a bijective mapping $W$ which associates with every situation and with $\mathcal{L}$ in a position to move the analogous situation with $\mathcal{R}$ having the right to move. Relations $Q$ and $S$ then clearly satisfy $WS = QW^{\mathsf{T}}$. (With rows and columns numbered suitably, the matrices of $Q$ and $S$ are equal.) In the example, $W(l_i) = r_i$ for $0 \leq i \leq 6$.

---

[6] The principle of changing the point of view exists in chess, too. However, one must not simply transfer the right of moving from one player to the other while leaving the situation on the board unchanged. If, e.g., White is being given check, then in the same board situation it cannot possibly be Black's turn to move. So there is no geometric symmetry for this type of change. For chess the operation $W$ is subtler: It consists of transferring the right to move to the other player, reflecting the configuration on the board about the middle line (between positions 4 and 5), and interchanging the colors of both pieces and squares.

In the sequel, we consider exclusively games with symmetric rules. There, each player can follow the same instructions, apart from the change of point of view (in contrast to Fig. 8.3.1).

For games of the kind described, one can identify situations related by $W$, as is done in Fig. 8.3.4 for the game from Fig. 8.3.2. Arrows $(l_5, r_4)$ and $(r_5, l_4)$, for example, are identified as arrow $(5, 4)$. The results are then referred to as *positions*. Situation $l_5$ is now replaced with the pair $(5, \mathcal{L})$, consisting of position 5 and player $\mathcal{L}$ having the right to move.

When moving along the graph in Fig. 8.3.4, it is no longer easy to see how the right to move switches from left to right, but on the other hand the graph itself is more clearly arranged. As a consequence of identifying left and right, it becomes unnecessary to distinguish the players as $\mathcal{L}$ and $\mathcal{R}$. Instead, we need only refer to the player $\xi$ ($\mathcal{L}$ or $\mathcal{R}$) and his opponent $\bar{\xi}$ ($\mathcal{R}$ or $\mathcal{L}$), the opponent $\bar{\bar{\xi}}$ of the latter being again $\xi$.

$$B = \begin{pmatrix} O & Q \\ S & O \end{pmatrix} =$$

| | $l_0$ | $l_1$ | $l_2$ | $l_3$ | $l_4$ | $l_5$ | $l_6$ | $r_0$ | $r_1$ | $r_2$ | $r_3$ | $r_4$ | $r_5$ | $r_6$ |
|---|---|---|---|---|---|---|---|---|---|---|---|---|---|---|
| $l_0$ | 0 | 0 | 0 | 0 | 0 | 0 | 0 | 0 | 0 | 0 | 0 | 0 | 0 | 0 |
| $l_1$ | 0 | 0 | 0 | 0 | 0 | 0 | 0 | 1 | 0 | 0 | 0 | 0 | 0 | 0 |
| $l_2$ | 0 | 0 | 0 | 0 | 0 | 0 | 0 | 1 | 1 | 0 | 0 | 0 | 0 | 0 |
| $l_3$ | 0 | 0 | 0 | 0 | 0 | 0 | 0 | 0 | 1 | 1 | 0 | 0 | 0 | 0 |
| $l_4$ | 0 | 0 | 0 | 0 | 0 | 0 | 0 | 0 | 0 | 1 | 1 | 0 | 0 | 0 |
| $l_5$ | 0 | 0 | 0 | 0 | 0 | 0 | 0 | 0 | 0 | 0 | 1 | 1 | 0 | 0 |
| $l_6$ | 0 | 0 | 0 | 0 | 0 | 0 | 0 | 0 | 0 | 0 | 0 | 1 | 1 | 0 |
| $r_0$ | 0 | 0 | 0 | 0 | 0 | 0 | 0 | 0 | 0 | 0 | 0 | 0 | 0 | 0 |
| $r_1$ | 1 | 0 | 0 | 0 | 0 | 0 | 0 | 0 | 0 | 0 | 0 | 0 | 0 | 0 |
| $r_2$ | 1 | 1 | 0 | 0 | 0 | 0 | 0 | 0 | 0 | 0 | 0 | 0 | 0 | 0 |
| $r_3$ | 0 | 1 | 1 | 0 | 0 | 0 | 0 | 0 | 0 | 0 | 0 | 0 | 0 | 0 |
| $r_4$ | 0 | 0 | 1 | 1 | 0 | 0 | 0 | 0 | 0 | 0 | 0 | 0 | 0 | 0 |
| $r_5$ | 0 | 0 | 0 | 1 | 1 | 0 | 0 | 0 | 0 | 0 | 0 | 0 | 0 | 0 |
| $r_6$ | 0 | 0 | 0 | 0 | 1 | 1 | 0 | 0 | 0 | 0 | 0 | 0 | 0 | 0 |

**Fig. 8.3.3** Associated relation of a game with symmetric rules

The positions of the game in Fig. 8.3.4 are numbered 0 through 6. But even when making 6 the starting position and stipulating that it is $\xi$'s turn to move there, giving the position does not, in general, determine the player who is to make the move. For example, position 4 can reached by "$\xi$ taking 2", but also by "$\xi$ taking 1", followed by "$\bar{\xi}$ taking 1". In the first case it will be $\bar{\xi}$'s turn, in the second, $\xi$'s turn.

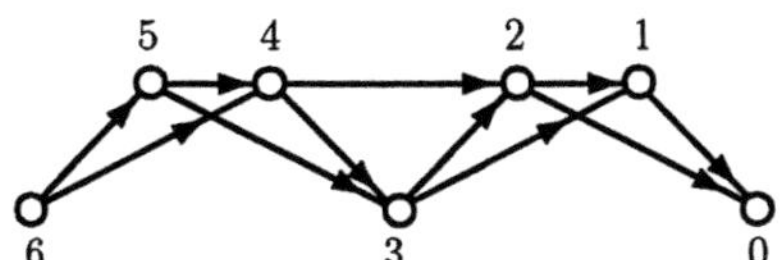


**Fig. 8.3.4** Positions in a game of matches

So positions, in contrast to situations, do *not* determine the player entitled to make the next move. One rather adopts the convention of *considering positions only with regard to the player about to move.*

Starting with position 6 and player $\xi$ to move, it is easy to see that player $\bar{\xi}$ can always manage to win. All he needs to do is never to draw horizontally, so as to have $\xi$ draw only from the lower row of vertices and eventually from 0. This rule can also be expressed as: The number of matches to be removed has to equal 3 minus the number last taken by the opponent. If $\bar{\xi}$ fails to follow this rule just once, then $\xi$ can draw from the upper row and, by following the rule in turn, can make $\bar{\xi}$ lose. The rule provides an a priori decision of how to move in any given position of the game, irrespective of whether the position will be reached in a play of the game, or not. Rules like this are called *strategies*, a fundamental notion of mathematical game theory introduced by H. Steinhaus in 1925.

So the player eager to win looks for the optimal strategy; its definition, however, does not refer to its quality.

**8.3.2 Definition.** A **strategy** is a mapping which assigns to every nonterminal position precisely one move. A **play of the game**, starting in position $x$ with player $\xi$ to move, is a path in the position graph starting from $x$, with moves determined alternately by $\xi$ and $\bar{\xi}$ (starting with $\xi$). $\qquad\qquad$ □

Once a starting position is determined, and each player has selected a strategy, the sequence of moves (play of the game) is uniquely determined. We can now make the assessment of positions more precise.

**8.3.3 Definition.** Given a position $x$ of the game graph, and given that it is player $\xi$'s turn to move from $x$, we call

$x$ **winning position** for $\xi$ $\quad:\Longleftrightarrow$
> There is a strategy for $\xi$ that, against every strategy of $\bar{\xi}$, results in a play starting in $x$ and ending in a position where $\bar{\xi}$ has no more move available.

$x$ **losing position** for $\xi$ $\quad:\Longleftrightarrow$
> There is a strategy for $\bar{\xi}$ that, against every strategy of $\xi$, results in a play starting in $x$ and ending in a position in which $\xi$ is to move but cannot.

$x$ **draw position** $\quad:\Longleftrightarrow$
> $x$ is neither a winning nor a losing position. $\qquad\qquad$ □

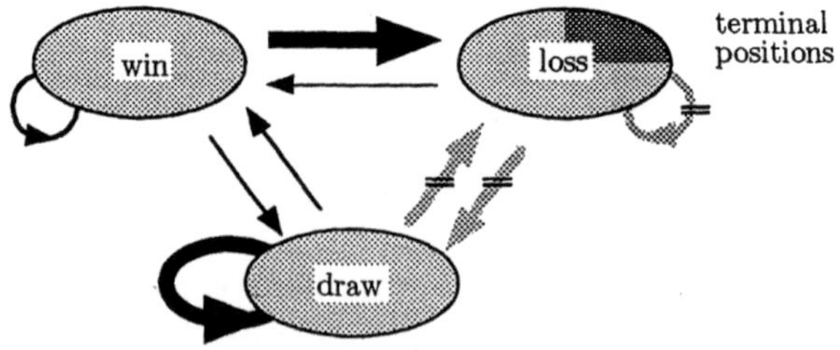

**Fig. 8.3.5** Win, loss, and draw (for the player about to move)

We give the following three obvious facts, keeping in mind that positions are being considered from the viewpoint of the player next to move. We omit the proof which relies on the notion of strategy and is not completely trivial.

- From a winning position there is at least one possible move into a losing position.

- From a losing position there is no move into a position of draw or a losing position.

- From a position of draw there is at least one move into some other position of draw, but no move into a losing position.

Therefore, any position of draw is the starting point of a path of infinite length in the graph of the game. Figure 8.3.5 shows the general pattern. Thick arrows indicate that there is at least one possible move from any position ("a good move"), crossed-out arrows mark the nonexistence of arrows, and simple arrows stand for possibly, but not necessarily, existing arrows. In the cases of "win→win", "win→draw", and "draw→win" such arrows correspond to a "poorly chosen" move, and to an "unavoidably bad" move in the case of "loss→win".

We compare this to the relationship between a kernel and its complement, represented in Fig. 8.3.6, which is obviously the special case with no positions of draw. This special case suggests looking for kernels in the graph in order to clarify the relationship between win, loss, and draw.

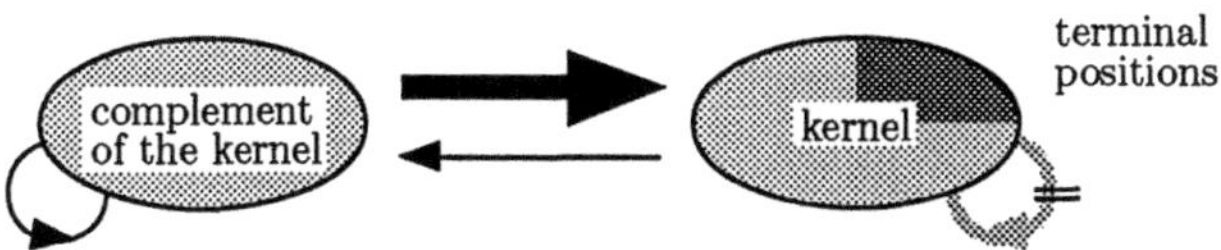


**Fig. 8.3.6** A kernel and its complement

The question arises as to what it is actually worth for a player to know whether the position he is about to move from is one of win, loss, or draw. That information alone need not yet mean much to him, for he would also want to know how to exploit a chance of winning or maintaining a draw. Thus many positions to be reached only later are equally important.

We now assume that the player knows about his initial position as well as all the other positions he can move into, whether they are ones of win, loss, or draw. In a winning position, this knowledge provides a strategy which avoids loss but does not necessarily secure win. The strategy says: "Move into a losing position". (Such a move is possible by hypothesis.) In order to achieve win, the player needs additional information about the layers $a_0, a_1, \ldots$ of the set $a$ of all losing positions. His strategy then reads: "Move into a losing position belonging to the smallest $a_i$ that can be reached".

In a position of draw there is a strategy for the player about to move which will secure a draw for him; it says: "Move into a draw position". (This move again exists by hypothesis.) Out of a losing position the player can move only into a winning position, if at all. If in addition he knows the sets $b_i$, then he can chose an *optimal defensive move* which has the effect of avoiding a terminal position for the longest possible period of time. Provided he is not yet caught in a terminal position, the strategy for that goal is the following: "Move into a winning position belonging to the greatest possible $\overline{b_i}$ that can be reached, but which does not belong to $\overline{b_{i-1}}$"(Compare Sect. 8.2.).

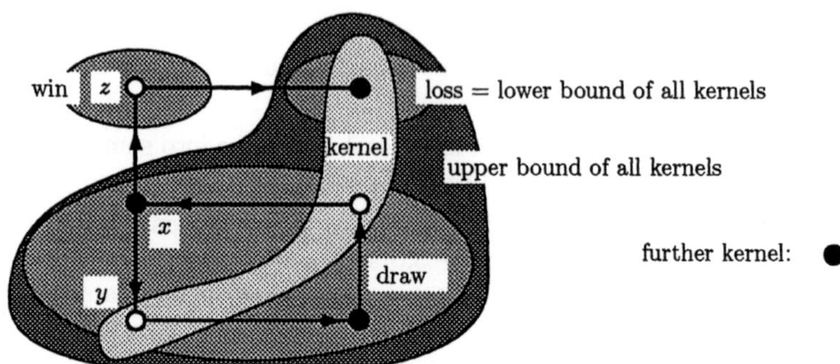


**Fig. 8.3.7** Kernel with positions of win, loss, and draw

The situation changes a little when the descriptive kernel bounds $a$ and $b$ are not known and one knows just some kernel which merely contains all the losing positions and possibly some draw positions. Figure 8.3.7 deals with this situation. The knowledge of some kernel provides the player moving from some position outside the kernel with a strategy that avoids loss but does not secure win. The strategy says: "Move into the kernel". After such a move the opponent either finds himself in a terminal position of the kernel or in some position from which he has to leave the kernel again, thus putting the first player in the same situation as before. On the other hand, in a position within the kernel, the knowledge of that kernel is not much help to the player about to move; he will not be able to maintain a draw, even when he is in a draw position. We consider player $\xi$ moving from position $x$ in Fig. 8.3.7. With every strategy prescribing to move from $x$ to $y$, player $\xi$ will maintain a draw by repeating certain moves indefinitely. Moving to $z$, however, means loss for $\xi$ because $\overline{\xi}$ can move to the only terminal position.


## Nim-Type Games

Games with progressively finite graphs have no draw positions, and, by Proposition 8.2.11, there is a unique kernel. Since progressive finiteness prevents games from continuing indefinitely, the kernel consists precisely of the losing positions for the player to move. Conversely, the winning positions are described by the complement of the kernel.

The game of nim is the prototype of a game with a progressively finite graph (C. L. Bouton, 1902). We illustrate this with a special example due to W. A. Wythoff (1907).

Two players alternately move a piece from one lattice point in the first quadrant to another, either in a direction parallel to one of the two axes towards the other, or towards the origin in a direction parallel to the diagonal. For example, from $(2, 1)$ a player can move to $(1, 1), (0, 1), (2, 0)$, or $(1, 0)$. The loser will be the player whose turn it is to move when the piece is in the origin, for there no further move is possible.

The player who moves from position $(i, k)$ will lose, if and only if $(i, k)$ belongs to the unique kernel of the graph of positions, defined on the point set $\{ (i, k) \mid 0 \le i, k \}$, which is progressively bounded, moreover.

The first points of the kernel are printed as squares in Fig. 8.3.8. Also, the iterative construction of the kernel is indicated (cf. Sect. 8.2).

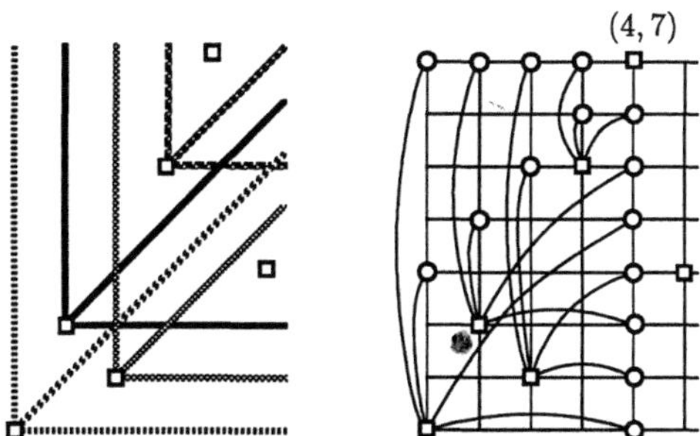


**Fig. 8.3.8** Kernel of the game of Wythoff

If player $\xi$ starts the game with the piece situated in the kernel (in position $(4, 7)$, say), then there is a strategy for his opponent $\bar{\xi}$, ensuring that $\xi$ will have to move, and hence lose, when the piece is in the origin. We have illustrated this for the point $(4, 7)$ by indicating at every point that can be reached by $\xi$ from $(4, 7)$, and which must lie outside the kernel, a move by which $\bar{\xi}$ can reenter the kernel and thus get closer to $(0, 0)$.

The kernel of this game can be given in a closed form by the method of "autogenous sequences", due to G. Aumann (1966). We use the following modified version. The starting point of the construction is the origin, which belongs to the kernel. Defining the following vectors of the lattice,

$$s := (2, 1) \quad \text{and} \quad k := (3, 2),$$

the sequence of kernel points below the diagonal begins with

$$(0,0) \quad (2,1) \quad (5,3) \quad (7,4) \quad (10,6) \quad (13,8) \quad (15,9)\ldots$$
$$s \qquad k \qquad s \qquad k \qquad k \qquad s \qquad k\ldots$$

It is then easy to see that the sequence of occurrences of $s$ and $k$ is "autogenous" in the following sense: The sequence starts as a one-element sequence $s$. It is then scanned from the left while appending $k$ or $sk$ to it from the right, according to the following "production principle" (cf. Fig. 8.3.9):

On scanning $s$ append $k$, on scanning $k$ append $sk$.

One will recognize the recursion principle for larger partial words to be attached. Starting with $w_1 := s$, $w_2 := k$, one has $w_h := w_{h-2} w_{h-1}$

$$s \mid k \mid sk \mid ksk \mid skksk \mid kskskksk \mid skkskkskskksk \mid$$
$$w_1 \quad w_2 \quad w_3 \quad w_4 \quad w_5 \qquad w_6 \qquad\qquad w_7 \qquad\qquad \cdots$$

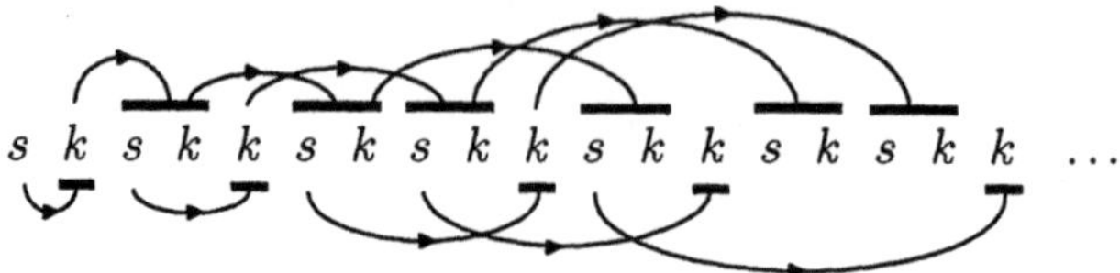

**Fig. 8.3.9** Autogenous kernel

The lengths of these partial words clearly constitute the Fibonacci sequence, and the lower half of the kernel is indeed represented by

$$\{\,(\lfloor \tfrac{3+\sqrt{5}}{2}\,n\rfloor,\ \lfloor \tfrac{1+\sqrt{5}}{2}\,n\rfloor)\ \mid\ n = 0,1,2,\dots\,\}.$$

Compare the discussion in COXETER 53. So the points of the kernel "almost" sit on a straight line.

## Endgames in Chess

There is a mathematical tradition of studying chess endgames[7]. Already E. Zermelo (1912) and D. König (1927) contributed to the mathematical analysis of finiteness problems in chess. Also, the world chess champions E. Lasker and M. Euwe were mathematicians.

Since there are only finitely many positions in a game of chess, one can, in principle, determine the winning positions, the losing positions, and the draw positions as in the previous examples. But, because of the huge number of positions, there is no hope of ever finding out by some *iterative process* whether the starting position in chess is one of winning or losing—notwithstanding that such a result might some day be proved in an ingenious way.

So it is impossible to successfully apply the above iterative process to the game of chess in its entirety. We now mention some results of computational analysis restricted to endgames of chess with a small number of chessmen.

| type of endgame | g. w. | n. e. | example | |
|---|---|---|---|---|
| (1) $wR$ | 16 | 121 | $w\,Ka1\,Rb2,$ | $b\,Kc3$ |
| (2) $wQ$ | 10 | 1 | $w\,Ka1\,Qb2,$ | $b\,Ke6$ |
| (3) $wR:bB$ | 18 | 28 | $w\,Ka4\,Rc3,$ | $b\,Ka7\,Ba6$ |
| (4) $wR:bN$ | 27 | 2 | $w\,Kd1\,Rh1,$ | $b\,Kb1\,Ng4$ |
| (5) $wQ:bR$ | 31 | 4 | $w\,Ka1\,Qa4,$ | $b\,Kf3\,Re1$ |
| (6) $wQ:bQ$ | 10 | 5 | $w\,Ke1\,Qg1,$ | $b\,Kb1\,Qa1$ |
| (7) $wQ:bR+bP(d2)$ | 29 | 10 | $w\,Kh8\,Qa4,$ | $b\,Kf8\,Rf2\,Pd2$ |
| (8) $wB+B:bN$ | 66 | | $w\,Ka8\,Bh1\,h6,$ | $b\,Kf3\,Ng2$ |

g. w. = greatest number of moves necessary to win
n. e. = number of extremal positions

**Fig. 8.3.10** Chess endings with 3, 4 or 5 pieces

---

[7] At a first glance the previous considerations do *not* apply to chess because there is the stalemate, a terminal position rated as a draw. But one can slightly modify the rules, which will not affect the game of chess proper, to the effect that out of a position of stalemate the game has to be continued indefinitely by alternately repeating certain moves. Then a "draw by position" becomes a "draw by repetition of moves" which can be dealt with as before.

Several endgames with three or four pieces were analyzed during 1967-1969 in Munich at the Leibniz-Rechenzentrum on the TR4 computer made by AEG-Telefunken. The iteration process had to be carefully refined in order to cope with the memory size and computing time available. As a result, it was possible to a large extent to work with machine words as bit vectors, in the sense that by one operation on a machine word all the positions encoded by the bits could be treated simultaneously. For that purpose the game graph had to be enlarged by many illegal positions (such as two pieces on a square) whose influence had to be eliminated afterwards. In a four-piece endgame opposing the white king and rook (no castling allowed) on the one hand, and the black king and one knight on the other, there were $64^4 = 16,777,216$ positions to be considered. After restricting to legal positions only, and exploiting all symmetries, there remained almost 2 million positions.

At that time the first five games from Table 8.3.10 were analyzed. Games (1) and (2) are published in STRÖHLEIN, ZAGLER 78. Games (6) and (7) were analyzed later in a master's thesis. Game (8) was calculated by K. Thompson. The numbers of moves given in Table 8.3.10 stand for different things. In games (3) to (5) the number of moves towards win is the number of moves necessary to beat the black king (checkmate + 1) or the other black chessman; in games (1), (2), (6), and (7) the number indicates the number of moves necessary for win by checkmate or by beating some black chessman.

| type of endgame | mostly | extremal case |
| --- | --- | --- |
| $wR + B : bB$ | draw | 59 moves or less for mate or reduction |
| $wR + R : bR$ | won | 31 moves or less for mate |
| $wQ + Q : bQ$ | won | 30 moves or less for mate |
| $wQ : bB + B$ | won | 71 moves or less for mate or reduction |
| $wQ : bB + N$ | won | 42 moves or less for mate or reduction |
| $wQ : bN + N$ | won | 63 moves or less for mate or reduction |
| $wQ + R : bQ$ | won | 67 moves or less for mate |
| $wQ + B : bQ$ | draw | 33 moves or less for mate |
| $wQ + N : bQ$ | draw | 41 moves or less for mate |
| $wR + N : bR$ | draw | 33 moves or less for mate or reduction |

**Fig. 8.3.11** Further chess endings with 5 pieces

In spring of 1986 the chess journals reported on further results in the iterative analysis of chess endgames obtained by Thompson at Bell Laboratories. We cite a few of these results in Fig. 8.3.11. The numbers indicated stand for the moves that are necessary in order to achieve one of the following effects, starting from the least favourable position:

- checkmate or
- a capturing move that results in reducing the endgame to one with fewer pieces in a configuration which is already known to lead to win.

In some cases, even the maximum number of moves necessary for checkmate was given. Today already some endgames with 6 pieces have been analyzed.

## 8.4 References

See also the references of Chap. 6.

AHRENS W: *Mathematische Unterhaltungen und Spiele.* Teubner, Leipzig, 1901.

AUMANN G: *Über autogene Folgen und die Konstruktion des Kerns eines Graphen.* Bayer. Akad. Wiss. Math.-Natur. Kl. Sitzungsber. 1966 (1967) 53–63.

BALL WWR, COXETER HSM: *Mathematical recreations and essays.* Macmillan, London, 1956.

BOUTON CL: *Nim, a game with a complete mathematical theory.* Ann. of Math. (2) **3** (1902) 35–39.

COXETER HSM: *The golden section, phyllotaxis, and Wythoff's game.* Scripta Math. **19** (1953) 135–143.

KRAÏTCHIK M: *La mathématique des jeux ou récréations mathématiques.* Stevens Frères, Brüssel, 1930.

RICHARDSON M: *Solutions of irreflexive relations.* Ann. of Math. (2) **58** (1953) 573–590.

ROTH AE: *A fixedpoint approach to stability in cooperative games.* In Karamardian S and Garcia CB (eds.): Fixed points—algorithms and applications, Proc. of a Conf., Clemson, SC, June 26-28, 1974, Academic Press, New York, 1977, pp. 165-180

SCHMIDT G, STRÖHLEIN T: *On kernels of graphs and solutions of games: A synopsis based on relations and fixpoints.* SIAM J. Algebraic Discrete Methods **6** (1985) 54–65.

STEINHAUS H: *Definitions for a theory of games and pursuit.* (Polish, 1925), Engl. reprint: Naval Res. Logist. Quart. **7** (1960) 105–108.

STRÖHLEIN T: *Untersuchungen über kombinatorische Spiele.* Diss., Techn. Univ. München, 1970.

STRÖHLEIN T, ZAGLER L: *Analyzing games by boolean matrix iteration.* Discrete Math. **19** (1977) 183–193.

STRÖHLEIN T, ZAGLER L: *Ergebnisse einer vollständigen Analyse von Schachendspielen— König und Turm gegen König—König und Turm gegen König und Läufer.* TUM-INFO-09-78-00-FBMA, 202 p., Institut für Informatik der Techn. Univ. München, 1978.

WYTHOFF WA: *A modification of the game of Nim.* Nieuw Arch. Wisk. (2) **7** (1905/07) 199–202.

# 9. Matchings and Coverings

The results of this chapter show the close interplay between the adjacency relation $\Gamma$ and the incidence $M$, and yield beautiful applications to combinatorics. Section 9.1 deals with independence of points in a set of points of a graph, i.e., with their mutually being non-neighbors of each other, which can easily be characterized in terms of the adjacency $\Gamma$. We shall be interested in the biggest possible independent sets.

In Sect. 9.2 we study point sets that "cover" all the edges by means of the incidence $M$, a property which, at a first glance, does not seem to have much to do with independence. Here we are interested in the smallest possible sets with this property. Since incidence and adjacency are related by the equation $\Gamma = \overline{I} \sqcap M^{\mathsf{T}} M$, we then see that in the best case a set is independent, if and only if its complement has the covering property.

For bipartite graphs, which will be treated in Sect. 9.3, independence and covering properties are still more closely related, as is expressed by the well-known theorems of König and Hall. We establish this relationship step by step. It may be subsumed, however, to a general lattice-theoretic setting described in Appendix A.3. Section 9.4 is devoted to starlikeness which is important for minimal coverings.

## 9.1 Independence

In Definition 8.1.1 we discussed absorptiveness for point sets of a directed graph, which involved the (not necessarily symmetric) associated relation $B$. In contrast, the property of a point set that no two distinct members of it are neighbors (i.e., independence) depends on the adjacency $\Gamma$.

These properties will be primarily discussed for simple graphs. The results can then easily be extended to hypergraphs and their adjacency $\Gamma = \overline{I} \sqcap M^{\mathsf{T}} M$. Furthermore, the concept of independence can be dualized for sets of edges, using the edge-adjacency $\mathrm{K} = \overline{I} \sqcap M M^{\mathsf{T}}$.

There is a typical applied problem the chef de protocole of an embassy is confronted with, when he has to seat guests at various tables. Of course, he will avoid seating two political enemies at the same table. A related problem is the **8-queens problem** attributed to Franz Nauck (1850) which was considered also by C. F. Gauss in the same year. (It is described in detail in the recreational mathematics classics by Ahrens, Kraïtchik, and Ball-Coxeter.)

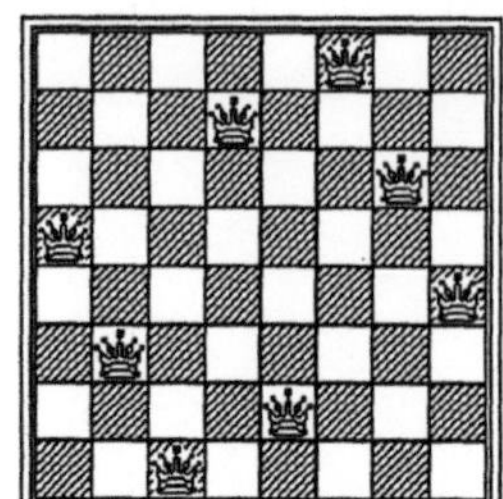

**Fig. 9.1.1** One solution of the
8-queens problem of Gauss

Consider a number of queens placed on a chessboard. Two queens are said to be enemies if they are threatening each other. The problem consists in "inviting" as many queens as possible and placing them on the board in such a way that no two of them are enemies. One of the 92 solutions to this problem is given in Fig. 9.1.1. The graph-theoretic setting is given by 64 points $(i, k)$ with $1 \leq i, k \leq 8$, representing the squares, joined by an adjacency $\Gamma$ representing the queens' moves:

$$\big((i, k), (j, l)\big) \in \Gamma \;:\!\Longleftrightarrow\; [(i \neq j) \vee (k \neq l)] \wedge [(i = j) \vee (k = l) \vee (|i - j| = |k - l|)].$$

The problem now translates into finding a set of mutually nonadjacent points. (The graph is too big to be reproduced here.)

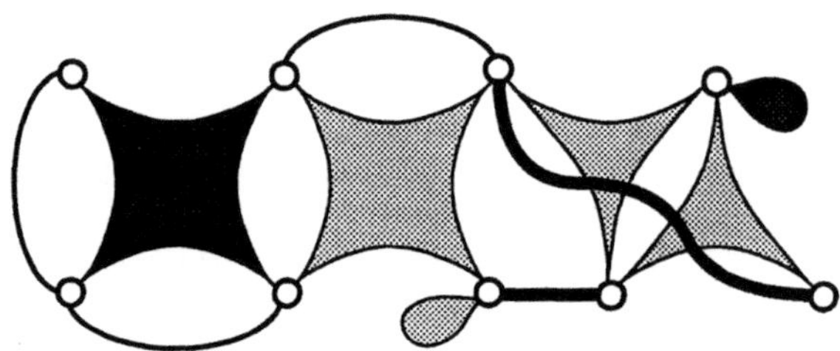

**Fig. 9.1.2** Independent hyperedges

The dual problem is illustrated in Fig. 9.1.2 by four mutually nonadjacent hyperedges.

Our preliminary observations lead to the following definition.

**9.1.1 Definition.** Let $\Gamma$ be the adjacency, and let $K$ be the edge-adjacency relation of a graph, and consider a point set $s$ and an edge set $r$. Then we call

i)   $s$ **independent**   $:\!\Longleftrightarrow$   $\Gamma s \subset \bar{s}$   $\Longleftrightarrow$   $ss^{\mathsf{T}} \subset \overline{\Gamma}$
     $\Longleftrightarrow$   No two points of $s$ are mutually adjacent.

ii)  $r$ **independent**[1]   $:\!\Longleftrightarrow$   $Kr \subset \bar{r}$   $\Longleftrightarrow$   $rr^{\mathsf{T}} \subset \overline{K}$
     $\Longleftrightarrow$   No two edges of $r$ are mutually adjacent.

Independent point sets of a hypergraph are sometimes called **strongly stable**.

iii)  $\alpha := \max\{\, |s| \in \mathbb{N} \mid \Gamma s \subset \bar{s} \,\}$   **point independence number;**

iv)  $\alpha^* := \max\{\, |r| \in \mathbb{N} \mid Kr \subset \bar{r} \,\}$   **edge independence number.**   $\square$

---

[1] An independent edge set is also called a **matching** or a couplage (in French).

For independent sets of edges, especially in bipartite graphs, the terms "correspondence" and "pairing" are also used. However, we reserve these for the quite similar situation described in Def. 9.1.5 and before Def. 9.3.5.

Independence is closely related to stability (Sect. 8.1) where loops play an additional role. For simple graphs, as well as for loop-free directed graphs, independence and stability mean the same. The general relationship follows from $\Gamma = \overline{I} \sqcap (B \sqcup B^{\mathsf{T}})$:

$$x \ \text{stable} \ \implies \ x \ \text{independent},$$
$$x \ \text{stable} \ \impliedby \ x \ \text{independent}, \quad \text{if } B \text{ is loop-free}.$$

Clearly, any subset of an independent set (of points or edges) is again independent. So the natural question arises for biggest possible independent sets. In Fig. 9.1.3.a two independent point sets $s_0$ and $s_1$ are given. Both are maximal with respect to inclusion, i.e., they cannot be enlarged by adding some point. But only $s_0$ contains the maximum number of elements, hence $\alpha = |s_0| = 2$. Similarly, in part b) of that figure the two "triangles" and the three "parallels" each form an independent set of edges which cannot be enlarged by adding more edges. One has $\alpha^* = |r_0| = 3$. It is more interesting to ask for independent edge sets with the maximum number of elements when all edges contain the same number of points (usually 2).

When $s$ increases then so does the left-hand side in $\Gamma s \subset \overline{s}$, while the right-hand side decreases. One will expect that the equation $\Gamma s = \overline{s}$ characterizes sets $s$ which can no longer be enlarged, i.e., which are maximal with respect to inclusion.

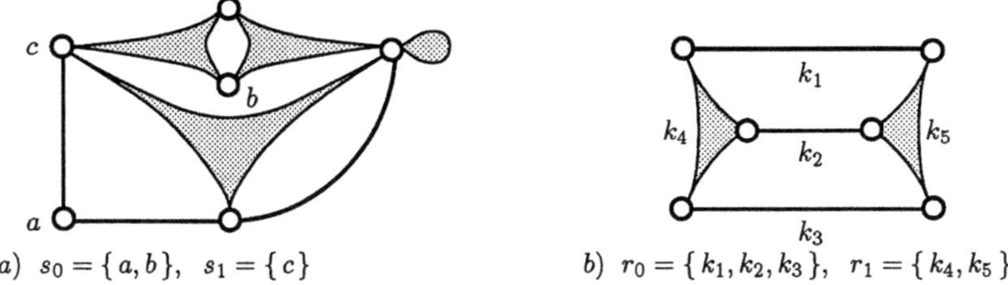

a) $s_0 = \{a, b\}, \quad s_1 = \{c\}$      b) $r_0 = \{k_1, k_2, k_3\}, \quad r_1 = \{k_4, k_5\}$

**Fig. 9.1.3** Independent sets

**9.1.2 Proposition.** i)   If $\Gamma$ is an adjacency relation and $s$ a point set, then

$$\Gamma s = \overline{s} \ \iff \ s \text{ is a maximal independent point set}.$$

ii)   If K is an edge-adjacency relation and $r$ is an edge set, then

$$\mathrm{K}r = \overline{r} \ \iff \ r \text{ is a maximal independent edge set}.$$

**Proof:** In both cases the proof is the same as in Props. 8.1.2 and 8.1.3.        □

This result yields a simple proof of some known lower bounds for the independence numbers $\alpha$ and $\alpha^*$.

**9.1.3 Corollary.** In a graph with $n$ points, $m$ edges, at most $h$ neighbors per point, and at most $k$ adjacent edges per edge the following statements hold:

i)    $s$ $\begin{array}{c}\text{inclusion-maximal}\\\text{independent point set}\end{array}$ $\implies$ $\dfrac{n}{h+1} \le |s| \le \alpha;$

ii)    $r$ $\begin{array}{c}\text{inclusion-maximal}\\\text{independent edge set}\end{array}$ $\implies$ $\dfrac{m}{k+1} \le |r| \le \alpha^*.$

**Proof:** i) $n - |s| = |\bar{s}| = |\Gamma s| \le h|s|$, and similarly for (ii).     $\square$

The bounds are indeed tight. For if $p$ denotes the smallest integer $\ge \frac{n}{h+1}$, then we can partition $n$ given points into $p$ disjoint, nonempty subsets containing each at most $(h+1)$ points. Consider then the graph on $n$ points where any two of them are joined by an edge, if and only if they lie in the same subset. It clearly contains no $p+1$ independent points. In the 8-queens problem, for example, one has $n = 64$ and $h = 27$ (the number of squares threatened by the queen in position $d4$) which gives the inequality $\alpha \ge 64/27$, and hence $\alpha \ge 3$ since $\alpha$ is an integer, whereas the exact value is $\alpha = 8$.

An equally loose upper bound is obtained by counting all related pairs.

**9.1.4 Proposition.** For a simple graph with $n$ points and $m$ edges the following inequality holds:

$$\alpha \le \sqrt{n^2 - 2m}.$$

**Proof:** Given an arbitrary independent point set $s$, we have $ss^\mathsf{T} \subset \bar{\Gamma}$. Now we interpret $\Gamma$ as a Boolean matrix, and we count the nonvanishing entries on both sides of this inclusion. We obtain $|s| \cdot |s| \le n^2 - 2m$, since for symmetric $\bar{\Gamma}$ precisely two entries per edge vanish.     $\square$

This bound is again tight (up to fractional parts) for every $n$ and $m \le \binom{n}{2}$, as can be seen from the $n$-clique with all edges removed between $\lfloor \sqrt{n^2 - 2m} \rfloor$ chosen points. As it is often the case with bounds for greatly varying domains, the present one can be very far-off. For the 8-queens problem we have $\alpha = 8$, but $\sqrt{n^2 - 2m} = \sqrt{64^2 - 2640} \approx 51.39$. The number $2m$ comes down to $2640 = 4 \times 27 + 12 \times 25 + 20 \times 23 + 28 \times 21$ because the queen threatens 27 squares from 4 positions, etc.

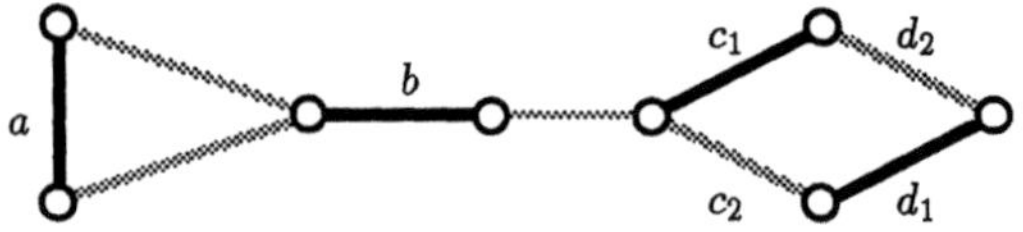


**Fig. 9.1.4** Rigid and flexible edges

When dealing with maximum-number matchings (this is an example where independent sets are often called matchings), one occasionally tries to classify the edges more precisely. In Fig. 9.1.4, for example, there are exactly two maximum-number matchings, $\{\,a,b,c_1,d_1\,\}$ and $\{\,a,b,c_2,d_2\,\}$. Consider the set $\mathcal{M}$ of all maximum-number matchings. One calls an edge **rigid** if it occurs in *every* matching of $\mathcal{M}$, and **inadmissible** (or a **pseudo-edge**) if it occurs in *no* matching of $\mathcal{M}$. All the other edges are termed **flexible** (or **free**). These properties are relevant for combinatorial problems, but they may require considerable effort to establish; see, e.g., SCHMIDT, STRÖHLEIN 76. In the example given, $a$ and $b$ are rigid, $c_1$, $c_2$, $d_1$ and $d_2$ are flexible, and the unnamed remaining edges are inadmissible. For inclusion-maximal matchings there is no simple classification as above.

## Pairings

An independent edge set is a set of some property which is defined in terms of the *edge*-adjacency $K$. We now consider an alternate construct that contains in essence the same information, but is defined by means of the *point*-adjacency $\Gamma$, and will be called a *pairing* of points. In order to show the relationship between both, independent edge sets and pairings, we assume that $M$, $\Gamma$, $K$ are all at our disposal. If an independent edge set $r$ is given as in Fig. 9.1.5, it is natural to consider the relation $\mu$ defined by the edges in $r$.

$$
M = \begin{array}{c} \\ p \\ q \\ w \\ s \\ t \\ u \\ v \end{array}
\begin{array}{ccccccc} a & b & c & d & e & f & g \\ \end{array}
\begin{pmatrix}
1 & 1 & 0 & 0 & 0 & 0 & 0 \\
0 & 1 & 1 & 0 & 0 & 0 & 0 \\
0 & 0 & 1 & 1 & 0 & 0 & 0 \\
0 & 1 & 0 & 0 & 1 & 0 & 0 \\
0 & 0 & 0 & 0 & 1 & 1 & 0 \\
0 & 0 & 0 & 1 & 0 & 1 & 0 \\
0 & 0 & 0 & 0 & 0 & 1 & 1
\end{pmatrix}
\qquad
r = \begin{pmatrix} 1 \\ 0 \\ 1 \\ 0 \\ 1 \\ 0 \\ 0 \end{pmatrix}
\qquad
K = \begin{array}{c} p & q & w & s & t & u & v \end{array}
\begin{pmatrix}
0 & 1 & 0 & 1 & 0 & 0 & 0 \\
1 & 0 & 1 & 1 & 0 & 0 & 0 \\
0 & 1 & 0 & 0 & 0 & 1 & 0 \\
1 & 1 & 0 & 0 & 1 & 0 & 0 \\
0 & 0 & 0 & 1 & 0 & 1 & 1 \\
0 & 0 & 1 & 0 & 1 & 0 & 1 \\
0 & 0 & 0 & 0 & 1 & 1 & 0
\end{pmatrix}
$$

$$
\Gamma = \begin{array}{c} a \\ b \\ c \\ d \\ e \\ f \\ g \end{array}
\begin{array}{ccccccc} a & b & c & d & e & f & g \\ \end{array}
\begin{pmatrix}
0 & 1 & 0 & 0 & 0 & 0 & 0 \\
1 & 0 & 1 & 0 & 1 & 0 & 0 \\
0 & 1 & 0 & 1 & 0 & 0 & 0 \\
0 & 0 & 1 & 0 & 0 & 1 & 0 \\
0 & 1 & 0 & 0 & 0 & 1 & 0 \\
0 & 0 & 0 & 1 & 1 & 0 & 1 \\
0 & 0 & 0 & 0 & 0 & 1 & 0
\end{pmatrix}
\qquad \supset \qquad
\mu = \begin{array}{c} a & b & c & d & e & f & g \end{array}
\begin{pmatrix}
0 & 1 & 0 & 0 & 0 & 0 & 0 \\
1 & 0 & 0 & 0 & 0 & 0 & 0 \\
0 & 0 & 0 & 1 & 0 & 0 & 0 \\
0 & 0 & 1 & 0 & 0 & 0 & 0 \\
0 & 0 & 0 & 0 & 0 & 1 & 0 \\
0 & 0 & 0 & 0 & 1 & 0 & 0 \\
0 & 0 & 0 & 0 & 0 & 0 & 0
\end{pmatrix}
$$

**Fig. 9.1.5** Independent edge set and pairing

The graph above seems to suggest that an independent edge set provides the same information as a univalent, injective, and symmetric relation $\mu$.

**9.1.5 Definition.** Given an adjacency relation $\Gamma$, we call[2] a relation

$$
\mu \ \textbf{pairing} \ :\Longleftrightarrow \ \mu \subset \Gamma, \quad \mu = \mu^{\mathsf{T}}, \quad \mu^2 \subset I. \qquad \square
$$

---

[2] If necessary, we speak of a $\Gamma$-pairing. A pairing is sometimes called a **correspondence**.

One would expect that the relationship between pairings and independent edge sets is easily established via the incidence $M$. The proof, however, is lengthy.

**9.1.6 Proposition.** In a simple graph with incidence $M$ we consider an edge set $r$ and a relation $\mu$ on the point set.

i)   $r$   independent edge set   $\implies$   $\mu_r := \overline{I} \sqcap (M \sqcap rL)^{\mathsf{T}}(M \sqcap rL)$   pairing.

ii)   $\mu$   pairing   $\implies$   $r_\mu := (M \sqcap M\mu)L$   independent edge set.

iii)   In this way pairings and independent edge sets are related bijectively to one another, i.e., $r = r_{\mu_r}$ and $\mu = \mu_{r_\mu}$.

**Proof** for (i) only: Since the graph is *simple*, we have $M = A \sqcup E$ as a union of two mappings $A$ and $E$ with $A \sqcap E = O$ (loop-freeness), $AA^{\mathsf{T}} \sqcap EE^{\mathsf{T}} \subset I$ (no parallel arcs), and $AE^{\mathsf{T}} \sqcap EA^{\mathsf{T}} = O$ (no pair of arcs in opposite direction between the same points). By definition, $\mu_r$ is symmetric and contained in $\Gamma$. Using univalence of $A$ and $E$, as well as $A \sqcap E = O \iff A^{\mathsf{T}}E \subset \overline{I} \iff AE^{\mathsf{T}} \subset \overline{I}$ we obtain

$$\mu_r = (A \sqcap rL)^{\mathsf{T}}(E \sqcap rL) \sqcup (E \sqcap rL)^{\mathsf{T}}(A \sqcap rL).$$

Now the inclusion $\mu_r^2 \subset \overline{I}$ is established for the four terms separately according to the following pattern

$$(A \sqcap rL)^{\mathsf{T}}\,[(E \sqcap rL)(A \sqcap rL)^{\mathsf{T}}](E \sqcap rL)$$
$$\subset (rL)^{\mathsf{T}}KrL \subset (rL)^{\mathsf{T}}\overline{r}L = (rL)^{\mathsf{T}}\overline{rL} \subset O,$$
$$(A \sqcap rL)^{\mathsf{T}}\,[(E \sqcap rL)(E \sqcap rL)^{\mathsf{T}}](A \sqcap rL)$$
$$\subset A^{\mathsf{T}}[EE^{\mathsf{T}} \sqcap rL(rL)^{\mathsf{T}}]A \subset A^{\mathsf{T}}[(I \sqcup K) \sqcap \overline{K}]A \subset A^{\mathsf{T}}A \subset I. \qquad \square$$

For bipartitioned graphs $(X, Y, Q)$ the situation is simpler. In order to define a pairing $\mu$ in this case, it suffices to give a relation $\lambda \subset Q$ between the left and the right point sets. We have decided to call $\lambda$ a *matching*, see Sect. 9.3. Relations $\lambda$ and $\mu$ satisfy

$$\mu = \begin{pmatrix} O & \lambda \\ \lambda^{\mathsf{T}} & O \end{pmatrix}.$$

Univalence and injectivity of $\lambda$ follow from $\mu^2 \subset I$ and symmetry:

$$\lambda^{\mathsf{T}}\lambda \subset I, \quad \lambda\lambda^{\mathsf{T}} \subset I.$$

# 9.2  Coverings

In Sect. 9.1 we defined independence for points or edges by means of adjacency as the property of not being neighbors of each other. The "complementary" concept of covering or transversal is based on incidence and applies to both, points as well as edges. The relationship between independence and covering for points is established in Proposition 9.2.2.

In applications one needs to find sets of points in a graph or hypergraph with the property that every edge is incident with at least one point of that set. The dual problem consists in finding sets of edges so that every point is incident with at least one such edge. For reasons of economy these sets should contain as few points or edges as possible. By successively removing points one obtains point sets of that kind which are minimal with respect to inclusion. The resulting incidence is then often starlike, a property which we shall study in detail in Sect. 9.4.

In the following definition we allow edges to be empty and points to be free, i.e., to be incident with no point or edge, respectively. As we have seen earlier in Sect. 5.3, $ML$ is the set of nonempty edges, and $M^{\mathsf{T}}L$ is the set of nonfree points.

**9.2.1 Definition.** Given an incidence $M$ together with a point set $u$, or an edge set $q$, we call

i)  $\quad u$ **covering** $\quad :\Longleftrightarrow \quad ML \subset Mu \quad \Longleftrightarrow \quad ML = Mu$

$\qquad\qquad \Longleftrightarrow \quad$ Every nonempty edge is incident with at least one point of $u$.

ii)  $\quad q$ **covering** $\quad :\Longleftrightarrow \quad M^{\mathsf{T}}L \subset M^{\mathsf{T}}q \quad \Longleftrightarrow \quad M^{\mathsf{T}}L = M^{\mathsf{T}}q$

$\qquad\qquad \Longleftrightarrow \quad$ Every nonfree point is incident with at least one edge of $q$.

Sometimes the term **transversal** is used instead of covering.

iii)  $\quad \tau \;:= \min\{\,|u| \in \mathbb{N} \mid ML \subset Mu\,\} \quad$ **point covering number,**

iv)  $\quad \tau^* := \min\{\,|q| \in \mathbb{N} \mid M^{\mathsf{T}}L \subset M^{\mathsf{T}}q\,\} \quad$ **edge covering number.** $\qquad \Box$

In many cases a slightly simpler form of Definition 9.2.1 suffices. If there are no free points, then clearly

$$q \text{ covering} \quad \Longleftrightarrow \quad L \subset M^{\mathsf{T}}q.$$

So this holds for surjective hypergraphs, 1-graphs without isolated points, and simple graphs without isolated points. If, conversely, there are no empty edges, then

$$u \text{ covering} \quad \Longleftrightarrow \quad L \subset Mu.$$

Simple graphs and 1-graphs contain no empty edges; they occur only in nontotal hypergraphs.

Finding a covering set of points or edges without any further properties poses no problem. Starting with an arbitrary point set $s$, the set $\overline{Ms}$ consists of all edges not covered by $s$. The point set $s \sqcup M^{\mathsf{T}}\overline{Ms}$ is then indeed covering, for

$$ML = Ms \sqcup (M\,\overline{s} \sqcap \overline{Ms}) \subset Ms \sqcup (M \sqcap \overline{Ms}\,\overline{s}^{\mathsf{T}})(\overline{s} \sqcap M^{\mathsf{T}}\overline{Ms})$$
$$\subset Ms \sqcup MM^{\mathsf{T}}\overline{Ms} = M(s \sqcup M^{\mathsf{T}}\overline{Ms}).$$

Thus $s$ has been enlarged by all those points which are contained in edges not incident with $s$.

It is clear from the definition that any set containing a covering set is again covering; in particular, the full sets of points or edges are covering. For economic reasons one is interested in covering sets with the minimum number of elements—these can be difficult to characterize—and also in covering sets that are minimal with respect to inclusion, i.e., which lose the covering property when some element is removed. In Fig. 9.2.1 the point sets $u_1, u_2$, and $u_3$ are covering, but only $u_1$ and $u_2$ are minimal. From $u_3$ point $x_4$ could be removed. Moreover, $u_1$ contains fewer points than $u_2$ and is indeed a covering set of smallest cardinality, as can easily be checked.

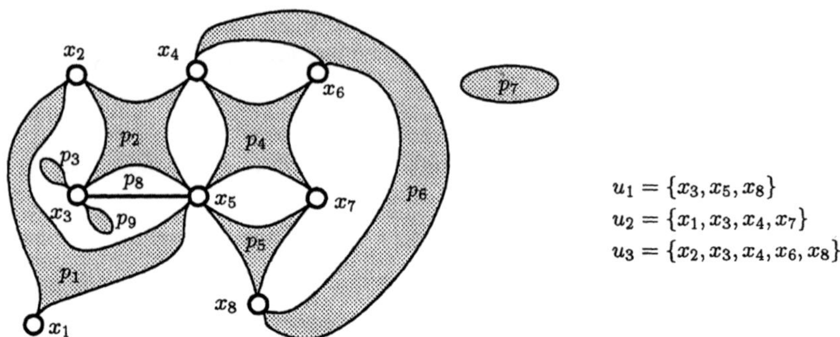


**Fig. 9.2.1** Covering point sets in a nontotal hypergraph

Dually, the set of edges $q := \{p_1, p_3, p_4, p_5\}$ is a covering of minimum cardinality, and hence is minimal.

Transversality derives from incidence, and independence derives from adjacency. We now study how these two properties are related. For this purpose we need that the graph comes with an incidence $M$ and related adjacency $\Gamma := \overline{I} \sqcap M^{\mathsf{T}}M$. One will observe that we are dealing with points only and do not ask for dual results on edges.

**9.2.2 Proposition.** i)   In a drop-free incidence we have for a point set $s$:

$$s \text{ independent} \quad \Longrightarrow \quad \overline{s} \text{ covering.}$$

ii)   With respect to an incidence of rank $\leq 2$ and a point set $s$ we have:

$$s \text{ independent} \quad \Longleftarrow \quad \overline{s} \text{ covering.}$$

iii)   In simple graphs and in loop-free total graphs the following holds:

$$s \text{ independent} \quad \Longleftrightarrow \quad \overline{s} \text{ covering.}$$

**Proof:** We will show

$$\Gamma s \subset \overline{s} \quad \Longrightarrow \quad ML \subset M\overline{s}, \quad \text{if} \quad M \subset M\overline{I}$$
$$\Gamma s \subset \overline{s} \quad \Longleftarrow \quad ML \subset M\overline{s}, \quad \text{if} \quad r(M) \leq 2.$$

"$\Longrightarrow$": $ML = Ms \sqcup M\overline{s}$ and $Ms = (M\overline{I} \sqcap M)s \subset (M \sqcap M\overline{I})(\overline{I} \sqcap M^{\mathsf{T}}M)s \subset M\Gamma s \subset M\overline{s}$.

"$\Longleftarrow$": If the rank of $M$ is not greater than 2, we may assume $M$ to be partitioned as $M = A \sqcup E$. Using $B = A^{\mathsf{T}}E$, Def. 2.2.2, and Prop. 5.1.7, we have $\Gamma = \overline{I} \sqcap (B \sqcup B^{\mathsf{T}})$. Now we may reason as follows

$$Bs = A^{\mathsf{T}}Es \subset A^{\mathsf{T}}[(As \sqcap Es) \sqcup \overline{As}] = A^{\mathsf{T}}\overline{\overline{As} \sqcup \overline{Es}} \sqcup A^{\mathsf{T}}\overline{As}$$
$$\subset A^{\mathsf{T}}\overline{\overline{As} \sqcup \overline{Es}} \sqcup \overline{s}, \qquad \text{since } A \text{ univalent implies } As \subset \overline{As}$$
$$= A^{\mathsf{T}}\overline{(A \sqcup E)s} \sqcup \overline{s} \subset M^{\mathsf{T}}\overline{Ms} \sqcup \overline{s}$$
$$\subset M^{\mathsf{T}}\overline{Ms} \sqcup \overline{s}, \qquad \text{since } Ms \subset ML \subset M\overline{s} \text{ by assumption}$$
$$\subset \overline{s} \sqcup \overline{s} = \overline{s}. \qquad\qquad\qquad \Box$$

According to the graphs mentioned in Proposition 9.2.2.iii, we have the equivalence

$$s \quad \text{covering} \quad \Longleftrightarrow \quad \Gamma\overline{s} \subset s$$

which is sometimes taken as a definition for a point set $s$ to be covering. As in Proposition 9.1.2, then

$$s \quad \text{inclusion-minimal covering} \quad \Longleftrightarrow \quad \Gamma\overline{s} = s.$$

The dualized notion $K\overline{v} \subset v$ concerning edges is not of interest; see Fig. 9.2.4, for example.

a)                                                     b)

**Fig. 9.2.2** Independence and transversality

Figure 9.2.2 explains the restrictions that had to be made in Proposition 9.2.2. The picture on the left contains a hyperedge of rank 3 which is responsible for the set $s := \{b, c, d\}$ *not* to be independent (since $c$ and $d$ are adjacent), whereas its complement $\overline{s} := \{a\}$ is a covering. In the picture on the right the point set $s := \{f\}$ is independent, but its complement $\overline{s} = \{e\}$ is not covering. The adjacency $\Gamma = \overline{I} \sqcap (B \sqcup B^{\mathsf{T}})$ contains no information about the loop of the directed graph at $f$. But it is precisely this loop which is not being covered by $\overline{s} = \{e\}$.

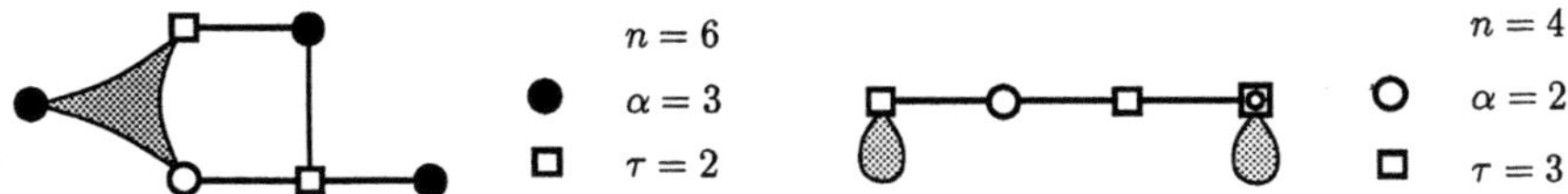


**Fig. 9.2.3** Point independence number and point covering number

Proposition 9.2.2 yields quantitative results which we first illustrate by the examples in Fig. 9.2.3.

**9.2.3 Corollary.** Let a graph with $n$ points be given. Then the following holds for the point independence number $\alpha$ and the point covering number $\tau$ :

| | | | |
|---|---|---|---|
| i) | drop-free incidence | $\Longrightarrow$ | $\alpha + \tau \leq n,$ |
| ii) | incidence of rank $\leq 2$ | $\Longrightarrow$ | $\alpha + \tau \geq n,$ |
| iii) | in simple graphs and in loop-free total graphs | $\Longrightarrow$ | $\alpha + \tau = n.$ |

**Proof:** i) Given a maximum independent set $s$, the complement $\bar{s}$ is covering (but possibly not minimum), therefore

$$\tau \leq |\bar{s}| = n - |s| = n - \alpha.$$

ii) Given a minimum covering set $t$, the complement $\bar{t}$ is independent (but possibly not maximum), therefore

$$\alpha \geq |\bar{t}| = n - |t| = n - \tau. \qquad \square$$

This result could be extended to the edge independence number $\alpha^*$, the edge covering number $\tau^*$, and the edge number $m$, but that would be of little interest because incidences of degree 2 are quite special. This would apply only to graphs consisting of several cycles as in Fig. 9.2.4.

Fig. 9.2.4 A covering and an independent edge set in a graph of degree 2

We have tried so far to use the independence of some point set in order to show that some other *point* set, actually its complement, is covering. But certain independent point sets, namely the maximal ones, also determine covering sets of *edges*. Figure 9.2.5 gives a hint in that direction.

Fig. 9.2.5 Transversality of the edges incident with a
maximal independent point set

**9.2.4 Proposition.** If $\Gamma = \bar{I} \sqcap M^{\mathsf{T}} M$ is the adjacency with respect to some incidence $M$ and if $s$ is a point set, then

$$s \begin{array}{l} \text{inclusion-maximal} \\ \text{independent point set} \end{array} \Longrightarrow Ms \text{ covering edge set.}$$

**Proof:** Obviously $M^\mathsf{T}L = M^\mathsf{T}(Ms \sqcup \overline{Ms}) = M^\mathsf{T}Ms \sqcup M^\mathsf{T}\overline{Ms}$. The containment $M^\mathsf{T}\overline{Ms} \subset \overline{s}$ always holds and may, by Proposition 9.1.2, be extended to $\overline{s} = \Gamma s \subset M^\mathsf{T}Ms$ for maximal $s$. Therefore, $M^\mathsf{T}L \subset M^\mathsf{T}Ms$.     $\square$

Let us explain the statement above in a less formal way. If $Ms$ were not incident with some nonfree point $x$, then $x$ could be added to $s$ and $s$ would remain independent, for $x$ is incident with no edge from $Ms$, i.e., with no edge incident with $s$.

Let us now discuss the reverse direction of Prop. 9.2.4. Assume that there are no free points and no empty edges, and let for some $s$ the set $Ms$ be a covering edge set. Then one has

$$s \sqcup \Gamma s = s \sqcup (\overline{I} \sqcap M^\mathsf{T}M)s \supset (I \sqcap M^\mathsf{T}M)s \sqcup (\overline{I} \sqcap M^\mathsf{T}M)s = M^\mathsf{T}Ms$$

$$\supset M^\mathsf{T}L, \quad \text{since } Ms \text{ is covering and there are no empty edges}$$

$$\supset L, \quad \text{since there are no free points,}$$

hence $\overline{s} \subset \Gamma s$. Together with Prop. 9.1.2 we have that an independent set $s$ with $Ms$ a covering set is also maximal.

From Proposition 9.2.4 the following relationship between the edge covering number $\tau^*$ and the point independence number $\alpha$ can be derived.

**9.2.5 Corollary.** For an incidence of degree $g$, we have

$$\tau^* \leq g \cdot \alpha.$$

**Proof:** We consider an independent point set $s$ which is maximal with respect to inclusion and with respect to cardinality. By Proposition 9.2.4, $Ms$ is a covering which satisfies

$$\tau^* \leq |Ms| \leq g \cdot |s| = g \cdot \alpha.     \square$$

Similarly, for an incidence of rank $r$ the inequality $\tau \leq r \cdot \alpha^*$ is obtained. It specializes to $\tau \leq 2\alpha^*$ if "proper" hypergraphs are excluded.

We now investigate minimal covering sets of edges $q$ in greater detail. If any edge is removed from $q$, then $q$ is no longer covering. To every edge in $q$ there is therefore a point being covered solely by that edge, and which is hence no longer covered when that edge is removed. In Fig. 9.2.6 $q = \{p_2, p_4, p_5\}$ is a minimal edge covering. Points $x_2$ and $x_3$, for example, are being covered by $p_2$, but not by $p_4$ or $p_5$.

So the points that are incident with just *one* edge in $q$ are of special interest. Similar examples have occurred earlier, and we described them by means of the univalent part $\mathrm{unp}(R)$ or the multivalent part $\mathrm{mup}(R)$ of some relation $R$.

Following Prop. 5.3.6, we decomposed the entire point set into the set of free points $\overline{M^\mathsf{T}L}$, the set of edge-connecting points $\mathrm{mup}(M^\mathsf{T})L$, and the set $\mathrm{unp}(M^\mathsf{T})L$ of peaks, i.e., points incident with precisely one edge. We now make use of our earlier observations while considering not the entire incidence $M$ but the part of it related to the set of edges $q$.

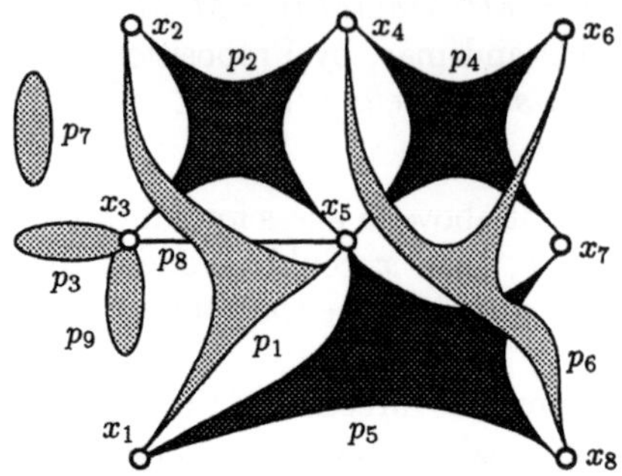

**Fig. 9.2.6** Reduction and boundary

**9.2.6 Definition.** Given an incidence $M$ and an edge set $q$, we consider their relationship and call

i)    $M_q := M \sqcap qL$        the **reduction** of $M$ by $q$,

ii)      $\delta := \mathrm{unp}(M_q^{\mathsf{T}})L$   the **boundary** of an edge set $q$.        □

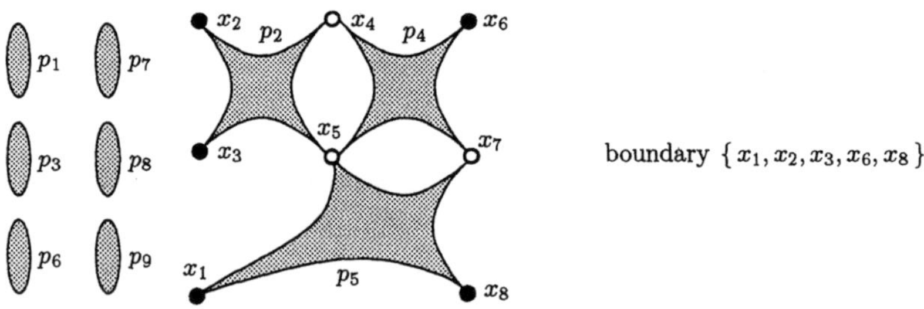

boundary $\{\, x_1, x_2, x_3, x_6, x_8 \,\}$

**Fig. 9.2.7** Reduction $M_q$ of the incidence $M$ of Fig. 9.2.6

Since the boundary $\delta$ depends on $M$ and $q$, we shall write $\delta_{M,q}$ instead, if necessary. Figure 9.2.7 refers to the incidence $M$ from Fig. 9.2.6. The reduction $M_q$ of $M$ is illustrated by the edge set $q = \{\, p_2, p_4, p_5 \,\}$. By Prop. 5.3.6, the boundary $\delta = \{\, x_1, x_2, x_3, x_6, x_8 \,\}$ of the set of edges $q$ consists of the peaks of the reduced incidence $M_q$ obtained by removing all edges not belonging to the set $q$.

Using these two new concepts we can characterize minimal edge coverings in terms of the incidence.

**9.2.7 Proposition.** Let an incidence $M$ and a covering edge set $q$ be given. If $M_q$ is the reduction of the incidence determined by $q$, and $\delta$ is the boundary of the edge set $q$, the following holds

$$q \text{ inclusion-minimal} \quad \Longleftrightarrow \quad q \subset M_q \delta.$$

**Proof:** For "$\Longleftarrow$" we assume a covering edge set $r \subset q$ to exist and argue as follows:

$$q \subset M_q\delta = [\mathrm{unp}\,(M_q^\mathsf{T})]^\mathsf{T}L, \quad \text{by Prop. 5.3.8.i}$$
$$\subset [\mathrm{unp}\,(M_r^\mathsf{T})]^\mathsf{T}L, \quad \text{by Prop. 4.2.9.iii, since} \quad M_q \supset M_r \text{ and}$$
$$M_r^\mathsf{T}L = M^\mathsf{T}r = M^\mathsf{T}L = M^\mathsf{T}q = M_q^\mathsf{T}L$$
$$\subset ([rL]^\mathsf{T})^\mathsf{T}L = rL = r.$$

So $r$ cannot be a proper subset of $q$.

We now prove "$\Longrightarrow$" assuming $q \not\subset M_q\delta$. Then there is an edge[3] $k \subset q \sqcap \overline{M_q\delta}$, and one obtains the properly contained covering edge set $q \sqcap \overline{k}$, contradicting minimality. In order to prove $M^\mathsf{T}L \subset M^\mathsf{T}(q \sqcap \overline{k})$, we handle the parts in $M^\mathsf{T}L = \mathrm{unp}\,(M_q^\mathsf{T})L \sqcup \mathrm{mup}\,(M_q^\mathsf{T})L$ separately:

$$\mathrm{unp}\,(M_q^\mathsf{T})L = \delta = M_q^\mathsf{T}L \sqcap \delta \subset (M_q^\mathsf{T} \sqcap \delta L)(L \sqcap M_q\delta)$$
$$\subset M_q^\mathsf{T}M_q\delta \subset M^\mathsf{T}(q \sqcap \overline{k})$$
$$\mathrm{mup}\,(M_q^\mathsf{T})L = \mathrm{mup}\,(M_q^\mathsf{T})\overline{k}, \qquad\qquad\qquad \text{Prop. 4.2.10.iv}$$
$$\subset M_q^\mathsf{T}\overline{k} = (M \sqcap qL)^\mathsf{T}\overline{k} = M^\mathsf{T}(q \sqcap \overline{k}), \qquad \text{Prop. 2.4.2.} \quad \square$$

From the definition $M_q = M \sqcap qL$ it is clear that condition $q \subset M_q\delta$ may equivalently be replaced by $q = M_q\delta$ or $q \subset M\delta$.

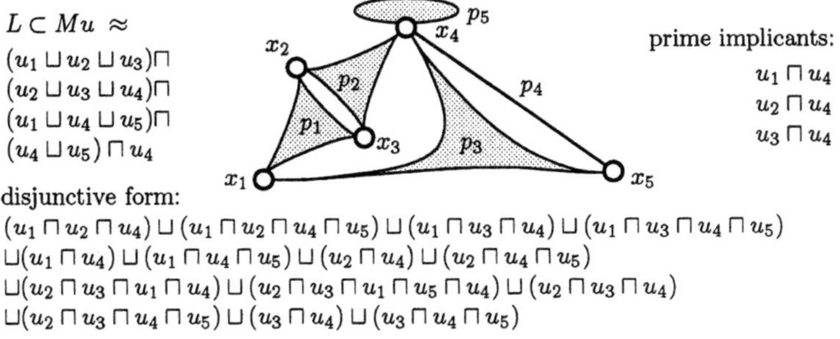


**Fig. 9.2.8** Transversals and implicants

We want to establish a connection with switching algebra, based on a systematic procedure for finding minimal transversals, suitable for small graphs (K. Maghout, 1959).

Suppose a total graph is given as in Fig. 9.2.8. We interpret the relation-algebraic condition $L \subset Mu$ for a transversal $u$ as a Boolean expression $\forall p\, \exists x : M_{px} \sqcap u_x$. This expression is to be put into disjunctive form according to the well-known procedure. Then to every conjunctive term (a so-called implicant) there corresponds in an obvious way a transversal, and to every prime implicant, a minimal transversal.

---

[3] Here, the point axiom is used.

## 9.3 Matching Theorems

We continue our study of independence and transversality and deal next with bipartite graphs for which there are stronger results, such as the two famous theorems of graph theory and combinatorics due to König and Hall.

### Pairings and Coverings for Bipartite Graphs

First we specialize our basic definitions and results from Sects. 9.1 and 9.2 to the bipartite case. This requires a change of notation which may make things look quite different. Also, some special language is used. Coverings, for example, are called **cuts** by some authors.

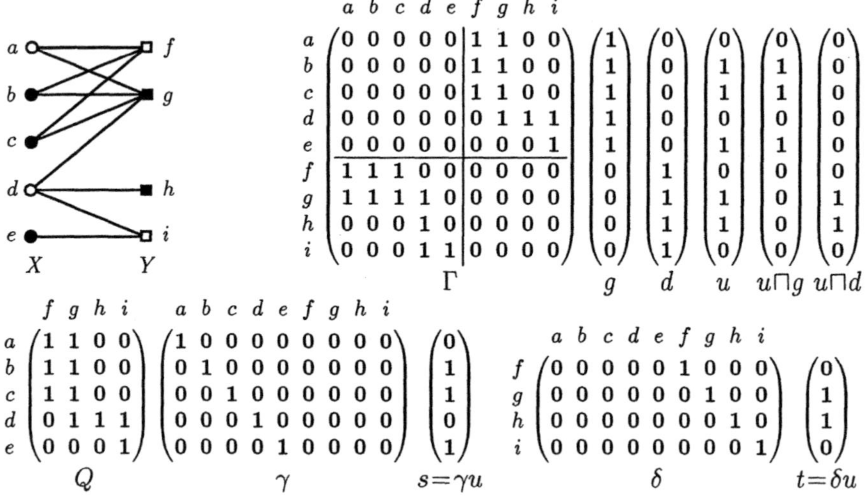

Fig. 9.3.1 Bipartitioning of an adjacency and of point sets

The bipartitioning of the total set of points of the graph into left-hand points $(g)$ and right-hand points $(d)$ applies to each point subset $u$ as illustrated in Fig. 9.3.1. A set of points $u$ can either be decomposed as in Sect. 4.1, $u = (u \sqcap g) \sqcup (u \sqcap d)$, or it can be represented, according to Sect. 5.2, as a pair $(s, t)$ with $s = \gamma u$ and $t = \delta u$. The complement $\overline{u}$ of $u$ then gives rise to the pair $(\overline{s}, \overline{t})$.

Definitions 9.1.1 and 9.2.1, together with Proposition 9.2.2, take the following form which respects the bipartition

$$(s, t) \text{ covering} \quad :\Longleftrightarrow \quad \begin{pmatrix} O & Q \\ Q^\mathsf{T} & O \end{pmatrix} \begin{pmatrix} \overline{s} \\ \overline{t} \end{pmatrix} \subset \begin{pmatrix} s \\ t \end{pmatrix}$$

$$(s, t) \text{ independent} \quad :\Longleftrightarrow \quad \begin{pmatrix} O & Q \\ Q^\mathsf{T} & O \end{pmatrix} \begin{pmatrix} s \\ t \end{pmatrix} \subset \begin{pmatrix} \overline{s} \\ \overline{t} \end{pmatrix}$$

According to Proposition 9.1.2, in certain cases one has equality instead of the inclusion. The essential part of the proof below has thus already been given.

**9.3.1 Proposition.** In a bipartitioned graph $(X, Y, Q)$ we call a bipartitioned point set

i)   $(s, t)$   covering   $\Longleftrightarrow$   $Q\bar{t} \subset s$   $\Longleftrightarrow$   $Q \subset sL \sqcup (tL)^{\mathsf{T}}.$

$(s, t)$   $\left. \begin{array}{c} \text{inclusion-minimal} \\ \text{covering} \end{array} \right\}$   $\Longleftrightarrow$   $Q\bar{t} = s$   and   $Q^{\mathsf{T}}\bar{s} = t.$

ii)  $(s, t)$   independent   $\Longleftrightarrow$   $Qt \subset \bar{s}$   $\Longleftrightarrow$   $Q \subset \overline{sL} \sqcup \overline{tL}^{\mathsf{T}}.$

$(s, t)$   $\left. \begin{array}{c} \text{inclusion-maximal} \\ \text{independent} \end{array} \right\}$   $\Longleftrightarrow$   $Qt = \bar{s}$   and   $Q^{\mathsf{T}}s = \bar{t}.$

**Proof:** Without loss of generality, it remains to prove the equivalence of the second and the first variant in (i).

"$\Longrightarrow$": It is trivial that $Q \sqcap sL \subset sL$, and because of Prop. 2.4.2.i also

$$Q \sqcap \overline{sL} = IQ \sqcap \overline{sL} = (I \sqcap \overline{sL})Q = (I \sqcap \overline{sL}^{\mathsf{T}})Q \subset \overline{sL}^{\mathsf{T}}Q = (Q^{\mathsf{T}}\overline{sL})^{\mathsf{T}} \subset (tL)^{\mathsf{T}}.$$

"$\Longleftarrow$":   $Q^{\mathsf{T}}\bar{s} \subset (sL)^{\mathsf{T}}\bar{s} \sqcup (tL)\bar{s} \subset O \sqcup tL = t.$   $\square$

One will observe that the conditions $Qt \subset \bar{s}$ and $Q^{\mathsf{T}}s \subset \bar{t}$ are equivalent by the Schröder equivalences, but $Qt = \bar{s}$ and $Q^{\mathsf{T}}s = \bar{t}$ are not.

Figure 9.3.2 indicates the variety of pairs of subsets that can be used for covering a given relation. Among these are the pairs $(L, O)$, $(O, L)$, and the somewhat more economic $(QL, O)$ and $(O, Q^{\mathsf{T}}L)$. When looking for a smallest possible covering one may start with the covering $(\{\, b, c, d, e\, \}, \{\, 2, 3, 4\, \})$ from the first part of Fig. 9.3.2 and shrink it to a covering with $\{\, b, c, d\, \}$ as the left-hand set. The contrary effect is then that $\{\, 2, 3, 4\, \}$ has to be enlarged at least to $\{\, 1, 2, 3, 4\, \}$, so as to cover the edge from $e$ to $1$ just vacated. It is this counter-running effect that makes it difficult to find *smallest* possible, or even just *minimal*, point coverings.

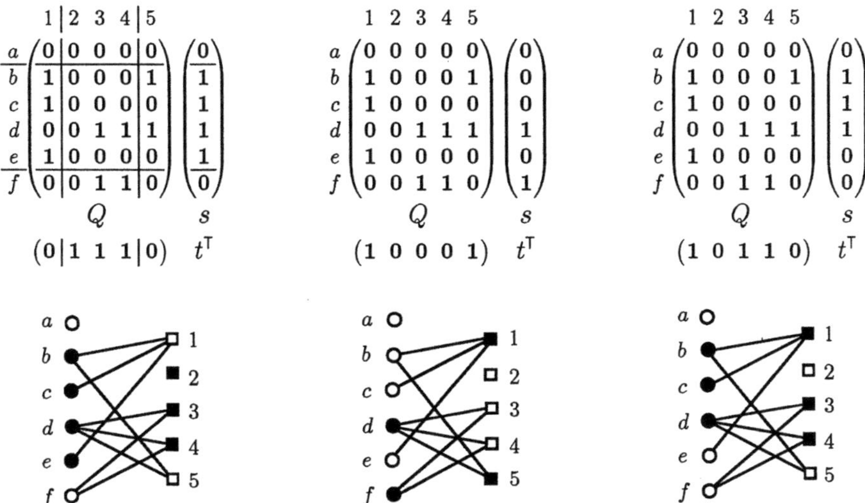

Fig. **9.3.2** Three different coverings

Nevertheless, the set of all covering point sets does carry some structure:

**9.3.2 Proposition** (*A. L. Dulmage, N. S. Mendelsohn, 1958*). Let us for the moment define two new operations for bipartitioned point sets $(s,t)$ and $(u,v)$ of a bipartitioned graph:

$$\text{union} \qquad (s,t) \perp (u,v) := (s \sqcup u, \, t \sqcap v) \quad \text{and}$$
$$\text{intersection} \qquad (s,t) \top (u,v) := (s \sqcap u, \, t \sqcup v).$$

For these operations and the corresponding infinite versions thereof, the following holds: If $(s,t)$ and $(u,v)$ are point coverings, then so is their $\perp$-union and $\top$-intersection. With these operations, the set of all point coverings constitutes a complete lattice with least element $(O, L)$ and greatest element $(L, O)$.

**Proof:** This is fairly simple; for instance:

$$Q \overline{(t \sqcup v)} = Q(\overline{t} \sqcap \overline{v}) \subset Q\overline{t} \sqcap Q\overline{v} \subset s \sqcap u. \qquad \square$$

We now consider the left-hand subsets and the right-hand subsets, both being ordered by "$\subset$". In order to investigate the counter-running effect described above, we define by

$$s \mapsto \sigma(s) := Q^{\top}\overline{s} \quad \text{resp.} \quad t \mapsto \tau(t) := Q\overline{t}$$

an antitone mapping $\sigma$ sending a left-hand subset into a right-hand subset, and also a mapping $\tau$ which does the converse. Obviously, one has

$$\tau(\sigma(s)) \subset s \quad \text{and} \quad \sigma(\tau(t)) \subset t.$$

Therefore, $\sigma$ and $\tau$ form a Galois correspondence (up to reversing both orderings which does not affect the antitonicity), cf. Sect. 3.3 and Appendix A.3. By the general theory, the left-hand subsets $s$ satisfying $\tau(\sigma(s)) = s$, as well as the right-hand subsets $t$ satisfying $\sigma(\tau(t)) = t$, in other words, the fixed points for the two composite mappings, each form a complete lattice, and these two lattices are isomorphic. The operations of these two lattices do not, in general, coincide with the operations $\perp, \top$, though.

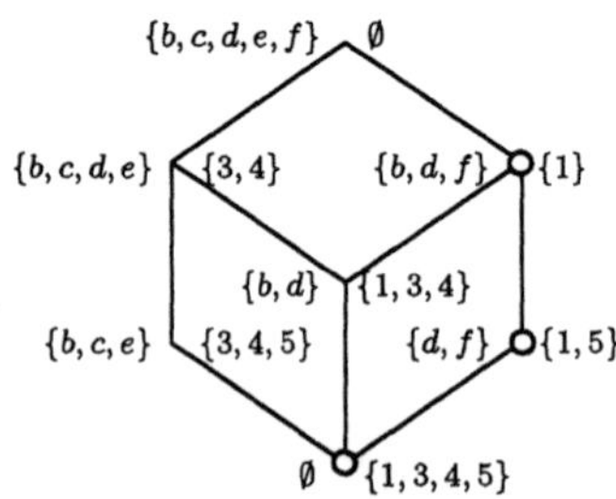

**Fig. 9.3.3** Lattice of inclusion-minimal coverings, where the minimum ones are highlighted

On the other hand, $\tau(\sigma(s)) = Q\overline{Q^{\top}\overline{s}} = s$ with $t := Q^{\top}\overline{s}$ implies $Q\overline{t} = Q\overline{Q^{\top}\overline{s}} = s$, and vice versa. Therefore, any such pair $(s,t)$ consisting of corresponding fixedpoints is a minimal covering. Figure 9.3.3 represents the lattice of minimal coverings for the example from Fig. 9.3.2, by indicating the pairs of corresponding sets.

One will observe that the minimal coverings $(s,t)=(\{\,b,c,e\,\},\ \{\,3,4,5\,\})$ and $(u,v) = (\{\,b,d\,\},\{\,1,3,4\,\})$ give rise to $(s,t)\ \top\ (u,v) = (\{\,b\,\},\ \{\,1,3,4,5\,\})$, a covering which is *not* minimal. Join and meet in the lattice of minimal coverings in Fig. 9.3.3 are therefore *not* the set-theoretic union and intersection in terms of which the operations $\bot, \top$ are defined. Conversely, the pairs $(s,t) = (\{\,b,c,e\,\}, \{\,3,4,5\,\})$ and $(u,v) = (\{\,d,f\,\},\{\,1,5\,\})$ provide an example of a $\bot$-union of minimal coverings which is not minimal.

Figure 9.3.4 shows another typical example of a minimal covering, using the sets $\{\,b,d\,\},\{\,1,3,4\,\}$ and $\{\,d,f\,\},\{\,1,5\,\}$ from Fig. 9.3.2 where points or rows and columns have been rearranged so as to make the presentation clearer.

The covering on the right is even of minimum cardinality. In the example on the left, one can remove points $b, d$ on the left, and add to the pair $(s,t)$ point 5 on the right, so as to obtain a new covering with fewer points, in contrast to the example on the right. So relation $Q$ associates with *every* left-hand subset of $s$ at least as many points in $\bar{t}$ on the right, rendering an exchange uninteresting.

We have thus motivated our next results which again deal with independence numbers and covering numbers. Since we have to assume certain finiteness conditions, we consider mainly finite bipartite graphs for the sake of simplicity.

**9.3.3 Definition.** In a finite graph, we define the **deficiency** of a subset $p$ with respect to a relation $Q$ and the deficiency of that relation itself by

$$\texttt{deficiency}(p) := |p| - |Q^{\mathsf{T}}p|;$$
$$\texttt{deficiency}(Q) := \texttt{max}_p\ \texttt{deficiency}(p).$$

Furthermore, we say that

$x$ satisfies the **Hall condition**

$\quad :\Longleftrightarrow$ For every subset $z \subset x$, the cardinality of $Q^{\mathsf{T}}z$ is not smaller than that of $z$.

$\quad \Longleftrightarrow$ For every subset $z \subset x$ there exists a relation $\rho$ satisfying $\rho\rho^{\mathsf{T}} \subset I$, $\rho^{\mathsf{T}}\rho \subset I$, $\rho L = z$, and $\rho^{\mathsf{T}}L \subset Q^{\mathsf{T}}z$.

(in case of a finite $Q$:

$\quad \Longleftrightarrow$ $|z| \leq |Q^{\mathsf{T}}z|$ for every subset $z \subset x$.) $\hfill \square$

If $x := L$ satisfies the Hall condition, then clearly $\texttt{deficiency}(Q) = 0$. Since $\texttt{deficiency}(O) = 0$ for the empty set, one always has that $\texttt{deficiency}(Q) \geq 0$. The point covering number and the deficiency are related by the formula below. It turns out that the edges emanating from the left-hand set $p$ in the case $|p| > |Q^{\mathsf{T}}p|$ can more economically be covered from the right-hand side.

**9.3.4 Proposition.** In a finite bipartitioned graph $(X,Y,Q)$

$$\tau = |X| - \texttt{deficiency}(Q).$$

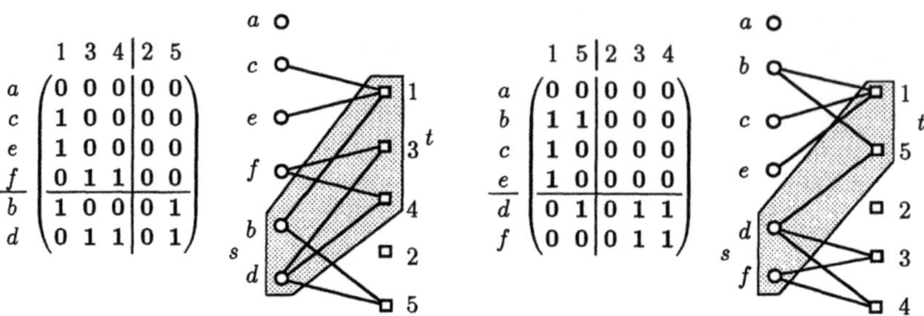

Fig. 9.3.4 Two inclusion-minimal coverings

**Proof:** The minimum $\tau$ of *all* cardinalities of point coverings can only be assumed by an inclusion-minimal one, so that by Proposition 9.3.1

$$\tau = \min\{\,|u| \mid u \subset X \cup Y \text{ covering}\,\}$$
$$= \min\{\,|s| + |t| \mid s \subset X,\; t \subset Y,\; (s,t) \text{ covering}\,\}$$
$$= \min\{\,|s| + |t| \mid s \subset X,\; t \subset Y,\; (s,t) \text{ inclusion-minimal covering}\,\}$$
$$= \min\{\,|s| + |Q^\mathsf{T}\bar{s}| \mid s \subset X\,\}.$$

Then we obtain by elementary arguments

$$\tau = \min\{\,|X| - |s'| + |Q^\mathsf{T}s'| \mid s' \subset X\,\}$$
$$= |X| - \max\{\,|s| - |Q^\mathsf{T}s| \mid s \subset X\,\} = |X| - \texttt{deficiency}(Q). \qquad \square$$

For a graph with $x := L$ satisfying the Hall condition, we obtain $\tau = |X|$. If $(s,t)$ is a minimum covering, then $\texttt{deficiency}(Q) = |X| - \tau = |X| - (|s| + |t|) = |\bar{s}| - |t|$. Because of the three minimum coverings of Fig. 9.3.3, $\texttt{deficiency}(Q) = |\bar{s}| - |t| = 2$.

## The Theorems of König and Hall

The deep results we are going to discuss next can be illustrated in terms of human relations as follows. Suppose the relation $Q$ is expressing feelings of sympathy between young men and girls of some village, referred to as belonging to set $X$ or $Y$, respectively. The bipartitioned graph in Fig. 9.3.5 represents men on the left

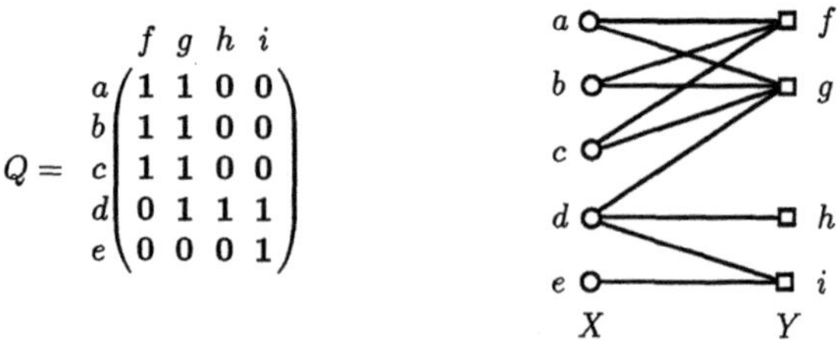

Fig. 9.3.5 Relation of sympathy

and girls on the right, two persons being joined by an edge when their feelings towards each other are marked by sympathy. We exclude nonreciprocal relations of that kind, as well as any within the two groups themselves. So the graph is symmetric and we work with the adjacency, i.e., with the bipartitioned graph $(X, Y, Q)$.

The problem consists in finding a matching relation $\lambda$ for $Q$, representing possible marriages. So $\lambda$ must be contained in $Q$, for a couple should marry only when they like each other. Moreover, $\lambda$ must be univalent and injective in order to avoid polygamy.

Clearly, $\lambda := O$ is such a relation, but would hardly satisfy the lovers involved. Indeed, $\lambda$ should lead to as many marriages as possible. Therefore, also the sets $\lambda L$ and $\lambda^{\mathsf{T}} L$, representing the males and females to be married, must be considered and actually maximized, either with respect to their cardinality or just with respect to set-theoretic inclusion.

If a subset $x$ of men is to be married simultaneously, then there have to be at least $|x|$ girls related to the fellows from $x$ by means of $\lambda$. And this condition must hold for every subset $x$ of $X$.

In graph-theoretic terms the problem consists in finding a pairing or a matching in a bipartitioned graph. We repeat our definition from the end of Sect. 9.1:

$$\lambda \ \textbf{matching} \quad :\Longleftrightarrow \quad \lambda \subset Q, \quad \lambda\lambda^{\mathsf{T}} \subset I, \quad \lambda^{\mathsf{T}}\lambda \subset I$$

$$\Longleftrightarrow \quad \lambda \text{ is a univalent and injective} $$
$$\text{relation contained in } Q.$$

We also need the following

**9.3.5 Definition.** Given a bipartitioned graph $(X, Y, Q)$, we say that a left-hand point set

$$x \ \text{ can be } \textbf{saturated} \quad :\Longleftrightarrow \quad \text{There exists a matching } \lambda \text{ with } \lambda L = x.$$

One may also say $x$ "can be **matched** into $Y$". When we want to specify the $Q$ considered, then we speak of a $Q$-matching. $\qquad\qquad\square$

For any $Q$ there is the trivial matching $O$ and also $\mathtt{unp}(\mathtt{unp}(Q)^{\mathsf{T}})^{\mathsf{T}}$ which is the injective part of the univalent part of $Q$. If the set $x$ can be saturated, then so, too, any subset $x' \subset x$, for if $\lambda$ is a matching with $\lambda L = x$, then $\lambda' := \lambda \sqcap x' L$ is a matching with $\lambda' L = x'$.

**Fig. 9.3.6** A set that cannot be saturated but satisfies the Hall condition

A first simple observation is

**9.3.6 Proposition.** $x$ can be saturated $\implies$ $x$ satisfies the Hall condition.

**Proof:** We define a mapping $\rho$ from $z$ into $Q^{\mathsf{T}}z$ by $\rho := \lambda \sqcap zL$. Now we easily see that $\rho L = (\lambda \sqcap zL)L = \lambda L \sqcap zL = xL \sqcap zL = zL$ and $\rho^{\mathsf{T}}L = (\lambda \sqcap zL)^{\mathsf{T}}L = \lambda^{\mathsf{T}}zL \subset Q^{\mathsf{T}}zL$. $\qquad\square$

In the infinite case the converse does not hold in general. To see this, we give in Fig. 9.3.6 a set which cannot be saturated but which does satisfy the Hall condition. The relation $Q$ in question is defined between $\mathbb{N} \cup \{\infty\}$ and $\mathbb{N}$ and associates with every $i$ all the $j$'s in $\mathbb{N}$ such that $j \leq i$.

Returning to the marriage problem, we consider Fig. 9.3.7. In case a) two couples have been married although, clearly, three marriages would be possible. Relation $Q$ is the same for cases b) and c). In all three cases there is clearly no matching $\lambda_1 \gneqq \lambda$. In case a), after all, there exists a matching $\lambda_1$ with $\lambda_1 L \gneqq \lambda L$. In cases b) and c) the matching cannot be enlarged so that $\lambda_1 \gneqq \lambda$, nor can it be replaced by some matching with $\lambda_1 L \gneqq \lambda L$. Despite the matchmaker's efforts, in villages b) and c) three persons from the upper left group of five will remain bachelors. Also, the marriages in the lower part are not at all uniquely determined by the maximality condition. Nevertheless, it is certain that the marriages must be formed within the lower part. Marrying $(x_0, y_1)$ would certainly leave an additional young man unmarried.

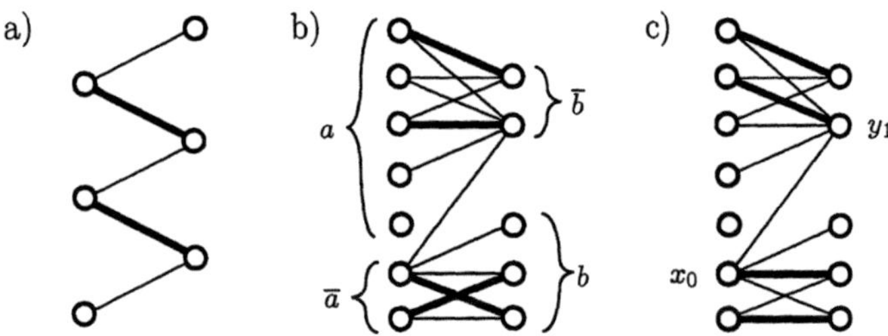

**Fig. 9.3.7** Matchings

In b) and c) one would expect a general subdivision $a \sqcup \bar{a}$ of the left-hand point set. The "uncritical" part $\bar{a}$ seems to consist of the two lower points with $\texttt{deficiency}(\bar{a}) = 0$ to which there correspond sufficiently many points on the right-hand side. The other part $a$ comprises the five upper points with $\texttt{deficiency}(a) = 3$. The two matchings indicated in b) and c) are inclusion-maximal with respect to $\lambda L$. There, the five upper points are claiming their two counterparts from the set $\bar{b}$ (who are too few), thus pushing the points from $\bar{a}$ towards their partners on the lower right who are sufficient in number.

On the other hand, the young men from the noncritical set $\bar{a}$ need not fear for their wedlock when changing to a matching $\lambda'$ with $\lambda' L \gneqq \lambda L$, since the fellows from $a$ are not interested in their spouses, anyway.

We now try to characterize the set $a$. It certainly contains all points not affected by the matching $\lambda$, so $a \supset \overline{\lambda L}$. It also contains all those points vying with the points just mentioned for the right-hand points from $\overline{b}$, the latter kind being the points from $\lambda Q^{\mathsf{T}} \overline{\lambda L}$, $(\lambda Q^{\mathsf{T}})^2 \overline{\lambda L}$, etc., from which $\overline{\lambda L}$ can be reached by alternating $\lambda$-$Q^{\mathsf{T}}$-paths. In Fig. 9.3.7 one cannot get from $\overline{a}$ to $\overline{\lambda L}$ in this way.

The following definition is based on our considerations above. There are, however, many subsets of the described sort out of which a suitable one has to be chosen.

**9.3.7 Definition.** In a bipartitioned graph $(X, Y, Q)$ we consider an arbitrary matching $\lambda$ and call a left-hand set

$$x \text{ \textbf{alternating}} \quad :\Longleftrightarrow \quad \overline{\lambda L} \sqcup \lambda Q^{\mathsf{T}} x \subset x \quad \Longleftrightarrow \quad \overline{\lambda L} \sqcup \lambda Q^{\mathsf{T}} x = x$$

$$\Longleftrightarrow \quad \overline{\lambda \overline{Q^{\mathsf{T}} x}} \subset x \quad \Longleftrightarrow \quad \overline{\lambda \overline{Q^{\mathsf{T}} x}} = x$$

$$\Longleftrightarrow \quad x \text{ comprises all the points matched}$$
$$\text{onto a } Q\text{-successor of } x.$$

We call a right-hand set

$$y \text{ \textbf{attracting}} \quad :\Longleftrightarrow \quad Qy \subset \lambda y \quad \Longleftrightarrow \quad Qy = \lambda y \quad \Longleftrightarrow \quad y \subset \overline{Q^{\mathsf{T}} \overline{\lambda y}}$$

$$\Longleftrightarrow \quad \text{All } Q\text{-predecessors of } y \text{ are matched}$$
$$\text{onto a point of } y.$$

If several relations $\lambda$ or $Q$ are in question, we will qualify the sets more precisely as $\lambda$-$Q^{\mathsf{T}}$-alternating and $\lambda$-$Q^{\mathsf{T}}$-attracting. Furthermore, we consider the sets

$$\mathcal{A} := \{\, x \mid x \text{ alternating} \,\}, \qquad \mathcal{B} := \{\, y \mid y \text{ attracting} \,\}$$

for the alternating or the attracting sets. We call their limits

$$a := \inf \mathcal{A} \ \textbf{concurring}, \qquad b := \sup \mathcal{B} \ \textbf{safe}. \qquad \square$$

Again we have presented several variants in the definition that are indeed equivalent. This follows by Proposition 4.2.2.v from $\lambda \overline{Q^{\mathsf{T}} x} = \lambda L \sqcap \overline{\lambda Q^{\mathsf{T}} x}$, and from $\lambda^{\mathsf{T}} x \subset Q^{\mathsf{T}} x \Longleftrightarrow \lambda \overline{Q^{\mathsf{T}} x} \subset \overline{x}$. Since $L \in \mathcal{A}$, certainly $\mathcal{A} \neq \emptyset$, and because of $O \in \mathcal{B}$ also $\mathcal{B} \neq \emptyset$. We will afterwards prove that the extrema indeed exist and that

$$a = \inf \{\, x \mid \overline{\lambda L} \sqcup \lambda Q^{\mathsf{T}} x \subset x \,\} = (\lambda Q^{\mathsf{T}})^* \overline{\lambda L},$$

i.e., that the concurring set consists of all those points from which $\overline{\lambda L}$ is reachable via $\lambda Q^{\mathsf{T}}$.

We are now aiming at showing that the infimum $a$ is alternating, hence contained in $\mathcal{A}$, and that it is thus the smallest alternating set. Similarly, we shall show that $b$ is attracting, i.e., contained in $\mathcal{B}$, and therefore the biggest attracting set. For that purpose we introduce the mappings

$$x \mapsto \sigma(x) := \overline{Q^{\mathsf{T}} x}$$
$$y \mapsto \pi(y) := \overline{\lambda y},$$

and have that $x$ is alternating precisely when $\pi\left(\sigma(x)\right) \subset x$, and $y$ is attracting precisely when $y \subset \sigma\left(\pi(y)\right)$.

We next establish some general properties of $\sigma$ and $\pi$ which are not symmetric because $\lambda$ is univalent but $Q$ is not.

**9.3.8 Proposition.** i)   $\sigma$ and $\pi$ are antitonic;

ii)   $\sigma(\sup_\iota x_\iota) = \inf_\iota \sigma(x_\iota)$;

iii)   $\pi(\sup_\iota y_\iota) = \inf_\iota \pi(y_\iota)$   and   $\pi(\inf_\iota y_\iota) = \sup_\iota \pi(y_\iota)$.

**Proof:** i) is trivial. ii) $\sigma(x \sqcup x') = \overline{Q^{\mathsf{T}}(x \sqcup x')} = \overline{Q^{\mathsf{T}}x \sqcup Q^{\mathsf{T}}x'} = \sigma(x) \sqcap \sigma(x')$.

iii) The left-hand part is proved the same way as was (ii). For univalent $\lambda$, the equation on the right follows from

$$\pi(y \sqcap y') = \overline{\lambda(y \sqcap y')} = \overline{\lambda y \sqcap \lambda y'} = \pi(y) \sqcup \pi(y').$$

The pattern of proof stays the same for unions and intersections which are not necessarily finite.    □

The following corollary is nonsymmetric, too, for $\pi\left(\sigma(.)\right)$ is expanding, but $\sigma\left(\pi(.)\right)$ is not contracting in general.

**9.3.9 Corollary.** The mappings

$$x \mapsto \pi\left(\sigma(x)\right) = \overline{\lambda\overline{Q^{\mathsf{T}}x}} \quad \text{and} \quad y \mapsto \sigma\left(\pi(y)\right) = \overline{Q^{\mathsf{T}}\overline{\lambda y}}$$

are isotonic and satisfy

$$\pi\left(\sigma(\sup_\iota x_\iota)\right) = \sup_\iota \pi\left(\sigma(x_\iota)\right) \quad \text{and} \quad \sigma\left(\pi(\inf_\iota y_\iota)\right) = \inf_\iota \sigma\left(\pi(y_\iota)\right).$$

Moreover, the mapping $x \mapsto \pi\left(\sigma(x)\right)$ is expanding: $x \subset \pi\left(\sigma(x)\right)$.

**Proof:** Isotonicity is trivial and, e.g.,

$$\pi\left(\sigma(\sup_\iota x_\iota)\right) = \pi\left(\inf_\iota \sigma(x_\iota)\right) = \sup_\iota \pi\left(\sigma(x_\iota)\right).$$

The lack of symmetry with regard to $\pi$ and $\sigma$ should be noted. Finally, we have
$x \subset \pi\left(\sigma(x)\right) \iff \lambda\overline{Q^{\mathsf{T}}x} \subset \overline{x} \iff \lambda^{\mathsf{T}}x \subset Q^{\mathsf{T}}x$.    □

We want to approximate the bounds

$$a = \inf \mathcal{A} = \inf\{\, x \mid \pi\left(\sigma(x)\right) \subset x \,\},$$
$$b = \sup \mathcal{B} = \sup\{\, y \mid y \subset \sigma\left(\pi(y)\right) \,\}$$

by lattice-theoretic methods which are listed in Appendix A.3. The composition $\sigma\left(\pi(.)\right)$ is isotonic, but in general it is not contracting. Starting from $b_0 = L$, however, we obtain the sequence of approximations

$$L = b_0 \supset b_1 = \sigma\left(\pi(b_0)\right) \supset b_2 \supset \dots$$

by isotonicity.

**9.3.10 Proposition.** For the approximation of the sets $a$ and $b$ the following hold:

i) $\quad a \in \mathcal{A}, \quad$ i.e., $\quad a = \overline{\lambda \overline{Q^\mathsf{T} a}} = \overline{\lambda L} \sqcup \lambda Q^\mathsf{T} a \quad$ or $\quad a = \pi\left(\sigma(a)\right);$

$\quad\ b \in \mathcal{B}, \quad$ i.e., $\quad b = \overline{Q^\mathsf{T} \overline{\lambda b}} \qquad\qquad\qquad$ or $\quad b = \sigma\left(\pi(b)\right).$

ii) The following iteration approximates $a = \sup_i a_i$ from below:

$$a_0 := O, \quad a_{i+1} := \pi\left(\sigma(a_i)\right) = \overline{\lambda \overline{Q^\mathsf{T} a_i}}, \quad i \geq 0.$$

iii) The following iteration approximates $b = \inf_i b_i$ from above:

$$b_0 := L, \quad b_{i+1} := \sigma\left(\pi(b_i)\right) = \overline{Q^\mathsf{T} \overline{\lambda b_i}}, \quad i \geq 0.$$

iv) Furthermore,

$$\overline{b} = Q^\mathsf{T} a, \quad \overline{a} = \lambda b.$$

**Proof:** i) For all $x \in \mathcal{A}$ we have $x = \pi\left(\sigma(x)\right)$ and $a \subset x$. Since the mappings are isotonic, $\pi\left(\sigma(a)\right) \subset \pi\left(\sigma(x)\right) = x$, so that $\pi\left(\sigma(a)\right)$ is one of the lower bounds of $\mathcal{A}$ and is, therefore, contained in $a$: $\pi\left(\sigma(a)\right) \subset a$. So $a \in \mathcal{A}$, and in addition $a = \pi\left(\sigma(a)\right)$. For all $y \in \mathcal{B}$ we have $y \subset \sigma\left(\pi(y)\right)$ and $y \subset b$. Again, because of the isotone mappings, $y \subset \sigma\left(\pi(y)\right) \subset \sigma\left(\pi(b)\right)$, so that $\sigma\left(\pi(b)\right)$ is one of the upper bounds and, therefore, contains $b$: $b \subset \sigma\left(\pi(b)\right)$. Now, $b$ belongs to $\mathcal{B}$. Since the mappings involved are isotonic, $\sigma\left(\pi(b)\right) \subset \sigma\left(\pi\left(\sigma\left(\pi(b)\right)\right)\right)$, and as a consequence $\sigma\left(\pi(b)\right) \in \mathcal{B}$. In contrast to the preceding case, however, we now have only $\sigma\left(\pi(b)\right) \subset b$ and $b = \sigma\left(\pi(b)\right)$. Equality does not hold for all $y \in \mathcal{B}$.

ii) and iii) We shall show simultaneously that $a_i$ is a monotone ascending sequence contained in $a$, and that $b_i$ is a monotone decreasing sequence containing $b$, and furthermore, that

$$a_{i+1} = \pi(b_i) \quad \text{and} \quad b_{i+1} = \sigma(a_{i+1}), \quad i \geq 0.$$

From (i) we obtain the starting situation for the iteration

$$a_0 = O \subset a_1 = \overline{\lambda L} = \pi(b_0) \subset a \quad \text{and} \quad b \subset b_1 = \overline{Q^\mathsf{T} \overline{\lambda L}} = \sigma(a_1) \subset b_0 = L.$$

Using the induction hypothesis,

$$a_i \subset a_{i+1} = \pi(b_i) \subset a \quad \text{and} \quad b \subset b_{i+1} = \sigma(a_{i+1}) \subset b_i,$$

we conclude that for isotone mappings we have

$$a_{i+1} = \pi(\sigma(a_i)) \subset a_{i+2} = \pi(\sigma(a_{i+1})) = \pi(\sigma(\pi(b_i))) = \pi(b_{i+1}) \subset \pi(\sigma(a)) = a,$$
$$b = \sigma(\pi(b)) \subset b_{i+2} = \sigma(\pi(b_{i+1})) = \sigma(\pi(\sigma(a_{i+1}))) = \sigma(a_{i+2}) \subset b_{i+1} = \sigma(\pi(b_i)).$$

Therefore, $\sup_i a_i \subset a$ and $b \subset \inf_i b_i$. From Prop. 9.3.8.i,ii it follows that

$$\sigma(\sup_i a_i) = \inf_i \sigma(a_i) = \inf_i b_i, \quad \pi(\inf_i b_i) = \sup_i \pi(b_i) = \sup_i a_i,$$

as well as

$$\pi\left(\sigma(\sup_i a_i)\right) = \pi(\inf_i b_i) = \sup_i a_i, \quad \sigma\left(\pi(\inf_i b_i)\right) = \sigma(\sup_i a_i) = \inf_i b_i.$$

The limit $\sup_i a_i$ thus belongs to $\mathcal{A}$, and $\inf_i b_i$ to $\mathcal{B}$, from which we conclude that they coincide with $a$ and $b$, respectively.

iv) We have $\pi(b) = a$, $\sigma(a) = b$, so that $\overline{\lambda \overline{b}} = a$ and $\overline{Q^\mathsf{T} a} = b$. $\qquad\square$

The iteration for $a$ starts as $a_0 = O$, $a_1 = \overline{\lambda L}$, $a_2 = \overline{\lambda L} \sqcup \lambda Q^\mathsf{T} \overline{\lambda L} = \overline{\lambda Q^\mathsf{T} \overline{\lambda L}}$, $a_3 = \overline{\lambda L} \sqcup \lambda Q^\mathsf{T} \overline{\lambda L} \sqcup \lambda Q^\mathsf{T} \lambda Q^\mathsf{T} \overline{\lambda L} = \overline{\lambda Q^\mathsf{T} \lambda Q^\mathsf{T} \overline{\lambda L}}, \ldots$ leading to $a = (\lambda Q^\mathsf{T})^* \overline{\lambda L}$. This shows the alternating character of the iteration. Algorithms using alternating chains are based on this lattice-theoretic principle.

Now the first connection with coverings emerges: $\overline{a} = \lambda b$ and $\overline{b} = Q^\mathsf{T} a$ yield the following corollary.

**9.3.11 Corollary.** For a matching $\lambda$ in a bipartitioned graph, the concurring set $a$ and the safe set $b$ satisfy $\overline{a} = Qb$, and the pair $(\overline{a}, \overline{b})$ of complements thus constitutes an inclusion-minimal covering.

**Proof:** Condition 9.3.1 for an inclusion-minimal covering applied to $(\overline{a}, \overline{b})$ gives $Qb = \overline{a}$ and $Q^\mathsf{T} a = \overline{b}$. While the second of these two conditions is part of 9.3.10.iv, the first follows from $\overline{a} = \lambda b \subset Qb$ and $Qb \subset \overline{a} \iff Q^\mathsf{T} a \subset \overline{b}$. $\qquad\square$

So far we have dealt with arbitrary matchings $\lambda$. In particular, $\lambda = O$ gives rise to $a = L$, $b = \overline{Q^\mathsf{T} L}$, and the minimal covering $(O, Q^\mathsf{T} L)$ which, however, has no optimal cardinality. In the sequel we assume a matching which is inclusion-maximal with respect to $\lambda L$, or is of maximum cardinality.

$$Q = \begin{array}{c} \\ a \\ \overline{a} \end{array} \begin{pmatrix} \overset{b}{O} & \overset{\overline{b}}{\otimes} \\ \times & * \end{pmatrix} \qquad \lambda = \begin{array}{c} \\ a \\ \overline{a} \end{array} \begin{pmatrix} \overset{b}{O} & \overset{\overline{b}}{\otimes} \\ \times & O \end{pmatrix} \qquad \begin{array}{l} \otimes \approx \text{surjective} \\ \times \approx \text{total} \end{array}$$

**Fig. 9.3.8** Partition generated by a concurring and a safe point set

We consider the decomposition of $Q$ and $\lambda$ in Fig. 9.3.8 resulting from $a$ and $b$. An easy calculation starting from $Q^\mathsf{T} a = \overline{b}$, $\overline{a} = Qb$ shows the properties of the submatrices of $Q$. From $\overline{a} = \lambda b$ follows totality in the lower left part of $\lambda$. From our discussion of Fig. 9.3.7 we expect that $\lambda$ contains a $O$ in the lower right rectangle (i.e., $\lambda \overline{b} \subset a$) and a surjective relation in the upper right. For the proof we need the point axiom and the intermediate point theorem.

**9.3.12 Proposition.** A matching $\lambda$, where $\lambda L$ is inclusion-maximal, satisfies

$$\overline{b} = \lambda^\mathsf{T} a.$$

**Proof:** As we have already Prop. 9.3.10.iv, only $\overline{b} \subset \lambda^\mathsf{T} a$ is a new result. Assuming $\overline{b} \not\subset \lambda^\mathsf{T} a$, we construct a matching $\lambda'$ satisfying $\lambda' L \not\subset \lambda L$ thus contradicting maximality. The starting point is

$$\overline{b} = Q^\mathsf{T} a = Q^\mathsf{T}(\lambda Q^\mathsf{T})^* \overline{\lambda L} = \sup_i Q^\mathsf{T}(\lambda Q^\mathsf{T})^i \overline{\lambda L}.$$

Therefore an $i$ exists such that $Q^\mathsf{T}(\lambda Q^\mathsf{T})^i\overline{\lambda L}\sqcap\overline{\lambda^\mathsf{T}a}\neq O$, so that we can find some point which we call $x_{2i+1}$. After that we obtain further points in $2i+1$ steps:

for $\quad i\geq j\geq 1$

take $\quad x_{2j}\quad$ with $\quad x_{2j+1}x_{2j}^\mathsf{T}\subset Q^\mathsf{T}\quad$ and $\quad x_{2j}\subset\lambda(Q^\mathsf{T}\lambda)^{j-1}Q^\mathsf{T}\overline{\lambda L}$

as well as $\quad x_{2j-1}:=\lambda^\mathsf{T}x_{2j}=\lambda^\mathsf{T}\lambda(Q^\mathsf{T}\lambda)^{j-1}Q^\mathsf{T}\overline{\lambda L}\subset(Q^\mathsf{T}\lambda)^{j-1}Q^\mathsf{T}\overline{\lambda L}$

and finally $x_0$ with $x_1x_0^\mathsf{T}\subset Q^\mathsf{T}$, $x_0\subset\overline{\lambda L}$.

We shall now prove that $\lambda':=(\lambda\sqcap\overline{\sup_j x_{2j}})\sqcup\sup_{0\leq k\leq i}x_{2k}x_{2k+1}^\mathsf{T}$ is a matching satisfying $\lambda'L\supsetneq\lambda L$.

First we need that $\lambda'\subset Q$, which follows from $\lambda\subset Q$ and $x_{2j}x_{2j+1}^\mathsf{T}\subset Q$. Univalence is obtained from $\overline{x_{2k}}^\mathsf{T}x_{2k}x_{2k+1}^\mathsf{T}\subset\overline{x_{2k}}^\mathsf{T}x_{2k}L\subset O$ and $x_{2k}^\mathsf{T}x_{2j}=O$ for $k\neq j$. Injectivity is proved analogously.

The proof is complete since $\lambda'L$ is properly enlarged:

$$\begin{aligned}
\lambda'L &= (\lambda\sqcap\overline{\sup_j x_{2j}})L\sqcup\sup_j x_{2j}x_{2j+1}^\mathsf{T}L\\
&= (\lambda L\sqcap\overline{\sup_j x_{2j}})\sqcup\sup_j x_{2j}L,\quad\text{because}\quad x_{2j+1}^\mathsf{T}L=(Lx_{2j+1}L)^\mathsf{T}=L.\\
&= \lambda L\sqcup\sup_j x_{2j}L,\\
&= \lambda L\sqcup x_0L,\quad\text{since }x_{2j}\subset\lambda(Q^\mathsf{T}\lambda)^{j-1}Q^\mathsf{T}\overline{\lambda L}\subset\lambda L\\
&\supsetneq\lambda L,\quad\text{since}\quad x_0L\subset\overline{\lambda L}.\qquad\qquad\qquad\qquad\square
\end{aligned}$$

Two famous theorems follow.

**9.3.13 Proposition** (*D. König, 1931*). In finite bipartite graphs the edge independence number and the point covering number coincide:

$$\alpha^*=\tau.$$

**Proof:** Choose a maximum independent edge set, consider its corresponding pairing $\mu$ as in Prop. 9.1.6, and obtain the matching $\lambda$ due to the bipartition so that $\alpha^*=|\lambda L|$. Now determine the $a$ and $b$ according Prop. 9.3.10 resulting in $\overline{a}=\lambda b$ which (since $\lambda$ is univalent) says that $\lambda$ joins no point from $\overline{a}$ to one from $\overline{b}$. Therefore, $|\lambda L|=|\overline{a}|+|\overline{b}|$. By Corollary 9.3.11, $(\overline{a},\overline{b})$ is an inclusion-minimal covering, so that $(\overline{a},\overline{b})$ is a covering of cardinality $\alpha^*$. This proves $\alpha^*\geq\tau$. On the other hand, at least that many points are necessary for a covering, since no two edges of the independent edge set can be covered by one point. $\qquad\square$

This theorem will be complemented by the similar result of Corollary 9.4.8. As a second consequence we have the well-known matching theorem.

**9.3.14 Proposition** (*Matching Theorem, P. Hall 1935*). In a finite bipartite graph, any set $x$ satisfying the Hall condition can be saturated.

**Proof:** Without loss of generality we may restrict ourselves to the case $Q\subset xL$. This eliminates all connections which do not meet $x$ on the left-hand side and which, of course, cannot contribute to the saturation of $x$. Assume $x$ cannot be saturated. Then $\lambda L \subsetneqq x$ for any matching $\lambda$. We choose a matching for which $\lambda L$ is inclusion-maximal, and determine $a$ and $b$. Now we show that $|Q^{\mathsf{T}}(a \sqcap x)| < |a \sqcap x|$, violating the Hall condition.

The relation defined as $\iota := \lambda \sqcap aL$ is univalent and injective, and, by Prop. 9.3.12, satisfies $\iota^{\mathsf{T}}L = (\lambda \sqcap aL)^{\mathsf{T}}(a \sqcup \overline{a}) = (\lambda \sqcap aL)^{\mathsf{T}}a \sqcup (\lambda \sqcap aL)^{\mathsf{T}}\overline{a} = \lambda^{\mathsf{T}}a \sqcup O = \overline{b}$ and $\iota L = (\lambda \sqcap aL)L = \lambda L \sqcap aL \subset x \sqcap a$. This gives an injection of the set $\overline{b}$ into $a \sqcap x$, but not onto $a \sqcap x$. The reason for this is that $a \sqcap x \subset \lambda L$ cannot hold since this together with $\overline{a} \sqcap x \subset \overline{a} = \lambda b \subset \lambda L$ would result in $x \subset \lambda L$, contradicting the assumption that $x$ is unsaturable. Therefore $|\overline{b}| < |a \sqcap x|$. As a consequence $|Q^{\mathsf{T}}(a \sqcap x)| \leq |Q^{\mathsf{T}}a| = |\overline{b}| < |a \sqcap x|$.         □

# 9.4  Starlikeness

There is an incidence of a special type by means of which inclusion-minimal or minimum cardinality transversals can be studied (under certain additional assumptions). This type of incidence is related to the Helly property, to confluence, and to inductive orderings. We will not discuss these relationships in detail.

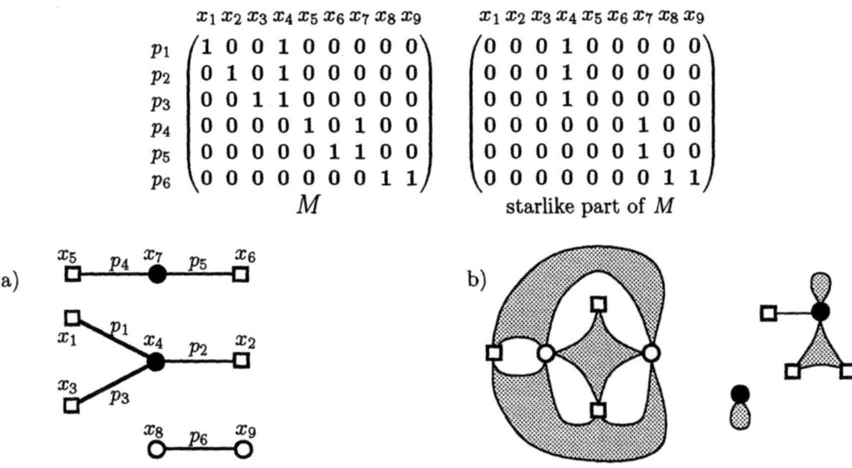

$$\begin{array}{c} x_1\ x_2\ x_3\ x_4\ x_5\ x_6\ x_7\ x_8\ x_9 \\ \begin{array}{c} p_1 \\ p_2 \\ p_3 \\ p_4 \\ p_5 \\ p_6 \end{array} \begin{pmatrix} 1 & 0 & 0 & 1 & 0 & 0 & 0 & 0 & 0 \\ 0 & 1 & 0 & 1 & 0 & 0 & 0 & 0 & 0 \\ 0 & 0 & 1 & 1 & 0 & 0 & 0 & 0 & 0 \\ 0 & 0 & 0 & 0 & 1 & 0 & 1 & 0 & 0 \\ 0 & 0 & 0 & 0 & 0 & 1 & 1 & 0 & 0 \\ 0 & 0 & 0 & 0 & 0 & 0 & 0 & 1 & 1 \end{pmatrix} \\ M \end{array} \qquad \begin{array}{c} x_1\ x_2\ x_3\ x_4\ x_5\ x_6\ x_7\ x_8\ x_9 \\ \begin{pmatrix} 0 & 0 & 0 & 1 & 0 & 0 & 0 & 0 & 0 \\ 0 & 0 & 0 & 1 & 0 & 0 & 0 & 0 & 0 \\ 0 & 0 & 0 & 1 & 0 & 0 & 0 & 0 & 0 \\ 0 & 0 & 0 & 0 & 0 & 0 & 1 & 0 & 0 \\ 0 & 0 & 0 & 0 & 0 & 0 & 1 & 0 & 0 \\ 0 & 0 & 0 & 0 & 0 & 0 & 0 & 1 & 1 \end{pmatrix} \\ \text{starlike part of } M \end{array}$$



Fig. 9.4.1 Starlike incidences

**9.4.1 Definition.** Let the incidence $M$ of a hypergraph be given together with its edge-adjacency $\mathsf{K} := \overline{I} \sqcap MM^{\mathsf{T}}$. We call

$$M \ \mathbf{starlike} \ :\Longleftrightarrow \ M \subset (M \sqcap \overline{\mathsf{K}M})L \ \Longleftrightarrow \ M \subset (M \sqcap \overline{MM^{\mathsf{T}}\overline{M}})L$$

$$\Longleftarrow \quad \text{Every nonempty edge } p \text{ is incident with a point,}$$
which in turn is incident with all edges edge-adjacent to $p$.         □

$M \sqcap \overline{\mathsf{K}M}$ will be called the starlike part of $M$. In Exercise 9.4.2 this part is shown to be equal to $M \sqcap \overline{MM^{\mathsf{T}}\overline{M}}$. Thus the conditions given in the above definition are equivalent. Graph a) in Fig. 9.4.1, whose relations are also indicated, has a starlike incidence. The same holds for the hypergraph b), but not for the hypergraph nor the graph in Fig. 9.4.2.

There is the important special case when $M$ is an ordering and hence a homogeneous relation. Instead of saying "$x$ is edge-adjacent to $y$", one then puts "$x$ and $y$ are distinct and have a common upper bound". Starlikeness for an ordering implies that all the elements having some upper bound in common with a given element $x$ also have a *common* upper bound. We therefore expect that this common upper bound depends unambiguously on $x$ and that the confluence condition holds in the direction towards this bound.

In view of these examples a starlike graph is characterized by its connected components being *stars*. All the edges of a star are pairwise adjacent. This is certainly true for the graph a) in Fig. 9.4.1. In the 3-star as well as in the 2-star the edges are pairwise adjacent. We shall give the general proof by showing that $MM^\mathsf{T}$ is a symmetric and transitive relation.

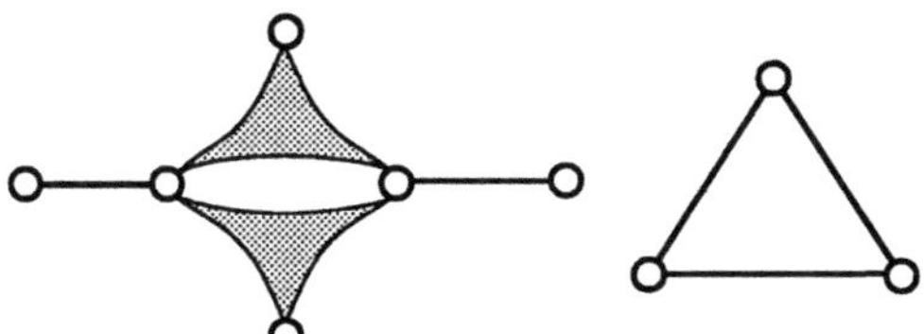

**Fig. 9.4.2** Nonstarlike incidence

First we prove two technical results concerning the starlike part of an arbitrary incidence.

**9.4.2 Proposition.** The following inclusions hold for the starlike part $M \sqcap \overline{K\overline{M}}$ of an arbitrary incidence $M$:

i)
$$MM^\mathsf{T}(M \sqcap \overline{K\overline{M}}) \subset M;$$

ii)
$$(M \sqcap \overline{K\overline{M}})M^\mathsf{T}(M \sqcap \overline{K\overline{M}}) \subset M \sqcap \overline{K\overline{M}}.$$

**Proof:** i) We prove instead the equivalent form

$$MM^\mathsf{T}\overline{M} \subset \overline{M} \sqcup K\overline{M},$$

using $MM^\mathsf{T}\overline{M} = (I \sqcap MM^\mathsf{T})\overline{M} \sqcup (\overline{I} \sqcap MM^\mathsf{T})\overline{M}$.

ii) Because of (i) we have "$\subset M$". Using (i) in a transposed form, we obtain $(M \sqcap \overline{K\overline{M}})^\mathsf{T}K \subset M^\mathsf{T}$ and, again using the above inclusion,

$$M(M \sqcap \overline{K\overline{M}})^\mathsf{T}K\overline{M} \subset MM^\mathsf{T}\overline{M} \subset \overline{M} \sqcup K\overline{M};$$

this is equivalent with "$\subset \overline{K\overline{M}}$".    □

Both assertions are elementary albeit somewhat cumbersome to formulate. In situation (i), when passing from an edge $p$ to some adjacent edge $q$ (possibly $p = q$), and then to some point $x$ of $q$ which is incident with each edge-adjacent to $q$, then $x$ is incident with $p$, too. Suppose $p$ also has some point $y$ which is incident with each edge-adjacent to $p$, as required in (ii). If the passage to $q$ described above is restricted to passing by such points as $y$, then $x$ is also a point of $p$ with this property. For all edges adjacent to $p$ are adjacent to $q$ via $y$, and are hence incident with $x$.

Specializing to starlike incidences yields the transitivity of $MM^\mathsf{T}$. The reflexive closure $I \sqcup MM^\mathsf{T}$ is then an equivalence relation on the set of edges. In addition, we get a decomposition of the point set.

**9.4.3 Corollary.** The following statements hold for every starlike incidence $M$:

i)    $MM^\mathsf{T} \subset (M \sqcap \overline{K\overline{M}})M^\mathsf{T}$,

ii)   $MM^\mathsf{T}$ is transitive,

iii)  $M^\mathsf{T}L = [\mathbf{unp}\,(M \sqcap \overline{K\overline{M}})]^\mathsf{T}L \sqcup [\mathbf{mup}\,(M \sqcap \overline{K\overline{M}})]^\mathsf{T}L \sqcup (M \sqcap K\overline{M})^\mathsf{T}L$
is a disjoint decomposition of the set of nonfree points.

**Proof:** i) $MM^\mathsf{T} = [(M \sqcap \overline{K\overline{M}})L \sqcap M]M^\mathsf{T}$
$$\subset (M \sqcap \overline{K\overline{M}} \sqcap ML)[L \sqcap (M \sqcap \overline{K\overline{M}})^\mathsf{T}M]M^\mathsf{T}$$
$$\subset (M \sqcap \overline{K\overline{M}})(M \sqcap \overline{K\overline{M}})^\mathsf{T}MM^\mathsf{T} \subset (M \sqcap \overline{K\overline{M}})M^\mathsf{T},$$
because of Prop. 9.4.2.i. Part ii) follows directly from (i) and Prop. 9.4.2.i:
$$MM^\mathsf{T}MM^\mathsf{T} \subset MM^\mathsf{T}(M \sqcap \overline{K\overline{M}})M^\mathsf{T} \subset MM^\mathsf{T}.$$

iii) As the union obviously gives $M^\mathsf{T}L$, we need only check its disjointness. This is done using the equivalence $X^\mathsf{T}L \sqcap Y^\mathsf{T}L = O \iff XY^\mathsf{T} = O$ which is easily established. The third part is disjoint from the first two, since

$$(M \sqcap \overline{K\overline{M}})(M \sqcap K\overline{M})^\mathsf{T} = (M \sqcap \overline{K\overline{M}})(\overline{M}^\mathsf{T}K \sqcap M^\mathsf{T})$$
$$\subset (M \sqcap \overline{K\overline{M}})[\overline{M}^\mathsf{T}M(M \sqcap \overline{K\overline{M}})^\mathsf{T} \sqcap M^\mathsf{T}], \text{ because of (i)}$$
$$\subset (M \sqcap \overline{K\overline{M}})[\overline{M}^\mathsf{T}M \sqcap M^\mathsf{T}(M \sqcap \overline{K\overline{M}})][(M \sqcap \overline{K\overline{M}})^\mathsf{T} \sqcap M^\mathsf{T}\overline{M}M^\mathsf{T}]$$
$$\subset [(M \sqcap \overline{MM^\mathsf{T}\,\overline{M}})\overline{M}^\mathsf{T}M \sqcap (M \sqcap \overline{K\overline{M}})M^\mathsf{T}(M \sqcap \overline{K\overline{M}})]L,$$
$$\text{since } M \sqcap \overline{MM^\mathsf{T}\,\overline{M}} = M \sqcap \overline{K\overline{M}}$$
$$\subset [\overline{MM^\mathsf{T}\,\overline{M}}\,\overline{M}^\mathsf{T}M \sqcap (M \sqcap \overline{K\overline{M}})]L, \text{ because of Prop. 9.4.2.ii}$$
$$\subset (\overline{MM^\mathsf{T}M} \sqcap M)L \subset (\overline{M} \sqcap M)L = O.$$

The first two parts are disjoint, because

$$\mathbf{unp}\,(M \sqcap \overline{K\overline{M}})[\mathbf{mup}\,(M \sqcap \overline{K\overline{M}})]^\mathsf{T} \subset \mathbf{unp}\,(M \sqcap \overline{K\overline{M}})L,$$

but also
$$\mathbf{unp}\,(M \sqcap \overline{K\overline{M}})[\mathbf{mup}\,(M \sqcap \overline{K\overline{M}})]^\mathsf{T}$$
$$\subset (M \sqcap \overline{K\overline{M}})[\overline{I}(M \sqcap \overline{K\overline{M}})^\mathsf{T} \sqcap (M \sqcap \overline{K\overline{M}})^\mathsf{T}]$$
$$\subset (M \sqcap \overline{K\overline{M}})[\overline{I} \sqcap (M \sqcap \overline{K\overline{M}})^\mathsf{T}(M \sqcap \overline{K\overline{M}})]$$
$$[(M \sqcap \overline{K\overline{M}})^\mathsf{T} \sqcap \overline{I}(M \sqcap \overline{K\overline{M}})^\mathsf{T}] \quad \text{via Dedekind rule}$$
$$\subset [(M \sqcap \overline{K\overline{M}})\overline{I} \sqcap (M \sqcap \overline{K\overline{M}})(M \sqcap \overline{K\overline{M}})^\mathsf{T}(M \sqcap \overline{K\overline{M}})]L$$
$$\subset [(M \sqcap \overline{K\overline{M}})\overline{I} \sqcap (M \sqcap \overline{K\overline{M}})]L, \text{ following Prop. 9.4.2.ii}$$
$$= \mathbf{mup}\,(M \sqcap \overline{K\overline{M}})L;$$
the product, therefore, vanishes by Proposition 4.2.11.    $\square$

The decomposition is shown in Fig. 9.4.1. Filled-in points are associated in an *unambiguous* way with edges so that they are incident with all adjacent edges. If an edge has several points incident with all adjacent edges, then so have all these adjacent edges. Such points are represented as circles. Points of an edge that are not incident with all adjacent edges are marked as squares—this property, too, refers to all edges incident with that point.

Incidentally, transitivity of $MM^\mathsf{T}$ does not imply that the incidence is starlike, not even for simple graphs. A counterexample is provided by the "triangle" in Fig. 9.4.2, where $MM^\mathsf{T} = L$ which is clearly transitive.

For incidences $M$ of rank 2 we shall now establish the announced relation between inclusion-minimal transversal edge sets and starlikeness. First we prove a preparatory result on incidences in which each nonempty edge is a cusp. Although the result seems to be geometrically clear, the algebraic proof of it is quite involved. We recall the definition of a cusp as an edge that is incident with at least one peak (a point incident with no other edge), cf. Definition 5.3.7 and Proposition 5.3.8.

**9.4.4 Proposition.** A hypergraph of rank $\leq 2$, all of whose nonempty edges are cusps, has starlike incidence.

**Proof:** We decompose the incidence $M$ into

$$M = [\mathbf{unp}\,(M^\mathsf{T}) \sqcup \mathbf{mup}\,(M^\mathsf{T})]^\mathsf{T} \subset [\mathbf{unp}\,(M^\mathsf{T})]^\mathsf{T} \sqcup [\mathbf{mup}\,(M^\mathsf{T})]^\mathsf{T}L$$
$$= \left([\mathbf{unp}\,(M^\mathsf{T})]^\mathsf{T} \sqcap \overline{[\mathbf{mup}\,(M^\mathsf{T})]^\mathsf{T}L}\right) \sqcup [\mathbf{mup}\,(M^\mathsf{T})]^\mathsf{T}L.$$

We therefore consider the incidence of edges having no other points than peaks separately. The remaining edges will meet at least one edge-connecting point.

In the first case we have

$$[\mathbf{unp}\,(M^\mathsf{T})]^\mathsf{T} \sqcap \overline{[\mathbf{mup}\,(M^\mathsf{T})]^\mathsf{T}L} \subset M \sqcap \overline{KL} \subset M \sqcap \overline{KM} \subset (M \sqcap \overline{KM})L,$$

since $KL = (MM^\mathsf{T} \sqcap \overline{I})L \subset (M \sqcap \overline{I}M)(M^\mathsf{T} \sqcap M^\mathsf{T}\overline{I})L \subset [\mathbf{mup}\,(M^\mathsf{T})]^\mathsf{T}L$.

For the second part we prove $[\mathbf{mup}\,(M^\mathsf{T})]^\mathsf{T} \subset M \sqcap \overline{KM}$. Obviously "$\subset M$". It remains to prove $M \sqcap \overline{I}M \subset \overline{KM}$ or, equivalently, $M \sqcap \overline{I}M \sqcap KM \subset O$. To this end we decompose the first $M$ from $K = \overline{I} \sqcap MM^\mathsf{T}$ into $M = (M \sqcap \overline{I}M) \sqcup (M \sqcap \overline{I}M)$. Now we may proceed with

$$\overline{I} \sqcap (M \sqcap \overline{I}M)M^\mathsf{T} \subset \overline{I} \sqcap \overline{I}MM^\mathsf{T} \subset \overline{I} \sqcap I = O,$$

or

$$[\overline{I} \sqcap (M \sqcap \overline{I}M)M^\mathsf{T}]\overline{M} \sqcap (M \sqcap \overline{I}M) \subset (M \sqcap \overline{I}M)M^\mathsf{T}\overline{M} \sqcap (M \sqcap \overline{I}M)$$
$$\subset (M \sqcap \overline{I}M)\overline{I} \sqcap (M \sqcap \overline{I}M) = \mathbf{mup}(M \sqcap \overline{I}M) = O.$$

The two hypotheses of the proposition occur in the very last conclusion only, where use is made of the fact that $M \sqcap \overline{I}M$ is univalent and therefore has no multivalent part: Indeed, each nonempty edge is a cusp by hypothesis and is thus incident with a peak by Proposition 5.3.8. Since the rank is $\leq 2$, there can be *at most* one more point in that edge which is edge-connecting, i.e., which lies in $\mathbf{mup}\,(M^\mathsf{T})L$; see Def. 5.3.5. Consequently, $M \sqcap [\mathbf{mup}\,(M^\mathsf{T})L]^\mathsf{T} = M \sqcap \overline{I}M$ is univalent. The nontrivial part of the latter equation follows from

$$[\text{mup}\,(M^{\mathsf{T}})L]^{\mathsf{T}} = L(M \sqcap \bar{I}M) = I(M \sqcap \bar{I}M) \sqcup \bar{I}(M \sqcap \bar{I}M) \subset \bar{I}M \sqcup \bar{I}M. \qquad \square$$

### Starlikeness of inclusion-minimal edge coverings

We return to the problem of finding inclusion-minimal transversal sets, discussed in Sect. 9.2.

Given an incidence $M$ together with an inclusion-minimal transversal edge set $q$, we consider the reduced incidence $M_q$ and the boundary $\delta$, cf. Definition 9.2.6. If the rank of $M$ is 2, then the ranks of the hyperedges are 0, 1, or 2. There are no empty edges in $q$, for such edges could have been deleted in the first place without affecting the covering property. An edge of rank 1 is incident with its point, and is the *only* edge in $q$ incident with this point, for else it could have been deleted. So edges of rank 1 satisfy the starlikeness condition. Edges of rank 2 meet the boundary $\delta$. So adjacency of such an edge with other edges in $q$ is established via *at most* one of its two points, namely the one not in $\delta$, which is then incident with all the adjacent edges in $q$. So starlikeness holds in this case, too.

The connection with Proposition 9.4.4 is now evident: Each nonempty edge of the reduced incidence $M_q$ is a cusp, for it meets some point of $\delta$ which is a peak with respect to $M_q$.

**9.4.5 Corollary.** The reduction $M_q$ of the incidence of a hypergraph of rank $\leq 2$ by an inclusion-minimal transversal edge set $q$ is a starlike incidence.

**Proof:** According to Prop. 9.2.7 we have $q \subset M_q\delta$. Using Prop. 5.3.8 we obtain

$$M_q = M \sqcap qL \subset qL = q \subset M_q\delta = M_q\,\text{unp}\,(M_q^{\mathsf{T}})L = [\text{unp}\,(M_q^{\mathsf{T}})]^{\mathsf{T}}L.$$

Definition 5.3.7 shows that indeed every nonempty edge of the reduction $M_q$ is a cusp (with respect to $M_q$). Since $M$ has rank 2, so has $M_q$, and applying Prop. 9.4.4 completes the proof. $\qquad \square$

We derive one more quantitative result from that corollary which enables us to relate the number of edges of a covering of minimum cardinality to the number of points.

**9.4.6 Proposition.** Let an inclusion-minimal transversal edge set $q$ be given in a hypergraph of rank $\leq 2$, together with the star decomposition. If $\sigma$ denotes the number of stars and $\rho$ the number of edges of rank 2, then

$$\rho + \sigma = |M^{\mathsf{T}}L|.$$

**Proof:** By Corollary 9.4.5, the reduction of $M$ by $q$ is starlike. Using Corollary

9.4.3, the nonfree points may be disjointly decomposed into

$$M^{\mathsf{T}}L = M^{\mathsf{T}}q = M_q^{\mathsf{T}}L$$
$$= [\mathtt{unp}\,(M_q \sqcap \overline{M_q M_q^{\mathsf{T}} \overline{M_q}})]^{\mathsf{T}}L \sqcup [\mathtt{mup}\,(M_q \sqcap \overline{M_q M_q^{\mathsf{T}} \overline{M_q}})]^{\mathsf{T}}L$$
$$\sqcup (M_q \sqcap M_q M_q^{\mathsf{T}} \overline{M_q})^{\mathsf{T}}L$$
$$=: a \sqcup b \sqcup c.$$

The first part $a$ of this union contains the points assigned *unambiguously* to their edges as the common point (filled points in Fig. 9.4.1.a). The second part contains precisely the *pairs* of points from isolated covering edges of rank 2 (circles in Fig. 9.4.1.a). The rest is made up of points that neither belong to an isolated covering edge nor are incident with *all* edges adjacent to the edges incident with them (squares in Fig. 9.4.1.a). The number of edges of rank 2 in $q$ and the number of stars are therefore

$$\rho = |c| + \tfrac{1}{2}|b|, \qquad \sigma = |a| + \tfrac{1}{2}|b|. \qquad \Box$$

This decomposition is illustrated in Fig. 9.4.3. Proposition 9.4.6 can be formulated more concisely when considering only coverings $q$ of minimum cardinality and excluding free points and drops (but allowing empty edges). The number $|M^{\mathsf{T}}L|$ of nonfree points then equals $n$, and the number of rank 2 edges in $q$ equals $\tau^*$. So one obtains $\tau^* = n - \sigma$.

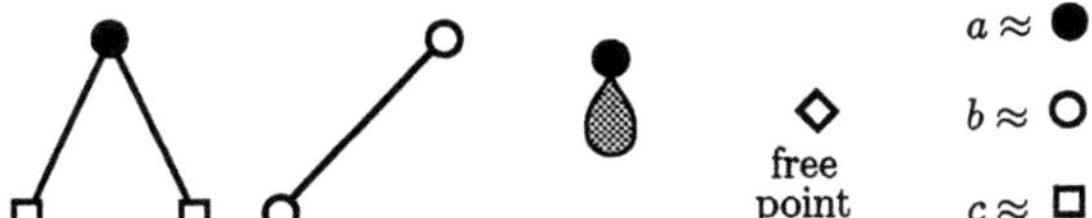


Fig. **9.4.3** Decomposition of the nonfree points in $a \sqcup b \sqcup c$

The interesting result below bears some similarity with Proposition 9.2.3.iii, of which it is no genuine dualization, for otherwise it would yield the number of edges and not the number of points.

**9.4.7 Proposition** (*T. Gallai; R. Z. Norman–M. O. Rabin, 1959*). In a simple graph with $n$ nonfree points the following holds:

$$\alpha^* + \tau^* = n.$$

**Proof:** "$\geq$": Any simple graph has rank 2; therefore, $\tau^* = n - \sigma$. From each of the $\sigma$ stars we now choose precisely one edge. These edges are pairwise nonadjacent, thus forming an independent set of edges such that $\sigma \leq \alpha^*$.

"$\leq$": Let an independent edge set $r$ be given which then obviously covers $2 \cdot |r|$ points. The rest of $n - 2 \cdot |r|$ (nonfree) points may be covered by at most $n - 2 \cdot |r|$ edges. So we have a covering with $n - |r| \geq \tau^*$ edges. Assume $r$ to be chosen maximum with regard to cardinality; then we have $n - \alpha^* \geq \tau^*$. $\qquad \Box$

The final corollary follows up on Proposition 9.3.13.

**9.4.8 Corollary.** In a bipartite graph without free points we have

$$\alpha = \tau^*.$$

**Proof:**   In the equation $\alpha = n - \tau$ from Corollary 9.2.3.iii, we may set $\tau = \alpha^*$, as a consequence of Proposition 9.3.13, and $n - \alpha^* = \tau^*$ because of Proposition 9.4.7.                                                                           $\square$

### Exercises

**9.4.1** Prove that a starlike incidence with an essential transposed relation is univalent.

**9.4.2** Prove that $M \sqcap \overline{MM^\mathsf{T}\,\overline{M}} = M \sqcap \overline{K\overline{M}}$ for arbitrary $M$ and $K = \overline{I} \sqcap MM^\mathsf{T}$.

**9.4.3** Prove that $M \sqcap [(M \sqcap \overline{K\overline{M}})^\mathsf{T}L^\mathsf{T}] = M \sqcap \overline{K\overline{M}}$ if $M$ is starlike.

**9.4.4** Prove that $\overline{MM^\mathsf{T}\,\overline{M}}$ is univalent if the incidence $M$ is reflexive, antisymmetric, and starlike.

**9.4.5** Prove the following identity for a reflexive, antisymmetric, and starlike incidence:
$$M \sqcap \overline{MM^\mathsf{T}\,\overline{M}} = M \sqcap \overline{(M \sqcap \overline{I})L}^\mathsf{T}.$$

**9.4.6** Prove that $M \sqcap \overline{MM^\mathsf{T}\,\overline{M}} = \mathsf{syq}(MM^\mathsf{T}, M) \sqcap ML$ for every $M$.

**9.4.7** Prove that starlikeness of an incidence implies the Helly property, and disprove the converse of this implication by giving a counterexample.

## 9.5 References

See also the references of Chap. 8.


DULMAGE AL, MENDELSOHN NS: *Coverings of bipartite graphs.* Canadian J. Math. **10** (1958) 517–534.

GALLAI T: *Über extreme Punkt- und Kantenmengen.* Ann. Univ. Sci. Budapest, Eötvös Sect. Math. **2** (1959) 133–138.

HALL P: *On representations of subsets.* J. London Math. Soc. **10** (1935) 26–30.

KÖNIG D: *Graphen und Matrizen.* Mat. Fiz. Lapok **38** (1931) 116–119.

MAGHOUT K: *Sur la détermination des nombres de stabilité et du nombre chromatique d'un graphe.* C. R. Acad. Sci. Paris **248** (1959) 3522–3523.

NORMAN RZ, RABIN MO: *An algorithm for a minimum cover of a graph.* Proc. Amer. Math. Soc. **10** (1959) 315–319.

SCHMIDT G, STRÖHLEIN T: *A boolean matrix iteration in timetable construction.* Linear Algebra Appl. **15** (1976) 27–51.

# 10. Programs: Correctness and Verification

Flowcharts relate programs and graphs. They serve for distinguishing the various steps of the program in the course of its execution. In the subsequent discussion we shall not take into account questions related to programming languages or the syntax of programs. When it becomes unavoidable in the sequel to deal with texts in progamming language, we shall use the following short expressions without specifying their syntax in a formal way:

- *sequential composition* $E; F$ of program steps,
- *branching* **if** $b$ **then** $E$ **else** $F$ **fi**,
- *nondeterministic guarded branching* $\lceil b : E [] c : F \rfloor$ with predicates $b$ and $c$ and program steps $E$ and $F$, where in case $b = c = $ **true** also $E [] F$ may be used,
- *loops* **while** $b$ **do** $E$ **od** and **repeat** $E$ **until** $b$,
- *jump command* **goto** $M$.

Programs with recursion have also been dealt with by relation-algebraic means, but including them here would lead too far.

In Sect. 10.1 we define the action and effect of flowchart programs in relation-algebraic terms. Then in Sect. 10.2 partial correctness of programs with regard to preconditions and postconditions is discussed. The verification rules for the various program constructs follow more or less immediately from the very general contraction theorem.

In Sect. 10.3 we develop three different concepts for investigating termination and total correctness. Two of these concepts coincide for finitely branching nondeterministic programs, and all three coincide for deterministic programs. In Sect. 10.4 the concept of total correctness is translated into applicable computational rules for weakest preconditions. We also discuss a connection with dynamic logic.

Section 10.5 features a special kind of program homomorphism based on the coverings introduced in Chap. 7. These homomorphisms affect only the execution sequence of the program steps, while leaving the individual steps of the program unchanged. We show how this concept can be used to establish semantic equivalence of programs.

## 10.1 Programs and Their Effect

The relation-algebraic concept of a program is based on flowcharts which describe the control flow of a program in terms of graphs, and indicate the individual steps of the program (often in a schematic fashion only). In Fig. 10.1.1 we describe three program constructs in this way by means of the underlying *flow*graphs whose arrows are "weighted" by relations. (The first two examples are also given in the usual box form for flowcharts. In the sequel, however, program steps will be associated with *arrows* rather than with boxes, i.e., with *points* of the graph.)

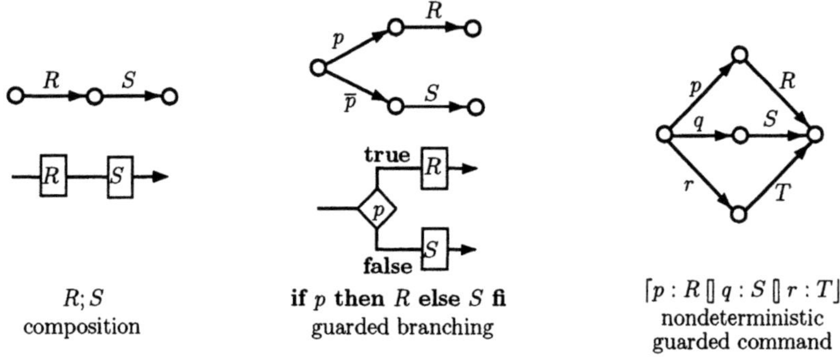



**Fig. 10.1.1** Programming constructs and their underlying flowgraphs

The graphs for these program constructs have the following associated relations (the points being arranged in an obvious way):

$$\begin{pmatrix} 0 & 1 & 0 \\ 0 & 0 & 1 \\ 0 & 0 & 0 \end{pmatrix} \qquad \begin{pmatrix} 0 & 1 & 1 & 0 & 0 \\ 0 & 0 & 0 & 1 & 0 \\ 0 & 0 & 0 & 0 & 1 \\ 0 & 0 & 0 & 0 & 0 \\ 0 & 0 & 0 & 0 & 0 \end{pmatrix} \qquad \begin{pmatrix} 0 & 1 & 1 & 1 & 0 \\ 0 & 0 & 0 & 0 & 1 \\ 0 & 0 & 0 & 0 & 1 \\ 0 & 0 & 0 & 0 & 1 \\ 0 & 0 & 0 & 0 & 0 \end{pmatrix}.$$

It is now convenient to replace the matrix entries $0$, $1$ by the operations executed during the program steps. So one obtains

$$\begin{pmatrix} O & R & O \\ O & O & S \\ O & O & O \end{pmatrix} \qquad \begin{pmatrix} O & p & \overline{p} & O & O \\ O & O & O & R & O \\ O & O & O & O & S \\ O & O & O & O & O \\ O & O & O & O & O \end{pmatrix} \qquad \begin{pmatrix} O & p & q & r & O \\ O & O & O & O & R \\ O & O & O & O & S \\ O & O & O & O & T \\ O & O & O & O & O \end{pmatrix}.$$

Thus, the program steps have been made the entries of a matrix. For the moment this is just another symbolic notation, marked also by the use of $O$ instead of $0$. The mathematical background is given in Appendix A.2.5 where we prove that matrices with entries in a homogeneous relation algebra again form a relation algebra. Before interpreting our new notation in terms of ordinary relations, we recall how the product of these matrices— $\begin{pmatrix} O & R & O \\ O & O & S \\ O & O & O \end{pmatrix} \begin{pmatrix} O & R & O \\ O & O & S \\ O & O & O \end{pmatrix} = \begin{pmatrix} O & O & RS \\ O & O & O \\ O & O & O \end{pmatrix}$ —is formed, cf. the discussion of reachability in Sect. 6.1.

So the double step is weighted by the product relation $RS$. Among the three or five positions, respectively, in the three graphs in Fig. 10.1.1, we next distinguish the input or output positions by

$$e_G = \begin{pmatrix} 1 \\ 0 \\ 0 \end{pmatrix} \quad a_G = \begin{pmatrix} 0 \\ 0 \\ 1 \end{pmatrix} \qquad e_G = \begin{pmatrix} 1 \\ 0 \\ 0 \\ 0 \\ 0 \end{pmatrix} \quad a_G = \begin{pmatrix} 0 \\ 0 \\ 0 \\ 1 \\ 1 \end{pmatrix} \qquad e_G = \begin{pmatrix} 1 \\ 0 \\ 0 \\ 0 \\ 0 \end{pmatrix} \quad a_G = \begin{pmatrix} 0 \\ 0 \\ 0 \\ 0 \\ 1 \end{pmatrix}.$$

For a somewhat bigger example we consider the following program scheme given in pseudo-code,

$$\begin{aligned}
&\textbf{if } p \textbf{ then} \\
&\qquad \textbf{while } q \textbf{ do } R \textbf{ od} \\
&\quad \textbf{else} \\
&\qquad S;\ \textbf{while } r \textbf{ do } T \textbf{ od} \\
&\quad \textbf{fi},
\end{aligned}$$

which can be interpreted concretely, and provided with input and output, as follows:

$$\begin{aligned}
&\textbf{nat } n; \\
&read\,(n); \\
&\textbf{if } n > 1 \quad \textbf{then} \\
&\qquad \textbf{while } n > 2 \textbf{ do } n := n - 3 \textbf{ od} \\
&\quad \textbf{else} \\
&\qquad n := n * n; \\
&\qquad \textbf{while } n > 0 \textbf{ do } n := n - 1 \textbf{ od} \\
&\quad \textbf{fi}; \\
&print\,(n)
\end{aligned}$$

We now strip this program (which happens to be absurd) of its syntactical representation in pseudo-code, and consider the totality of its possible execution sequences arising from different inputs. We must distinguish between the actual value of the variable $n$, to be understood as a "machine state", and the actual position in the flowgraph of the program.

The *position* in the graph, together with the *state*, describe the situation in which the concrete execution of the program may be at a given moment. This cannot be achieved by the state alone given by the value of the variable since, for example, $n = 4$ can result from entering the number 4 or from entering the number 10 and twice subtracting 3 by the first **while**-loop.

In Fig. 10.1.2 we sketch the (infinite) situation graph on top of the underlying flowgraph, so as to give all possible situations. It indicates, above each position of the flowgraph, all possible values for $n$, i.e., all possible states. The arrows in the situation graph correspond to the program steps given in the flowgraph. Input and output are indicated by $read\,(n)$ or $print\,(n)$, respectively. In the figure they are expressed by letting the set of states (enclosed in an oval) before entering into, or after running, the program correspond to the situation graph (dotted lines).

We now give the definition of our relation-algebraic program concept whose characteristic features are already included in the example above. It is characteristic of our point of view to distinguish between the following:

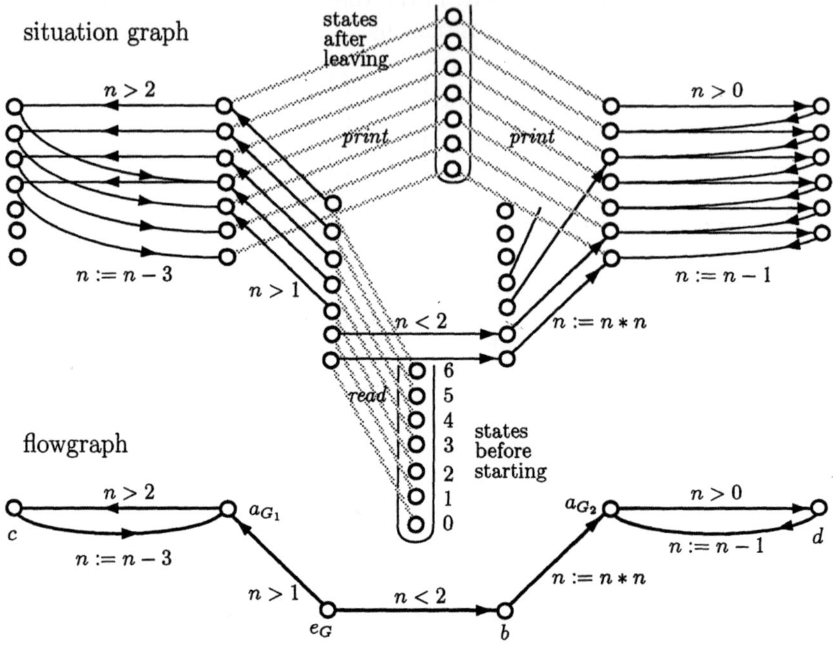



**Fig. 10.1.2** Flowgraph and situation graph

- *states* before starting the program,
- *states* after running the program,
- *situations* in the course of the program,
- *positions* in the flowgraph,

and the indication, in a schematic way, of the various relations holding between these.

The following definition is therefore based on two graphs. These are inter-related in a simple way by requiring that transitions in the situation graph are possible only when permitted by the flowgraph. In addition, it will be indicated which situations are to be associated with states before entering into the program, and which situations with states after running it.

**10.1.1 Definition.** We call the quintuple $\mathcal{P} = (G, S, \Theta, e, a)$ a **(flowdiagram) program**, provided

i)  $S = (V, B)$ is a graph, called the **situation graph**. Its points are called **situations.**

ii)  $G = (V_G, B_G)$ is a graph, called the underlying **flowgraph**. Its points are called **positions.**

iii)  $\Theta: V \longrightarrow V_G$ is a surjective graph homomorphism of $S$ onto $G$; therefore, $\Theta$ is subject to:

$$\Theta\Theta^{\mathsf{T}} \supset I, \quad \Theta^{\mathsf{T}}\Theta = I, \quad B\Theta \subset \Theta B_G.$$

| | | $e_G$ | | | | $b$ | | | | $c$ | | | | $a_{G_1}$ | | | | $d$ | | | | $a_{G_2}$ | | | |
|---|---|---|---|---|---|---|---|---|---|---|---|---|---|---|---|---|---|---|---|---|---|---|---|---|---|
| | | 0 | 1 | 2 | 3... | 0 | 1 | 2 | 3... | 0 | 1 | 2 | 3... | 0 | 1 | 2 | 3... | 0 | 1 | 2 | 3... | 0 | 1 | 2 | 3... |
| $e_G$ | 0 | 0 | 0 | 0 | 0... | 1 | 0 | 0 | 0... | 0 | 0 | 0 | 0... | 0 | 0 | 0 | 0... | 0 | 0 | 0 | 0... | 0 | 0 | 0 | 0... |
| | 1 | 0 | 0 | 0 | 0... | 0 | 1 | 0 | 0... | 0 | 0 | 0 | 0... | 0 | 0 | 0 | 0... | 0 | 0 | 0 | 0... | 0 | 0 | 0 | 0... |
| | 2 | 0 | 0 | 0 | 0... | 0 | 0 | 0 | 0... | 0 | 0 | 0 | 0... | 0 | 0 | 1 | 0... | 0 | 0 | 0 | 0... | 0 | 0 | 0 | 0... |
| | 3 | 0 | 0 | 0 | 0 | 0 | 0 | 0 | 0 | 0 | 0 | 0 | 0 | 0 | 0 | 0 | 1 | 0 | 0 | 0 | 0 | 0 | 0 | 0 | 0 |
| $b$ | 0 | 0 | 0 | 0 | 0... | 0 | 0 | 0 | 0... | 0 | 0 | 0 | 0... | 0 | 0 | 0 | 0... | 0 | 0 | 0 | 0... | 1 | 0 | 0 | 0... |
| | 1 | 0 | 0 | 0 | 0... | 0 | 0 | 0 | 0... | 0 | 0 | 0 | 0... | 0 | 0 | 0 | 0... | 0 | 0 | 0 | 0... | 0 | 1 | 0 | 0... |
| | 2 | 0 | 0 | 0 | 0... | 0 | 0 | 0 | 0... | 0 | 0 | 0 | 0... | 0 | 0 | 0 | 0... | 0 | 0 | 0 | 0... | 0 | 0 | 0 | 0 1 |
| | 3 | 0 | 0 | 0 | 0 | 0 | 0 | 0 | 0 | 0 | 0 | 0 | 0 | 0 | 0 | 0 | 0 | 0 | 0 | 0 | 0 | 0 | 0 | 0 | 0 |
| $c$ | 0 | 0 | 0 | 0 | 0... | 0 | 0 | 0 | 0... | 0 | 0 | 0 | 0... | 0 | 0 | 0 | 0... | 0 | 0 | 0 | 0... | 0 | 0 | 0 | 0... |
| | 1 | 0 | 0 | 0 | 0... | 0 | 0 | 0 | 0... | 0 | 0 | 0 | 0... | 0 | 0 | 0 | 0... | 0 | 0 | 0 | 0... | 0 | 0 | 0 | 0... |
| | 2 | 0 | 0 | 0 | 0... | 0 | 0 | 0 | 0... | 0 | 0 | 0 | 0... | 0 | 0 | 0 | 0... | 0 | 0 | 0 | 0... | 0 | 0 | 0 | 0... |
| | 3 | 0 | 0 | 0 | 0 | 0 | 0 | 0 | 0 | 0 | 0 | 0 | 0 | 1 0 0 0 / 0 1 | | | | 0 | 0 | 0 | 0 | 0 | 0 | 0 | 0 |
| $a_{G_1}$ | 0 | 0 | 0 | 0 | 0... | 0 | 0 | 0 | 0... | 0 | 0 | 0 | 0... | 0 | 0 | 0 | 0... | 0 | 0 | 0 | 0... | 0 | 0 | 0 | 0... |
| | 1 | 0 | 0 | 0 | 0... | 0 | 0 | 0 | 0... | 0 | 0 | 0 | 0... | 0 | 0 | 0 | 0... | 0 | 0 | 0 | 0... | 0 | 0 | 0 | 0... |
| | 2 | 0 | 0 | 0 | 0... | 0 | 0 | 0 | 0... | 0 | 0 | 0 | 0... | 0 | 0 | 0 | 0... | 0 | 0 | 0 | 0... | 0 | 0 | 0 | 0... |
| | 3 | 0 | 0 | 0 | 0 | 0 | 0 | 0 | 0 | 0 | 0 | 0 | 1 | 0 | 0 | 0 | 0 | 0 | 0 | 0 | 0 | 0 | 0 | 0 | 0 |
| $d$ | 0 | 0 | 0 | 0 | 0... | 0 | 0 | 0 | 0... | 0 | 0 | 0 | 0... | 0 | 0 | 0 | 0... | 0 | 0 | 0 | 0... | 0 | 0 | 0 | 0... |
| | 1 | 0 | 0 | 0 | 0... | 0 | 0 | 0 | 0... | 0 | 0 | 0 | 0... | 0 | 0 | 0 | 0... | 0 | 0 | 0 | 0... | 1 | 0 | 0 | 0... |
| | 2 | 0 | 0 | 0 | 0... | 0 | 0 | 0 | 0... | 0 | 0 | 0 | 0... | 0 | 0 | 0 | 0... | 0 | 0 | 0 | 0... | 0 | 1 | 0 | 0... |
| | 3 | 0 | 0 | 0 | 0 | 0 | 0 | 0 | 0 | 0 | 0 | 0 | 0 | 0 | 0 | 0 | 0 | 0 | 0 | 0 | 0 | 0 | 0 | 1 | 0 |
| $a_{G_2}$ | 0 | 0 | 0 | 0 | 0... | 0 | 0 | 0 | 0... | 0 | 0 | 0 | 0... | 0 | 0 | 0 | 0... | 0 | 0 | 0 | 0... | 0 | 0 | 0 | 0... |
| | 1 | 0 | 0 | 0 | 0... | 0 | 0 | 0 | 0... | 0 | 0 | 0 | 0... | 0 | 0 | 0 | 0... | 0 | 1 | 0 | 0... | 0 | 0 | 0 | 0... |
| | 2 | 0 | 0 | 0 | 0... | 0 | 0 | 0 | 0... | 0 | 0 | 0 | 0... | 0 | 0 | 0 | 0... | 0 | 0 | 1 | 0... | 0 | 0 | 0 | 0... |
| | 3 | 0 | 0 | 0 | 0 | 0 | 0 | 0 | 0 | 0 | 0 | 0 | 0 | 0 | 0 | 0 | 0 | 0 | 0 | 0 | 1 | 0 | 0 | 0 | 0 |

$$
B = \begin{array}{c} e_G \\ b \\ c \\ a_{G_1} \\ d \\ a_{G_2} \end{array}
\begin{array}{cccccc}
e_G & b & c & a_{G_1} & d & a_{G_2} \\
O & \overline{p} & O & p & O & O \\
O & O & O & O & O & S \\
O & O & O & R & O & O \\
O & O & q & O & O & O \\
O & O & O & O & O & T \\
O & O & O & O & r & O
\end{array}
\qquad
B_G = \begin{array}{cccccc}
e_G & b & c & a_{G_1} & d & a_{G_2} \\
0 & 1 & 0 & 1 & 0 & 0 \\
0 & 0 & 0 & 0 & 0 & 1 \\
0 & 0 & 0 & 1 & 0 & 0 \\
0 & 0 & 1 & 0 & 0 & 0 \\
0 & 0 & 0 & 0 & 0 & 1 \\
0 & 0 & 0 & 0 & 1 & 0
\end{array}
$$

$$
e = \begin{array}{c} e_G \\ b \\ c \\ a_{G_1} \\ d \\ a_{G_2} \end{array}
\begin{pmatrix} I \\ O \\ O \\ O \\ O \\ O \end{pmatrix}
\quad
e_G = \begin{pmatrix} 1 \\ 0 \\ 0 \\ 0 \\ 0 \\ 0 \end{pmatrix}
\quad
a = \begin{pmatrix} O \\ O \\ O \\ I \\ O \\ I \end{pmatrix}
\quad
a_G = \begin{pmatrix} 0 \\ 0 \\ 0 \\ 1 \\ 0 \\ 1 \end{pmatrix}
\;;
\quad
I = \begin{array}{c} 0 \\ 1 \\ 2 \\ 3 \\ \vdots \end{array}
\begin{array}{c} \begin{array}{cccc} 0 & 1 & 2 & 3\ldots \end{array} \\
\begin{pmatrix}
1 & 0 & 0 & 0\ldots \\
0 & 1 & 0 & 0\ldots \\
0 & 0 & 1 & 0\ldots \\
0 & 0 & 0 & 1 \\
\vdots & \vdots & \vdots & & \ddots
\end{pmatrix}
\end{array}
$$

**Fig. 10.1.3** Relations belonging to the flowgraph and situation graph of Fig. 10.1.2

iv)   $e$ is a so-called **input relation**, defined between situations and **states** before starting the program, satisfying:

$$e^{\mathsf{T}}e = I, \quad ee^{\mathsf{T}} \subset I, \quad eL = \Theta\Theta^{\mathsf{T}}eL, \quad \Theta^{\mathsf{T}}e = \Theta^{\mathsf{T}}eL.$$

v)   $a$ is a so-called **output relation** (with $aL \sqcap eL = O$) that relates situations and states after leaving the program, and satisfies:

$$a^{\mathsf{T}}a = I, \quad aa^{\mathsf{T}} \sqcap \Theta\Theta^{\mathsf{T}} \subset I, \quad aL = \Theta\Theta^{\mathsf{T}}aL.$$

If $B$ is univalent, we call the program $\mathcal{P}$ **deterministic**. If the underlying flowgraph is a rooted *tree*, we call it a **rooted tree program**.     □


This program concept is based on a few simple components, but its range of applications is surprisingly broad. It can serve for deriving correctness conditions and results on semantic equivalence. By means of this concept, problems concerning termination can be studied even for recursive and nondeterministic programs. Results from dynamic logic can be better understood; also, certain repetitive and recursive programs and their corresponding iterative versions can be shown to have the same effect.

It can be shown that

$$e_G := \Theta^{\mathsf{T}}eL \quad \text{and} \quad a_G := \Theta^{\mathsf{T}}aL$$

represent a single point and a nonempty point set, respectively, in the flowgraph, by verifying that

$$O \neq e_G = e_G L, \quad e_G e_G^{\mathsf{T}} \subset I, \quad O \neq a_G = a_G L.$$

This can be elucidated as follows. Consider the underlying flowgraph $(V_G, B_G)$ only, with an input position $e_G$, a set $a_G$ of output positions, and a set $Z$ of machine states. At each position then consider the totality of all machine states, so as to get the set of situations $V := V_G \times Z$. Situations are then specified by indicating a position in the flowgraph and the state holding there. In this special case $\Theta: V_G \times Z \longrightarrow V_G$ is the projection onto the component $V_G$. A transition along $B$ from a situation in $\{x\} \times Z \subset V_G \times Z$ to a situation in $\{y\} \times Z \subset V_G \times Z$ can take place only when there is an arrow from $x$ to $y$ in the flowgraph.

The vector for the input position $e_G$ in the flowgraph determines the input relation $e = (O \ldots I \ldots O)^{\mathsf{T}} \subset (V_G \times Z) \times Z$ with the identity $I$ in the place of $\mathbf{1}$. The vector $a_G$ of the output positions in the flowgraph determines the output relation $a = (O \ldots I \ldots I \ldots O)^{\mathsf{T}} \subset (V_G \times Z) \times Z$ with the identity $I$ in all places where there was a $\mathbf{1}$.

Figure 10.1.2 shows the situation graph and the flowgraph of the program given above; it also illustrates how the projection $\Theta$ acts. In addition, we have the associated relations $B_G \subset V_G \times V_G$ of the flowgraph and $B \subset (V_G \times Z) \times (V_G \times Z)$ of the situation graph. Passing to the matrix representations we have $B_G \in \mathbb{B}^{V_G \times V_G}$ and $B \in \mathbb{B}^{(V_G \times Z) \times (V_G \times Z)}$. But $B$ can also be interpreted as an element in $(\mathbb{B}^{Z \times Z})^{V_G \times V_G}$, thus making clear that arrows in the program graph are being associated with relations in $\mathbb{B}^{Z \times Z}$. The algebraic treatment is not affected by the fact that the coefficient relations in Fig. 10.1.2 are infinite.

## Effect and Action

Programs written in pseudo-code will usually be compiled and then applied to input data (or interpreted in connection with data). So the question arises as to the semantics of such programs, i.e., which input data yield which results (along which control sequence). We are allowing here also nondeterministic programs, for which the answer is much harder to find than for deterministic programs. One will also observe that we did not require the output situations in our program concept to be terminal situations.

Primarily, we are going to use reachability and terminating reachability from Sect. 6.1, with the terminology adjusted to the present topic.

**10.1.2 Definition.** In a program with input relation $e$, output relation $a$, and associated relation $B$ of the situation graph, we call

$$B^* \qquad \textbf{action,}$$
$$C := B^* \sqcap \overline{BL}^{\mathsf{T}} \quad \textbf{terminating action,}$$
$$\Sigma := e^{\mathsf{T}} C a \quad \textbf{effect.} \qquad \square$$

The action of a program is thus the reachability in the situation graph. It links a given situation $s$ with all those that can arise in a computation sequence starting with $s$. The terminating action, on the other hand, associates with a given situation only the last situations reachable from it, and these need not be situated above the distinguished output positions of the flowgraph (abortion). Operations $e$ and $a$ extract the part leading from input to output positions. The effect of the program is the relation that relates the states before entering into the program to those after exiting it, without keeping track of any details of the situation graph traversed.

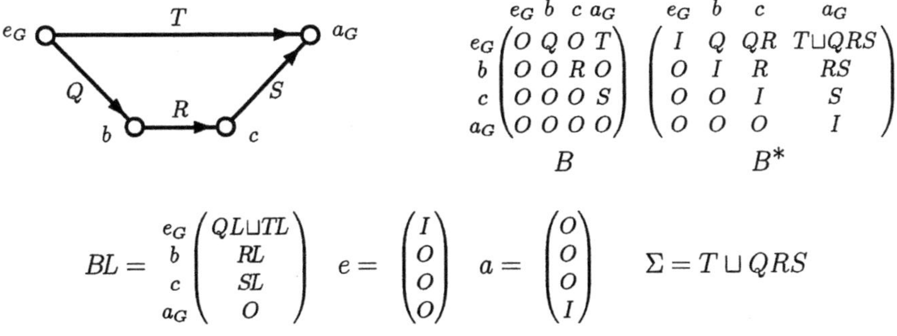

$$
B = \begin{array}{c} \\ e_G \\ b \\ c \\ a_G \end{array}
\begin{array}{cccc} e_G & b & c & a_G \\ \left(\begin{array}{cccc} O & Q & O & T \\ O & O & R & O \\ O & O & O & S \\ O & O & O & O \end{array}\right) \end{array}
\qquad
B^* = \begin{array}{cccc} e_G & b & c & a_G \\ \left(\begin{array}{cccc} I & Q & QR & T\sqcup QRS \\ O & I & R & RS \\ O & O & I & S \\ O & O & O & I \end{array}\right) \end{array}
$$

$$
BL = \begin{array}{c} e_G \\ b \\ c \\ a_G \end{array}\left(\begin{array}{c} QL\sqcup TL \\ RL \\ SL \\ O \end{array}\right)
\qquad
e = \left(\begin{array}{c} I \\ O \\ O \\ O \end{array}\right)
\qquad
a = \left(\begin{array}{c} O \\ O \\ O \\ I \end{array}\right)
\qquad
\Sigma = T \sqcup QRS
$$

**Fig. 10.1.4** Action and effect of $\lceil T \rceil\!\lceil Q; R; S \rfloor\rfloor$

Let us explain everything by means of the little program from Fig. 10.1.4. Two different routes can be taken (in a nondeterministic fashion) from input position $e_G$ to output position $a_G$, according to whether program step $T$ or the sequence of steps $QRS$ will be executed. So we expect the effect of the program to be $T \sqcup QRS$. Reasoning formally, we take the associated relation $B$, compute its transitive closure $B^*$, and determine the terminal situations $\overline{BL}$. Extracting $e^{\mathsf{T}}Ca$ from the terminating action $C$, we obtain indeed $T \sqcup QRS$ as the effect.

One will observe that in Fig. 10.1.4 we indicated neither the set of states nor the relations $T, Q, R$, and $S$ that should hold between them. So far, only a program *scheme* is given, and we have retained a high degree of generality. In the following two examples,

$$T = \begin{pmatrix} 1 & 0 \\ 0 & 0 \end{pmatrix} \quad Q = \begin{pmatrix} 0 & 0 \\ 1 & 0 \end{pmatrix} \quad R = \begin{pmatrix} 0 & 1 \\ 1 & 0 \end{pmatrix} \quad S = \begin{pmatrix} 1 & 1 \\ 1 & 0 \end{pmatrix} \quad \text{or}$$

$$T = \begin{pmatrix} 0 & 1 & 1 \\ 0 & 0 & 0 \\ 0 & 0 & 0 \end{pmatrix} \quad Q = \begin{pmatrix} 0 & 0 & 1 \\ 1 & 1 & 0 \\ 0 & 0 & 1 \end{pmatrix} \quad R = \begin{pmatrix} 0 & 1 & 0 \\ 0 & 0 & 1 \\ 1 & 0 & 0 \end{pmatrix} \quad S = \begin{pmatrix} 0 & 0 & 0 \\ 1 & 0 & 0 \\ 0 & 0 & 1 \end{pmatrix},$$

the effect is

$$\Sigma = \begin{pmatrix} 1 & 0 \\ 1 & 0 \end{pmatrix} \quad \text{or} \quad \Sigma = \begin{pmatrix} 0 & 1 & 1 \\ 1 & 0 & 1 \\ 0 & 0 & 0 \end{pmatrix},$$

respectively. The example also shows that one can use $B^*$ instead of $C$ for computing the effect, provided the situations above the output positions in the flowgraph are terminal. Next we show by relation-algebraic arguments that the effect relation $\Sigma$ for a deterministic program is univalent.


**10.1.3 Proposition.** If $\mathcal{P}$ is a program with associated relation $B$, input relation $e$, output relation $a$, and effect $\Sigma$, we have

i) $\qquad\qquad\qquad a \subset \overline{BL} \qquad \Longrightarrow \qquad \Sigma = e^{\mathsf{T}}B^*a;$

ii) $\qquad\qquad\qquad \mathcal{P} \text{ deterministic} \quad \Longrightarrow \quad \Sigma \text{ univalent.}$

**Proof:** i) $\Sigma = e^{\mathsf{T}}Ca = e^{\mathsf{T}}(B^* \sqcap \overline{BL}^{\mathsf{T}})a = e^{\mathsf{T}}B^*(\overline{BL} \sqcap a) = e^{\mathsf{T}}B^*a$, because of Prop. 2.4.2.ii. ii) From $B^{\mathsf{T}}B \subset I$ follows that $B^{*\mathsf{T}}B^* = B^{*\mathsf{T}} \sqcup B^* \subset I \sqcup (BL)^{\mathsf{T}} \sqcup BL$. Using this, we first prove the univalence of $C$,

$$C^{\mathsf{T}}C = (B^{*\mathsf{T}} \sqcap \overline{BL})(B^* \sqcap \overline{BL}^{\mathsf{T}})$$

$$= B^{*\mathsf{T}}B^* \sqcap \overline{BL}^{\mathsf{T}} \sqcap \overline{BL} \subset (I \sqcup (BL)^{\mathsf{T}} \sqcup BL) \sqcap \overline{BL}^{\mathsf{T}} \sqcap \overline{BL} \subset I,$$

and then univalence of $\Sigma$:

$$\Sigma^{\mathsf{T}}\Sigma = a^{\mathsf{T}}C^{\mathsf{T}}ee^{\mathsf{T}}Ca \subset a^{\mathsf{T}}C^{\mathsf{T}}Ca \subset a^{\mathsf{T}}a = I. \qquad\qquad \square$$


According to Exercise 7.1.1, condition (i) holds in particular when the output positions in the flowgraph are terminal, as is usually required for flowdiagrams. (We have not imposed this restriction in order to be able to treat recursion.)

Later we will compare the effect of programs that are different but have the same program steps as building blocks, according to how similar their executions are. For this we need a concept of homomorphy for programs.

Two programs will be called **homomorphic** if the following holds: Their flowgraphs can be mapped homomorphically into one another in such a way that input and output positions are preserved and that each program step associated with some arrow is contained, as a relation, in the program step associated with the image arrow. The precise definition will be given in Sect. 10.5 where homomorphisms are used for investigating coverings.


## 10.2 Partial Correctness and Verification

One of the most important problems of programming is to ensure that a given program actually processes the input data for which it is to be used in the intended way. In this case the program is said to be *correct*. We shall define correctness mathematically as a condition on the *effect* of the program.

The effect of a program contains the transitions from states *before* entering into the program to states *after* executing it. But often only some part of this relation is relevant. The initial and final states one is interested in are selected by means of predicates termed *precondition* and *postcondition*. A program is said to be correct, if all computation sequences terminating in output positions, which start in states satisfying the precondition, also end in states satisfying the postcondition.


**10.2.1 Definition.** For a program $\mathcal{P}$ with effect $\Sigma$ and predicates on the set of states, called pre- and postcondition $v$ and $n$, we define

$$\{v\}\mathcal{P}\{n\} \quad :\Longleftrightarrow \quad \Sigma^{\mathsf{T}}v \subset n;$$

if this holds, we say $\mathcal{P}$ is **partially correct** with respect to $v$ and $n$. $\qquad\square$


The curly brackets and variants thereof are commonly used in the literature.

An equivalent form of the condition, $\Sigma\overline{n} \subset \overline{v}$, reads as follows: If the effect of the program leads to some state contradicting the postcondition, then the precondition was already violated. Condition $\Sigma \subset \overline{v\overline{n}^{\mathsf{T}}}$, finally, can be interpreted in the sense that it is impossible, by executing the program, to get from states satisfying the precondition $v$ to states violating the postcondition $n$.

If only a postcondition is given by $n$, then $v_{\mathsf{sup}} := \overline{\Sigma\overline{n}}$ is the "weakest" precondition for $n$ (i.e., it contains as many states as possible) in the sense that correctness of the program with respect to $n$ and a given $v$, i.e., $\Sigma^{\mathsf{T}}v \subset n$, is always equivalent to $\Sigma\overline{n} \subset \overline{v}$ and hence to $v \subset \overline{\Sigma\overline{n}} = v_{\mathsf{sup}}$.

If on the other hand only a precondition $v$ is given, then $n_{\mathsf{inf}} := \Sigma^{\mathsf{T}}v$ is the "strongest" postcondition for $v$ (i.e., it contains as few states as possible), since correctness, $\Sigma^{\mathsf{T}}v \subset n$, always implies $n \supset n_{\mathsf{inf}}$.

Clearly, one would like to check correctness of the program more directly by looking at its associated relation $B$ rather than at the complicated effect $\Sigma = e^{\mathsf{T}}(B^{*} \sqcap \overline{BL}^{\mathsf{T}})a$. The essential ideas in this direction are contained in the *method of inductive assertions* by R. W. Floyd (1967) and in the axiomatic arguments by C. A. R. Hoare (1969) which were extended to the  recursive case by J. W. de Bakker and L. G. L. T. Meertens (1974). The core of these methods is contained in the following relation-algebraic theorem.

**10.2.2 Proposition** (*Contraction Theorem*). Let $\mathcal{P}$ be a program with associated relation $B$, input relation $e$, and output relation $a$. If $v$ is a precondition and $n$ is a postcondition, the following holds:

$$\{v\}\mathcal{P}\{n\} \quad\Longleftrightarrow\quad \begin{cases} B^\mathsf{T}q \subset q, \quad v \subset e^\mathsf{T}q, \quad a^\mathsf{T}(q \sqcap \overline{BL}) \subset n \\ \text{for some predicate } q \text{ on the situations.} \end{cases}$$

**Proof:** "$\Longrightarrow$": Define $q := B^{*\mathsf{T}}ev$. Then obviously $q = qL$ and $B^\mathsf{T}q \subset q$ hold as well as $v = e^\mathsf{T}ev \subset e^\mathsf{T}B^{*\mathsf{T}}ev = e^\mathsf{T}q$. Furthermore, $a^\mathsf{T}(q \sqcap \overline{BL}) = a^\mathsf{T}(B^{*\mathsf{T}}ev \sqcap \overline{BL}) = a^\mathsf{T}(B^{*\mathsf{T}} \sqcap \overline{BL})ev = a^\mathsf{T}C^\mathsf{T}ev = \Sigma^\mathsf{T}v \subset n$. "$\Longleftarrow$": $\Sigma^\mathsf{T}v \subset \Sigma^\mathsf{T}e^\mathsf{T}q = a^\mathsf{T}C^\mathsf{T}ee^\mathsf{T}q \subset a^\mathsf{T}C^\mathsf{T}q = a^\mathsf{T}(B^{*\mathsf{T}} \sqcap \overline{BL})q = a^\mathsf{T}(B^{*\mathsf{T}}q \sqcap \overline{BL}) = a^\mathsf{T}(q \sqcap \overline{BL}) \subset n$. $\qquad\square$

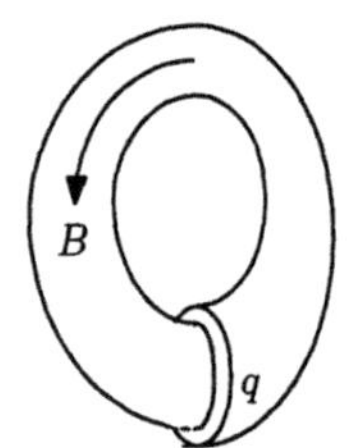

**Fig. 10.2.1** Contraction

We have thus separated the concept of correctness from the effect $\Sigma$ of the program which depends strongly on the termination of the program. So it becomes possible to consider programs that are not intended to terminate, such as operating systems. The formula $B^\mathsf{T}q \subset q$ characterizes sets $q$ of situations which cannot be left when applying $B$, as ought to be the case for the set of well-defined operating system situations.

Our starting point, however, has been a set of programs whose termination is of interest to us. We stress again that $v$ and $n$ mark states before entering into the program or after exiting it, respectively. On the other hand, the predicate $q$ is defined on the situations, and thus represents some kind of "interpolation" between $v$ and $n$. The restriction of $q$ to input situations over $e_G$ must contain $v$, the restriction of $q$ to terminal output situations must be contained in $n$, and the associated relation $B$ must be "contracted" by $q$, that is $B^\mathsf{T}q \subset q$.

From this general contraction theorem the verification rules concerning partial correctness of elementary program constructs can be deduced. In Fig. 10.2.2 six simple relational programs are given together with their usual notation in pseudo-code.

The semantics of the versions in programming language are understood to be defined by the relational programs for which these versions will serve as abbreviations in the sequel. Incidentally, the **if-then-else-fi** program has been arbitrarily given *two* output positions instead of one, but this is permitted by the definition of programs 10.1.1.

We had originally defined correctness by means of the effect $\Sigma$. By the contraction theorem we can now regard it as a property of the program, namely its associated relation $B$. The verification rules for each of the constructs just introduced can now be obtained by elementary calculation.

**10.2.3 Proposition** (*Elementary verification rules*). For a program consisting of several program steps partial correctness with respect to some precondition

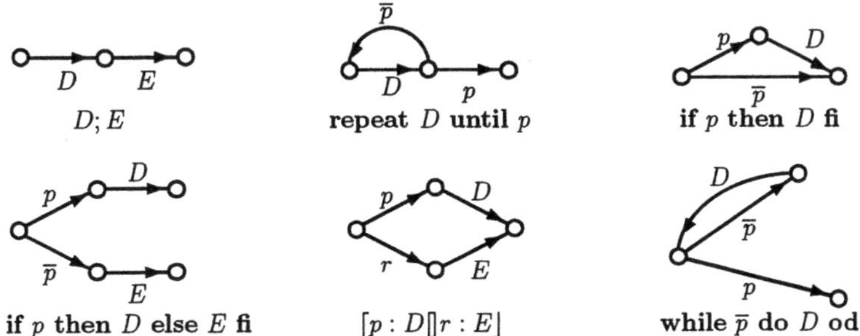

**Fig. 10.2.2** Composition, branching, and loops

$v$ and postcondition $n$ may be traced back to properties of the single program steps. In particular, we have for

i)   composition

$$\{v\}\, D; E\, \{n\} \quad \Longleftrightarrow \quad \begin{array}{l} \{v\}D\{y\},\ \{y\}E\{n\} \\ \text{for some predicate } y; \end{array}$$

ii)   guarded branching

$$\{v\}\ \textbf{if } p \textbf{ then } D \textbf{ fi } \{n\} \quad \Longleftrightarrow \quad \{p \sqcap v\}D\{n\},\ \overline{p} \sqcap v \subset n,$$
$$\{v\}\ \textbf{if } p \textbf{ then } D \textbf{ else } E \textbf{ fi } \{n\} \quad \Longleftrightarrow \quad \{p \sqcap v\}D\{n\},\ \{\overline{p} \sqcap v\}E\{n\};$$

iii)   nondeterministic guarded branching

$$\{v\}\ \lceil p : D \,[\!]\, r : E\rfloor\, \{n\} \quad \Longleftrightarrow \quad \{v \sqcap p\}D\{n\},\ \{v \sqcap r\}E\{n\}.$$

**Proof** for (i) and (iii); (ii) is a special case of (iii).

i) Let the program consist of two program steps $D$ and $E$ to be performed one after the other. Then its associated relation $B$, the input and output relations $e$, $a$, together with the predicate $q$ on the situations (assumed to be arbitrary) look as follows:

$$B = \begin{pmatrix} O & D & O \\ O & O & E \\ O & O & O \end{pmatrix} \qquad q = \begin{pmatrix} x \\ y \\ z \end{pmatrix} \qquad e = \begin{pmatrix} I \\ O \\ O \end{pmatrix} \qquad a = \begin{pmatrix} O \\ O \\ I \end{pmatrix}.$$

Using the contraction theorem, we have partial correctness with respect to $v$ and $n$ precisely when there is a predicate $q$ (given by its components $x, y, z$) satisfying

$$D^{\mathsf{T}}x \subset y, \quad E^{\mathsf{T}}y \subset z, \quad v \subset x \quad \text{and} \quad z \subset n.$$

One will observe that $x$ occurs on the smaller as well as on the greater side. Shrinking $x$ to $v$ will thus not affect solvability. Likewise, we may blow up $z$ to $n$ without affecting solvability. If there is a solution $q = \begin{pmatrix} x \\ y \\ z \end{pmatrix}$, there will be a solution of the form $q_1 = \begin{pmatrix} v \\ y \\ n \end{pmatrix}$.

That is, $q$ exists precisely when some $y$ exists satisfying

$$D^{\mathsf{T}}v \subset y \quad \text{and} \quad E^{\mathsf{T}}y \subset n.$$

iii) Writing $p$ instead of $I \sqcap pL$ for reasons of space, the relations look as follows:

$$B = \begin{pmatrix} O & p & r & O \\ O & O & O & D \\ O & O & O & E \\ O & O & O & O \end{pmatrix} \qquad e = \begin{pmatrix} I \\ O \\ O \\ O \end{pmatrix} \qquad a = \begin{pmatrix} O \\ O \\ O \\ I \end{pmatrix} \qquad q = \begin{pmatrix} w \\ x \\ y \\ z \end{pmatrix}.$$

Because of the contraction theorem, partial correctness is equivalent to

$$(I \sqcap pL)w \subset x, \quad (I \sqcap rL)w \subset y, \quad D^{\mathsf{T}}x \sqcup E^{\mathsf{T}}y \subset z, \quad v \subset w, \quad z \subset n.$$

Again, $w$ may be shrunk to $v$ and afterwards $x$ to $p \sqcap v$ and $y$ to $r \sqcap v$. Furthermore, $z$ may be enlarged to $n$ without affecting the existence of a $q$. Indeed, we obtain $D^{\mathsf{T}}(p \sqcap v) \subset n$, $E^{\mathsf{T}}(r \sqcap v) \subset n$. $\qquad\square$

**10.2.4 Proposition** (*Verification rules for loops*).

i) **repeat**-loop

$$\{v\} \textbf{ repeat } D \textbf{ until } p \: \{n\} \quad \Longleftrightarrow \quad \begin{array}{c} \{x\}D\{y\}, \quad p \sqcap y \subset n, \\ \overline{p} \sqcap y \subset x, \quad v \subset x \\ \text{for suitable predicates } x, y; \end{array}$$

ii) **while**-loop

$$\{v\} \textbf{ while } \overline{p} \textbf{ do } D \textbf{ od } \{n\} \quad \Longleftrightarrow \quad \begin{array}{c} \{x\}D\{y\}, \quad p \sqcap y \subset n, \\ \overline{p} \sqcap y \subset x, \quad v \subset y \\ \text{for suitable predicates } x, y. \end{array}$$

**Proof** of i): The relations to be considered are (writing again $p$ instead of $I \sqcap pL$ for reasons of space)

$$B = \begin{pmatrix} O & D & O \\ \overline{p} & O & p \\ O & O & O \end{pmatrix} \qquad e = \begin{pmatrix} I \\ O \\ O \end{pmatrix} \qquad a = \begin{pmatrix} O \\ O \\ I \end{pmatrix} \qquad q = \begin{pmatrix} x \\ y \\ z \end{pmatrix}.$$

From the contraction theorem we obtain

$$\underbrace{(I \sqcap \overline{pL})^{\mathsf{T}}y}_{= \overline{p} \sqcap y} \subset x, \quad D^{\mathsf{T}}x \subset y, \quad \underbrace{(I \sqcap pL)^{\mathsf{T}}y}_{= p \sqcap y} \subset z, \quad v \subset x, \quad z \subset n.$$

Enlarging $z$ to $n$ gives the result. (Note that we could not shrink $x$ to $v$!) $\qquad\square$

For the iterative cases one commonly uses a weaker form which we record as a corollary.

**10.2.5 Corollary** (*Conditions sufficient for partial correctness*).

i) **repeat**-loop

$$\{v\} \textbf{ repeat } D \textbf{ until } p \: \{n\} \quad \Longleftarrow \quad \{\overline{p} \sqcap n\}D\{n\}, \quad \{v\}D\{n\};$$

ii) **while**-loop

$$\{v\} \textbf{ while } \overline{p} \textbf{ do } D \textbf{ od } \{v \sqcap p\} \quad \Longleftarrow \quad \{v \sqcap \overline{p}\}D\{v\}.$$

**Proof:** i) Replace $x := v \sqcup (\overline{p} \sqcap n)$ and $y := n$ in Prop. 10.2.4.i.
ii) Choose $x := \overline{p} \sqcap v$, $y := v$ in Prop. 10.2.4.ii. $\qquad\square$

In both cases any predicate $q$ with components $(x, y, z)$ satisfying those sufficient conditions is called an *invariant* of the loop. When proving partial correctness one typically tries to find such an invariant. Often it is closely connected with the problem, and sometimes one can therefore guess it.

However, the invariant yields a proof of partial correctness only for pre- and postconditions that are related to $p$. Since this relationship ensures that the conditions are "adequate", one is usually content with the simplified verification rules. One will observe that in the **repeat**-case $x$ can be strictly greater than $v$. To see that the simplified verification rules are indeed only sufficient, consider the following program

$$\textbf{repeat } i := i + 1 \textbf{ until } i \geq 4$$

together with the precondition $i \leq 1$ (denoted by $v$) and the postcondition $i > 3$ (denoted by $n$). The correctness proof with respect to $v$ and $n$ must then also take into account the cases $i = 2$ and $i = 3$ in which we have not been interested initially.

We have so far only clarified how to achieve verification by decomposing the program into smaller constituents, such as breaking up compositions of program steps, branchings, or loops. Since the program cannot be cut up indefinitely, we have to provide for at least one "atomic" program step.

Since our programming style is state- and execution-oriented, we choose the assignment statement for the atomic step and assume that an assignment has no side effects. To give an example, consider the assignment $y := x + 3$, the precondition $x \geq 7$, and the postcondition $y \geq 9$. The assertion

$$\{ x \geq 7 \} \, y := x + 3 \, \{ y \geq 9 \}$$

is then seen to be valid, for tracing the postcondition by means of the assignment step leads to $x + 3 \geq 9$ which in turn follows from the precondition.

**10.2.6 Definition** (*Assignment Rule*). The correctness statement

$$\{ S_E^y n \} \, y := E \, \{ n \},$$

holds, where $S_E^y n$ denotes the conjunction of all predicates arising from $n$ by substituting for $y$ one of the values that are results of the (not necessarily deterministic) program step $E$. □

One has to be careful, though, not to formulate the predicates in too liberal a manner. Take for instance two variables, $x$ and $y$, declared as **real** and **integer**, respectively. Since the correctness assertion

$$\{ y \text{ is of type } \textbf{integer} \} \, x := y \, \{ x \text{ is of type } \textbf{integer} \}$$

cannot be accepted as true, the syntax of the language used for formulating the conditions has to be made more precise. But such syntactic considerations are not our purpose here.

Abortion or nontermination may occur when evaluating $E$, as can be seen from the following two correctness assertions which are valid by Definition 10.2.1:

$$\{\, x > 0 \,\} \lceil \textbf{ while } x > 1 \textbf{ do } x := 2 \textbf{ od; } y := 2 \rfloor \{\, y = 2 \,\},$$
$$\{\, |y| < 4 \,\} \, y := \sqrt{y} \, \{\, y \le 2 \,\}.$$

We now show the effect of the contraction principle on a slightly bigger example where the interaction of the predicates is more apparent. Consider the following program, whose effect cannot be understood immediately:

$$
\begin{aligned}
&read(x); \\
&y := x - 1; \\
&z := x; \\
M1: \quad &\textbf{if } y = 0 \textbf{ then goto } M3 \textbf{ fi;} \\
&u := x; \\
M2: \quad &\textbf{if } u \ge y \textbf{ then } u := u - y; \textbf{ goto } M2 \textbf{ fi;} \\
&\textbf{if } u = 0 \textbf{ then } z := z - y \textbf{ fi;} \\
&y := y - 1; \\
&\textbf{goto } M1; \\
M3: \quad &\textbf{if } z = 0 \textbf{ then } w := \text{``yes''} \\
&\qquad\qquad \textbf{else } \ w := \text{``no''} \textbf{ fi;} \\
&print(w).
\end{aligned}
$$

It does not seem to help much knowing that the program decides whether a given number $x$ is *perfect*, i.e., equal to the sum of its proper divisors, such as $6 = 1 + 2 + 3$. At any rate one can try to set up a flowchart program. The complete decomposition into single steps yields the diagram in Fig. 10.2.3 where we also indicate the interpretation of these steps in the state space $Z$; this state space consists of quintuples of the form $\textbf{nat } x, y, u, \textbf{int } z, \textbf{string } w$.

Our translation of texts in programming language into flowdiagrams is an informal one because we have not given a definition of the syntax and semantics. In BERGHAMMER, ZIERER 86 it is shown for applicative languages how to formally define their relational semantics.

The assertion about the effect of program $\mathcal{P}$ reads as follows

$$\{\, x \ge 1 \,\} \quad \mathcal{P} \quad \{\, x = \sum_{d|x,\, d<x} d \implies w = \text{``yes''}, \quad x \ne \sum_{d|x,\, d<x} d \implies w = \text{``no''} \,\}.$$

In other words, if the natural number $x$ entered is positive and the program terminates, then the answer is "yes" precisely when $x$ is perfect. (It is not claimed that the program terminates, for this is a question of total correctness which cannot be answered in terms of partial correctness.)

First we consider the program scheme only and identify the precondition and the post-condition,

$$v := \langle\!\langle x \ge 1 \rangle\!\rangle,$$
$$n := \langle\!\langle F(x,0) = 0 \implies w = \text{``yes''}, \quad F(x,0) \ne 0 \implies w = \text{``no''} \rangle\!\rangle,$$

using

$$F(x,y) = x - \sum_{d|x,\, y<d<x} d.$$

By introducing another parameter $y$ we have generalized the problem of verification, a method called *embedding* in the development and verification of programs. We now consider not only the sum of all divisors but the sum of divisors greater than $y$, the original problem corresponding to the case $y = 0$.

$$
B = \bordermatrix{
  & a & b & c & d & e & f & g & h & i & j & k & l & m \cr
a & O & O & O & O & O & O & O & O & O & O & O & O & O \cr
b & O & O & K & O & O & O & O & O & O & O & O & O & O \cr
c & O & O & O & \bar{p} & O & O & O & O & O & O & p & O & O \cr
d & O & O & O & O & O & O & C & O & O & O & O & O & O \cr
e & O & A & O & O & O & O & O & O & O & O & O & O & O \cr
f & O & O & O & O & O & O & D & O & O & O & O & O & O \cr
g & O & O & O & O & O & t & O & \bar{t} & O & O & O & O & O \cr
h & O & O & O & O & O & O & O & O & \bar{r} & r & O & O & O \cr
i & O & O & F & O & O & O & O & O & O & O & O & O & O \cr
j & O & O & O & O & O & O & O & O & E & O & O & O & O \cr
k & O & O & O & O & O & O & O & O & O & O & O & \bar{s} & s \cr
l & G & O & O & O & O & O & O & O & O & O & O & O & O \cr
m & H & O & O & O & O & O & O & O & O & O & O & O & O \cr
}
\qquad
q = \begin{pmatrix} a \\ b \\ c \\ d \\ e \\ f \\ g \\ h \\ i \\ j \\ k \\ l \\ m \end{pmatrix}
$$

$$
\begin{aligned}
A &\approx y := x - 1 \\
K &\approx z := x \\
C &\approx u := x \\
D &\approx u := u - y \\
E &\approx z := z - y \\
F &\approx y := y - 1 \\
G &\approx w := \text{``no''} \\
H &\approx w := \text{``yes''} \\
p &:= \langle\!\langle y = 0 \rangle\!\rangle \\
t &:= \langle\!\langle u \geq y \rangle\!\rangle \\
r &:= \langle\!\langle u = 0 \rangle\!\rangle \\
s &:= \langle\!\langle z = 0 \rangle\!\rangle
\end{aligned}
$$

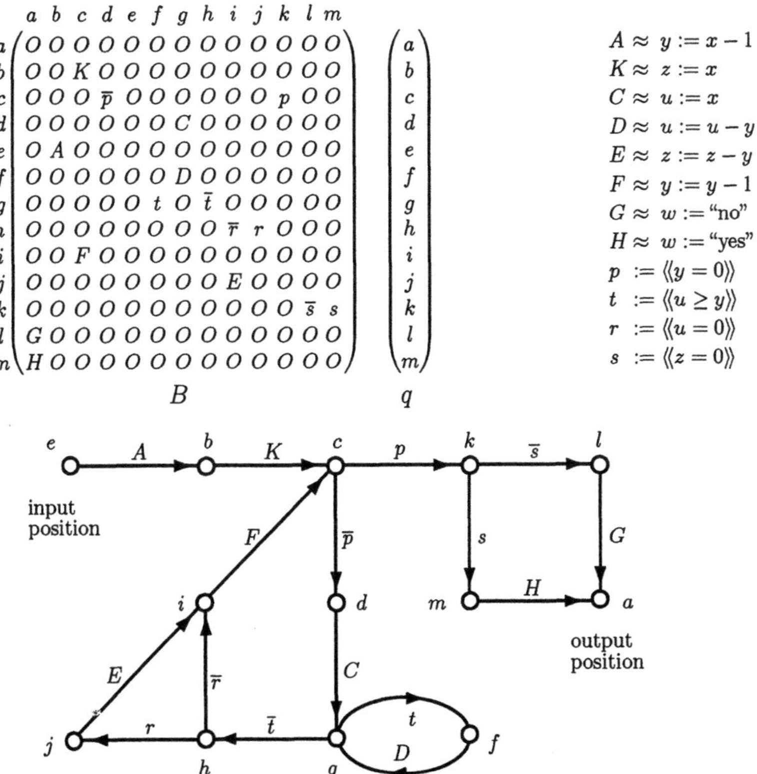


**Fig. 10.2.3** Interpreted program scheme for checking
numbers for perfectness



**Fig. 10.2.4** Contracted predicate for Fig. 10.2.3

By the contraction theorem, the correctness assertion $\{v\}\mathcal{P}\{n\}$ holds if and only if the predicate $q$ on the situations can be found. We consider a general predicate of the type required and denote the predicates for the situations over the points $a,\ldots,m$ again by $a,\ldots,m$, for the sake of simplicity.

From Proposition 10.2.2 then follow the conditions

$$
\begin{array}{lll}
G^{\mathsf{T}}l \sqcup H^{\mathsf{T}}m \subset a, & A^{\mathsf{T}}e \subset b, & K^{\mathsf{T}}b \sqcup F^{\mathsf{T}}i \subset c, \\
\overline{p} \sqcap c \subset d, & t \sqcap g \subset f, & C^{\mathsf{T}}d \sqcup D^{\mathsf{T}}f \subset g, \\
\overline{t} \sqcap g \subset h, & (\overline{r} \sqcap h) \sqcup E^{\mathsf{T}}j \subset i, & r \sqcap h \subset j, \\
p \sqcap c \subset k, & \overline{s} \sqcap k \subset l, & s \sqcap k \subset m,
\end{array}
$$

together with $v \subset e$ and $a \subset n$. (Again one could let $e$ shrink to $v$ and enlarge $a$ to $n$ in order to reduce the problem somewhat.)

Finding the other predicates amounts to finding a loop invariant. Since we have not systematically developed the syntax for formulating those predicates, we shall not pursue this finding process; in addition to the formal relationships established, it requires some intuition. Figure 10.2.4 indicates these predicates by inscribing them in the box corresponding to the respective position.

In order to complete the proof of the correctness statement $\{v\}\mathcal{P}\{n\}$ one has to verify all the above conditions between those predicates. (Incidentally, it is irrelevant for correctness to hold whether the predicates $a,\ldots,m$ have been calculated or guessed.) Checking the conditions presents various degrees of difficulty; we demonstrate three cases:

The first condition $G^{\mathsf{T}}l \sqcup H^{\mathsf{T}}m \subset a$, together with $a \subset n$, can be split up into $G^{\mathsf{T}}l \subset n$ and $H^{\mathsf{T}}m \subset n$, which can be checked separately and retranslated into the correctness statements $\{l\}G\{n\}$ and $\{m\}H\{n\}$ for steps $G$ and $H$. In the following interpreted form the validity of these assertions can be seen. In order to facilitate the check we have indicated by underbracing what program step and postcondition yield according to the assignment rule 10.2.6.

$$
\{\,F(x,0) \neq 0\,\} \quad \underbrace{w := \text{``no''} \quad \left\{
\begin{array}{l}
F(x,0) = 0 \implies w = \text{``yes''}, \\
F(x,0) \neq 0 \implies w = \text{``no''}
\end{array}
\right\}}_{\begin{array}{l} F(x,0) = 0 \implies \text{``no''} = \text{``yes''}, \\ F(x,0) \neq 0 \implies \text{``no''} = \text{``no''} \end{array}}
$$

$$
\{\,F(x,0) = 0\,\} \quad \underbrace{w := \text{``yes''} \quad \left\{
\begin{array}{l}
F(x,0) = 0 \implies w = \text{``yes''}, \\
F(x,0) \neq 0 \implies w = \text{``no''}
\end{array}
\right\}}_{\begin{array}{l} F(x,0) = 0 \implies \text{``yes''} = \text{``yes''}, \\ F(x,0) \neq 0 \implies \text{``yes''} = \text{``no''} \end{array}}
$$

Indeed, every precondition implies the conjunction underbraced.

We now check $K^{\mathsf{T}}b \sqcup F^{\mathsf{T}}i \subset c$ in the same way:

$$
\{\,x - 1 = y \geq 0\,\} \quad \underbrace{z := x \quad \{\,x > y \geq 0,\ z = F(x,y)\,\}}_{x > y \geq 0,\quad x = F(x,y)}
$$

$$\left\{ \begin{array}{c} z = F(x,y),\ y \mid x - u,\ y > u > 0 \\ \text{or} \\ z = F(x,y) - y,\ y \mid x,\ y \geq 1 \end{array} \right\} \underbrace{y := y - 1\ \{\, x > y \geq 0,\ z = F(x,y)\,\}.}_{x > y - 1 \geq 0,\quad z = F(x, y - 1)}$$

Again the result of applying the assignment rule at the underbraced place is indicated below. In the first case we have indeed

$$F(x, x - 1) = x - \sum_{d \mid x,\ x - 1 < d < x} d = x - 0 = x.$$

If in the second case $y$ divides $x$ with remainder $0 < u < y$, we have

$$F(x, y) = x - \sum_{d \mid x,\ y < d < x} d = x - \sum_{d \mid x,\ y \leq d < x} d = F(x, y - 1);$$

if, on the other hand, $y$ divides $x$, then $F(x, y) = F(x, y - 1) + y$.

Finally, we check $(\bar{r} \sqcap h) \sqcup E^{\mathsf{T}} j \subset i$ where two different forms occur together:

$$\begin{array}{c} u > 0,\ z = F(x,y),\ y \geq 1, \\ y > u,\ y \mid x - u \end{array} \implies \begin{array}{c} z = F(x,y),\ y \mid x - u,\ y > u > 0 \\ \text{or} \\ z = F(x,y) - y,\ y \mid x,\ y \geq 1 \end{array}$$

$$\left\{ \begin{array}{c} z = F(x,y),\ y \geq 1, \\ u = 0,\ y \mid x - u \end{array} \right\} \quad z := z - y \quad \left\{ \begin{array}{c} z = F(x,y),\ y \mid x - u,\ y > u > 0 \\ \text{or} \\ z = F(x,y) - y,\ y \mid x,\ y \geq 1, \end{array} \right\}$$

$$\underbrace{\phantom{z := z - y \left\{ z = F(x,y),\ y \mid x - u,\ y > u > 0 \right\}}}$$

$$\begin{array}{c} z - y = F(x,y),\ y \mid x - u,\ y > u > 0 \\ \text{or} \\ z - y = F(x,y) - y,\ y \mid x,\ y \geq 1 \end{array}$$

In the first case, the upper variant on the right comes to bear, while in the second case, the lower.

## Exercises

**10.2.1**  Prove Proposition 10.2.3.ii.

**10.2.2**  Prove Proposition 10.2.4.ii.

**10.2.3**  Check the remaining cases of Fig. 10.2.4 in order to establish a contracted predicate.

# 10.3 Total Correctness and Termination

Partial correctness alone does not yet suffice to assess programs with precision, since the effect $\Sigma$ does not take into account possible infinite loops or premature termination (abortion). The statement $\{v\}\mathcal{P}\{n\}$ means that

> "*if* program $\mathcal{P}$ is started in some state with precondition $v$
> *and* if the computation sequence terminates normally,
> *then* the final state satisfies postcondition $n$".

Of greater interest would be the following:

> "*If* program $\mathcal{P}$ is started in some state with precondition $v$,
> *then* all computation sequences terminate normally
> *and* the final states satisfy postcondition $n$".

Questions of the latter kind can generally no longer be treated by means of the effect. But we recall the concepts developed in connection with progressive finiteness and boundedness (Sect. 6.3), in particular the initial part (cf. Def. 6.3.2.i). They serve for characterizing those states before starting the program for which each computation sequence (we are studying the general nondeterministic case) has finite length, or for which all such sequence lengths have a common upper bound. Also, a condition is to be found ensuring that the terminating action $C$ terminates only in output states satisfying postcondition $n$. The effect cannot be used for this purpose because of the possibility of abortion.

With Definition 6.3.2 in mind we first specify the concepts of termination and correctness.

**10.3.1 Definition.** Let the program $\mathcal{P} = (G, S, \Theta, e, a)$ be given with associated relation $B$, terminating action $C := B^* \sqcap \overline{BL}^\mathsf{T}$, initial part $J(B)$, and effect $\Sigma := e^\mathsf{T} Ca$. If $v$ is a pre- and $n$ is a postcondition, we call $\mathcal{P}$ **totally correct with respect to** $v$ **and** $n$

> **of type 1,** if $v \subset e^\mathsf{T} \sup_{h \geq 0} \overline{B^h L} \sqcap e^\mathsf{T} \overline{C \overline{an}}$;
>
> **of type 2,** if $v \subset e^\mathsf{T} J(B) \sqcap e^\mathsf{T} \overline{C \overline{an}}$;
>
> **of type 3,** if $v \subset \Sigma n$.

We say that $\mathcal{P}$ **terminates** for $v$ **of type** $i$, if $\mathcal{P}$ is totally correct of type $i$ with respect to $v$ and $L$.     □

Saying that $\mathcal{P}$ is correct of type 1 with respect to $v$ and $n$ means the following: Given a state satisfying $v$ before entering into the program, there are no computation sequences of arbitrary length. Moreover, with the terminating action $C$ only states that satisfy postcondition $n$ after output by $a$ can be reached (because abortion is avoided and no state with $\overline{n}$ can be reached).

Total correctness of type 2 does not require an upper bound for the lengths of all computation sequences starting with an input state satisfying $v$. Each individual computation sequence must be finite, however.

For correctness of type 3, it suffices that for each input state with $v$ there is *at least* one terminating computation sequence leading to an output state with $n$. Therefore, $\Sigma L$ describes the domain of definition of the effect of the program.

In order to distinguish termination of type 1 or type 2 from termination of type 3, "supernatural" expressions have been used. Nondeterminism of type 3 is called **angelic** because it selects only the terminating computation sequences and discards others that abort or enter a loop. Nondeterminism of types 1 or 2, on the other hand, is called **demonic** because it would seize any opportunity of terminating prematurely or repeating indefinitely.

Types 1 and 2 will not be distinguished by such striking names. The difference between total correctness of type 1 and type 2 is often overlooked. According to Proposition 6.3.5, both types coincide for finite branching—which is usually the case in our programs. Only for programs with nonfinite branching do the two concepts differ. Consider the following program $\mathcal{P}$; it contains **some nat** and therefore the nonfinitely branching process of choosing some natural number.

$$m := \textbf{some nat};$$
$$\textbf{while } m > 0 \textbf{ do } m := m - 1 \textbf{ od}.$$

Clearly, $\mathcal{P}$ stops after finitely many steps, but there is no bound for the number of steps. So it is totally correct of type 2, but *not* of type 1, with respect to $v := L$ and $n := \langle\!\langle m = 0 \rangle\!\rangle$.

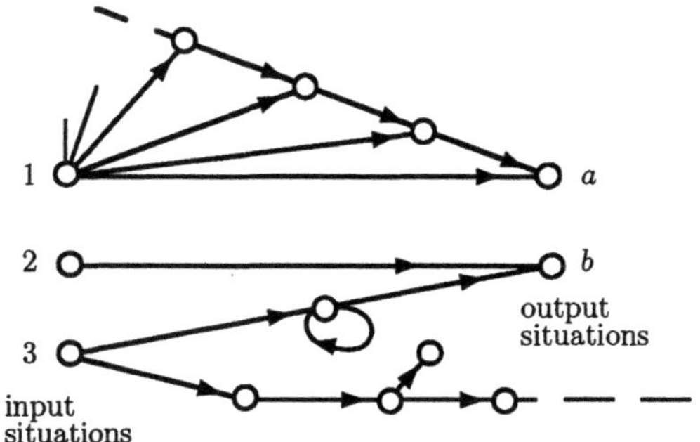


**Fig. 10.3.1** Total correctness

Total correctness is illustrated in Fig. 10.3.1. The graph offers the possibility of abortion, of nonfinite branching, and of nonterminating runs, due to loops, or simply by following an infinite path over an infinite sequence of different points. Input and output situations have been indicated so as to make this graph the situation graph of a program.

To give some examples, the program is totally correct with respect to the precondition $v$ and the postcondition $n \approx \{a, b\}$ in the following cases

| | |
|---|---|
| $v \approx \{2\}$ | of type 1 |
| $v \approx \emptyset, \quad v \approx \{1\}, \quad v \approx \{2\}, \quad v \approx \{1,2\}$ | of type 2 |
| $v \approx \{1,2,3\}$ | of type 3 |

For type 1, point 1 is excluded because it is not progressively bounded; for type 2, point 3 is excluded because it is not progressively finite. In the case of type 2, the set $\{1,2\}$ is the biggest of the four sets $v$ and corresponds to the *weakest* precondition such that $\mathcal{P}$ is totally correct for $v$ and $n$. This aspect will be discussed in Sect. 10.4. We prove two elementary properties.

**10.3.2 Proposition.** i)   For every program $\mathcal{P}$ we have:

$\mathcal{P}$ is totally correct with respect to $v$ and $n$

of type 1   $\implies$   of type 2   $\implies$   of type 3.

ii)   If the program $\mathcal{P}$ is deterministic, then all three types are equivalent:

$\mathcal{P}$ is totally correct with respect to $v$ and $n$

of type 1   $\iff$   of type 2   $\iff$   of type 3.

**Proof:** i) By Proposition 6.3.3.i we have

$$e^\mathsf{T} \sup\nolimits_{h\geq 0} \overline{B^h L} \sqcap e^\mathsf{T} \overline{C\,\overline{an}} \subset e^\mathsf{T} J(B) \sqcap e^\mathsf{T} \overline{C\,\overline{an}} \subset e^\mathsf{T} B^* \overline{BL} \sqcap e^\mathsf{T} \overline{C\,\overline{an}},$$

where the last expression is contained in $e^\mathsf{T} Can = \Sigma n$ since we may conclude, using Prop. 2.4.2, that

$$L = \overline{CL} \sqcup CL = \overline{(B^* \sqcap \overline{BL}^\mathsf{T})L \sqcup C\,\overline{an}} \sqcup Can = \overline{B^* \overline{BL} \sqcap \overline{C\,\overline{an}}} \sqcup Can.$$

ii) Applying Proposition 6.3.3.ii we immediately obtain the first equivalence. For the second to hold, only "$\Longleftarrow$" must be shown: $Can \subset CL = B^* \overline{BL}$ and, $C^\mathsf{T} C \subset I \implies C^\mathsf{T} C\,\overline{an} \subset \overline{an} \iff Can \subset \overline{C\,\overline{an}}$.  □

Total correctness of type 3 is the least complicated to define, and it applies to deterministic programs for which all three variants indeed coincide by Proposition 10.3.2. Nondeterministic programs are much harder to handle, as is reflected by the three differing concepts of total correctness.

We next investigate when total correctness amounts to the conjunction of partial correctness and termination. This is, however, only the case for total correctness of types 1 and 2. Total correctness of type 3 behaves differently from the two other types.

**10.3.3 Proposition.** For $i = 1, 2$, we have

$$\begin{array}{ccc} \mathcal{P} \text{ totally correct of type } i & & \mathcal{P} \text{ partially correct w. r. t. } v \text{ and } n, \\ \text{with respect to } v \text{ and } n & \Longleftrightarrow & \text{and } \mathcal{P} \text{ terminates of type } i \text{ for } v. \end{array}$$

This does *not* hold for $i = 3$ where we only have "$\Longleftarrow$".

The **proof** will be given simultaneously for both cases. We show

$$v \subset e^\mathsf{T} \overline{C\,\overline{an}} \quad \Longleftrightarrow \quad v \subset \overline{\Sigma \overline{n}} \text{ and } v \subset e^\mathsf{T} \overline{CaL}$$

by proving

$$e^\mathsf{T} \overline{C\,\overline{an}} = \overline{\Sigma \overline{n}} \sqcap e^\mathsf{T} \overline{CaL}.$$

By Proposition 4.2.4, the mapping $e^\mathsf{T}$ may be shifted underneath the bar from the left. So negating the preceding line yields

$$e^\mathsf{T} C\,\overline{an} = e^\mathsf{T} Ca\overline{n} \sqcup e^\mathsf{T} CaL.$$

This, however, is a true statement because for any univalent $a$ one has $\overline{an} = a\overline{n} \sqcup \overline{aL}$ by Proposition 4.2.2.v. For $i = 3$ we have in general

$$v \subset \Sigma n \quad \Longleftarrow \quad v \subset \overline{\Sigma \overline{n}} \text{ and } v \subset \Sigma L.$$

According to Proposition 4.2.2.v, the converse of that implication is a consequence of the univalence of $\Sigma$, which does not hold in general.  □

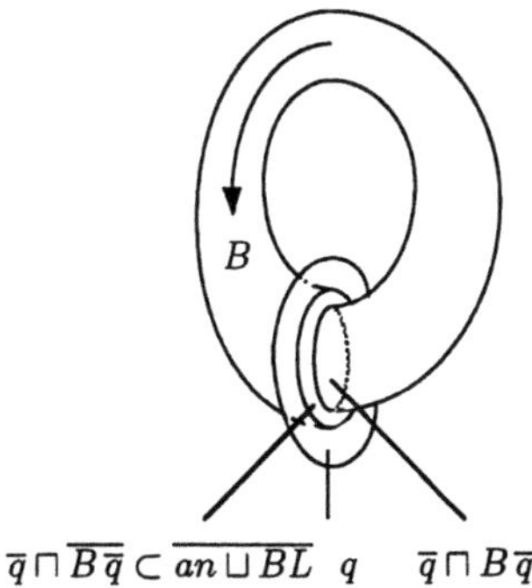

$$\overline{q} \sqcap \overline{B\overline{q}} \subset \overline{an \sqcup BL} \quad q \quad \overline{q} \sqcap B\overline{q}$$

**Fig. 10.3.2** Complement-expansion

As for partial correctness we ask whether total correctness can be directly read off the program, i.e., the associated relation $B$, or whether the derived construct $e^{\mathsf{T}} J(B) \sqcap e^{\mathsf{T}} \overline{C\,\overline{an}}$ of Def. 10.3.1 is unavoidable. What we are looking for is an analog, for total correctness, of the contraction theorem 10.2.2. This will be established in Props. 10.3.4 and 10.3.5 and used in Sect. 10.4 for investigating various constructs occurring in programming languages.

**10.3.4 Proposition** (*Complement-Expansion Theorem*). Let $\mathcal{P}$ be a program with associated relation $B$, input relation $e$, and output relation $a$. If $v$ is a pre- and $n$ is a postcondition, the following statements are equivalent:

$$\begin{array}{c} \mathcal{P} \text{ totally correct of type 2} \\ \text{with respect to } v,\, n \end{array} \iff \begin{array}{c} \text{For all predicates } q, \\ \overline{q} \subset B\overline{q} \sqcup \overline{an \sqcup BL} \text{ implies } v \subset e^{\mathsf{T}} q. \end{array}$$

**Proof:** The crucial point of the proof is the identity

$$J(B) \sqcap \overline{C\,\overline{an}} = \inf\{\, q \mid \overline{q} \subset B\overline{q} \sqcup \overline{an \sqcup BL}\,\}$$

which we shall prove afterwards. Applying it to Definition 10.3.1, we have total correctness for $\mathcal{P}$ of type 2 with respect to $v$ and $n$ precisely when

$$v \subset e^{\mathsf{T}} \inf\{\, q \mid \overline{q} \subset B\overline{q} \sqcup \overline{an \sqcup BL}\,\}.$$

Since $e^{\mathsf{T}}$ is univalent, multiplication by it from the left is $\sqcap$-distributive. Therefore, we have total correctness for $\mathcal{P}$ of type 2 with respect to $v$ and $n$ precisely when $v \subset e^{\mathsf{T}} q$ for *all* predicates $q$ of the selected kind.

Abbreviating $F := \overline{an \sqcup BL}$, we now prove the identity we have used above:

$$\inf\{\, x \mid \overline{x} \subset B\overline{x}\,\} \sqcap \overline{C\,\overline{an}} = \inf\{\, q \mid \overline{q} \subset B\overline{q} \sqcup F\,\}.$$

For any $x$ satisfying $\overline{x} \subset B\overline{x}$ the predicate $q := x \sqcap \overline{C\,\overline{an}} = x \sqcap \overline{B^*F}$ is contained in $x$ and satisfies $B\overline{q} \sqcup F = B\overline{x} \sqcup BB^*F \sqcup F \supset \overline{x} \sqcup B^*F = \overline{q}$; therefore "$\supset$". On the other hand, for any $q$ with $\overline{q} \subset B\overline{q} \sqcup F$ we may find an $x := \overline{B^*(\overline{q} \sqcap \overline{B^*F})}$, such that

$$\overline{x} = B^*(\overline{q} \sqcap \overline{B^*F}) = (\overline{q} \sqcap \overline{B^*F}) \sqcup BB^*(\overline{q} \sqcap \overline{B^*F})$$

$$\subset [(B\overline{q} \sqcup F) \sqcap \overline{B^*F}] \sqcup B\overline{x} = (B\overline{q} \sqcap \overline{B^*F}) \sqcup B\overline{x}$$

$$= ([B(\overline{q} \sqcap \overline{B^*F}) \sqcup B(\overline{q} \sqcap B^*F)] \sqcap \overline{B^*F}) \sqcup B\overline{x}$$

$$\subset [BB^*(\overline{q} \sqcap \overline{B^*F}) \sqcap \overline{B^*F}] \sqcup (B^+F \sqcap \overline{B^*F}) \sqcup B\overline{x}$$

$$\subset BB^*(\overline{q} \sqcap \overline{B^*F}) \sqcup O \sqcup B\overline{x} = B\overline{x}.$$

Because $x = \overline{B^*(\overline{q} \sqcap \overline{B^*F})} \subset \overline{I(\overline{q} \sqcap \overline{B^*F})} = q \sqcup B^*F$, we have $x \sqcap \overline{C\,\overline{an}} = x \sqcap \overline{B^*F} \subset q$ and therefore "$\subset$". □

We compare this theorem to the contraction theorem 10.2.2:

| $\mathcal{P}$ partially correct with respect to $v$, $n$ | $\Longleftrightarrow$ | There exists a predicate $q$ with $B^\mathsf{T}q \subset q$, $a^\mathsf{T}(q \sqcap \overline{BL}) \subset n$, and $v \subset e^\mathsf{T}q$. |

| $\mathcal{P}$ totally correct of type 2 with respect to $v$, $n$ | $\Longleftrightarrow$ | For all predicates $q$, $\overline{q} \subset B\overline{q} \sqcup \overline{an} \sqcup \overline{BL}$ implies $v \subset e^\mathsf{T}q$. |

When $B\overline{q} \subset \overline{q}$ (which is equivalent to $B^\mathsf{T}q \subset q$), then $q$ is being contracted. For total correctness one has to consider the reverse inclusion (with an exceptional rule for terminal points) and change the quantifiers.

We recall Proposition 6.3.4 which indicates how to get the progressively bounded points by some exhaustion process. There, *all* the terminal points have to be taken into account, whereas here only those terminal output states play a role which satisfy the postcondition $n$ after execution of the program.

**10.3.5 Proposition.** Let $\mathcal{P}$ be a program with associated relation $B$, input relation $e$, and output relation $a$. Given a precondition $v$ and a postcondition $n$, the following are equivalent:

| $\mathcal{P}$ totally correct of type 1 with respect to $v$, $n$ | $\Longleftrightarrow$ | The limit of the iteration $\overline{q_{-1}} := L;\quad \overline{q_i} := B\overline{q_{i-1}} \sqcup \overline{an} \sqcup \overline{BL},\quad i \geq 0$ satisfies $v \subset e^\mathsf{T} \sup_{i \geq 0} q_i$. |

**Proof:** As in Proposition 10.3.4 we just prove the identity

$$\sup\nolimits_{h \geq 0} \overline{B^h L} \sqcap \overline{C\,\overline{an}} = \sup\nolimits_{i \geq 0} q_i.$$

By Definition 10.3.1, $\mathcal{P}$ is then totally correct of type 1 with respect to $v$ and $n$ precisely when $v \subset e^\mathsf{T} \sup_{i \geq 0} q_i$.

Again, we abbreviate $F := \overline{an \sqcup BL}$ and show that for arbitrary $r \geq -1$

$$\overline{C\,\overline{an}} = \overline{B^* \overline{an \sqcup BL}} = \overline{B^* F} = \overline{\sup\nolimits_{0 \leq m \leq r} B^m F \sqcup B^{r+1}B^* F}$$
$$= \inf\nolimits_{0 \leq m \leq r} \overline{B^m F} \sqcap \overline{B^{r+1}C\,\overline{an}}.$$

So we have the following decomposition

$$\sup\nolimits_{h \geq 0} \overline{B^h L} \sqcap \overline{C\,\overline{an}} = \sup\nolimits_{s \geq -1}(\overline{B^{s+1}L} \sqcap \overline{C\,\overline{an}})$$
$$= \sup\nolimits_{s \geq -1}(\overline{B^{s+1}L} \sqcap \inf\nolimits_{0 \leq m \leq s} \overline{B^m F} \sqcap \overline{B^{s+1}C\,\overline{an}})$$
$$= \sup\nolimits_{s \geq -1}(\overline{B^{s+1}L} \sqcap \inf\nolimits_{0 \leq m \leq s} \overline{B^m F}).$$

This is a monotone ascending sequence whose parts coincide with $q_s$. This is obviously true for $q_{-1} = O$; furthermore, we obtain by induction,

$$\overline{q_{i+1}} = B\overline{q_i} \sqcup F = B(B^{i+1}L \sqcup \sup\nolimits_{0 \leq m \leq i} B^m F) \sqcup F = B^{i+2}L \sqcup \sup\nolimits_{0 \leq h \leq i+1} B^h F.$$

$$\square$$

We illustrate this result by means of the example of perfect numbers from Sect. 10.2.

Correctness of type 2 holds when the following conditions, which differ from those in Sect. 10.2, are satisfied:

$$\overline{a} \subset \overline{n} \qquad\qquad \overline{h} \subset (\overline{r} \sqcap \overline{i}) \sqcup (r \sqcap \overline{j}) \sqcup O$$

$$\overline{b} \subset K\overline{c} \sqcup \overline{KL} \qquad\qquad \overline{i} \subset F\overline{c} \sqcup \overline{FL}$$

$$\overline{c} \subset (\overline{p} \sqcap \overline{d}) \sqcup (p \sqcap \overline{k}) \sqcup O \qquad\qquad \overline{j} \subset E\overline{i} \sqcup \overline{EL}$$

$$\overline{d} \subset C\overline{g} \sqcup \overline{CL} \qquad\qquad \overline{k} \subset (\overline{s} \sqcap \overline{l}) \sqcup (s \sqcap \overline{m}) \sqcup O$$

$$\overline{e} \subset A\overline{b} \sqcup \overline{AL} \qquad\qquad \overline{l} \subset G\overline{a} \sqcup \overline{GL}$$

$$\overline{f} \subset D\overline{g} \sqcup \overline{DL} \qquad\qquad \overline{m} \subset H\overline{a} \sqcup \overline{HL}$$

$$\overline{g} \subset (t \sqcap \overline{f}) \sqcup (\overline{t} \sqcap \overline{h}) \sqcup O.$$

According to the interpretation of the program scheme indicated in Fig. 10.2.3 we must prove that for *all* predicates with these properties we have $v \subset e$. Knowing the sole predicate $q$ from Fig. 10.2.4, clearly, is not sufficient for completing the proof. It requires results from the next section.


## Dynamic Logic

Dynamic logic is a branch of modal logic. In works on this subject related to programming, the operators $\square$ (box) and $\Diamond$ (diamond) are used; see MANNA 82 and HAREL 79, for example. Given the associated relation $B$ of a program $\mathcal{P}$ with effect $\Sigma$, together with a postcondition $n$, one defines (cf. BERGHAMMER, SCHMIDT 82)

$[\mathcal{P}]n := \overline{\Sigma\overline{n}},$ to characterize those states before entering the program for which each terminating but nonaborting execution of the (nondeterministic) program leads to states satisfying $n$,

$<\mathcal{P}>n := \Sigma n,$ to characterize those states before entering the program for which there is at least one terminating and nonaborting execution of the program leading to $n$,

$\mathrm{loop}_{\mathcal{P}} := e^{\mathsf{T}} \overline{J(B)},$ for characterizing those states which are the starting points of infinite computation sequences,

$\mathrm{fail}_{\mathcal{P}} := e^{\mathsf{T}} \overline{CaL},$ to characterize those states which can lead to abortion of the program.

Using quantifiers, the first two cases can also be written as

$$s \in [\mathcal{P}]n \;\; := \;\; \forall t : (s,t) \in \Sigma \to t \in n,$$
$$s \in <\mathcal{P}>n \;\; := \;\; \exists t : (s,t) \in \Sigma \wedge t \in n.$$

We give a typical relationship between these constructs. Its proof and interpretation are straightforward.

**10.3.6 Proposition.** i)    $<\mathcal{P}> n = [\mathcal{P}]n \sqcap <\mathcal{P}> L,$    if $\mathcal{P}$ is deterministic.

ii)    $$[\mathcal{P}; \mathcal{Q}]n = [\mathcal{P}]([\mathcal{Q}]n). \qquad\qquad \square$$

The treatment of recursion by relation-algebraic methods can be found in, e.g., SANDERSON 80. Jumps and **goto** are investigated in ARBIB, ALAGIĆ 79, and CLINT, HOARE 72.

## Exercises

**10.3.1**   Prove Proposition 10.3.6.

**10.3.2**   Show that $\text{fail}_{\mathcal{P}} \subset \text{fail}_{\mathcal{P};\mathcal{Q}}$.

## 10.4 Weakest Preconditions

In Sect. 10.2 we defined the assertion $\{v\}\mathcal{P}\{n\}$ with regard to partial correctness as $\Sigma^{\mathsf{T}} v \subset n$. Given $v$ and $n$, we then considered $n_{\text{inf}} := \Sigma^{\mathsf{T}} v$, the strongest postcondition for $v$, and $v_{\text{sup}} := \overline{\Sigma \overline{n}}$, the weakest precondition for $n$, under which partial correctness still holds.

When dealing with total correctness one typically tries to find weakest preconditions in the presence of a given postcondition. There are three possibilities of doing this, according to the three types of total correctness which we termed types 1, 2, and 3. We shall not consider the last variant and rather concentrate on the first two. We incorporate the results of Propositions 10.3.4 and 10.3.5 into the following definition.

**10.4.1 Definition.**   For a program $\mathcal{P} = (G, S, \Theta, e, a)$ with associated relation $B$ and some postcondition $n$ we define as the **weakest precondition** of $\mathcal{P}$ with respect to $n$

i)   **of type 1,**   $\text{wp}_1(\mathcal{P}, n) := e^{\mathsf{T}} \sup_{h \geq 0} \overline{B^h L} \sqcap e^{\mathsf{T}} \overline{C \overline{an}} = e^{\mathsf{T}} \sup_{i \geq 0} q_i,$

$$\text{where } \overline{q_{-1}} := L \text{ and } \overline{q_i} := B \overline{q_{i-1}} \sqcup \overline{an} \sqcup \overline{BL} \text{ for } i \geq 0;$$

ii)   **of type 2**   $\text{wp}_2(\mathcal{P}, n) := e^{\mathsf{T}} J(B) \sqcap e^{\mathsf{T}} \overline{C \overline{an}}$

$$= e^{\mathsf{T}} \inf \{ q \mid \overline{q} \subset B \overline{q} \sqcup \overline{an} \sqcup \overline{BL} \}. \qquad \square$$

$\text{wp}_1(\mathcal{P}, L)$ and $\text{wp}_2(\mathcal{P}, L)$ describe just those states before entering the program from which all computation sequences terminate and none aborts. The terminology $\text{wp}_1$ and $\text{wp}_2$ is to indicate that the expressions have the properties of a weakest precondition in the sense of E. W. Dijkstra. We write $\text{wp}$ for $\text{wp}_1$ and $\text{wp}_2$ in case these concepts coincide or their proofs follow the same pattern.

Next we calculate the weakest preconditions for a number of program constructs. The simplest case consists of a single program step $D$, and the associated relation of the program is then $B = \left( \begin{smallmatrix} O & D \\ O & O \end{smallmatrix} \right)$. Clearly, types 1 and 2 of total correctness coincide in this case, and one obtains the following weakest precondition:

$$\mathtt{wp}(D,n) = (I\ O)\,\inf\left\{\begin{pmatrix} x \\ y \end{pmatrix} \,\middle|\, \begin{pmatrix} \overline{x} \\ \overline{y} \end{pmatrix} \subset \begin{pmatrix} O & D \\ O & O \end{pmatrix} \begin{pmatrix} \overline{x} \\ \overline{y} \end{pmatrix} \sqcup \overline{\begin{pmatrix} O \\ n \end{pmatrix}} \sqcup \overline{\begin{pmatrix} DL \\ O \end{pmatrix}}\right\}$$

$$= (I\ O)\,\inf\left\{\begin{pmatrix} x \\ y \end{pmatrix} \,\middle|\, \begin{pmatrix} \overline{D\overline{y}} \sqcap DL \\ n \end{pmatrix} \subset \begin{pmatrix} x \\ y \end{pmatrix}\right\} = (I\ O)\begin{pmatrix} \overline{D\overline{n}} \sqcap DL \\ n \end{pmatrix}$$

$$= \overline{D\overline{n}} \sqcap DL.$$

As one would expect, the program step $D$ is totally correct with respect to the weakest precondition $\mathtt{wp}(D,n)$ and $n$, provided $D$ is defined for each state with $\mathtt{wp}(D,n)$ and does not lead to some state violating $n$. This has already been the proof of part (i) of the following proposition.

**10.4.2 Proposition** (*Weakest Precondition for Composition*).

i)   If there is just one step $D$ in a program, we have
$$\mathtt{wp}(D,n) = \overline{D\overline{n}} \sqcap DL.$$

ii)   If the program consists of two program steps $D$ and $E$, we have
$$\mathtt{wp}\,((D;E),n) = \mathtt{wp}\,(D,\mathtt{wp}(E,n)).$$

**Proof** of ii): If program steps $D$ and $E$ are to be executed one after the other, $B = \begin{pmatrix} O & D & O \\ O & O & E \\ O & O & O \end{pmatrix}$ is the associated relation, and one obtains

$$\mathtt{wp}\,((D;E),n) =$$

$$= (I\ O\ O)\,\inf\left\{\begin{pmatrix} x \\ y \\ z \end{pmatrix} \,\middle|\, \begin{pmatrix} \overline{x} \\ \overline{y} \\ \overline{z} \end{pmatrix} \subset \begin{pmatrix} O & D & O \\ O & O & E \\ O & O & O \end{pmatrix} \begin{pmatrix} \overline{x} \\ \overline{y} \\ \overline{z} \end{pmatrix} \sqcup \overline{\begin{pmatrix} O \\ O \\ n \end{pmatrix}} \sqcup \overline{\begin{pmatrix} DL \\ EL \\ O \end{pmatrix}}\right\}$$

$$= (I\ O\ O)\,\inf\left\{\begin{pmatrix} x \\ y \\ z \end{pmatrix} \,\middle|\, \begin{pmatrix} \overline{D\overline{y}} \sqcap DL \\ \overline{E\overline{z}} \sqcap EL \\ n \end{pmatrix} \subset \begin{pmatrix} x \\ y \\ z \end{pmatrix}\right\} = (I\ O\ O)\begin{pmatrix} \overline{D\overline{E\overline{n}} \sqcap EL} \sqcap DL \\ \overline{E\overline{n}} \sqcap EL \\ n \end{pmatrix}$$

$$= \overline{D\overline{E\overline{n}} \sqcap EL} \sqcap DL = \overline{D\overline{\mathtt{wp}(E,n)}} \sqcap DL = \mathtt{wp}\,(D,\mathtt{wp}(E,n)). \qquad \square$$

It is now easily seen that for several program steps, performed one after the other, the weakest precondition can be determined step by step in the reverse order, starting from the end.

Our next result is the set of rules for weakest preconditions postulated by Dijkstra for $\mathtt{wp}$.

**10.4.3 Proposition.** For weakest preconditions of type 1 and type 2 we have

i)    $\mathtt{wp}\,(skip,n) = n;$
ii)   $\mathtt{wp}\,(abort,n) = O;$
iii)  $\mathtt{wp}\,(\mathcal{P},O) = O;$
iv)   $\mathtt{wp}\,(\mathcal{P},n_1) \sqcap \mathtt{wp}\,(\mathcal{P},n_2) = \mathtt{wp}\,(\mathcal{P},n_1 \sqcap n_2);$
v)    $\mathtt{wp}\,(\mathcal{P},n_1) \sqcup \mathtt{wp}\,(\mathcal{P},n_2) \subset \mathtt{wp}\,(\mathcal{P},n_1 \sqcup n_2),$ with "=", if $\mathcal{P}$ is deterministic;
vi)   $n_1 \subset n_2 \implies \mathtt{wp}\,(\mathcal{P},n_1) \subset \mathtt{wp}\,(\mathcal{P},n_2).$

**Proof:** "*skip*" stands for a program with associated relation $B = \begin{pmatrix} O & I \\ O & O \end{pmatrix}$, and "*abort*" for a program with $B = O$. Applying Def. 10.4.1, we obtain (i, ii, iii); (iv, v, vi) follow directly from the definition. $\qquad\square$

We will also give the rules for weakest preconditions for the conditional statement and for the **while**-loop. Figure 10.4.1 represents the distinction between cases as a flowchart diagram and also in terms of explicit relations.

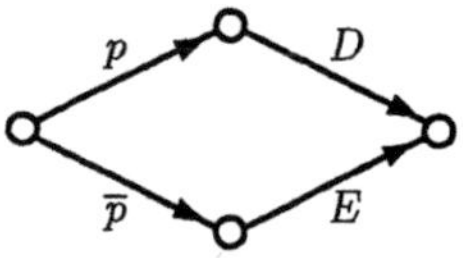

$$B = \begin{pmatrix} O & p & \bar{p} & O \\ O & O & O & D \\ O & O & O & E \\ O & O & O & O \end{pmatrix} \qquad BL = \begin{pmatrix} L \\ DL \\ EL \\ O \end{pmatrix} \qquad a = \begin{pmatrix} O \\ O \\ O \\ I \end{pmatrix}$$

**Fig. 10.4.1** Weakest precondition for the conditional statement

According to $\bar{q} \subset B\bar{q} \sqcup \overline{an} \sqcup \overline{BL}$ we begin with

$$\begin{pmatrix} \bar{x} \\ \bar{y} \\ \bar{z} \\ \bar{w} \end{pmatrix} \subset \begin{pmatrix} (p \sqcap \bar{y}) \sqcup (\bar{p} \sqcap \bar{z}) \\ \overline{D\bar{w}} \sqcup \overline{DL} \\ \overline{E\bar{w}} \sqcup \overline{EL} \\ \bar{n} \end{pmatrix}.$$

From this we obtain the four inclusions

$$(\bar{p} \sqcup y) \sqcap (p \sqcup z) \subset x, \quad \overline{D\bar{w}} \sqcap DL \subset y, \quad \overline{E\bar{w}} \sqcap EL \subset z, \quad n \subset w,$$

where for the components $x, y, z$, and $w$ of $q$ the infimum is to be found. Obviously we may set $w := n$, thus reducing $\overline{D\bar{w}}$ and $\overline{E\bar{w}}$ still further. Then we proceed with $z := \overline{E\bar{n}} \sqcap EL = \mathbf{wp}(E, n)$, $y := \overline{D\bar{n}} \sqcap DL = \mathbf{wp}(D, n)$, and arrive at $x := [\bar{p} \sqcup \mathbf{wp}(D, n)] \sqcap [p \sqcup \mathbf{wp}(E, n)]$. Because of $p \sqcup \bar{p} = L$, this is part (i) of the following theorem which we present using pseudo-code again.

**10.4.4 Proposition** (*Weakest precondition for conditional statements*).

i)   $\mathbf{wp}(\textbf{if } p \textbf{ then } D \textbf{ else } E \textbf{ fi}, n) = \big(p \sqcap \mathbf{wp}(D, n)\big) \sqcup \big(\bar{p} \sqcap \mathbf{wp}(E, n)\big)$;

ii)   $\mathbf{wp}(\textbf{if } p \textbf{ then } D \textbf{ fi}, n) = \big(p \sqcap \mathbf{wp}(D, n)\big) \sqcup (\bar{p} \sqcap n)$;

iii)   $\mathbf{wp}(\lceil p : D \rfloor r : E \rfloor, n) = (p \sqcup r) \sqcap \big(\bar{p} \sqcup \mathbf{wp}(D, n)\big) \sqcap \big(\bar{r} \sqcup \mathbf{wp}(E, n)\big)$.

**Proof:** ii) Interpreting  **if $p$ then $D$ fi**  as **if $p$ then $D$ else** *skip* **fi**, which means $B = \begin{pmatrix} O & p & \bar{p} \\ O & O & b \\ O & O & O \end{pmatrix}$, we obtain (ii) from (i) using (10.4.3).

iii) Passing in Fig. 10.4.1 from $p$, $\bar{p}$, and $L = p \sqcup \bar{p}$ to $p$, $r$, and $p \sqcup r$ we obtain

$$(p \sqcup r) \sqcap (\bar{p} \sqcup y) \sqcap (\bar{r} \sqcup z) \subset x, \quad \overline{D\bar{w}} \sqcap DL \subset y, \quad \overline{E\bar{w}} \sqcap EL \subset z, \quad n \subset w.$$

Now the proof can be completed as before, but without using $p \sqcup \bar{p} = L$ and $p \sqcap \bar{p} = O$. So one obtains $w := n$, $z := \mathbf{wp}(E, n)$, $y := \mathbf{wp}(D, n)$, $x := (p \sqcup r) \sqcap (\bar{p} \sqcup \mathbf{wp}(D, n)) \sqcap (\bar{r} \sqcup \mathbf{wp}(E, n))$, only. $\qquad\square$

We now consider the diagram and the relations for the **while**-loop in Fig. 10.4.2.

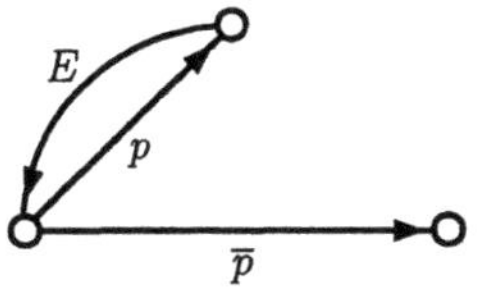

$$B = \begin{pmatrix} O & p & \bar{p} \\ E & O & O \\ O & O & O \end{pmatrix} \qquad BL = \begin{pmatrix} L \\ EL \\ O \end{pmatrix} \qquad a = \begin{pmatrix} O \\ O \\ I \end{pmatrix}$$

**Fig. 10.4.2** Weakest precondition for the **while**-loop

With regard to Def. 10.4.1, we again start from $\bar{q} \subset B\bar{q} \sqcup \overline{an \sqcup BL}$ or

$$\begin{pmatrix} \bar{x} \\ \bar{y} \\ \bar{z} \end{pmatrix} \subset \begin{pmatrix} (p \sqcap \bar{y}) \sqcup (\bar{p} \sqcap \bar{z}) \\ E\bar{x} \sqcup \overline{EL} \\ \bar{n} \end{pmatrix},$$

and obtain the three inclusions

$$(\bar{p} \sqcup y) \sqcap (p \sqcup z) \subset x, \quad \overline{E\bar{x}} \sqcap EL \subset y, \quad n \subset z,$$

where the smallest triple $x, y, z$ is to be found. Although part of the solution, $z := n$, is obtained directly, it remains to minimize $x$ and $y$ simultaneously. The result is part (i) of the proposition below. It is rather unwieldy, however, because there is still an infimum to be formed. Therefore, one often tries to avoid establishing rules for the total correctness of the **while**-construct in the presence of an arbitrary postcondition $n$. Rather, one considers pre- and postconditions which are closely related to each other, and to the loop condition, and satisfy $n = v \sqcap \bar{p}$. One then prefers a rule which is only sufficient and is actually a rule for partial correctness; it uses the fact that by Proposition 10.3.3 total correctness holds if and only if both termination and partial correctness do.

**10.4.5 Proposition** (*Weakest precondition for loops*).

i)   $\text{wp}(\textbf{while } p \textbf{ do } E \textbf{ od}, n) = \inf\{x \mid [p \sqcap \text{wp}(E, x)] \sqcup [\bar{p} \sqcap n] \subset x\};$

ii)   Assuming partial correctness $\{v \sqcap p\}E\{v\}$ to hold, we have
$$v \sqcap \text{wp}(\textbf{while } p \textbf{ do } E \textbf{ od}, L) \subset \text{wp}(\textbf{while } p \textbf{ do } E \textbf{ od}, v \sqcap \bar{p}).$$

**Proof** of ii): $\text{wp}(\textbf{while } p \textbf{ do } E \textbf{ od}, L)$ characterizes the input states from which the program terminates. Using Proposition 10.3.3, it suffices to show that

$$\{v\} \textbf{ while } p \textbf{ do } E \textbf{ od } \{v \sqcap \bar{p}\};$$

however, this follows from Corollary 10.2.5.ii.     $\square$

It is still unclear whether this infimum can be obtained as the supremum of an ascending iterated sequence. In general, the functional **wp** is not continuous, and therefore one will get only a lower bound, but not the infimum itself; cf. Appendix A.3.

We have thus shown how to obtain weakest preconditions for a number of program constructs by means of such conditions for their components. In Proposition 10.4.3 these conditions could be given directly for a number of special cases. It remains to give such a **wp**-formula directly for some nontrivial "atomic" construct which can be decomposed no further. For this we choose the assignment again.

In contrast to Def. 10.2.6, it is now important to know whether evaluating the expression to be assigned always terminates.

**10.4.6 Definition.** The weakest precondition for the assignment statement is the following:

i)   for a nonaborting, terminating, deterministic expression:

$$\mathbf{wp}\,(y := E, n) := S_E^y n.$$

ii)   for a nondeterministic expression:

$$\mathbf{wp}\,(y := E, n) = S_E^y n \sqcap \left\langle\!\left\langle \begin{array}{l} \text{Every evaluation path for } E, \text{ starting from} \\ \text{this state, terminates and does not abort.} \end{array} \right\rangle\!\right\rangle$$

Here $S_E^y n$ is the predicate, formed as the conjunction of all predicates arising from $n$ by replacing $y$ by some value of $E$ obtained nondeterministically.        $\Box$

So, for instance,

$$\mathbf{wp}\,(y := |x - 1|, \langle\!\langle y \le 7 \rangle\!\rangle) = \langle\!\langle |x - 1| \le 7 \rangle\!\rangle,$$
$$\mathbf{wp}\,(y := y/0, \langle\!\langle y < 1 \rangle\!\rangle) = O,$$
$$\mathbf{wp}\,(n := \lceil 3 [\!] 5 \rfloor, \langle\!\langle n \ge k \rangle\!\rangle) = \langle\!\langle 3 \ge k,\ 5 \ge k \rangle\!\rangle,$$
$$\mathbf{wp}\,(y := \lceil M :\ \textbf{if } x > 2 \textbf{ then } 2 \textbf{ else goto } M \textbf{ fi} \rfloor, \langle\!\langle y = 2 \rangle\!\rangle)$$
$$= \langle\!\langle 2 = 2,\ x > 2 \rangle\!\rangle = \langle\!\langle x > 2 \rangle\!\rangle;$$

the latter holds in contrast to an example following Def. 10.2.6.

We are now in a position to complete the problem of finding perfect numbers, begun in Sect. 10.2 and continued in Sect. 10.3.

Using the concept of weakest precondition, the network of relationships between the predicates established in Sect. 10.3 can be formulated as follows:

$$
\begin{array}{ll}
a \supset n & h \supset (r \sqcup i) \sqcap (\bar{r} \sqcup j) \\
b \supset \mathbf{wp}(K, c) & \quad = (r \sqcap j) \sqcup (\bar{r} \sqcap i) \\
c \supset (p \sqcup d) \sqcap (\bar{p} \sqcup k) & i \supset \mathbf{wp}(F, c) \\
\quad = (p \sqcap k) \sqcup (\bar{p} \sqcap d) & j \supset \mathbf{wp}(E, i) \\
d \supset \mathbf{wp}(C, g) & k \supset (s \sqcup l) \sqcap (\bar{s} \sqcup m) \\
e \supset \mathbf{wp}(A, b) & \quad = (s \sqcap m) \sqcup (\bar{s} \sqcap l) \\
f \supset \mathbf{wp}(D, g) & l \supset \mathbf{wp}(G, a) \\
g \supset (\bar{t} \sqcup f) \sqcap (t \sqcup h) & m \supset \mathbf{wp}(H, a) \\
\quad = (\bar{t} \sqcap h) \sqcup (t \sqcap f) &
\end{array}
$$

Given such a network, i.e., given an arbitrary predicate $q$ satisfying $\bar{q} \subset B\bar{q} \sqcup (an \sqcup BL)$, the condition $v \subset e$ must hold (or $v \subset e^\mathsf{T} q$, respectively; unfortunately, there is a collision of notation, for the letter $e$ stands for both, the input relation and the predicate over the input relation).

We now give the proof, whose technique is completely different from that in Sect. 10.2. The interpreted form of the network of conditions starts as follows:

$$a \supset \langle\!\langle F(x,0) = 0 \Longrightarrow w = \text{``yes''}, \quad F(x,0) \neq 0 \Longrightarrow w = \text{``no''} \rangle\!\rangle,$$

from which, using Def. 10.4.6 and Prop. 10.4.3.vi, iv, we obtain for $l$, $m$:

$$l \supset \langle\!\langle F(x,0) \neq 0 \rangle\!\rangle, \qquad\qquad m \supset \langle\!\langle F(x,0) = 0 \rangle\!\rangle.$$
$$k \supset (\langle\!\langle z = 0 \rangle\!\rangle \sqcap \langle\!\langle F(x,0) = 0 \rangle\!\rangle) \sqcup (\langle\!\langle z \neq 0 \rangle\!\rangle \sqcap \langle\!\langle F(x,0) \neq 0 \rangle\!\rangle)$$
$$= (\langle\!\langle z = 0 \text{ precisely when } F(x,0) = 0 \rangle\!\rangle)$$
$$c \supset (\langle\!\langle y = 0 \rangle\!\rangle \sqcap \langle\!\langle z = 0 \text{ precisely when } F(x,0) = 0 \rangle\!\rangle) \sqcup (\langle\!\langle y > 0 \rangle\!\rangle \sqcap d).$$

Direct evaluation of the weakest precondition fails here for the first time; because of the loops in the program there is self-reference among the conditions. For $D \approx u := u - y$ and $F \approx y := y - 1$, abortion, which might arise from the declaration **nat** $u, y$, has to be taken into consideration by introducing the conditions $u \geq y$ and $y \geq 1$.

$$i \supset \mathbf{wp}(F, c) = S_{y-1}^{y} c \sqcap \langle\!\langle y > 0 \rangle\!\rangle,$$
$$j \supset \mathbf{wp}(E, i) = S_{z-y}^{z} S_{y-1}^{y} c \sqcap \langle\!\langle y > 0 \rangle\!\rangle,$$
$$h \supset (\langle\!\langle u = 0 \rangle\!\rangle \sqcap j) \sqcup (\langle\!\langle u > 0 \rangle\!\rangle \sqcap i)$$
$$= (\langle\!\langle u = 0 \rangle\!\rangle \sqcap S_{z-y}^{z} S_{y-1}^{y} c \sqcap \langle\!\langle y > 0 \rangle\!\rangle)$$
$$\sqcup (\langle\!\langle u > 0 \rangle\!\rangle \sqcap S_{y-1}^{y} c \sqcap \langle\!\langle y > 0 \rangle\!\rangle).$$

For $g$ and $f$ we find another loop.

$$g \supset (\langle\!\langle u < y \rangle\!\rangle \sqcap h) \sqcup (\langle\!\langle u \geq y \rangle\!\rangle \sqcap f),$$
$$f \supset \mathbf{wp}(D, g) = S_{u-y}^{u} g \sqcap \langle\!\langle u \geq y \rangle\!\rangle$$
$$= (\langle\!\langle 0 \leq u - y < y \rangle\!\rangle \sqcap S_{u-y}^{u} h) \sqcup (\langle\!\langle u - y \geq y \rangle\!\rangle \sqcap S_{u-y}^{u} f).$$

Now we use that the declaration was **nat** $u$; so by descending finitely many times from $u$ to $u - y$ we arrive at $0 \leq u - y < y$. This allows us to estimate $f$ as

$$f \supset S_{\mathbf{mod}(u,y)}^{u} h.$$

As usual, the function **mod** denotes integer division remainder. This yields

$$g \supset (\langle\!\langle u < y \rangle\!\rangle \sqcap h) \sqcup \left( \langle\!\langle u \geq y \rangle\!\rangle \sqcap S_{\mathbf{mod}(u,y)}^{u} h \right)$$
$$= S_{\mathbf{mod}(u,y)}^{u} h, \qquad \text{since} \quad u < y \Longrightarrow u = \mathbf{mod}(u, y),$$

together with

$$d \supset \mathbf{wp}(C, g) = S_{x}^{u} g \supset S_{x}^{u} S_{\mathbf{mod}(u,y)}^{u} h = S_{\mathbf{mod}(x,y)}^{u} h$$
$$= \left( \langle\!\langle \mathbf{mod}(x,y) = 0 \rangle\!\rangle \sqcap S_{\mathbf{mod}(x,y)}^{u} S_{z-y}^{z} S_{y-1}^{y} c \sqcap \langle\!\langle y > 0 \rangle\!\rangle \right)$$
$$\sqcup \left( \langle\!\langle \mathbf{mod}(x,y) > 0 \rangle\!\rangle \sqcap S_{\mathbf{mod}(x,y)}^{u} S_{y-1}^{y} c \sqcap \langle\!\langle y > 0 \rangle\!\rangle \right)$$
$$= \left( \langle\!\langle y \mid x \rangle\!\rangle \sqcap S_{\mathbf{mod}(x,y)}^{u} S_{z-y}^{z} S_{y-1}^{y} c \sqcap \langle\!\langle y > 0 \rangle\!\rangle \right)$$
$$\sqcup \left( \langle\!\langle y \nmid x \rangle\!\rangle \sqcap S_{\mathbf{mod}(x,y)}^{u} S_{y-1}^{y} c \sqcap \langle\!\langle y > 0 \rangle\!\rangle \right).$$

At this point we obtain a recurrent condition for $c$:

$$c \supset \left( \langle\!\langle y = 0 \rangle\!\rangle \sqcap \langle\!\langle z = 0 \ \text{ precisely when} \ \ F(x,0) = 0 \rangle\!\rangle \right)$$
$$\sqcup \left( \langle\!\langle y > 0 \rangle\!\rangle \sqcap \langle\!\langle y \mid x \rangle\!\rangle \sqcap S^u_{\mathbf{mod}(x,y)} S^z_{z-y} S^y_{y-1} c \right)$$
$$\sqcup \left( \langle\!\langle y > 0 \rangle\!\rangle \sqcap \langle\!\langle y \nmid x \rangle\!\rangle \sqcap S^u_{\mathbf{mod}(x,y)} S^y_{y-1} c \right).$$

The first of the three terms of the estimate for $c$ is given explicitly, whereas the
other two again refer to $c$. So $c$ contains at least the first term and everything
resulting from the second and third term by iterated substitutions. One obtains

$$c \supset \langle\!\langle z = 0 \ \text{ precisely when} \ \ F(x,y) = 0 \rangle\!\rangle,$$

where, among others, the properties of $F(x,y)$ have been used that are known
from the end of Sect. 10.2. Applying the assignment rule,

$$b \supset \mathbf{wp}(K,c) = S^z_x c = \langle\!\langle x = 0 \ \text{ precisely when} \ \ F(x,y) = 0 \rangle\!\rangle,$$

we remember that at program step $A \approx y := x - 1$ an abortion has to be taken
into account for $\mathbf{nat}\ x = 0$. We finally obtain

$$e \supset \mathbf{wp}(A,b) = S^y_{x-1} b \sqcap \langle\!\langle x \geq 1 \rangle\!\rangle$$
$$= \langle\!\langle x = 0 \ \text{ precisely when} \ \ F(x,x-1) = 0 \rangle\!\rangle \sqcap \langle\!\langle x \geq 1 \rangle\!\rangle = \langle\!\langle x \geq 1 \rangle\!\rangle.$$

In the proof of total correctness just finished, essential use has been made of the
fact that a natural number can be (properly) decreased only finitely many times.
Such a proof may not have worked if the program scheme from Fig. 10.2.3 had
been interpreted differently.


### Exercise

**10.4.1** Prove for demonic composition $R \,;;\, S := RS \sqcap \overline{\overline{R}S L}$ (see Exercise 2.4.4)
that $\mathbf{wp}(R, \mathbf{wp}(S,n)) = \mathbf{wp}(R \,;;\, S, n)$.


## 10.5 Coverings of Programs

Our aim is to compare the effect of different programs which are made up of the
same program steps, by studying similarities in their execution. For this purpose
we need a concept of homomorphism for programs.

Two programs will be called homomorphic, if their flowgraphs can be mapped
homomorphically onto one another in such a way that input and output positions are preserved and that each program step associated with some arrow is
contained, as a relation, in the program step associated with the image arrow.


**10.5.1 Definition.** If $\mathcal{P} = (G, S, \Theta, e, a)$, $\mathcal{P}' = (G', S', \Theta', e', a')$ are programs,
we call the pair $(\eta, \xi)$ a **(program-)homomorphism** of $\mathcal{P}$ into $\mathcal{P}'$, if

i)   $\eta: G \longrightarrow G'$ and $\xi: S \longrightarrow S'$ are homomorphisms of the flowgraph and the
     situation graph.

ii)   $\xi^\mathsf{T} e \subset e', \quad a \subset \xi a'.$

iii)  $\xi^\mathsf{T}\Theta = \Theta'\eta^\mathsf{T}.$     □

In Fig. 10.5.1 an example of a program homomorphism is given. It is assumed that $R_1 \sqcup R_2 \subset R$ and $p_1 \sqcup p_2 \subset p$ hold.

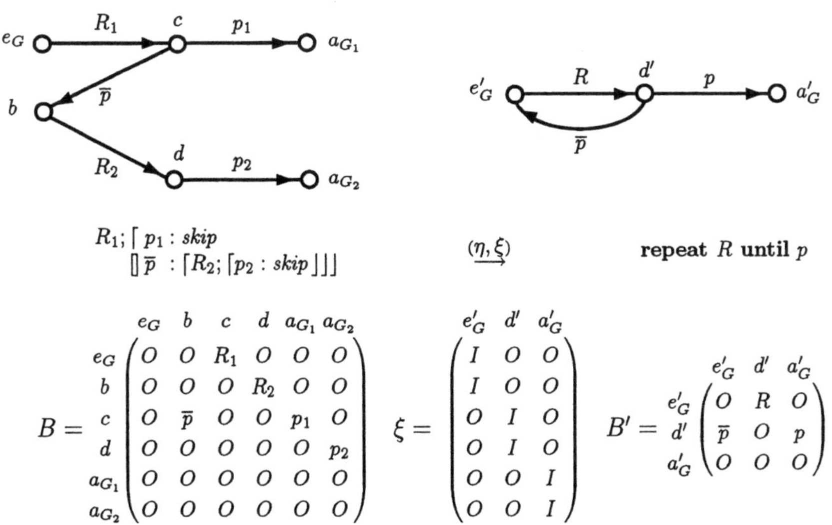

$$B = \begin{array}{c} \\ e_G \\ b \\ c \\ d \\ a_{G_1} \\ a_{G_2} \end{array} \begin{array}{c} \begin{array}{cccccc} e_G & b & c & d & a_{G_1} & a_{G_2} \end{array} \\ \begin{pmatrix} O & O & R_1 & O & O & O \\ O & O & O & R_2 & O & O \\ O & \overline{p} & O & O & p_1 & O \\ O & O & O & O & O & p_2 \\ O & O & O & O & O & O \\ O & O & O & O & O & O \end{pmatrix} \end{array} \qquad \xi = \begin{array}{c} \begin{array}{ccc} e'_G & d' & a'_G \end{array} \\ \begin{pmatrix} I & O & O \\ I & O & O \\ O & I & O \\ O & I & O \\ O & O & I \\ O & O & I \end{pmatrix} \end{array} \qquad B' = \begin{array}{c} \\ e'_G \\ d' \\ a'_G \end{array} \begin{array}{c} \begin{array}{ccc} e'_G & d' & a'_G \end{array} \\ \begin{pmatrix} O & R & O \\ \overline{p} & O & p \\ O & O & O \end{pmatrix} \end{array}$$

Fig. 10.5.1 A program homomorphism

By univalence and totality of $\xi$, the two conditions in (ii) are equivalent, although presented differently. The two different forms have been chosen in view of Def. 10.5.3 to come. Since $\xi$ is total and $\eta$ is univalent, we obtain from (iii) that $\Theta\eta \subset \xi\xi^\mathsf{T}\Theta\eta = \xi\Theta'\eta^\mathsf{T}\eta \subset \xi\Theta'$; therefore, if the left diagram is commutative, so is the right one. Because of Prop. 4.2.2.iv and $\Theta\eta L = \Theta L = L$, there is equality: $\Theta\eta = \xi\Theta'$. So (iii) means that the diagrams of Fig. 10.5.2 are *both* commutative, and the left diagram implies the right one. However, from the diagram on the right *alone* one cannot conclude that the diagram on the left is commutative, too.

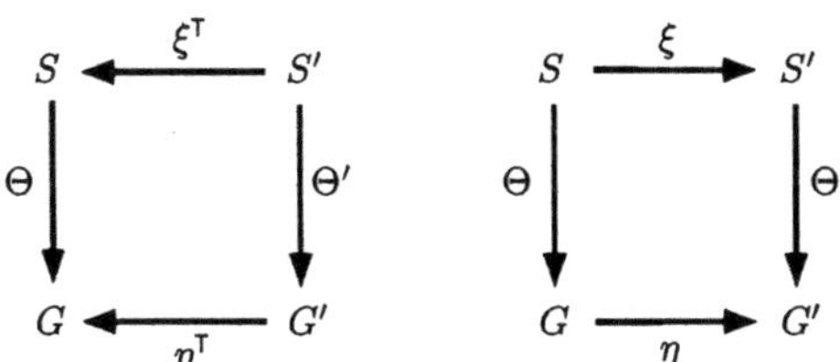

Fig. 10.5.2 Commutative diagrams of program homomorphisms

The diagram on the right can be interpreted as follows: Finding the position, during the execution, that corresponds to the image situation, amounts to giving the image of the position of the current situation. In addition, the diagram on the left says the following: If the position of some situation in $S'$ has an inverse image in $G$, then the situation considered also has an inverse image in $S$.

If $S'$ is the homomorphic image of a graph $S$, then by Exercise 7.1.1 the inverse image situations of a terminal situation are terminal, too. If one supposes the converse to hold for a program homomorphism, so that both programs terminate "simultaneously", then it is to be expected, by homomorphy, that one effect is contained in the other.

**10.5.2 Proposition.** If the program $\mathcal{P}$ is mapped homomorphically into the program $\mathcal{P}'$, and if the mapping $\xi$ of the situation graphs satisfies $\overline{BL} \subset \xi\overline{B'L}$, then the respective effects satisfy $\Sigma \subset \Sigma'$.

**Proof:** Using Exercise 7.1.1, one obtains for the terminating action $C$ that

$$C\xi = (B^* \cap \overline{BL}^\mathsf{T})\xi \subset (B^* \cap \overline{B'L}^\mathsf{T}\xi^\mathsf{T})\xi$$
$$= B^*\xi \cap \overline{B'L}^\mathsf{T} \subset \xi B'^* \cap \overline{B'L}^\mathsf{T} = \xi(B'^* \cap \overline{B'L}^\mathsf{T}) = \xi C',$$

and, thus, for the effect

$$\Sigma = e^\mathsf{T}Ca \subset e^\mathsf{T}C\xi a' \subset e^\mathsf{T}\xi C'a' \subset e'^\mathsf{T}C'a' = \Sigma'. \qquad \square$$

### Flow-Equivalence of Programs

We study special program homomorphisms for which it can be shown that their effects are equivalent, $\Sigma = \Sigma'$. The three programs in Fig. 10.5.3 differ with regard to their execution, but are built by the same program steps, namely $R$, $I \sqcap pL$ (denoted by $p$ for brevity), and $I \sqcap \overline{p}L$ (denoted by $\overline{p}$). One would generally agree that all three programs do the same.

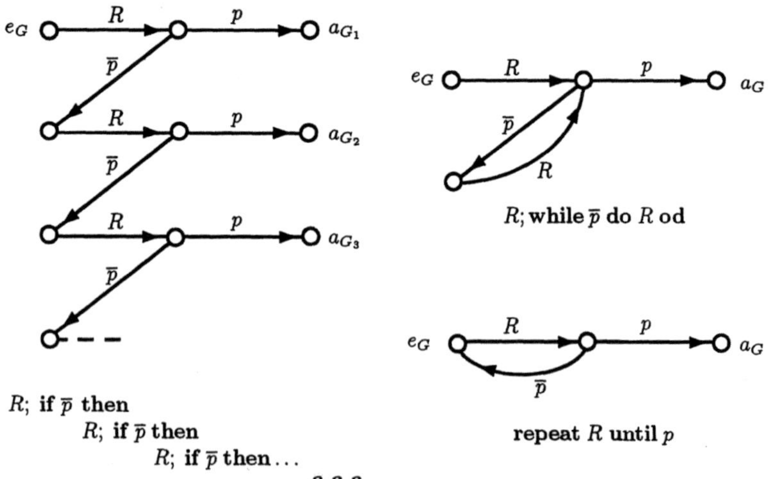


Fig. 10.5.3 Three programs with the same effect

It is often considered as self-evident that programs related in this way have the same effect. We attempt to give a formal proof for this, based on the identity of the individual program steps and on the "similarity" of the execution. For this purpose we take up the theme of Sect. 7.3 and study coverings, not merely for *graphs* but also for *programs*.

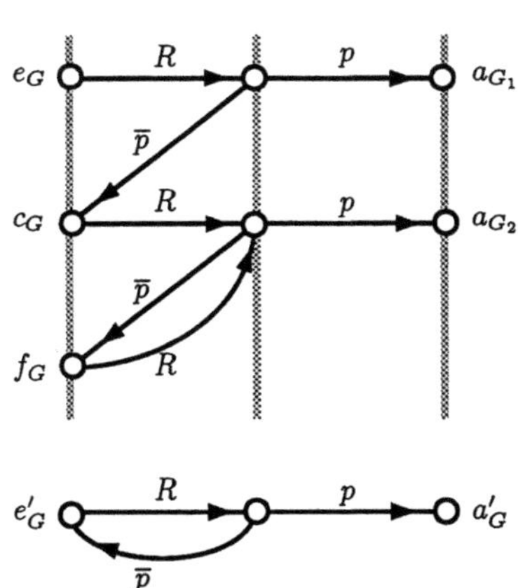


**Fig. 10.5.4** A covering
of programs

According to Def. 10.1.1 a program is given by a flowgraph and a situation graph, the latter being mapped homomorphically into the former. Coverings are used in a parallel way for both graphs. So we consider program homomorphisms that behave like graph coverings (Def. 7.3.1). In Fig. 10.5.4, the associated relations for the flowgraph and the situation graph can easily be given.

We call $\mathcal{P}$ a covering of $\mathcal{P}'$ if the associated flowgraphs cover each other in such a way that the same program steps are associated with arrows lying above one another. Furthermore, the input position of $\mathcal{P}$ is to lie above that of $\mathcal{P}'$, and the output positions of $\mathcal{P}$ must be precisely those positions which lie above output positions of $\mathcal{P}'$.

**10.5.3 Definition.** The homomorphism $(\eta, \xi)$ of the program $\mathcal{P} = (G, S, \Theta, e, a)$ into the program $\mathcal{P}' = (G', S', \Theta', e', a')$ is called a (program) **covering**[1] of $\mathcal{P}$ onto $\mathcal{P}'$, precisely when

i)    $\eta \colon G \to G'$ and $\xi \colon S \to S'$ are (graph) coverings,

ii)   $e' \subset \xi^{\mathsf{T}} e, \quad \xi a' \subset a,$

iii)  $\Theta \Theta^{\mathsf{T}} \sqcap \xi \xi^{\mathsf{T}} \subset I.$

If the coverings are both universal, we will speak of a covering rooted tree program, or a **universal covering** of the program.     □

Because of Def. 10.5.1.ii one has actually equality in (ii). But one will observe that the conditions on $e$ and $a$ are different. The input positions of $\mathcal{P}'$ are related by $\xi^{\mathsf{T}}$ only to those input positions of $\mathcal{P}$ lying above them in the sense given by the covering, not to others which may equally lie above. On the other

[1] These coverings correspond to the concept of flow-equivalence defined by GOGUEN 74 as a pair consisting of a special functor from the path category of the flowgraph of the first program into the corresponding category of the second, together with a natural transformation. Our definition is not based on some path category (which is infinite in general), but on the flowgraph itself.

hand, $a$ must be the entire $\xi$-fiber of $a'$. Thus, in Fig. 10.5.4, $a_{G_1}$ and $a_{G_2}$ must be output positions, whereas $c_G$ and $f_G$ need not be input positions.

Condition (iii) ensures that situations are uniquely determined by their positions and image situations. One has equality in (iii), too. So two situations belonging to the same position in $G$, and lying above the same situation in $S'$, are the same.

As a first application of this definition we prove for programs the result established for graphs in Exercise 7.3.4.

**10.5.4 Proposition.** The composition of two program coverings is again a program covering.

**Proof:** Let $\mathcal{P} = (G, S, \Theta, e, a)$ be a covering onto $\mathcal{P}' = (G', S', \Theta', e', a')$ by $(\eta, \xi)$, and let $(\eta', \xi')$ be a covering of $\mathcal{P}'$ onto $\mathcal{P}'' = (G'', S'', \Theta'', e'', a'')$. Then $\eta\eta'$ and $\xi\xi'$ are (graph) coverings of $G$ onto $G''$ and of $S$ onto $S''$, while $(\eta\eta', \xi\xi')$ is a (program) homomorphism. Furthermore, $e'' \subset \xi'^{\mathsf{T}}e' \subset \xi'^{\mathsf{T}}\xi^{\mathsf{T}}e = (\xi\xi')^{\mathsf{T}}e$ and $\xi\xi'a'' \subset \xi a' \subset a$. Because of $\Theta\Theta^{\mathsf{T}} \subset \Theta\eta\eta^{\mathsf{T}}\Theta^{\mathsf{T}} = \xi\Theta'\Theta'^{\mathsf{T}}\xi^{\mathsf{T}}$ we finally obtain

$$\Theta\Theta^{\mathsf{T}} \sqcap \xi\xi'\xi'^{\mathsf{T}}\xi^{\mathsf{T}} = \Theta\Theta^{\mathsf{T}} \sqcap \xi\Theta'\Theta'^{\mathsf{T}}\xi^{\mathsf{T}} \sqcap \xi\xi'\xi'^{\mathsf{T}}\xi^{\mathsf{T}}$$
$$= \Theta\Theta^{\mathsf{T}} \sqcap \xi(\Theta'\Theta'^{\mathsf{T}} \sqcap \xi'\xi'^{\mathsf{T}})\xi^{\mathsf{T}} \subset \Theta\Theta^{\mathsf{T}} \sqcap \xi\xi^{\mathsf{T}} \subset I. \qquad \square$$

But our main goal is to generalize Proposition 7.3.2 for programs. The situation is being illustrated by the diagrams in Fig. 10.5.5 below.

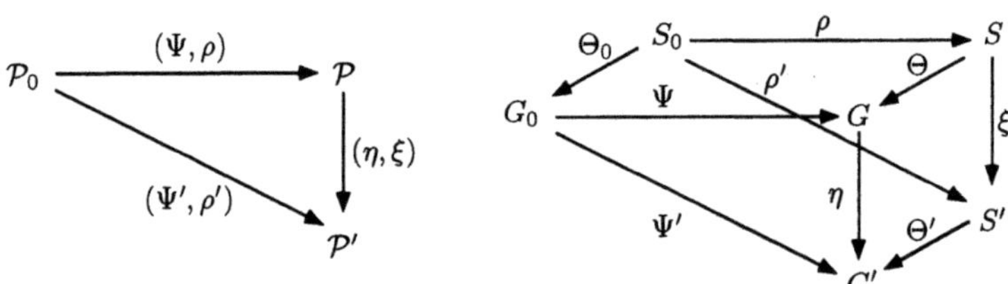

**Fig. 10.5.5** Lifting of program homomorphisms

**10.5.5 Proposition** (*Lifting of Computation Sequences*). Let $(\eta, \xi)$ be a covering of the program $\mathcal{P}$ onto the program $\mathcal{P}'$. If a rooted tree program $\mathcal{P}_0$ and a homomorphism $(\Psi', \rho')$ of $\mathcal{P}_0$ into $\mathcal{P}'$ are given, there exists a homomorphism $(\Psi, \rho)$ of $\mathcal{P}_0$ into $\mathcal{P}$ such that the composition of $(\Psi, \rho)$ and $(\eta, \xi)$ gives $(\Psi', \rho')$.

**Proof:** For the flowgraphs we have the configuration of Proposition 7.3.2 since it is easily shown that $\eta^{\mathsf{T}}e_G = e'_G$ and $\Psi'^{\mathsf{T}}e_{G_0} = e'_G$.

So there exists $\Psi$ such that $\Psi^{\mathsf{T}}e_{G_0} = e_G$ and that the front surface of the diagram of Fig. 10.5.5 is commutative: $\Psi' = \Psi\eta$. This $\Psi$ will now be used in an obvious manner to define $\rho$ as the intersection of the two intended commutativity connections for the rear surface and for the top surface:

$$\rho := \rho'\xi^{\mathsf{T}} \sqcap \Theta_0\Psi\Theta^{\mathsf{T}}.$$

Using relation-algebraic methods it is now possible to prove the following (in the indicated order): the factorization condition $\rho\xi = \rho'$, the homomorphism condition 10.5.1.iii $\rho^\mathsf{T}\Theta_0 = \Theta\Psi^\mathsf{T}$, and, furthermore, totality and univalence of $\rho$.

The proof just sketched ends by proving the homomorphism condition $B_0\rho \subset \rho B$ and conditions 10.5.1.ii, $\rho^\mathsf{T}e_0 \subset e$ and $a_0 \subset \rho a$. $\qquad\square$

In the case when the flowgraphs are also rooted graphs we have a stronger result: If $(\Psi', \rho')$ is a covering, then so is $(\Psi, \rho)$.

**10.5.6 Corollary.** If in Proposition 10.5.5 the flowgraph belonging to the program $\mathcal{P}_0$ is a rooted graph and $(\Psi', \rho')$ is a covering, then $(\Psi, \rho)$ is again a covering.

**Proof:** We may already use $\Psi B \subset B_{G_0}\Psi$ and $B_{G_0}^\mathsf{T} B_{G_0} \sqcap \Psi\Psi^T \subset I$ since $\Psi$ is a covering by Corollary 7.3.3. Then one shows, in that order, $\rho B \subset B_0\rho$, surjectivity of $\rho$, $e \subset \rho^\mathsf{T}e_0$, $\rho a \subset a_0$, and finally $\Theta_0\Theta_0^\mathsf{T} \sqcap \rho\rho^\mathsf{T} \subset I$. $\qquad\square$

For the reason given above, a covering *rooted tree* program is essentially unique, as two such covering rooted tree programs of a program $\mathcal{P}'$ are coverings of each other and are therefore isomorphic. So speaking of *the* covering rooted tree program in Definition 10.5.3 is justified.

Within a given relation algebra the universal covering for a program need not exist, although it can easily be constructed for programs over a rooted tree in the classical sense. Therefore, the following result is important, which establishes the semantic equivalence of programs covering each other *without* using universal coverings. Output positions may even be nonterminal.

**10.5.7 Proposition.** The effects of two programs covering each other coincide, i.e., $\Sigma = \Sigma'$.

**Proof:** We start with $B\xi = \xi B'$ which by Def. 7.3.1 is a consequence of the covering condition for $\xi$. This implies $B^*\xi = \xi B'^*$. The mapping $\xi$ satisfies $\xi\overline{B'L}\xi^\mathsf{T} = \overline{BL}$, and for the terminating action it follows that

$$C\xi = (B^* \sqcap \overline{BL}^\mathsf{T})\xi = (B^* \sqcap \xi\overline{B'L}^\mathsf{T}\xi^\mathsf{T})\xi = B^*\xi \sqcap \xi\overline{B'L}^\mathsf{T}$$
$$= \xi B'^* \sqcap \xi\overline{B'L}^\mathsf{T} = \xi(B'^* \sqcap \overline{B'L}^\mathsf{T}) = \xi C'.$$

So, $\Sigma = e^\mathsf{T}Ca = e^\mathsf{T}C\xi a' = e'^\mathsf{T}\xi C'a' = e'^\mathsf{T}C'a = \Sigma'$. $\qquad\square$

# 10.6 References



ARBIB MH, ALAGIĆ S: *Proof rules for gotos.* Acta Informat. **11** (1979) 139–148.

BAKKER JW DE, MEERTENS LGLT: *On the completeness of the inductive assertion method.* J. Comput. Syst. Sci. **11** (1975) 323–357.

BERGHAMMER R, KEMPF P, SCHMIDT G, STRÖHLEIN T: *Relation Algebra and Logic of Programs.* - In: Andréka H, Monk JD, Németi I (eds.): *Algebraic Logic.* Proc. of a Coll., Budapest, August 8-14, 1988, (Colloq. Math. Soc. J. Bolyai **54**) North-Holland Publ. Co., Amsterdam, 1991, 37-58.

BERGHAMMER R, SCHMIDT G: *A relational view on gotos and dynamic logic.* In: Schneider HJ, Göttler H (eds.): Proc. 8th Conf. on Graphtheoretic Concepts in Computer Science (WG 82), Neunkirchen a. Br. (Germany), June 16–18, 1982, Hanser, München, 1982, 13-24.

BERGHAMMER R, ZIERER H: *Relational algebraic semantics of deterministic and nondeterministic programs.* Theoret. Comput. Sci. **43** (1986) 123–147.

CLINT M, HOARE CAR: *Program proving: Jumps and functions.* Acta Informat. **1** (1972) 214–224.

FLOYD RW: *Assigning meaning to programs.* In: Schwartz JT (ed.): Mathematical Aspects of Computer Science. Amer. Math. Soc. Proc. Symp. in Appl. Math. XIX 1966, Providence, R. I. 1967, 19–32.

GOGUEN JA: *On homomorphisms, correctness, termination, unfoldments and equivalence of flow diagram programs.* J. Comput. System Sci. **8** (1974) 333–365.

HAREL D: *First-order dynamic logic.* Lecture Notes in Computer Science **68**, Springer, Berlin, 1979.

HOARE CAR: *An axiomatic basis for computer programming.* Comm. ACM **12** (1969) 576–580, 583.

MANNA Z: *Verification of sequential programs: Temporal axiomatization.* In: Broy M, Schmidt G (eds.): Theoretical foundations of programming methodology. Lecture Notes of an International Summer School at Marktoberdorf, 1981, Reidel, Dordrecht, 1982, 53-102.

MILI A: *A relational approach to the design of deterministic programs.* Acta Informat. **20** (1983) 315–328.

NGUYEN TT: *A relational model of demonic nondeterministic programs.* Internat. J. Found. Comput. Sci. **2** (1991) 101–131.

DE ROEVER WP: *Recursive program schemes: semantics and proof theory.* Ph. D.-Thesis, Math. Centrum Tracts, Amsterdam, 1974.

SANDERSON JG: *A relational theory of computing.* Lecture Notes in Computer Science **82**, Springer, Berlin, 1980.

SCHMIDT G: *Programme als partielle Graphen.* Habil. Thesis, Techn. Univ. München, 1977.

# Appendix

## A.1 Boolean Algebra

The simplest example of a Boolean lattice, also called a Boolean algebra, is the
**lattice of truth-values** $\mathbb{B} := \{\,0,1\,\}$ whose operations, negation $^-$, disjunction
$\vee$, and conjunction $\wedge$ are defined as follows:

$$
\begin{array}{c|cc}
^- & 0 & 1 \\\hline
 & 1 & 0
\end{array}
\qquad
\begin{array}{c|cc}
.\vee. & 0 & 1 \\\hline
0 & 0 & 1 \\
1 & 1 & 1
\end{array}
\qquad
\begin{array}{c|cc}
.\wedge. & 0 & 1 \\\hline
0 & 0 & 0 \\
1 & 0 & 1
\end{array}
$$

The algebraic structure of a Boolean lattice will be introduced by successively specializing the
concept of lattice, an approach quite different from G. Boole's in his *Mathematical Analysis
of Logic* of 1847. The system of axioms for a lattice, as it is being used today, is based on
R. Dedekind's "dual group" which, however, E. Schröder used already in 1897. During the
intermediate period of fifty years, notably C. Peirce and E. Schröder contributed to the subject
with treatises on the "algebra of logic".

**A.1.1 Definition.** If $X$ is a nonempty set on which two operations $\vee$ and $\wedge$
are defined satisfying the laws of

| | | |
|---|---|---|
| **associativity** | $(x \vee y) \vee z = x \vee (y \vee z),$ | $(x \wedge y) \wedge z = x \wedge (y \wedge z),$ |
| **commutativity** | $x \vee y = y \vee x,$ | $x \wedge y = y \wedge x,$ |
| **absorption** | $x \vee (x \wedge y) = x,$ | $x \wedge (x \vee y) = x,$ |

we call $(X, \vee, \wedge)$ a **lattice**. The operations $\vee$ and $\wedge$ are then called **join** and
**meet**. $\qquad\qquad\square$

The two laws in each line are "dual" to one another in the sense that one can be
obtained from the other by interchanging $\vee$ and $\wedge$ .

By the associative law for $\vee$ and $\wedge$ one can dispense with brackets in the
following definition of finite joins and meets

$$
\sup_{1 \le i \le n} x_i := \bigvee_{i=1}^{n} x_i := x_1 \vee x_2 \vee \ldots \vee x_n, \qquad
\inf_{1 \le i \le n} x_i := \bigwedge_{i=1}^{n} x_i := x_1 \wedge x_2 \wedge \ldots \wedge x_n.
$$

In Fig. A.1.1 we give two lattice structures on four elements by the respective
tables for $\vee$ and $\wedge$. There are no other four-element lattices.

The absorption laws relate the two lattice operations and are essential for deriving further properties.

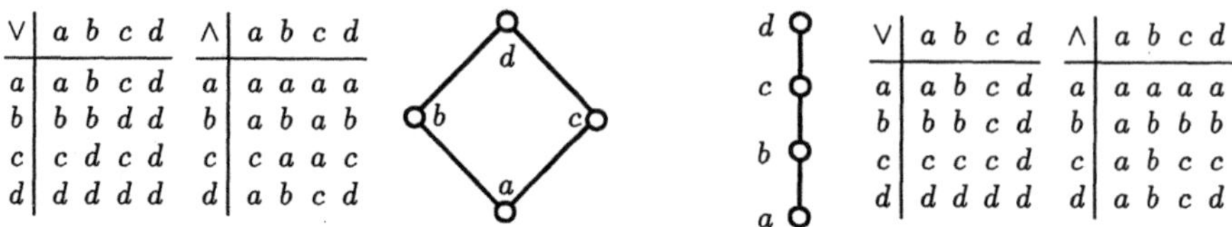

| $\vee$ | $a$ | $b$ | $c$ | $d$ |
|---|---|---|---|---|
| $a$ | $a$ | $b$ | $c$ | $d$ |
| $b$ | $b$ | $b$ | $d$ | $d$ |
| $c$ | $c$ | $d$ | $c$ | $d$ |
| $d$ | $d$ | $d$ | $d$ | $d$ |

| $\wedge$ | $a$ | $b$ | $c$ | $d$ |
|---|---|---|---|---|
| $a$ | $a$ | $a$ | $a$ | $a$ |
| $b$ | $a$ | $b$ | $a$ | $b$ |
| $c$ | $c$ | $a$ | $a$ | $c$ |
| $d$ | $a$ | $b$ | $c$ | $d$ |

| $\vee$ | $a$ | $b$ | $c$ | $d$ |
|---|---|---|---|---|
| $a$ | $a$ | $b$ | $c$ | $d$ |
| $b$ | $b$ | $b$ | $c$ | $d$ |
| $c$ | $c$ | $c$ | $c$ | $d$ |
| $d$ | $d$ | $d$ | $d$ | $d$ |

| $\wedge$ | $a$ | $b$ | $c$ | $d$ |
|---|---|---|---|---|
| $a$ | $a$ | $a$ | $a$ | $a$ |
| $b$ | $a$ | $b$ | $b$ | $b$ |
| $c$ | $a$ | $b$ | $c$ | $c$ |
| $d$ | $a$ | $b$ | $c$ | $d$ |

**Fig. A.1.1** Two lattices with four elements

**A.1.2 Proposition.** Every lattice $(X, \vee, \wedge)$ satisfies the laws of

**idempotence** $$x \vee x = x, \quad x \wedge x = x.$$

The **proof** follows easily from the absorption laws. $\qquad\square$

Clearly, the proofs of the idempotent laws for $\vee$ and $\wedge$ are completely analogous. This is an instance of the general duality principle of lattice theory which is a consequence of the symmetry in the laws (Def. A.1.1) and is already present in the work of E. Schröder. The **duality principle** says: If a theorem in lattice theory can be derived from the axioms A.1.1, then so can the dual theorem, obtained by interchanging $\vee$ and $\wedge$.

From the two absorption laws it follows that both equations, $x \vee y = y$ and $x \wedge y = x$, are equivalent. If such a pair of equations holds, this is notationally abbreviated as follows:

**A.1.3 Definition.** In a lattice we define for elements $x$ and $y$

$$x \subset y \quad :\Longleftrightarrow \quad x \wedge y = x,$$

and say that "$x$ **is contained in** (or equal to) $y$". The notation $y \supset x$ is understood to be equivalent to $x \subset y$. The relation $\subset$ is called the **inclusion (ordering)** of the lattice. $\qquad\square$

Indeed, $\subset$ is an ordering: Following Prop. A.1.2, we have $x \subset x$ for all $x$ (reflexivity). From $x \subset y$ and $y \subset z$ it follows that $x \wedge z = (x \wedge y) \wedge z = x \wedge (y \wedge z) = x \wedge y = x$, i.e., $x \subset z$ (transitivity). From $x \subset y$ and $y \subset x$, finally, $x = y$ (antisymmetry) is easily deduced. Furthermore, we have for $\vee$ and $\wedge$

**isotonicity** $$x \subset y \quad \Longrightarrow \quad x \vee z \subset y \vee z \quad \text{and} \quad x \wedge z \subset y \wedge z.$$

The algebraic Definition A.1.1 of a lattice turns out to be equivalent to an ordering $\subset$ on a set $M$ being given with respect to which any two elements $x, y$ of $M$ have a supremum $\sup\{x, y\} = x \vee y$ and an infimum $\inf\{x, y\} = x \wedge y$. We show in the next proposition that $x \vee y$ is indeed the least element greater or equal to both $x$ and $y$. Similarly, $x \wedge y$ is the greatest element less than or equal to both $x$ and $y$.

**A.1.4 Proposition.** For a lattice $(X, \vee, \wedge)$ with ordering $\subset$ the following holds:

i) $$x_1 \wedge x_2 \subset x_i \subset x_1 \vee x_2 \quad \text{for } i = 1, 2;$$

ii) $$u \subset x_i \subset v \text{ for } i = 1, 2 \quad \Longleftrightarrow \quad u \subset x_1 \wedge x_2 \text{ and } x_1 \vee x_2 \subset v. \qquad \square$$

The following inclusions can be shown to hold in any lattice

$$x \wedge (y \vee z) \supset (x \wedge y) \vee (x \wedge z), \quad x \vee (y \wedge z) \subset (x \vee y) \wedge (x \vee z).$$

The distributive law requires that these inclusions be equalities:

**A.1.5 Definition.** A lattice is said to be **distributive,** provided

$$x \vee (y \wedge z) = (x \vee y) \wedge (x \vee z), \quad x \wedge (y \vee z) = (x \wedge y) \vee (x \wedge z). \qquad \square$$

It can be shown that any one such distributive law implies the other. This symmetry is typical of lattices and is not shared by the distributive laws for other algebraic structures.

By induction, Def. A.1.5 generalizes for finite $n$ and $m$ as

$$\bigvee_{i=1}^{n} x_i \wedge \bigvee_{j=1}^{m} y_j = \bigvee_{i=1}^{n} \bigvee_{j=1}^{m} x_i \wedge y_j, \quad \bigwedge_{i=1}^{n} x_i \vee \bigwedge_{j=1}^{m} y_j = \bigwedge_{i=1}^{n} \bigwedge_{j=1}^{m} x_i \vee y_j.$$

**A.1.6 Definition.** Let $(X, \vee, \wedge)$ be a lattice. An element $O \in X$ is called the **zero element** (least element) and an element $L \in X$ is called the **universal element** (greatest element) of the lattice, if for any other element $x$ one has

$$O \vee x = x, \ O \wedge x = O \quad \text{resp.} \quad L \vee x = L, \ L \wedge x = x. \qquad \square$$

The lattices on four elements in Fig. A.1.1 have $a$ as the zero element and $d$ as the universal element. Clearly a lattice can have at most one such element. If $O$ and $L$ exist, then one has $O \subset x \subset L$ for all elements $x$ of the lattice.

**A.1.7 Definition.** An element $a$ of a lattice with zero element $O$ is called an **atom,** if $a \neq O$ and, furthermore,

$$O \neq x \subset a \quad \Longrightarrow \quad x = a.$$

A lattice is called **atomic,** if every element unequal to $O$ contains at least one atom. $\qquad \square$

Atoms are thus "indivisible" in the sense that they have no proper nonzero parts. (There are varying definitions of atomic lattices in the literature.) Accordingly, one has

$$a, b \text{ atoms}, \ a \neq b \quad \Longrightarrow \quad a \wedge b = O.$$

For, otherwise, $O \neq a \wedge b \subset a$ would imply $a \wedge b = b$, i.e., $O \neq b \subset a$ and therefore $b = a$, contradicting the assumption $a \neq b$.

In a distributive lattice one has the useful property of atoms that they cannot be contained in a union without being contained in one of its parts:

$$a \text{ atom, } a \subset x \vee y \implies a \subset x \text{ or } a \subset y.$$

This is because $a \not\subset x$ implies $a \wedge x \neq a$, hence $a \wedge x = O$, and similarly $a \not\subset y$ implies $a \wedge y = O$; putting these together, the distributive law $a \wedge (x \vee y) = (a \wedge x) \vee (a \wedge y) = O \vee O = O$ would produce the contradiction $a \not\subset x \vee y$. More generally,

$$a \text{ atom, } a \subset \sup \mathcal{A} \implies a \subset y \text{ for at least one } y \in \mathcal{A}.$$

If zero and universal elements exist, one may want to find a smallest possible element whose join with $x$ is $L$.

**A.1.8 Definition.** Let an element $x$ be given in a lattice $(X, \vee, \wedge)$ with zero element $O$ and universal element $L$. The element $c$ is called a **complement** of the element $x$, if

$$x \vee c = L, \quad x \wedge c = O.$$

A lattice $(X, \vee, \wedge)$ with zero element $O$ and universal element $L$ is called **complementary,** if for any element $x$ at least one complement exists.     $\square$

Complements which are uniquely determined occur mainly in the following situation.

**A.1.9 Proposition.** In a distributive lattice there is at most one complement for every element.

**Proof:** By the absorptive law and commutativity, the weaker assumption $x \vee c = x \vee c'$ and $x \wedge c = x \wedge c'$ already implies that $c = c'$:

$$c = c \wedge (c \vee x) = c \wedge (c' \vee x) = (c \wedge c') \vee (c \wedge x)$$
$$= (c \wedge c') \vee (c' \wedge x) = (x \vee c) \wedge c' = (x \vee c') \wedge c' = c'.$$     $\square$

If there exists exactly one complement for an element $x$, it is usually denoted by $\bar{x}$. In complementary, distributive lattices there is therefore the additional unary operation $^{-}$ of complementation. Sometimes, $O$ and $L$ are interpreted as nullary operations.

**A.1.10 Definition.** A complementary, distributive lattice where complementation, zero element, and universal element are denoted by $^{-}, O, L$, respectively, is called a **Boolean lattice** $\mathcal{B} = (X, \vee, \wedge, ^{-}, O, L)$. (A Boolean lattice is also called a Boolean algebra.)     $\square$

The ordering $\subset$ in a Boolean lattice has additional properties. Some frequently used ones are:

$$x \subset y \iff L \subset \overline{x} \vee y \iff x \wedge \overline{y} \subset O;$$
$$x \subset y \vee z \iff x \wedge \overline{y} \subset z \iff \overline{y} \subset \overline{x} \vee z \iff L \subset \overline{x} \vee y \vee z.$$

We prove the first equivalence only; the others follow from de Morgan's laws which we establish below. Direction "$\Longrightarrow$" follows from isotonicity, $L = \overline{x} \vee x \subset \overline{x} \vee y$. For "$\Longleftarrow$" we have $x \subset L \subset \overline{x} \vee y$, so that distributivity and the complementation properties yield $x = x \wedge (\overline{x} \vee y) = (x \wedge \overline{x}) \vee (x \wedge y) = O \vee (x \wedge y) = x \wedge y$ and hence $x \subset y$.

The forming of the complement is an involutionary, dual automorphism:

**A.1.11 Proposition.** In a Boolean lattice the following laws are satisfied:

| | |
|---|---|
| **involution** | $\overline{\overline{x}} = x;$ |
| **de Morgan** | $\overline{x \vee y} = \overline{x} \wedge \overline{y}, \quad \overline{x \wedge y} = \overline{x} \vee \overline{y}.$ |

**Proof:** The law of involution follows from the symmetry which holds in Definition A.1.8 with respect to $x$ and $\overline{x}$. Of the two de Morgan laws we deduce only the second from the equations

$$(x \wedge y) \wedge (\overline{x} \vee \overline{y}) = (x \wedge y \wedge \overline{x}) \vee (x \wedge y \wedge \overline{y}) = O \vee O = O,$$
$$(x \wedge y) \vee (\overline{x} \vee \overline{y}) = (x \vee \overline{x} \vee \overline{y}) \wedge (y \vee \overline{x} \vee \overline{y}) = L \wedge L = L,$$

which the complement of $x \wedge y$ has to satisfy. $\qquad\qquad\square$

We now want to allow infinite operations as well. In this situation it has become customary to work with the order-theoretic definition of a lattice instead of the algebraic. In other words, we now require suprema and infima to exist for *arbitrary* subsets.

**A.1.12 Definition.** A lattice $(X, \vee, \wedge)$ is called a **complete** lattice, if for arbitrary subsets $\mathcal{A} \subset X$ the supremum $\sup \mathcal{A}$ and the infimum $\inf \mathcal{A}$ exist. (Note that $\mathcal{A}$ may be empty.) $\qquad\qquad\square$

The general form of the de Morgan laws is:

$$\overline{\bigvee_i x_i} = \bigwedge_i \overline{x_i}, \qquad \overline{\bigwedge_i x_i} = \bigvee_i \overline{x_i},$$
$$\overline{\sup\{x \mid x \in S\}} = \inf\{y \mid \overline{y} \in S\}, \quad \overline{\inf\{x \mid x \in S\}} = \sup\{y \mid \overline{y} \in S\}.$$

Proposition A.1.4 thus takes the following form:

$$\inf \mathcal{A} \subset x \quad \text{and} \quad x \subset \sup \mathcal{A} \quad \text{for all } x \in \mathcal{A},$$
$$u \subset x \subset v \text{ for all } x \in \mathcal{A} \iff u \subset \inf \mathcal{A} \text{ and } \sup \mathcal{A} \subset v.$$

In a complete Boolean lattice $(X, \vee, \wedge)$ the zero element $O$ is the infimum of all the elements in $X$, and the universal element $L$ is their supremum. It is also consistent with the definitions to consider $O$ as the supremum of the empty set and $L$ as its infimum. In a complete Boolean lattice the distributive laws also hold for infinite sets:

**A.1.13 Proposition.** A complete Boolean lattice is **infinitely distributive,** i.e., for arbitrary subsets $\mathcal{A} \subset X$

$$x \wedge \sup \mathcal{A} = \sup \{\, x \wedge a \mid a \in \mathcal{A} \,\}, \quad x \vee \inf \mathcal{A} = \inf \{\, x \vee a \mid a \in \mathcal{A} \,\}.$$

**Proof:** In view of the duality principle we prove only the first identity. For empty $\mathcal{A}$ it is clearly satisfied, since by the above convention we have $O$ on both sides. In a complete lattice one has $\sup \mathcal{A} \supset a$ for all $a \in \mathcal{A}$, therefore $x \wedge \sup \mathcal{A} \supset x \wedge a$, and hence we have "$\supset$". Now we abbreviate the right-hand side by $u := \sup \{\, x \wedge a \mid a \in \mathcal{A} \,\}$. For any $a \in \mathcal{A}$ clearly $x \wedge a \subset u$, and therefore, since we may form complements and use distributivity also

$$a \subset a \vee \overline{x} = L \wedge (a \vee \overline{x}) = (x \vee \overline{x}) \wedge (a \vee \overline{x}) = (x \wedge a) \vee \overline{x} \subset u \vee \overline{x},$$

yielding $\sup \mathcal{A} \subset u \vee \overline{x}$ and $x \wedge \sup \mathcal{A} \subset u$. $\qquad\square$

The structure problem of Boolean lattices was in essence solved by M. H. Stone in 1936, who proved that any Boolean lattice is isomorphic to a field of sets. Thus, a finite Boolean lattice has a number of elements of the form $2^n$, and is isomorphic to the power set, ordered by inclusion, of a set of cardinality $n$. So, up to isomorphism, there is exactly one Boolean lattice with $1, 2, 4, 8, \ldots$ elements, respectively. The Boolean lattice with 2 elements is the lattice of truth-values.

## A.2  Abstract Relation Algebra

In Chap. 2 we laid the foundations for the *calculus* of relations and discussed axiomatic aspects. Here, we want to deepen our understanding of the abstract structure and of models thereof. As the standard model for a relation algebra one can take the set of $n \times n$ Boolean matrices which corresponds to all relations *on* a set. A set of relations, which is closed with respect to the usual operations, is commonly called an **algebra of (concrete) relations**. By contrast, an **(abstract) relation algebra** is an algebraic object in which all identities and rules known for relations are valid.

So the following "representation problem of relation algebra", formulated in 1940/41 by J. C. C. McKinsey and A. Tarski, naturally arises: Can every relation algebra be represented as an algebra of (concrete) relations? Or, worded differently, can one still do the usual calculations even when not dealing with ordinary relations on a set, i.e., are there nonstandard models? If such other models exist, then one will ask for their practical relevance, for instance with regard to the algebraic treatment of parallel processes. In order to analyze models we must establish a set of precise axioms for relation algebras by completing the considerations from Chap. 2.

These questions were first investigated in MCKINSEY 40 and TARSKI 41. Important contributions were also made by R. C. Lyndon and B. Jónsson. The chapter on lattice-ordered semigroups of the third edition of BIRKHOFF 67 contains some information. More detailed references to the literature can be found in SCHMIDT, STRÖHLEIN 85 and especially in NÉMETI 91.

The following definition, due to CHIN, TARSKI 51, applies to homogeneous relations, in which case all operations are total:

A **(homogeneous abstract) relation algebra** $(\mathcal{R}, \sqcup, \sqcap, {}^-, \circ, {}^\top)$ consists of a nonempty set $\mathcal{R}$, whose elements are called relations, such that

- $(\mathcal{R}, \sqcup, \sqcap, {}^-)$ is a complete, atomic Boolean algebra with zero element $O$, universal element $L$, and ordering $\subset$,

- $(\mathcal{R}, \circ)$ is a semigroup with precisely one unit element $I$, and

- the Schröder equivalences and the Tarski rule hold.

The second postulate can be weakened inasmuch as only a right unit element is needed. The Schröder equivalences are sometimes attributed to LUCE 52, but J. Riguet observed already in 1948 that these equivalences are variants of a rule given by E. Schröder. In MADDUX 90 it is pointed out that A. de Morgan proved this equivalence as "Theorem K" already in 1864. There are other systems of axioms for relation algebras which do not postulate the validity of the Tarski rule which implies the "simplicity" of the structure. By postulating the Tarski rule we exclude *proper* homomorphisms between relation algebras; there are only the identity homomorphism and the zero homomorphism which annihilates everything. The independence of these axioms has been investigated, among others, by KAMEL 54, WOOYENAKA 59, and BRINK 79B.

When dealing with heterogeneous relations, $\sqcup, \sqcap$ and $\circ$ are *partial* operations. But when considering a relation algebra we want to say as little as possible about the question of whether relations can be joined, intersected, or multiplied. In practical problems, these operations will be permitted if they can be deduced from some corresponding union, intersection, or product given in the first place. One can let oneself be guided by the calculus of $m \times n$-matrices. (Category theory would provide an appropriate formal set-up; however, we have avoided categorical formalism throughout the book.) The product $R \circ S$ will be written simply as $RS$.

In Fig. A.2.1 we give a heterogeneous relation algebra which contains 10 relations. The set $\mathcal{R}_{\sqcup,e}$ of relations joinable with $e$ is $\{e, f, g, h\}$. From the viewpoint of the operations $\sqcup, \sqcap, {}^-$ this is a Boolean lattice in which $e$ is the least and $f$ is the greatest element.

This relation algebra can be put in correspondence with that from Fig. 7.5.5 (and that from Fig. A.2.2) as follows: associate with the elements $a, \ldots, j$, in their order, the Boolean matrices $O_1$, $L_1$, $O_{12}$, $L_{12}$, $O_2$, $L_2$, $I_2$, $\overline{I}_2$, $O_{21}$ and $L_{21}$. (A pair of indices $ij$ denotes the numbers of rows and columns of a matrix, a single index denotes the number of rows of a square matrix.)

With this example in mind, we now give an axiomatic definition of a heterogeneous relation algebra.

**A.2.1 Definition.** A **(heterogeneous abstract) relation algebra** $(\mathcal{R}, \sqcup, \sqcap, {}^-, \circ, {}^\top)$ consists of a nonempty set $\mathcal{R}$ together with (totally defined) unary operations ${}^-, {}^\top$ and (partially defined) binary operations $\sqcup, \sqcap, \circ$. We denote by $\mathcal{R}_{\sqcup,R}$ those elements $S \in \mathcal{R}$ for which the union $R \sqcup S$ is defined and demand $R \in \mathcal{R}_{\sqcup,R}$ for every element $R \in \mathcal{R}$. The operations satisfy the following rules:

272    Appendix

i) $(\mathcal{R}_{\sqcup,R},\sqcup,\sqcap,{}^{-})$ is a complete atomic Boolean lattice with zero element $O_R$, universal element $L_R$, and ordering $\subset$. (Often we just write $O$ and $L$.)

ii) If the products $QS$ and $RS$ are defined, so is $QR^{\mathsf{T}}$. If the products $QR$ and $QS$ are defined, so is $R^{\mathsf{T}}S$. If $QR$ exists, so does $QS$ for all $S \in \mathcal{R}_{\sqcup,R}$.

iii) The multiplication is associative: $\quad Q(RS) = (QR)S$.

iv) There are elements ${}^{R}I$ and $I^{R}$ associated to every relation $R \in \mathcal{R}$. ${}^{R}I$ behaves as a right identity and $I^{R}$ as a left identity for $\mathcal{R}_{\sqcup,R}$. (In most cases we will write $I$ for both.)

v) The Schröder rule holds: $\quad QR \subset S \iff Q^{\mathsf{T}}\overline{S} \subset \overline{R} \iff \overline{S}R^{\mathsf{T}} \subset \overline{Q}$.

vi) The Tarski rule is satisfied: $\quad R \neq O \implies LRL = L$. $\qquad\qquad\Box$

We first deduce the computational rules in Sects. 2.2 and 2.3 from these axioms alone. For the multiplication rules this was already done in Proposition 2.3.6.

| $\sqcup$ | $a$ | $b$ | $c$ | $d$ | $e$ | $f$ | $g$ | $h$ | $i$ | $j$ |
|---|---|---|---|---|---|---|---|---|---|---|
| $a$ | $a$ | $b$ | | | | | | | | |
| $b$ | $b$ | $b$ | | | | | | | | |
| $c$ | | | $c$ | $d$ | | | | | | |
| $d$ | | | $d$ | $d$ | | | | | | |
| $e$ | | | | | $e$ | $f$ | $g$ | $h$ | | |
| $f$ | | | | | $f$ | $f$ | $f$ | $f$ | | |
| $g$ | | | | | $g$ | $f$ | $g$ | $f$ | | |
| $h$ | | | | | $h$ | $f$ | $f$ | $h$ | | |
| $i$ | | | | | | | | | $i$ | $j$ |
| $j$ | | | | | | | | | $j$ | $j$ |

| $\sqcap$ | $a$ | $b$ | $c$ | $d$ | $e$ | $f$ | $g$ | $h$ | $i$ | $j$ |
|---|---|---|---|---|---|---|---|---|---|---|
| $a$ | $a$ | $a$ | | | | | | | | |
| $b$ | $a$ | $b$ | | | | | | | | |
| $c$ | | | $c$ | $c$ | | | | | | |
| $d$ | | | $c$ | $d$ | | | | | | |
| $e$ | | | | | $e$ | $e$ | $e$ | $e$ | | |
| $f$ | | | | | $e$ | $f$ | $g$ | $h$ | | |
| $g$ | | | | | $e$ | $g$ | $g$ | $e$ | | |
| $h$ | | | | | $e$ | $h$ | $e$ | $h$ | | |
| $i$ | | | | | | | | | $i$ | $i$ |
| $j$ | | | | | | | | | $i$ | $j$ |

| $\circ$ | $a$ | $b$ | $c$ | $d$ | $e$ | $f$ | $g$ | $h$ | $i$ | $j$ |
|---|---|---|---|---|---|---|---|---|---|---|
| $a$ | $a$ | $a$ | $c$ | $c$ | | | | | | |
| $b$ | $a$ | $b$ | $c$ | $d$ | | | | | | |
| $c$ | | | | | $c$ | $c$ | $c$ | $c$ | $a$ | $a$ |
| $d$ | | | | | $c$ | $d$ | $d$ | $d$ | $a$ | $b$ |
| $e$ | | | | | $e$ | $e$ | $e$ | $e$ | $i$ | $i$ |
| $f$ | | | | | $e$ | $f$ | $f$ | $f$ | $i$ | $j$ |
| $g$ | | | | | $e$ | $f$ | $g$ | $h$ | $i$ | $j$ |
| $h$ | | | | | $e$ | $f$ | $g$ | $h$ | $i$ | $j$ |
| $i$ | $i$ | $i$ | $e$ | $e$ | | | | | | |
| $j$ | $i$ | $j$ | $e$ | $f$ | | | | | | |

| | $.^{\mathsf{T}}$ | $\overline{.}$ |
|---|---|---|
| $a$ | $a$ | $b$ |
| $b$ | $b$ | $a$ |
| $c$ | $i$ | $d$ |
| $d$ | $j$ | $c$ |
| $e$ | $e$ | $f$ |
| $f$ | $f$ | $e$ |
| $g$ | $g$ | $h$ |
| $h$ | $h$ | $g$ |
| $i$ | $c$ | $j$ |
| $j$ | $d$ | $i$ |

| $R$ | $O_R$ | $L_R$ | $I^R$ | ${}^{R}I$ |
|---|---|---|---|---|
| $a$ | $a$ | $b$ | $b$ | $b$ |
| $b$ | $a$ | $b$ | $b$ | $b$ |
| $c$ | $c$ | $d$ | $b$ | $g$ |
| $d$ | $c$ | $d$ | $b$ | $g$ |
| $e$ | $e$ | $f$ | $g$ | $g$ |
| $f$ | $e$ | $f$ | $g$ | $g$ |
| $g$ | $e$ | $f$ | $g$ | $g$ |
| $h$ | $e$ | $f$ | $g$ | $g$ |
| $i$ | $i$ | $j$ | $g$ | $b$ |
| $j$ | $i$ | $j$ | $g$ | $b$ |

**Fig. A.2.1** A 10-element heterogeneous relation algebra

**A.2.2 Proposition.** In every relation algebra it is possible to prove the following rules without applying the Tarski rule:

**Rules of multiplication:**

i) $OR = RO = O$;

ii) $R \subset S \implies QR \subset QS,\ RQ \subset SQ$ $\qquad\qquad$ (monotonicity);

iii) $Q(R \sqcap S) \subset QR \sqcap QS$ $\quad (R \sqcap S)Q \subset RQ \sqcap SQ$ $\quad$ ($\sqcap$-subdistributivity),

$\quad Q(R \sqcup S) = QR \sqcup QS$ $\quad (R \sqcup S)Q = RQ \sqcup SQ$ $\quad$ ($\sqcup$-distributivity);

**Rules of transposition:**

iv) $(R^{\mathsf{T}})^{\mathsf{T}} = R$;

v) $(RS)^{\mathsf{T}} = S^{\mathsf{T}}R^{\mathsf{T}}$;

vi) $R \subset S \iff R^{\mathsf{T}} \subset S^{\mathsf{T}}$;

vii) $\overline{R}^{\mathsf{T}} = \overline{R^{\mathsf{T}}}$;

viii) $(R \sqcup S)^{\mathsf{T}} = R^{\mathsf{T}} \sqcup S^{\mathsf{T}}$;

$\quad\ (R \sqcap S)^{\mathsf{T}} = R^{\mathsf{T}} \sqcap S^{\mathsf{T}}$.

In a homogeneous relation algebra we have in addition:

$$I = I^{\mathsf{T}}, \quad O = O^{\mathsf{T}}, \quad L = L^{\mathsf{T}}.$$

**Proof:** Cases (i, ii, iii) have already been proved relation-algebraically in Proposition 2.3.6.

iv) $RI \subset R \iff R^{\mathsf{T}}\overline{R} \subset \overline{I} \iff R^{\mathsf{T}\mathsf{T}}I \subset \overline{R} \iff R^{\mathsf{T}\mathsf{T}} \subset R$;
$\quad R^{\mathsf{T}\mathsf{T}}I \subset R^{\mathsf{T}\mathsf{T}} \iff R^{\mathsf{T}}\overline{R^{\mathsf{T}\mathsf{T}}} \subset \overline{I} \iff RI \subset R^{\mathsf{T}\mathsf{T}} \iff R \subset R^{\mathsf{T}\mathsf{T}}$.

v) $(RS)^{\mathsf{T}}I \subset (RS)^{\mathsf{T}} \iff RS\overline{(RS)^{\mathsf{T}}} \subset \overline{I} \iff R^{\mathsf{T}}I \subset \overline{S\overline{(RS)^{\mathsf{T}}}} \iff$
$\quad S\overline{(RS)^{\mathsf{T}}} \subset \overline{R^{\mathsf{T}}} \iff S^{\mathsf{T}}R^{\mathsf{T}} \subset (RS)^{\mathsf{T}}$;
$\quad S^{\mathsf{T}}R^{\mathsf{T}} \subset S^{\mathsf{T}}R^{\mathsf{T}} \iff SS^{\mathsf{T}}R^{\mathsf{T}} \subset \overline{R^{\mathsf{T}}} \iff R^{\mathsf{T}}I \subset \overline{SS^{\mathsf{T}}R^{\mathsf{T}}} \iff$
$\quad RS\overline{S^{\mathsf{T}}R^{\mathsf{T}}} \subset \overline{I} \iff (RS)^{\mathsf{T}} \subset S^{\mathsf{T}}R^{\mathsf{T}}$.

vi) $S^{\mathsf{T}}I \subset S^{\mathsf{T}} \iff S\overline{S^{\mathsf{T}}} \subset \overline{I} \iff I\overline{S^{\mathsf{T}}}^{\mathsf{T}} \subset \overline{S} \implies I\overline{S^{\mathsf{T}}}^{\mathsf{T}} \subset \overline{R} \iff$
$\quad R\overline{S^{\mathsf{T}}} \subset \overline{I} \iff R^{\mathsf{T}}I \subset S^{\mathsf{T}} \iff R^{\mathsf{T}} \subset S^{\mathsf{T}}$. "$\impliedby$" is proved dually.

vii) $RI \subset R \iff R^{\mathsf{T}}\overline{R} \subset \overline{I} \iff I\overline{R}^{\mathsf{T}} \subset \overline{R^{\mathsf{T}}} \iff \overline{R}^{\mathsf{T}} \subset \overline{R^{\mathsf{T}}}$. Reasoning this way with $R^{\mathsf{T}}$ instead of $R$ yields $\overline{R^{\mathsf{T}}}^{\mathsf{T}} \subset \overline{R}$, and therefore $\overline{R^{\mathsf{T}}} \subset \overline{R}^{\mathsf{T}}$.

viii) From $Q := R^{\mathsf{T}} \sqcup S^{\mathsf{T}}$ we get $I\overline{R^{\mathsf{T}}} \subset Q \iff \overline{Q}R \subset \overline{I}$ and $I\overline{S^{\mathsf{T}}} \subset Q \iff$
$\quad \overline{Q}S \subset \overline{I}$; therefore, $\overline{Q}(R \sqcup S) = \overline{Q}R \sqcup \overline{Q}S \subset \overline{I} \iff I(R \sqcup S)^{\mathsf{T}} \subset Q \iff$
$\quad (R \sqcup S)^{\mathsf{T}} \subset R^{\mathsf{T}} \sqcup S^{\mathsf{T}}$. "$\supset$" results from (vi) and (iv).

For homogeneous relations we have: $I\overline{I} = \overline{I} \implies I\overline{I} \subset \overline{I} \implies I^{\mathsf{T}} = I^{\mathsf{T}}I \subset I$. Transposition yields $I = I^{\mathsf{T}\mathsf{T}} \subset I^{\mathsf{T}}$. In any case $O \subset O^{\mathsf{T}}$, and $OL = O \subset \overline{I}$ implies $O^{\mathsf{T}} = O^{\mathsf{T}}I \subset \overline{L} = O$. From that we obtain $L = \overline{O} = \overline{O^{\mathsf{T}}} = \overline{O}^{\mathsf{T}} = L^{\mathsf{T}}$.  $\square$

The relation-algebraic proof of the Dedekind rule,

$$QR \sqcap S \subset (Q \sqcap SR^{\mathsf{T}})(R \sqcap Q^{\mathsf{T}}S),$$

was given as Exercise 2.3.3. This rule appeared in Lorenzen 54 and Gericke 63 as the "Bund-Axiom". It holds in every relation algebra, and so do the other relation-algebraically derived results of Chap. 2. (Note that the existence of points (Prop. 2.4.6) could not be established in this way.)

For the sake of simplicity let us now turn to homogeneous relation algebras again where we no longer face the problem of applicability of the partially defined operations such as multiplication in partial algebras.

The investigation of the structure of homogeneous relation algebras can be achieved by "classical" mathematical means such as "representing" an algebra in a ring of matrices. Since a homogeneous relation algebra is in particular a complete, atomic lattice, its structure theory is based on Stone's results as described in Sect. A.1.

| .⊔. | a | b | c | d |
|---|---|---|---|---|
| a | a | b | c | d |
| b | b | b | d | d |
| c | c | d | c | d |
| d | d | d | d | d |

| .П. | a | b | c | d |
|---|---|---|---|---|
| a | a | a | a | a |
| b | a | b | a | b |
| c | a | a | c | c |
| d | a | b | c | d |

| .o. | a | b | c | d |
|---|---|---|---|---|
| a | a | a | a | a |
| b | a | b | c | d |
| c | a | c | b | d |
| d | a | d | d | d |

| | .⊤ | ⁻ |
|---|---|---|
| a | a | d |
| b | b | c |
| c | c | b |
| d | d | a |

**Fig. A.2.2**  Nonconcrete relation algebra

There are different approaches in the literature to solving the representation problem.  In essence, an underlying set has to be constructed for which the

| .⊔. / .П. | O | I | a | b | c | u | v | w | x | y | z | Ī | ā | b̄ | c̄ | L |
|---|---|---|---|---|---|---|---|---|---|---|---|---|---|---|---|---|
| O | O | I | a | b | c | u | v | w | x | y | z | Ī | ā | b̄ | c̄ | L |
| I | O | I | u | v | x | u | v | c̄ | x | b̄ | ā | L | ā | b̄ | c̄ | L |
| a | O | O | u | w | y | u | c̄ | w | b̄ | y | Ī | Ī | L | b̄ | c̄ | L |
| b | O | O | O | b | z | c̄ | v | w | ā | Ī | z | Ī | ā | L | c̄ | L |
| c | O | O | O | O | c | b̄ | ā | Ī | x | y | z | Ī | ā | b̄ | L | L |
| u | O | I | a | O | O | u | c̄ | c̄ | b̄ | b̄ | L | L | L | b̄ | c̄ | L |
| v | O | I | O | b | O | I | v | c̄ | ā | L | ā | L | ā | L | c̄ | L |
| w | O | O | a | b | O | a | b | w | L | Ī | Ī | Ī | L | L | c̄ | L |
| x | O | I | O | O | c | I | I | O | x | b̄ | ā | L | ā | b̄ | L | L |
| y | O | O | a | O | c | a | O | a | c | y | Ī | Ī | L | b̄ | L | L |
| z | O | O | O | b | c | O | b | b | c | c | z | Ī | ā | L | L | L |
| Ī | O | O | a | b | c | a | b | w | c | y | z | Ī | L | L | L | L |
| ā | O | I | O | b | c | I | v | b | x | c | z | z | ā | L | L | L |
| b̄ | O | I | a | O | c | u | I | a | x | y | c | y | x | b̄ | L | L |
| c̄ | O | I | a | b | O | u | v | w | I | a | b | w | v | u | L | L |
| L | O | I | a | b | c | u | v | w | x | y | z | Ī | ā | b̄ | c̄ | L |

| | .⊤ | ⁻ |
|---|---|---|
| O | L | O |
| I | Ī | I |
| a | ā | c |
| b | b̄ | c |
| c | c̄ | a |
| u | z | x |
| v | y | v |
| w | x | z |
| x | w | u |
| y | v | y |
| z | u | w |
| Ī | Ī | I |
| ā | a | c̄ |
| b̄ | b | b̄ |
| c̄ | c | ā |
| L | O | L |

| .o. | O | I | a | b | c | u | v | w | x | y | z | Ī | ā | b̄ | c̄ | L |
|---|---|---|---|---|---|---|---|---|---|---|---|---|---|---|---|---|
| O | O | O | O | O | O | O | O | O | O | O | O | O | O | O | O | O |
| I | O | I | a | b | c | u | v | w | x | y | z | Ī | ā | b̄ | c̄ | L |
| a | O | a | a | w | L | a | w | w | L | L | L | L | L | L | w | L |
| b | O | b | w | b̄ | z | w | L | L | z | Ī | L | L | L | Ī | L | L |
| c | O | c | L | w | c | L | z | L | c | L | z | L | z | L | L | L |
| u | O | u | a | w | L | u | c̄ | w | L | L | L | L | L | L | c̄ | L |
| v | O | v | w | L | z | c̄ | L | L | ā | Ī | L | L | L | L | L | L |
| w | O | w | w | L | L | w | L | L | L | L | L | L | L | L | L | L |
| x | O | x | L | z | c | L | ā | L | x | L | z | L | ā | L | L | L |
| y | O | y | L | Ī | L | L | Ī | L | L | L | L | L | L | L | L | L |
| z | O | z | L | L | z | L | L | L | z | L | L | L | L | L | L | L |
| Ī | O | Ī | L | L | L | L | L | L | L | L | L | L | L | L | L | L |
| ā | O | ā | L | L | z | L | L | L | ā | L | L | L | L | L | L | L |
| b̄ | O | b̄ | L | Ī | L | L | L | L | L | L | L | L | L | L | L | L |
| c̄ | O | c̄ | w | L | L | c̄ | L | L | L | L | L | L | L | L | L | L |
| L | O | L | L | L | L | L | L | L | L | L | L | L | L | L | L | L |

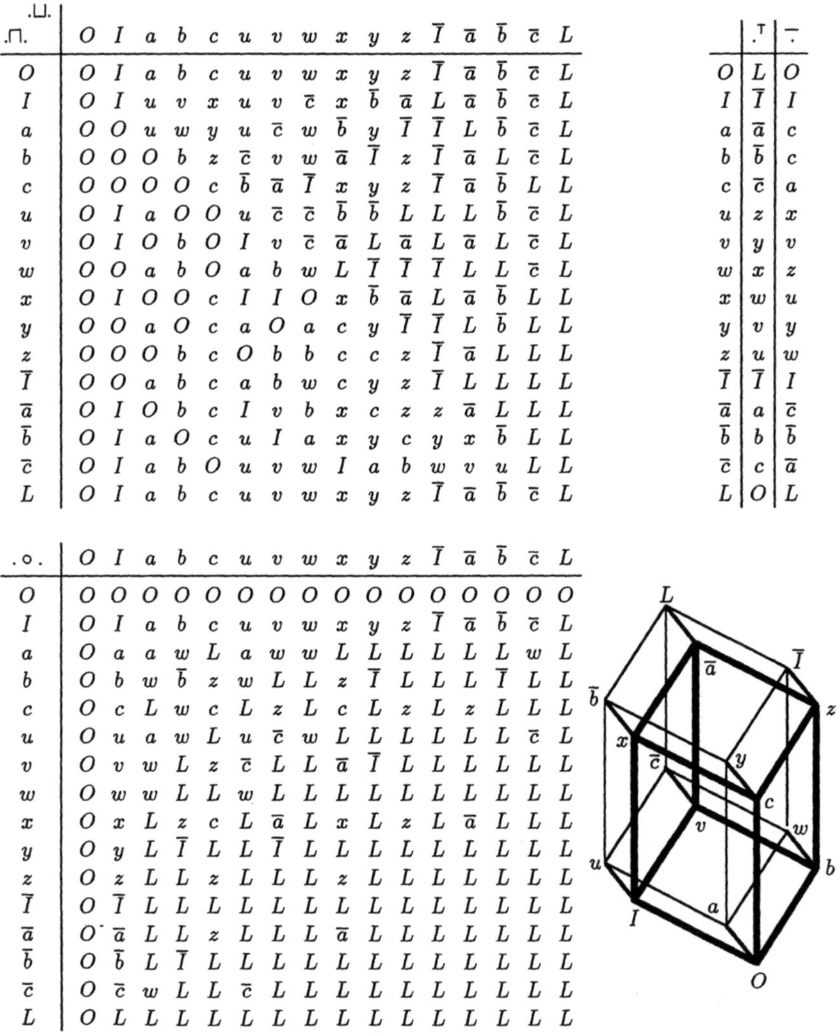


**Fig. A.2.3**  McKenzie's nonrepresentable relation algebra

relation algebra can be interpreted as relations in the classical sense. This cannot be achieved without making additional assumptions: The relation algebra in Fig. A.2.2, for instance, has cardinality 4 ($\neq 2^{n \times n}$) and therefore cannot be the algebra of all relations on some set. One also sees that $a$ is the zero element, $d$ the universal element, and $b, c$ are the elements $I$ and $\overline{I}$, which can equally be read off the representation by Boolean $2 \times 2$-matrices in Fig. 7.5.5. None of these four elements is a point because $a = O$, $b \neq bL$, $c \neq cL$ and $dd^{\mathsf{T}} \not\subseteq I$.

With regard to McKenzie's relation algebra of Fig. A.2.3, the situation is even worse. There are four elements which might be called generators, namely $I, a, b, c$; they are the upper neighbors of $O$. Each of the 16 elements can be expressed as a $\sqcup$-union of a subset of these four, so that a Boolean lattice is established giving a four-dimensional cube. So $O$ corresponds to the union of none and $L$ to the union of all of them, $a$ to the union of just one, and so forth. For instance, $\overline{a}$ means that all but $a$ are united. Furthermore, we have used the abbreviations

$$u := I \sqcup a, \quad v := I \sqcup b, \quad w := a \sqcup b,$$

$$x := I \sqcup c, \quad y := a \sqcup c, \quad z := b \sqcup c.$$

It is a little bit tedious to check that all the laws of a relation algebra are satisfied. This may be slightly facilitated by giving the set

$$I^{\mathsf{T}} = I, \quad a^{\mathsf{T}} = c, \quad b^{\mathsf{T}} = b, \quad c^{\mathsf{T}} = a, \quad a^2 = a, \quad b^2 = \overline{b}, \quad c^2 = c,$$

$$ac = ca = L, \quad ab = ba = a \sqcup b, \quad cb = bc = c \sqcup b$$

of basic compositions, from which the rest of the tables can be elaborated.

Assume now that there exists a representation for the elements of this relation algebra as a set of 16 Boolean $n \times n$-matrices. We will prove that this assumption leads to a contradiction, so that such a representation cannot exist. As $a$ satisfies $a^2 = a \subset \overline{I}$ and $a \sqcap a^{\mathsf{T}} = a \sqcap c = O$, it corresponds to a transitive and irreflexive matrix which is therefore a strict-ordering. By definition, $c$ is the converse of $a$, so that $b$ gives the relationship of incomparability with respect to this strict-ordering $a$ since $\overline{b} = I \sqcup a \sqcup c$. From $aa^{\mathsf{T}} = L$ it follows that any pair of elements has a common upper bound which is different from both of them. On the other hand, $\overline{b} = b^2$ demands that any pair of comparable elements has some element to which they are both incomparable.

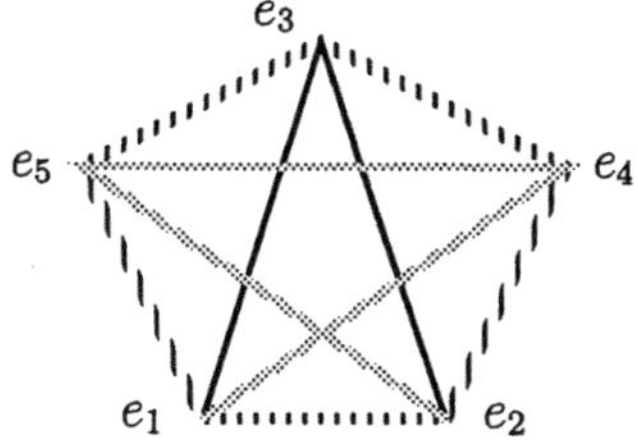


**Fig. A.2.4** Case analysis

Now we switch to the usual notation $<$ for the strict-ordering $a$. This is the crucial step. We are no longer satisfied with the operation tables of Fig. A.2.3, but are looking for some set together with a strict-ordering satisfying the properties just mentioned. Since $b \neq O$, there exist at least two incomparable elements $e_1, e_2$ in the set. They have some common upper bound $e_3$ with $e_1 < e_3, e_2 < e_3$, so that elements $e_4, e_5$ must exist with all the pairs $(e_1, e_5), (e_3, e_5), (e_2, e_4)$, $(e_3, e_4)$ incomparable. However, from consecutive incomparability we may then conclude that $(e_2, e_5), (e_1, e_4), (e_4, e_5)$ are comparable pairs of elements. Now a case analysis leads to a contradiction: First we have that $e_1 < e_4, e_2 < e_5$, since $e_4, e_5$ must not be less than $e_3$. Then $e_4 \neq e_5$ since, by assumption, $e_1, e_2$ are incomparable which would contradict $(e_1, e_5), (e_2, e_4)$ (with $e_4 = e_5$) being incomparable. In this situation there is no possibility to decide whether $e_4 < e_5$ or $e_5 < e_4$, without introducing the forbidden relationships $e_1 < e_5$ or $e_2 < e_4$ by transitivity.

The problem just encountered again amounts to finding rows and columns of a matrix or, equivalently, to finding points in the sense discussed earlier. In an abstract relation algebra there may not be enough points. Additional assumptions are necessary to ensure the existence of sufficiently many points.

Under relatively strong such assumptions MCKINSEY 48 gave a construction of a representation of relation algebras. LYNDON 50/56 gave a complicated example showing that the answer is negative in general (but compare VELOSO 77/79). There is a construction method for non-representable relation algebras based on their similarity with projective algebras. According to MONK 64 the class of representable relation algebras is not finitely axiomatizable.

We now turn to our approach to the representation of a relation algebra. In Sect. 2.4 we introduced the algebraic concept of points. They are atoms among the vectors (see Prop. 2.4.5.i), and they also describe in essence the atoms among the relations (see Exercise 2.4.2). As a replacement for Prop. 2.4.6 we now postulate the existence of sufficiently many points.

**A.2.3 Definition.** A relation algebra satisfies the **point axiom**, if

$$R \neq O \quad \Longrightarrow \quad \text{There exist points } x, y \text{ with } xy^{\mathsf{T}} \subset R. \qquad \square$$

This point axiom has already been used in several proofs, in particular for establishing Props. 2.4.7, 2.4.8, and 6.1.3. From it follows the decomposition of the identity

$$I = \sup\{\, xx^{\mathsf{T}} \mid x \text{ point} \,\}.$$

We now give the solution to the representation problem for relation algebras satisfying the point axiom. A similar result was independently published by MADDUX 78 and MADDUX, TARSKI 76 and, with a simplified proof, by SCHMIDT, STRÖHLEIN 85. However, a very general investigation was already carried out by JÓNSSON, TARSKI 52.

**A.2.4 Proposition.** Let $\mathcal{R}$ be a relation algebra satisfying the point axiom, and denote the set of its points by $\mathcal{P}$. Then $\mathcal{R}$ is isomorphic to the algebra of relations on the set $\mathcal{P}$.

**Proof:** We refer to Prop. 2.4.5, as well as to Exercises 2.4.2 and 2.4.3. Because of the point axiom all the atoms in the relation algebra may be written as a product $xy^{\mathsf{T}}$; this representation is, in addition, unique.

Now we compare the relation algebra $\mathcal{R}$ with $\mathbf{2}^{\mathcal{P} \times \mathcal{P}}$ by defining the mapping $\mu \colon \mathcal{R} \longrightarrow \mathbf{2}^{\mathcal{P} \times \mathcal{P}}$ by

$$\mu(R) := \{\, (x,y) \in \mathcal{P} \times \mathcal{P} \mid xy^{\mathsf{T}} \subset R \,\}.$$

The mapping $\mu$ assigns to every relation $R$ of $\mathcal{R}$ the set of all atoms (= pairs of points) contained in $R$.

As one knows from lattice theory, such a mapping establishes an isomorphism for the structure of a complete atomic Boolean lattice.

So it remains to prove that $\mu(RS) = \mu(R)\mu(S)$ and $\mu(R^{\mathsf{T}}) = [\mu(R)]^{\mathsf{T}}$, which we do using the intermediate point theorem 2.4.8 and applying Prop. 2.4.4.i several times:

$$\mu(RS) = \{\, (x,y) \mid xy^{\mathsf{T}} \subset RS \,\} = \{\, (x,y) \mid \exists z : xz^{\mathsf{T}} \subset R \text{ and } zy^{\mathsf{T}} \subset S \,\}$$
$$= \{\, (x,y) \mid xy^{\mathsf{T}} \subset R \,\}\{\, (x,y) \mid xy^{\mathsf{T}} \subset S \,\} = \mu(R)\mu(S).$$
$$\mu(R^{\mathsf{T}}) = \{\, (x,y) \mid xy^{\mathsf{T}} \subset R^{\mathsf{T}} \,\} = \{\, (y,x) \mid xy^{\mathsf{T}} \subset R \,\} = [\mu(R)]^{\mathsf{T}}. \qquad \square$$

Matrices with entries in a relation algebra, again form a relation algebra as is easily verified:

**A.2.5 Proposition.** If $\mathcal{R}$ is a homogeneous relation algebra, then the $n \times n$-matrices with coefficients in $\mathcal{R}$ form again a relation algebra, if one defines $\sqcup$, $\sqcap$, $^{-}$, $O$, $L$ componentwise and defines transposition and multiplication, in contrast to Definition 2.3.1, by

$$S^{\mathsf{T}}(i,k) := [S(k,i)]^{\mathsf{T}};$$
$$RS(i,k) := \mathbf{sup}\{\, R(i,j)S(j,k) \mid 1 \le j \le n \,\}. \qquad \square$$

Our treatment, in Chap. 10, of graphs with arrows weighted by relations as relation algebras is therefore justified.

We add a few historical remarks. De Morgan and Peirce were the first to systematically develop such algebras (see, e.g., BRINK 79A and MICHAEL 79). Their work was continued by E. Schröder with his *Algebra der Logik* where he studied operations applied to relations without using that "these relations hold between individuals". This algebraic treatment was continued, notably by L. Löwenheim. The question has arisen to which extent an algebraic formulation is possible for relationships which one would otherwise express in terms of predicate logic. The answer has been made precise by MADDUX 83: he proved that an equation is valid in every relation algebra, if and only if its predicate-logic form can be deduced using at most 4 variables.

Tarski (1941) and Riguet (1948) founded the modern "calculus of relations". A theory of "projective algebra" was developed independently, starting with EVERETT, ULAM 46. Subtle refinements concerning "cylindrical algebras" are discussed extensively in HENKIN, MONK, TARSKI, ANDRÉKA, NÉMETI 81. The interrelationships between these types of algebras were soon recognized, see CHIN, TARSKI 48, McKINSEY 48, LYNDON 61, MONK 61 or BEDNAREK, ULAM 78. Completeness and decidability were investigated by KALICKI, SCOTT 55, TARSKI 56 and SCHÖNFELD 79. WADGE 76 gave a complete natural deduction system for the relational calculus. The consequences of weakening the associativity law in a relation algebra to semi-associativity has been studied by MADDUX 90.

# A.3 Fixedpoint Theorems and Antimorphisms

At several occasions we determined a relation $X$ as a solution to a fixedpoint equation $X = \rho(X)$. In each case the questions as to existence and computability by iteration were decided according to the same pattern, depending on the lattice-theoretic properties of $\rho$ but not so much on the theory of relations. The first results in this direction are due to A. Tarski and B. Knaster.

A mapping $\rho: V \longrightarrow W$ between complete lattices[1] is called **isotonic** (or monotonic), if the following conditions, which are easily seen to be equivalent, hold (as usual, $\rho(S) = \{\, \rho(s) \mid s \in S \,\}$):

$$\rho(x) \subset \rho(y) \qquad \text{if } x, y \in V \text{ and } x \subset y;$$
$$\sup \rho(S) \subset \rho(\sup S) \quad \text{for all } S \subset V;$$
$$\rho(\inf S) \subset \inf \rho(S) \quad \text{for all } S \subset V.$$

**A.3.1 Proposition** (*Existence of least fixedpoints of isotone mappings*). In a complete lattice $V$ every isotone mapping $\rho: V \longrightarrow V$ has a (uniquely determined) least fixedpoint $a$. It may also be described as the least **contracted element** (**post-fixedpoint**)

$$a = \inf \{\, x \mid \rho(x) = x \,\} = \inf \{\, x \mid \rho(x) \subset x \,\}.$$

**Proof:** Consider the set $\mathcal{A} := \{\, x \mid \rho(x) \subset x \,\}$ of all contracted elements of $V$. This set is nonempty since the greatest element of the lattice is obviously contracted. We define $a := \inf \mathcal{A}$, i.e., we use the second characterization as a definition; in the next paragraph, it will be shown that $a$ thus defined is a fixedpoint. It is then the least fixedpoint, since it belongs to $\mathcal{A}$.

For all $x \in \mathcal{A}$ we have by definition, that $a \subset x$, and hence $\rho(a) \subset \rho(x)$ because $\rho$ is isotonic. Since the elements $x$ considered are all contracted elements this may be continued by $\rho(x) \subset x$, so that $\rho(a)$ is another lower bound for $\mathcal{A}$; it is necessarily less than or equal to the *greatest* lower bound $a$; i.e., $\rho(a) \subset a$. This, however, makes $a$ a contracted element which therefore belongs to $\mathcal{A}$. By isotonicity we have $\rho(\rho(a)) \subset \rho(a)$, therefore $\rho(a) \in \mathcal{A}$, and hence $a \subset \rho(a)$. Combining this with the preceding result we obtain indeed that $\rho(a) = a$.    $\square$

Replacing $\subset$ by $\supset$, we have the dual result:

**A.3.2 Proposition** (*Existence of greatest fixedpoints of isotone mappings*). In a complete lattice $V$ every isotone mapping $\rho: V \longrightarrow V$ has a (uniquely determined) greatest fixedpoint $a$. It may also be described as the greatest **expanded element** (**pre-fixedpoint**)

$$b = \sup \{\, x \mid x = \rho(x) \,\} = \sup \{\, x \mid x \subset \rho(x) \,\}.$$    $\square$

---

[1] Some of the results of this section may also be proved in a slightly more general setting, for instance in complete partial orderings.

It should be noted, however, that the amount of symmetry holding in lattices does not carry over fully when these are applied to relation algebra. For instance, the first of the two propositions above is of markedly greater relevance when the mapping $\rho$ is of the form $\rho(x) = Bx$. In the condition $Bx \subset x$ the product is situated on the lesser side and therefore lends itself more readily to algebraic manipulation; whereas the position of the product on the greater side in $x \subset Bx$ has proved to render transformations more difficult.

So far, the least and the greatest fixedpoints of $\rho$ have been established by the usual mathematical means in a *descriptive* way. This does not yet help much for actually *computing* the fixedpoints, apart from the case of a finite lattice where one can always check all possibilities. But for infinite lattices one has to find ways of approximating the fixedpoints or computing them by iteration. Unfortunately, iteration yields, in general, only a lower bound for the least fixedpoint, or an upper bound for the greatest. Additional conditions on the isotone mapping $\rho$ are necessary to ensure that the respective fixedpoint can be approximated.

**A.3.3 Proposition** (*Iterative bounds for the least fixedpoint of an isotone mapping*). Let $\rho\colon V \longrightarrow V$ be an isotone mapping of a complete lattice $V$ into itself, and let $a_0$ be an element of $V$ satisfying $a_0 \subset \rho(a_0) \subset a$ (usually chosen as $a_0 := O$). Then the iteration

$$a_{i+1} := \rho(a_i) \quad \text{for } i \geq 0$$

yields a lower bound $a_\infty := \sup_i a_i$ for the least fixedpoint $a$ of $\rho$.

**Proof:** We know already that a least fixedpoint $a$ exists. For the iteration sequence we prove

$$a_0 \subset a_1 \subset a_2 \subset \ldots \subset a.$$

We have $a_0 \subset a$ by hypothesis. Using that $\rho$ is isotonic, we obtain $a_1 = \rho(a_0) \subset \rho(a) = a$. Together we have the initial step $a_0 \subset a_1 \subset a$ of an inductive argument. Assuming $a_{i-1} \subset a_i \subset a$, we conclude for isotone $\rho$ that $a_i = \rho(a_{i-1}) \subset a_{i+1} = \rho(a_i) \subset \rho(a) = a$. The least fixedpoint $a$ is therefore an upper bound for the monotone increasing sequence and hence contains the least upper bound $a_\infty$. $\qquad\Box$

The dual result is:

**A.3.4 Proposition** (*Iterative bounds for the greatest fixedpoint of isotone mappings*). Let $\rho\colon V \longrightarrow V$ be an isotone mapping of a complete lattice into itself. Then, starting from any element $b_0$ satisfying $b \subset \rho(b_0) \subset b_0$ (usually chosen to be $b_0 := L$), the iteration

$$b_{i+1} := \rho(b_i) \quad \text{for } i \geq 0$$

yields an upper bound $b_\infty := \inf_i b_i$ for the greatest fixedpoint $b$. $\qquad\Box$

With isotone $\rho$ we began in the starting condition

$$a_0 \subset \rho(a_0) \subset a,$$

and concluded that the fixedpoint $a$ exists, and that it is bounded from below by the iterative sequence

$$a_0 \subset a_1 \subset \ldots \subset a_\infty \subset a.$$

Since $a_i \subset a_\infty \subset a$, we have moreover $a_{i+1} = \rho(a_i) \subset \rho(a_\infty) \subset \rho(a) = a$, so that the above estimate can be refined to

$$a_0 \subset a_1 \subset \ldots \subset a_\infty \subset \rho(a_\infty) \subset a.$$

In general, one will not have equality at any place in this chain. The iteration could be started afresh with $a_\infty$ since it satisfies the starting condition, and one would hope to get still nearer to the fixedpoint $a$. We will not, however, pursue this "transfinite continuation" of the iteration process beyond the first accumulation point $a_\infty$ because this is "not accesssible to computation". Instead, we want to find conditions ensuring that $a = a_\infty$, i.e., that the iteration process approximates $a$. So far we have only

$$\rho(a_\infty) = \rho(\sup_{i \geq 0} a_i) \supset \sup_{i \geq 0} \rho(a_i) = \sup_{i \geq 0} a_{i+1} = a_\infty.$$

In the following definition we postulate that equality holds in this chain.


**A.3.5 Definition.** The mapping $\rho \colon V \longrightarrow V$ of a complete lattice into itself is called **(upward) continuous**, if for any subset $S \subset V$

$$\rho(\sup S) = \sup \rho(S). \qquad \square$$


Continuous mappings are isotonic: Consider the two-element set $S = \{a, b\}$ with $a \subset b$. One has $\sup S = b$, hence $\rho(b) = \rho(\sup S) = \sup(\rho(S)) = \sup\{\rho(a), \rho(b)\}$, and therefore $\rho(a) \subset \rho(b)$.


**A.3.6 Corollary** (*Approximation of the least fixedpoint of a continuous mapping*). If $\rho$ is not only isotonic but continuous, then the lower bound given by the iteration is the least fixedpoint $a$, i.e.,

$$a_0 \subset a_1 \subset \ldots \subset a_\infty = a. \qquad \square$$


There is a dual version to this, which, however, requires continuity with respect to the *downward* direction as opposed to continuity upward used before. Indeed, for a **downward continuous** mapping $\rho$, i.e., one which satisfies

$$\rho(\inf S) = \inf \rho(S),$$

the greatest fixedpoint can be approximated:

$$b = b_\infty \subset \ldots \subset b_1 \subset b_0.$$

## Fixedpoints Determined by Pairs of Antitone Mappings

In many instances the isotone mapping $\rho$ of some complete lattice into itself is given as the composite of two antitone mappings. In this situation more special fixedpoint theory can be developed. Examples have already come up in Sect. 6.3 on the initial part, in Sect. 8.2 on the existence of kernels for graphs, and in Sect. 9.3 on correspondences in bipartite graphs.

The situation is the following: There are two complete lattices given (which may coincide), as well as two antitone mappings between them (which may equally be the same).

**A.3.7 Definition.** A mapping $\sigma\colon V \longrightarrow W$ of a complete lattice $V$ into a complete lattice $W$ is called **antitonic**, provided one of the following equivalent statements holds:

i) $\qquad\qquad\qquad \sigma(v') \subset \sigma(v) \qquad$ if $v \subset v'$;

ii) $\qquad\qquad\quad \sigma(\sup S) \subset \inf \sigma(S) \quad$ for all $S \subset V$;

iii) $\qquad\qquad\quad \sup \sigma(S) \subset \sigma(\inf S) \quad$ for all $S \subset V$. $\qquad\qquad\quad$ $\square$

So the *general assumption* is that two antitone mappings between complete lattices, $\sigma\colon V \longrightarrow W$ and $\pi\colon W \longrightarrow V$, are given with their composites $\rho\colon V \longrightarrow V$ and $\varphi\colon W \longrightarrow W$ defined by $\rho(v) := \pi\left(\sigma(v)\right)$ and $\varphi(w) := \sigma\left(\pi(w)\right)$. Clearly, $\rho$ and $\varphi$ are isotone mappings of the complete lattice $V$ or $W$ into itself, respectively, to which we can apply our previous results on the existence of fixedpoints, their approximation, and bounds. (We do *not* suppose that $v \subset \rho(v)$ and $w \subset \varphi(w)$ always hold, as for Galois correspondences.)

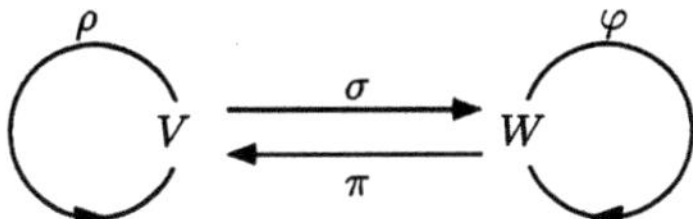

**Fig. A.3.1** Pair of antitone mappings

**A.3.8 Proposition** (*Existence of extremal fixedpoints for a pair of antitone mappings*). Under the general assumption the following holds:

i) $\rho$ and $\varphi$ are isotonic.

ii) $\rho$ has a uniquely determined least fixedpoint $a$, representable also as the least contracted element,
$$a = \inf\left\{\, v \mid \rho(v) \subset v \,\right\} = \inf\left\{\, v \mid \rho(v) = v \,\right\}.$$

iii) $\varphi$ has a uniquely determined greatest fixedpoint $b$, representable also as the greatest expanded element,
$$b = \sup\left\{\, w \mid w \subset \varphi(w) \,\right\} = \sup\left\{\, w \mid w = \varphi(w) \,\right\}.$$

iv) The extremal fixedpoints are related by
$$b = \sigma(a), \quad \pi(b) = a.$$

**Proof:** (i), (ii), (iii) are clear from the preceding remarks. So it remains to prove the statement (iv) about the position of the extremal fixedpoints relative to one another. First, we prove only one half of the result:

$$\sup\{\,w \mid \varphi(w) = w\,\} = b \subset \sigma(a), \quad \pi(b) \subset a = \inf\{\,v \mid \rho(v) = v\,\}.$$

For the first case, we pick one of the $w$'s considered. It satisfies $w = \varphi(w) = \sigma\left(\pi(w)\right)$ and, in addition, $\pi(w) = \pi\left(\varphi(w)\right) = \pi\left(\sigma\left(\pi(w)\right)\right) = \rho\left(\pi(w)\right)$, so that $\pi(w)$ is a fixedpoint of $\rho$. Since $a$ is the least fixedpoint of $\rho$, we have $a \subset \pi(w)$, from which we obtain, by antitonicity of $\sigma$, that $w = \varphi(w) = \sigma\left(\pi(w)\right) \subset \sigma(a)$. All $w$'s considered, as well as their supremum $b$, are therefore less than or equal to $\sigma(a)$. The second case is handled dually.

Now applying $\sigma$ to $\pi(b) \subset a$, and $\pi$ to $b \subset \sigma(a)$, yields $b = \varphi(b) = \sigma\left(\pi(b)\right) \supset \sigma(a)$ and $\pi(b) \supset \pi\left(\sigma(a)\right) = \rho(a) = a$ which completes the proof.  $\square$

Again, we have given the solution in a descriptive form only, which cannot be used for computation by iteration. But as in Proposition A.3.3, there are iterative bounds by which the extremal fixedpoints can be approximated, provided certain continuity conditions are met as in Corollary A.3.6.

**A.3.9 Proposition** (*Iterative bounds for the extremal fixedpoints of a pair of antitone mappings*). Under the general assumption the two-sided iteration

$$a_{i+1} := \pi(b_i), \qquad b_{i+1} := \sigma(a_i), \quad i \geq 0$$
$$a_\infty := \sup{}_i\, a_i, \qquad b_\infty := \inf{}_i\, b_i,$$

starting from two elements $a_0, b_0$ satisfying $a_0 \subset \pi(b_0) \subset a$ and $b \subset \sigma(a_0) \subset b_0$ (usually chosen as $a_0 := O$ and $b_0 := L$), produces the following iterative bounds for the extremal fixedpoints $a$ and $b$:

$$a_0 \subset a_1 \subset \ldots \subset a_\infty \subset \pi(b_\infty) \subset a, \quad b \subset \sigma(a_\infty) \subset b_\infty \subset \ldots \subset b_1 \subset b_0.$$

**Proof:** First we prove that choosing $a_0 := O$, $b_0 := L$ subsumes under the general case: Only $\pi(b_0) \subset a$ and $b \subset \sigma(a_0)$ are nontrivial; this, however, follows from applying the antitone functions to $b_0 \supset \sigma(a)$, $\pi(b) \supset a_0$ and using that $a = \pi\left(\sigma(a)\right)$ and $b = \sigma\left(\pi(b)\right)$.

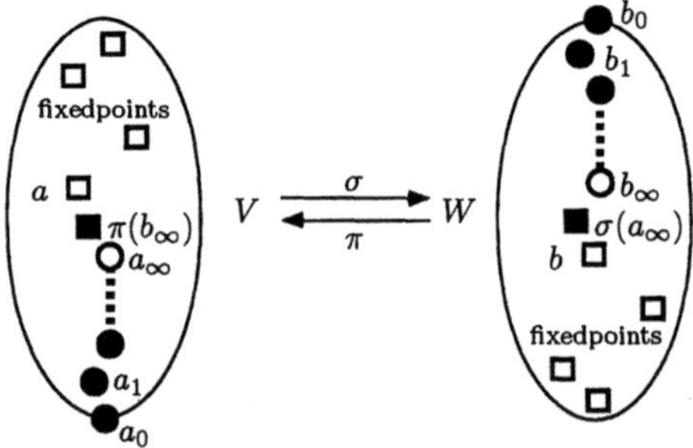


**Fig. A.3.2** Bounds and fixedpoints of antitone mappings

Since $\rho$ and $\varphi$ are isotonic, we are in the situation of Proposition A.3.8, and there is the least fixedpoint $a$ of $\rho$ and the greatest fixedpoint $b$ of $\varphi$. So we start the iteration in the following situation:

$$a_0 \subset a_1 = \pi(b_0) \subset a_2 = \pi(b_1) \subset a, \quad b \subset b_2 = \sigma(a_1) \subset b_1 = \sigma(a_0) \subset b_0.$$

This incorporates already that from

$$a_0 \subset a_1 \subset a, \qquad\qquad\qquad b \subset b_1 \subset b_0$$

one obtains

$$\sigma(a_0) \supset \sigma(a_1) \supset \sigma(a), \qquad\qquad \pi(b) \supset \pi(b_1) \supset \pi(b_0),$$

or $b_1 \supset b_2 \supset b$ and $a \supset a_2 \supset a_1$. Note that there are two nested chains

$$
\begin{array}{ccccccc}
a_0 & \subset & a_2 & \subset & a_4 & \subset & \cdots \\
 & \diagdown\diagup & & \diagdown\diagup & & \diagdown\diagup & \\
 & a_1 & \subset & a_3 & \subset & a_5 & \subset \quad \cdots
\end{array}
$$

of which the first starts from $a_0$ and is generated applying $\rho$, and the second from $a_1 = \pi(b_0)$ again by applying $\rho$; a similar situation prevails for the $b$-chain. For the inductive argument we need therefore two starting conditions. Assume

$$a_{i-1} \subset a_i \subset a \quad \text{and} \quad b \subset b_i \subset b_{i-1} \quad \text{for } i \geq 1.$$

Applying the antitone functions, one obtains

$$\sigma(a_{i-1}) = b_i \supset \sigma(a_i) = b_{i+1} \supset \sigma(a) = b,$$
$$\pi(b) = a \supset \pi(b_i) = a_{i+1} \supset \pi(b_{i-1}) = a_i.$$

So we get indeed with

$$a_0 \subset a_1 \subset a_2 \subset \ldots \subset \sup_i a_i =: a_\infty \subset a$$
$$b_0 \supset b_1 \supset b_2 \supset \ldots \supset \inf_i b_i =: b_\infty \supset b$$

the iterative bounds $a_\infty$ and $b_\infty$ for the two fixedpoints. Applying the antitone functions again, one has

$$a_\infty = \sup_i \pi(b_i) \subset \pi(\inf_i b_i) = \pi(b_\infty), \quad b_\infty = \inf_i \sigma(a_i) \supset \sigma(\sup_i a_i) = \sigma(a_\infty),$$

as well as $\pi(b_\infty) \subset \pi(b) = a$, and $\sigma(a_\infty) \supset \sigma(a) = b$ where it is used that $b_\infty \supset b$ and $a_\infty \subset a$.                 $\square$

We have seen how the general assumption that $\sigma$ and $\pi$ are antitonic, together with the starting conditions, led to the existence of the fixedpoints $a$ and $b$, as well as to their iterative bounds,

$$a_0 \subset a_1 \subset \ldots \subset a_\infty \subset \pi(b_\infty) \subset a, \quad b \subset \sigma(a_\infty) \subset b_\infty \subset \ldots \subset b_1 \subset b_0.$$

In general, there is no equality in these chains of inclusions. If $a_\infty$ and $b_\infty$ are obtained, then one can start the process all over again because this pair satisfies the starting conditions, thus possibly getting closer to $a$ and $b$. But we are more interested in finding conditions ensuring that $a_\infty = a$ and $b = b_\infty$, so that, as before, the fixedpoints can be approximated by the iteration process.

Galois correspondences provide a first result. If $v \subset \rho(v)$ and $w \subset \varphi(w)$ hold, then the decreasing iteration $b_{i+1} = \sigma\left(\pi(b_{i-1})\right) = \varphi(b_{i-1}) \subset b_{i-1}$ becomes stationary with $b_2 = b_1 = b_0$ already, the increasing iteration, therefore, with $a_0 \subset a_1 = a_2$. (Note that the results for Galois correspondences are no longer dual as far as equality is concerned.)

The general case will be treated somewhat indirectly, since we first prove a result on $a_\infty$ and $b_\infty$, not on the entire set $a$, $a_\infty$, $b$, $b_\infty$. Basically, the premises in (i) and (ii) are what we desire.

**A.3.10 Proposition** (*Approximation of extremal fixedpoints of a "continuously antitone" pair of mappings*).

i)    $\sup \pi(T) = \pi(\inf T)$    for all $T \subset W$    $\Longrightarrow$    $a_\infty = \pi(b_\infty)$;

ii)    $\sigma(\sup S) = \inf \sigma(S)$    for all $S \subset V$    $\Longrightarrow$    $b_\infty = \sigma(a_\infty)$;

iii)    Assuming both (i) and (ii) we obtain $a = a_\infty$,    $b = b_\infty$.

**Proof:** For (i) and (ii) consider the last displayed line of formulas in the proof of A.3.9. For (iii) this may be continued as follows: $a_\infty = \pi(b_\infty) = \pi\left(\sigma(a_\infty)\right) = \rho(a_\infty)$, $b_\infty = \sigma(a_\infty) = \sigma\left(\pi(b_\infty)\right) = \varphi(b_\infty)$, establishing $a_\infty$, $b_\infty$ as fixedpoints of $\rho$ and $\varphi$, respectively. However, $a_\infty, b_\infty$ cannot lie strictly below or above the respective least or greatest fixedpoint.    $\square$

Premises (i) and (ii) are sharpened versions of the antitonicity property (Definition A.3.7), but are so in *different* directions for $\sigma$ and $\pi$. It can be seen from the following considerations that usually *both* premises are not satisfied simultaneously; sometimes other properties can be used instead.

**A.3.11 Corollary** (*Approximation of the extremal fixedpoints of a pair of antitone mappings in the finite case*). Let the complete lattices be finite. Then $a_\infty = a$, $b_\infty = b$.

**Proof:** The approximating sequences become stationary,

$$a_0 \subset \dots \subset a_m = a_\infty \subset \pi(b_\infty) \subset a, \quad b \subset \sigma(a_\infty) \subset b_\infty = b_n \subset \dots \subset b_0,$$

for reasons of finiteness. Now $\pi(b_\infty) = \pi(b_n) = \pi\left(\sigma(a_{n-1})\right) = a_{n+1} \subset a_\infty$ gives $\pi(b_\infty) = a_\infty$; $\sigma(a_\infty) = b_\infty$ is obtained analogously. For the rest of the proof proceed as in that of Prop. A.3.10.iii.    $\square$

It can happen that two conditions of the kind required in Prop. A.3.10 are satisfied, such as Prop. A.3.10.i holding for both $\pi$ *and* $\sigma$, or, similarly with Prop. A.3.10.ii. This would not be the situation required and would not readily yield approximation. However, we next study one case, important for applications, where two such counter-running conditions are given and approximation is still possible. For that purpose, one of the two conditions has to be "flipped in the right direction". This can be achieved for complete lattices which in addition are *Boolean*, so that one can work with complements according to the usual rules; see Sect. A.1.

**A.3.12 Proposition** (*Approximation of the extremal fixedpoints of pairs of antitone mappings in complete* Boolean *lattices*). Let a continuous mapping $f$ be given. Assume $\pi$ to have the properties

$$\pi(\sup T) = \inf \pi(T) \qquad \text{for all } T \subset W,$$
$$\pi(w) = f(\overline{\pi(\overline{w})}) \qquad \text{for all } w \in W,$$

and $\sigma$ to have the properties

$$\sigma(\sup S) = \inf \sigma(S) \qquad \text{for all } S \subset V.$$

Then $a_\infty = a$, $b_\infty = b$.

**Proof:** Clearly, $\sigma$ satisfies the condition of Prop. A.3.10.ii. Using the negation formula (and abbreviating $\overline{T} = \{\overline{x} \mid x \in T\}$) we flip the condition on $\pi$ so as to have it in the pattern of Prop. A.3.10.i:

$$\pi(\inf T) = f(\overline{\pi(\overline{\inf T})}) = f(\overline{\pi(\sup \overline{T})}) = f(\overline{\inf \pi(\overline{T})})$$
$$= f(\sup \overline{\pi(\overline{T})}) = \sup f(\overline{\pi(\overline{T})}) = \sup \pi(T). \qquad \square$$

For the mapping $f$ one may take $f(x) := x \sqcup d$ with $d$ a constant; in our applications, $f(x) := x \sqcup \pi(L)$. The dual result is given in the following proposition.

**A.3.13 Proposition** (*Approximation of extremal fixedpoints of pairs of antitone functions in complete* Boolean *lattices*). Let a downward continuous mapping $g$ be given. Furthermore, let $\sigma$ satisfy

$$\sup \sigma(S) = \sigma(\inf S) \qquad \text{for all } S \subset V,$$
$$\sigma(v) = g(\overline{\sigma(\overline{v})}) \qquad \text{for all } v \in V,$$

and let $\pi$ satisfy

$$\sup \pi(T) = \pi(\inf T) \qquad \text{for all } T \subset W.$$

Then $a_\infty = a$, $b_\infty = b$. $\qquad \square$

A typical mapping $g$ is the intersecting with a constant $c$, $g(x) = x \sqcap c$, especially with $c = \sigma(O)$.

We end with five examples which show how central these theorems on antimorphisms are. The general theory developed here could have been used in each case to considerably shorten the respective arguments in the previous chapters.

## The Initial Part as a Fixedpoint of a Pair of Antitone Mappings

In Sect. 6.3 we considered the complete lattice given by the power set of the point set $V$ of a graph. We then defined in Def. 6.3.2 the initial part as

$$J(B) = \inf \{ x \mid \overline{x} \subset B\overline{x} \},$$

and considered the sequence obtained by iteration,

$$z_{-1} := O, \quad z_{i+1} := \overline{B\,\overline{z_i}}, \quad i \geq -1.$$

In the framework of our present general theory we can start instead with the antitone mappings $\sigma, \pi: 2^V \longrightarrow 2^V$ defined by

$$\sigma(x) := B\,\overline{x}, \quad \pi(x) := \overline{x}.$$

Then $\rho(x) = \pi(\sigma(x)) = \overline{B\,\overline{x}}$ and $\varphi(y) = \sigma(\pi(y)) = By$ (where $\rho$ need not be continuous). As a consequence, there exist two extremal fixedpoints $a, b$, where $a = J(B)$, and, since $a = \pi(b)$, also $\overline{b} = J(B) = a$. Finally we have $b = \sigma(a)$, thus establishing the relationship $\overline{J(B)} = BJ(B)$. One will easily see that Def. 6.3.2.i follows the pattern of Prop. A.3.8.ii, iii. The iteration then reads:

$$a_0 := O, \quad a_{i+2} := \pi(\sigma(a_i)) = \rho(a_i) = \overline{B\,\overline{a_i}}, \quad i \geq 0;$$

consequently, $a_\infty$ corresponds to $\sup_h \overline{B^h L}$. (Every step of the iteration of Prop. 6.3.4 is counting for two.) The iterations start as follows:

$$O = a_0 = a_1 = O \subset a_2 = \overline{BL} = a_3 = \overline{BL} \subset a_4 = \overline{B^2L} \subset \ldots \subset a_\infty = \sup_h \overline{B^h L};$$
$$L = b_0 \supset b_1 = BL = b_2 = BL \supset b_3 = B^2L = b_4 = B^2L \supset \ldots \supset b_\infty = \inf_h B^h L.$$

Finally, we consider the approximation of the fixedpoints. On one side we have

$$\sup \pi(T) = \sup\{\pi(t) \mid t \in T\} = \sup\{\overline{t} \mid t \in T\} = \overline{\inf\{t \mid t \in T\}} = \pi(\inf T)$$

and therefore $a_\infty = \pi(b_\infty)$, because of Prop. A.3.10.i. In general, however, we will not reach the limit, as only

$$\sigma(\sup S) = B\,\overline{\sup S} = B \inf\{\overline{v} \mid v \in S\}$$
$$\subset \inf\{B\overline{v} \mid v \in S\} = \inf\{\sigma(v) \mid v \in S\} = \inf \sigma(S),$$

can be proved; so only $\sigma(a_\infty) \subset b_\infty$ holds, corresponding to $B \inf_h B^h L \subset \inf_h B^h L$.

It is due to Corollary A.3.11 that we were, nevertheless, able to prove equality, and hence approximation, for finite graphs (Prop. 6.3.3.v). Moreover, Prop. 6.3.3.ii can be deduced from Prop. A.3.13 because by Prop. 4.2.2 $B$ is univalent if and only if $B\overline{v} = \overline{Bv} \sqcap BL$ holds, i.e., $\sigma(v) = \sigma(\overline{v}) \sqcap \sigma(O)$, and because the $\sqcup$-distributive law implies

$$\sup \sigma(S) = \sup\{\sigma(v) \mid v \in S\} = \sup\{B\overline{v} \mid v \in S\} = B\sup\{\overline{v} \mid v \in S\}$$
$$= B\,\overline{\inf\{v \mid v \in S\}} = \sigma(\inf S).$$

## Kernels as Fixedpoints of a Pair of Antitone Mappings

In Sect. 8.2 we defined kernels of a graph as point sets $s$ with the property $\overline{Bs} = s$. The sequence generated by the iteration,

$$a_0 := O, \quad a_{i+1} := \overline{B\,\overline{Ba_i}}, \quad i \geq 0$$

could be used for establishing iterative bounds. Clearly, we have here again a pair of antitone mappings which are both given by

$$\sigma(x) := \pi(x) := \overline{Bx},$$

with the corresponding mappings $\rho(x) = \varphi(x) = \overline{B\overline{Bx}}$ equally coinciding. Both iterations take place in the same complete Boolean lattice $\mathbf{2}^V$, producing the greatest lower bound in the first instance and the least upper bound, in the second. By Prop. A.3.8 we know that these fixedpoints $a$ and $b$ exist and satisfy $a = \pi(b)$ (here $a = \overline{Bb}$) and $\sigma(a) = b$ (here $\overline{Ba} = b$). By Prop. A.3.9, the iterative bounds $a_\infty$ and $b_\infty$ satisfy

$$a_\infty \subset \overline{Bb_\infty} \subset a = \overline{Bb} \subset b = \overline{Ba} \subset \overline{Ba_\infty} \subset b_\infty.$$

Remember that, concerning the last inclusion, in Sect. 8.2 a more precise result has been obtained: $\overline{Ba_\infty} = b_\infty$, or $\sigma(a_\infty) = b_\infty$. By Prop. A.3.10.ii this is a consequence of $\sqcup$-distributivity,

$$\sigma(\sup{}_i a_i) = \overline{B \sup{}_i a_i} = \overline{\sup{}_i Ba_i} = \inf{}_i \overline{Ba_i} = \inf{}_i \sigma(a_i).$$

## Concurring and Safe Sets as Fixedpoints
## of a Pair of Antitone Mappings

We now describe how the mechanism developed here applies to determining the concurring and safe set in a bipartite graph (Def. 9.3.7). The two sets $V$ and $W$ need no longer be the same. The antitone mappings are now

$$\sigma(v) := \overline{Q^\mathsf{T} v} \quad \text{and} \quad \pi(w) := \overline{\lambda w}.$$

As a consequence, on the left-hand side we have to find the least fixedpoint of $\rho(v) = \overline{\lambda \overline{Q^\mathsf{T} v}}$ and on the right-hand side, the greatest fixedpoint of $\varphi(w) := \overline{Q^\mathsf{T} \overline{\lambda w}}$. Both must satisfy the relationship $\overline{\lambda b} = a$, $b = \overline{Q^\mathsf{T} a}$. From Prop. A.3.12 it follows that approximation is possible. For

$$\sigma(\sup S) = \overline{Q^\mathsf{T} \sup S} = \overline{\sup\{Q^\mathsf{T} s \mid s \in S\}}$$
$$= \inf\{\overline{Q^\mathsf{T} s} \mid s \in S\} = \inf\{\sigma(s) \mid s \in S\} = \inf \sigma(S),$$

because of $\sqcup$-distributivity of the relational product, and finally

$$\pi(\sup T) = \inf \pi(T), \quad \text{as well as} \quad \pi(w) = \overline{\lambda w} = \overline{\lambda \overline{w} \sqcup \lambda L} = \overline{\pi(\overline{w})} \sqcup \pi(L).$$

## Minimal Point Coverings as Fixedpoints
## of a Pair of Antitone Mappings

Pairs of antitone mappings appeared yet in another place, namely when investigating point coverings of bipartite graphs $(X, Y, Q)$ in Sect. 9.3. There, we defined $\sigma(s) := \overline{Q^\mathsf{T} \overline{s}}$ and $\tau(t) := \overline{Q\overline{t}}$ with the composites being $\rho(s) := \tau(\sigma(s)) = \overline{Q\overline{Q^\mathsf{T} \overline{s}}}$ and $\varphi(t) := \sigma(\tau(t)) = \overline{Q^\mathsf{T} \overline{Q\overline{t}}}$. Figure 9.3.3 illustrates how the (left-hand) fixedpoints of $\rho$ are in (a "counter-running") correspondence with the (right-hand) fixedpoints of $\varphi$, the least left-hand fixedpoint $a \approx \emptyset$ corresponding to the greatest right-hand $b \approx \{1,3,4,5\}$. In this case, both fixedpoints are furnished by the cross-iteration

$$a_0 := O, \quad a_{i+1} := \tau(b_i); \qquad b_0 := L, \quad b_{i+1} := \sigma(a_i), \quad i \geq 0.$$
$$O = a_0 = a_1 = a_2 = a_\infty = a; \qquad L = b_0 \supset b_1 = Q^\mathsf{T} L = b_\infty = b.$$

They clearly satisfy $a = \tau(b)$, here $O = Q\overline{Q^\mathsf{T} L}$, and $b = \sigma(a)$, here $Q^\mathsf{T} L = Q^\mathsf{T} L$. This iteration is of course a short one, since, due to the Galois property, it stops after the first step.

## Upper and Lower Bounds as Fixedpoints of a Pair of Antitone Mappings

In Sect. 3.3 we studied the functionals ubd and lbd which associate with each subset of an ordered set the set of all upper and lower bounds, respectively. These are clearly antitone mappings $\mathrm{lbd}, \mathrm{ubd} \colon 2^V \longrightarrow 2^V$.

They satisfy

$$\mathrm{ubd}\,(\sup T) = \inf\,(\mathrm{ubd}\,(T)), \quad \mathrm{lbd}\,(\sup S) = \inf\,(\mathrm{lbd}\,(S))$$

and also

$$\mathrm{lbd}\,(\mathrm{ubd}\,(T)) \supset T, \quad \mathrm{ubd}\,(\mathrm{lbd}\,(S)) \supset S,$$

making them a Galois correspondence. Here again the iteration comes to an end quickly: $\mathrm{ubd}\,(T) = \mathrm{ubd}\,(\mathrm{lbd}\,(\mathrm{ubd}\,(T)))$, $\mathrm{lbd}\,(S) = \mathrm{lbd}\,(\mathrm{ubd}\,(\mathrm{lbd}\,(S)))$.

The fixedpoints for $T \mapsto \mathrm{lbd}\,(\mathrm{ubd}\,(T))$ and $S \mapsto \mathrm{ubd}\,(\mathrm{lbd}\,(S))$ are the upward or downward directed cones, respectively. The least fixedpoint of $T \mapsto \mathrm{lbd}\,(\mathrm{ubd}\,(T))$ is either $\emptyset$, if no least element exists in the ordered set, or $\{\mathrm{lea}\,(V)\}$ if it exists.

# A.4 References

See also the references of Chapts. 2 and 4.

BEDNAREK AR, ULAM S: *Projective algebra and the calculus of relations.* J. Symbolic Logic **43** (1978) 56–64.

BIRKHOFF G: *Lattice Theory.* Amer. Math. Soc. Colloq. Publ. Vol. XXV, Providence, R. I., 1967.

BRINK C: *The algebra of relatives.* Notre Dame J. Formal Logic **20** (1979) 900–908.

BRINK C: *Two axiom systems for relation algebras.* Notre Dame J. Formal Logic **20** (1979) 909–914.

BRINK C: *Boolean modules.* J. Algebra **71** (1981) 291–313.

CHIN LH, TARSKI A: *Remarks on projective algebras.* (Abstract 90) Bull. Amer. Math. Soc. **54** (1948) 80–81.

EVERETT CJ, ULAM S: *Projective algebra I.* Amer. J. Math. **68** (1946) 77–88.

HENKIN L, MONK JD, TARSKI A, ANDRÉKA H, NÉMETI I: *Cylindric set algebras.* Lecture Notes in Mathematics **883**, Springer, Berlin, 1981.

HENKIN L, MONK JD, TARSKI A: *Cylindric algebras, Part II.* North-Holland Publ. Co, Amsterdam, 1985.

JÓNSSON B, TARSKI A: *Boolean algebras with operators. Part II.* Amer. J. Math. **74** (1952) 127–162.

KALICKI J, SCOTT D: *Equational completeness of abstract algebras.* Indag. Math. **17** (1955) 650–659.

KAMEL H: *Relational algebra and uniform spaces.* J. London Math. Soc. **29** (1954) 342–344.

KNASTER B: *Un théorème sur les fonctions d'ensembles.* Ann. Société Polonaise de Mathématique **6** (1927) 133–134.

LYNDON RC: *The representation of relational algebras.* Ann. of Math. (2) **51** (1950) 707–729.

LYNDON RC: *The representation of relation algebras. II.* Ann. of Math. (2) **63** (1956) 294–307.

LYNDON RC: *Relation algebras and projective geometries.* Michigan Math. J. **8** (1961) 21–28.

MADDUX RD, TARSKI A: *A sufficient condition for the representability of relation algebras.* Notices Amer. Math. Soc. **23** (1976) A-477.

MADDUX RD: *Some sufficient conditions for the representability of relation algebras.* Algebra Universalis **8** (1978) 162–172.

MADDUX RD: *The origin of relation algebras in the development and axiomatization of the calculus of relations.* Studia Logica **50** (1991) 421–455.

MCKENZIE R: *Representations of integral relation algebras.* Michigan Math. J. **17** (1970) 279–287.

MCKINSEY JCC: *Postulates for the calculus of binary relations.* J. Symbolic Logic **5** (1940) 85–97.

MCKINSEY JCC: *On the representation of projective algebras.* Amer. J. Math. **70** (1948) 375–384.

MICHAEL E: *A note on Peirce on Boole's algebra of logic.* Notre Dame J. Formal Logic 20 (1979) 636–638.

MONK JD: *Relation algebras and cylindric algebras.* Notices Amer. Math. Soc. **8** (1961) 358.

MONK JD: *On representable relation algebras.* Michigan Math. J. **11** (1964) 207–210.

NÉMETI I: *Algebraizations of quantifier logics, an introductory overview.* 10th version, Oct. 1991, full manuscript underlying the article in a forthcoming special volume of Studia Logica devoted to Algebraic Logic (Ed.: Blok WJ, Pigozzi D).

SCHMIDT G, STRÖHLEIN T: *Relation algebras: Concept of points and representability.* Discrete Math. **54** (1985) 83–92.

SCHÖNFELD W: *An undecidability result for relation algebras.* J. Symbolic Logic **44** (1979) 111–115.

TARSKI A: *A lattice-theoretical fixpoint theorem and its applications.* Pacific J. Math. **5** (1955) 285–309.

TARSKI A: *Equationally complete rings and relation algebras.* Indag. Math. **18** (1956) 39–41.

VELOSO PAS: *The history of an error in the theory of representations of relation algebras.* J. Symbolic Logic **42** (1977) 473, **44** (1979) 466.

WADGE W: *A complete natural deduction system for the relational calculus.* Theory of computation report 5, University of Warwick 1976.

WOOYENAKA Y: *On postulate-sets for relation algebras.* Notices Amer. Math. Soc. **6** (1959) 534–535.

# General References


See also the references at the end of every chapter.

AHO AV, HOPCROFT JE, ULLMANN JD: *The design and analysis of computer algorithms.* Addison-Wesley, Reading, Mass., 1974.

ANDRÉKA H, MONK JD, NÉMETI I (eds.): *Algebraic Logic.* Proc. of a Coll., Budapest, August 8-14, 1988, (Colloq. Math. Soc. J. Bolyai; **54**) North-Holland Publ. Co., Amsterdam, 1991.

BACKHOUSE RC, DE BRUIN PJ, HOOGENDIJK P, MALCOLM G, VAN DER WOUDE J: *Polynomial relators.* Computing Science Notes, Dept. Mathematics and Computing Science, Eindhoven Univ. of Technology, May 1991.

BAUER FL, WÖSSNER H: *Algorithmic language and program development.* Springer, Berlin, 1982.

BERGE C: *The theory of graphs and its applications.* Methuen, London, 1962.

BERGE C: *Graphs and hypergraphs.* North-Holland Publ. Co., Amsterdam, 1973.

BERGHAMMER R, SCHMIDT G: *Relational specifications.* to appear in: Proc. 38th Stefan Banach Semester, *Algebraic methods in logic and their computer science applications*, Warszaw 1991.

DESHARNAIS J, MADHAVJI NH: *Abstract relational specifications.* - In: Broy M, Jones CB (eds.): *Programming concepts and methods.* Proc. IFIP WG 2.2/2.3 Sea of Galilee, Israel, April 2–5 1990, North-Holland Publ. Co., Amsterdam, 1990, 267-284.

HAEBERER AM, VELOSO PAS: *Partial relations for program derivation: Adequacy, inevitability and expressiveness.* - In: Möller B (ed.): *Constructing programs from specifications.* Proc. IFIP TC 2/WG 2.1 Pacific Grove, CA, USA, May 13–16 1991, North-Holland Publ. Co., Amsterdam, 1991, 319-371.

KIM KH: *Boolean matrix theory and applications.* Dekker, New York, 1982.

KÖNIG D: *Theorie der endlichen und unendlichen Graphen.* Akademische Verlagsgesellschaft, Leipzig, 1936 (Engl. Reprint: Chelsea, New York).

MÖLLER B: *Relations as a program development language.* - In: Möller B (ed.): *Constructing programs from specifications.* Proc. IFIP TC 2/WG 2.1 Pacific Grove, CA, USA, May 13–16 1991, North-Holland Publ. Co., Amsterdam, 1991, 373-397.

RUDEANU S: *Boolean functions and equations.* North-Holland Publ. Co., Amsterdam, 1974.

SIKORSKI R: *Boolean Algebras.* Ergebnisse der Mathematik und ihrer Grenzgebiete **NF 25**, Springer, Berlin, 1964.

TARSKI A, GIVANT S: *A formalization of set theory without variables.* Amer. Math. Soc. Colloq. Publ. **41** 1987.

ZIERER H: *Relation algebraic domain constructions.* Theoret. Comput. Sci. **87** (1991) 163–188.

# Name Index

# Table of Symbols

### Sets

| | | |
|---|---|---|
| $\{\,x \mid E(x)\,\}$ | set | 1 |
| $x \in M$ | element | 1 |
| $|x|$ | number of elements | 25, 58 |
| $M \subset X$ | subset | 1 |
| $\mathbf{2}^X$ | power set | 1 |
| $\emptyset$ | empty set | 1 |
| $M \cup N,\ M \cap N$ | union, intersection | 1 |
| $\overline{M}$ | complement (if ground set is tacitly given) | 1 |
| $M + N$ | symmetric difference | 3 |
| $M \times N$ | Cartesian product | 4 |

### Logic

| | | |
|---|---|---|
| $\Longrightarrow$ | metalanguage consequence | 1 |
| $\Longleftrightarrow$ | metalanguage equivalence | 1 |
| $:\Longleftrightarrow$ | metalanguage definition | 8 |
| $:=$ | definitional equality | 2 |
| $\mathbb{B} = \{\,\mathbf{0}, \mathbf{1}\,\}$ | truth-values | 3 |
| $\wedge,\ \vee$ | "and" resp. "or" in propositional logic | 3 |
| $\rightarrow$ | "if—then" in propositional logic | 7 |
| $\leftrightarrow$ | "precisely when" in propositional logic | 7 |
| $\exists$ | existential quantifier | 7 |
| $\forall$ | universal quantifier | 7 |

### Relations

| | | | |
|---|---|---|---|
| $R \sqcup S$ | union | | 7 |
| $\sup \mathcal{A},\ \sup\{\,R \mid R \in \mathcal{A}\,\}$ | supremum | | 2, 7 |
| $R \sqcap S$ | intersection | | 7 |
| $\inf \mathcal{A},\ \inf\{\,R \mid R \in \mathcal{A}\,\}$ | infimum | | 2, 7 |
| $\overline{R}$ | complement | | 7 |
| $R \subset S$ | containment | | 7 |
| $O$ | empty relation | where the | 7, 51 |
| $L$ | universal relation | ground set | 7, 51 |
| $I$ | identity | is tacitly given | 8, 51 |
| $R^{\mathsf{T}}$ | transposed relation | | 9 |
| $R \circ S,\ RS,\ R\,;S$ | product | | 13, 19 |
| $S \dagger R$ | | | 19 |
| $S\ .\cdot\ R,\ S\ \cdot.\ R$ | residuation | | 19 |
| $\mathsf{syq}(R, S)$ | symmetric quotient | | 19, 71 |
| $\mathsf{unp}(R)$ | univalent part | | 60 |

| | | |
|---|---|---|
| $\mathrm{mup}(R)$ | multivalent part | 60 |
| $H_B$ | nontransitive part, Hasse diagram | 139 |
| $R^+$ | transitive closure | 35 |
| $R^*$ | reflexive-transitive closure | 35 |
| $h_{\mathrm{equiv}}(R)$ | equivalence closure | 35 |
| $J(R)$ | initial part | 121 |

## Graphs

| | | |
|---|---|---|
| $\|R\|$ | maximal cardinality of a row | 58 |
| $d_p(x),\ d_s(x)$ | predecessor and successor degree | 25 |
| $d_0(x)$ | simple degree | 27 |
| $\chi$ | connectivity number | 118 |
| $\beta$ | absorption number | 173 |
| $\alpha$ | point independence number | 198 |
| $\alpha^*$ | edge independence number | 198 |
| $\tau$ | point covering number | 203 |
| $\tau^*$ | edge covering number | 203 |
| $g,\ \gamma$ | left-hand side (gauche) in a bipartition | 52, 88 |
| $d,\ \delta$ | right-hand side (droit) in a bipartition | 52, 88 |

## Orderings

| | | |
|---|---|---|
| $E$ | "less or equal" as a relation | 32 |
| $C$ | "less" as a relation | 32 |
| $\max(t),\ \min(t)$ | maxima, minima | 42 |
| $\mathrm{ubd}(t),\ \mathrm{lbd}(t)$ | majorants, minorants | 43 |
| $\mathrm{gre}(t),\ \mathrm{lea}(t)$ | greatest and least points | 45 |
| $\mathrm{lub}(t),\ \mathrm{glb}(t)$ | least upper, greatest lower bounds | 46 |
| $\lfloor x \rfloor$ | greatest integer number $\le x$ | 194 |

## Programs

| | | |
|---|---|---|
| $\lceil \ldots \rfloor$ | **begin…end** | 230 |
| $E; F$ | composition | 229 |
| **if** $b$ **then** $E$ **else** $F$ **fi** | conditional | 229 |
| $E[]F,\ [b:E[]c:F]$ | nondeterministic guarded command | 229 |
| **while** $b$ **do** $E$ **od** | | 229 |
| **repeat** $E$ **until** $b$ | | 229 |
| **goto** $M$ | | 229 |
| $\{v\}\,\mathcal{P}\,\{n\}$ | partially correct | 237 |
| $\langle\langle x \ge 1 \rangle\rangle$ | proposition as a relation | 242 |
| $\square,\ [\mathcal{P}]$ | box | 251 |
| $\Diamond,\ <\mathcal{P}>$ | diamond | 251 |
| $\mathrm{wp}(\mathcal{P}, n)$ | weakest precondition | 251 |
| $R/S$ | weakest prespecification | 19 |

# Subject Index